XXI Winter School on Continuous Media Mechanics 2019

IOP Conference Series: Materials Science and Engineering
Volume 581

Perm, Russia
18 – 22 February 2019

ISBN: 978-1-5108-9449-5
ISSN: 1757-8981

Printed from e-media with permission by:

Curran Associates, Inc.
57 Morehouse Lane
Red Hook, NY 12571

Some format issues inherent in the e-media version may also appear in this print version.

Institute of Physics
Dirac House, Temple Back
Bristol BS1 6BE UK

Phone: 44 1 17 929 7481
Fax: 44 1 17 920 0979

techtracking@iop.org

Additional copies of this publication are available from:

Curran Associates, Inc.
57 Morehouse Lane
Red Hook, NY 12571 USA
Phone: 845-758-0400
Fax: 845-758-2633
Email: curran@proceedings.com
Web: www.proceedings.com

Table of contents

Volume 581

XXI Winter School on Continuous Media Mechanics

18–22 February 2019, Perm, Russian Federation

Accepted papers received: 2 July 2019
Published online: 29 July 2019

Preface

Papers

IOP Conf. Series: Materials Science and Engineering **581** (2019) 011001 doi:10.1088/1757-899X/581/1/011001

PREFACE

The XXI-th Winter School on Continuous Media Mechanics, a scientific event traditionally organized by the Institute of Continuous Media Mechanics UB RAS. It is a biennial meeting, which attracts researchers in the field of mechanics. Most participants are regular attendees of this scientific event.

The Schools are also attended by our foreign colleagues, who are involved in active collaboration with the scientists of the Institute of Continuous Media Mechanics UB RAS. Therefore, our Winter School has established itself as a recognized international event, yet most participants certainly come to the School from the universities and institutes of the city of Perm. About 350 people from 23 cities of 7 countries (Russia, Austria, Germany, Poland, France, Israel, Italy, Belarus) took part in the XXI Winter School on Continuum Mechanics. Russian participants represented 39 organizations in 23 cities (Perm, St. Petersburg, Moscow, Yekaterinburg, Krasnoyarsk, Novosibirsk, Kazan, Samara, Rostov-on-Don, etc.).

The Scope of the Conference traditionally covers the following topics:

- computational continuum mechanics;
- coupled problems of solid mechanics;
- physics and mechanics of meso- and nano-structured systems ;
- mechanics of functional materials;
- convection, hydrodynamic stability and turbulence;
- hydrodynamics of non-Newtonian liquids and liquids with special characteristics;
- interdisciplinary studies.

All these topics are among the central themes of modern continuum mechanics, which is evidenced by the increasing number of publications in leading Russian and foreign journals.

During the conference reports were made on fundamental and applied problems of continuum mechanics, including invited reports of well-known experts in the relevant branches of mechanics (academicians, corresponding members of the Russian Academy of Sciences, doctors of science), as well as representatives of enterprises producing high-tech products (JSC «UEC-Aviadvigatel», JSC Perm Scientific-Production Instrument-Making Company, etc.).This issue contains articles on the part of the reports presented at the School.

V.P. Matveenko
Chairman of the XXI Winter School
on Continuous Media Mechanics,
Member of the Russian Academy of Sciences,
Director of the Institute of Continuous Media Mechanics
of the Ural Branch of the Russian Academy of Sciences

Peer review statement

All papers published in this volume of *IOP Conference Series: Materials Science and Engineering* have been peer reviewed through processes administered by the proceedings Editors. Reviews were conducted by expert referees to the professional and scientific standards expected of a proceedings journal published by IOP Publishing.

IOP Conf. Series: Materials Science and Engineering **581** (2019) 012001 doi:10.1088/1757-899X/581/1/012001

Viscoelastic ferrocolloid modelled as the Jeffreys fluid: dynamic magnetic susceptibility in the presence of a bias field

V V Rusakov[1,2], **Yu L Raikher**[1]

[1] Institute of Continuous Media Mechanics, 614013 Perm, Russia

[2] Perm National Research Polytechnic University, 614000 Perm, Russia

E-mail: vvr@icmm.ru

Abstract. Orientational Brownian motion of a solid nanoparticle embedded in a viscoelastic medium modelled by the Jeffreys fluid is studied. The particle bears a magnetic moment and is subjected to a combination of a constant (bias) and a weak (probing) AC magnetic fields. First, the set of Langevin equations of the problem is presented and then on that basis a corresponding kinetic equation (KE) is derived. The solution of the KE that is linear with respect to the AC field amplitude is worked out. This solution is virtually exact since it is cast as a series that enables one to obtain the result with arbitrary accuracy. In that framework, a closed expression for the dynamic magnetic susceptibility of a statistical ensemble of magnetic particles in a viscoelastic environment is found, analyzed and compared to the previously existed approximate solution. It is shown that: (*i*) the dynamic elasticity of the medium has a substantial effect on the magnetic spectra; (*ii*) to correctly describe this effect, one has to retain a considerable number of terms in the series that presents the KE solution.

1. Introduction

For a long time, the Jeffreys model [1] works as a very helpful tool in phenomenological rheiology. Comprising only three parameters, it is able to mimic a wide range of polymer media: from weak solutions to gel-like systems. That is why we chose it to investigate the essential features of magnetic nanosuspensions with viscoelastic carriers. This approach enables one to deal with the challenging problem: to advance the theory of magnetodynamic response of ferrofluids beyond the case of ordinary systems (colloids in Newtonian liquids) to a wide variety of magnetic nanosuspensions whose base behaves viscoelastically. The problem is especially important since the modern trend is to use field-driven magnetic nanoparticles as drug delivery agents [2, 3, 4, 5], heat mediators [6, 7, 8, 9], immunoassay sensors [10, 11, 12, 13] and nanosurgery tools [14, 15] in biotechnology and medicine by embedding them in, for example, analytical animal / human tissue samples, blood, lymph, cell cytosol, etc., whose rheology is far from that of simple liquids.

Another aspect of application of magnetic nanoparticles ii their use in laboratory technique for active microrheology, the method where the particles are introduced in small quantities to the media which rheology is unknown or needs to be time-monitored in the course of the chemical processes undergoing in them. In this method the nanoparticles are set to motion by means of an applied magnetic field (e.g., rotary motion), and the induced dynamic magnetization signal is used for analyzing the mechanical properties of the particle environment [16, 17, 18, 19, 20].

IOP Conf. Series: Materials Science and Engineering **581** (2019) 012001 doi:10.1088/1757-899X/581/1/012001

The common feature of all those approaches is that the magnetic response of nanoparticles is strongly affected by Brownian motion which is the more intense the smaller the particles. Therefore, any correct theory ought to take this factor into account.

Given the above-presented considerations, in this paper we investigate the orientational motion of a single-domain ferromagnet particle suspended in the Jeffreys fluid. The particle with typical size about $10 - 20\,\text{nm}$ is assumed to be magnetically hard that means that its magnetic moment $\vec{\mu}$ has a constant length μ and is "frozen" in the particle body. Because of that, unit vector $\vec{e} = \vec{\mu}/\mu$ in below is used as the marker of the particle orientational motion.

The particle is surrounded by the Jeffreys fluid whose mechanical properties are schematized by the Newtonian and Maxwellian rheological elements connected in parallel [1], see Fig. 1. The Newtonian element has viscosity coefficient η_N; the Maxwellian one is a chain comprising the viscous element with coefficient η_M and a spring that models shear elasticity with modulus G. On a particle of hydrodynamic volume V, upon its motion, this fluid exerts an elastic restoring torque proportional to G and two interacting viscous torques proportional to η_N and η_M.

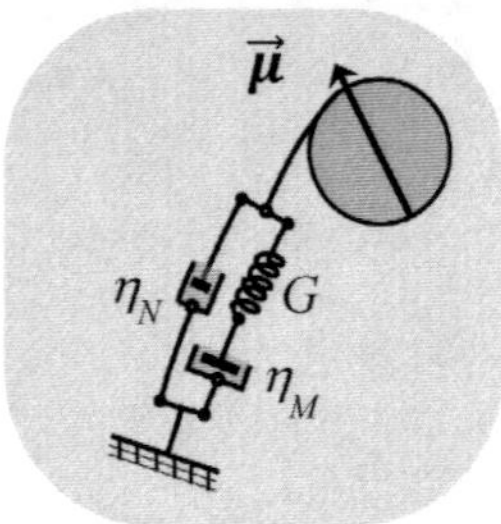

Figure 1. Schematic representation of a single-domain particle in a Jeffreys fluid.

2. Equations of motion and kinetic equation

On the nanoscale, the inertia effects are negligible, and, thus, the set of Langevin equations that governs the orientational motion of an embedded nanoparticle has the form [21]:

$$
\begin{aligned}
&\frac{\partial}{\partial t}\vec{e} = \left(\vec{\Omega} \times \vec{e}\right),\\
&\frac{\partial}{\partial t}\vec{\Omega} = \frac{1}{\zeta_N}\left[-\hat{\vec{L}}U + \vec{Q} + \vec{y}_N(t)\right], \quad \hat{\vec{L}} = \left(\vec{e} \times \frac{\partial}{\partial \vec{e}}\right),\\
&\left(1 + \tau_M \frac{\partial}{\partial t}\right)\vec{\Omega} = -\zeta_M \vec{\Omega} + \vec{y}_M(t).
\end{aligned}
\tag{1}
$$

Here $\vec{e}$ is unit vector of the particle orientation, $\vec{\Omega}$ its angular velocity, U the energy of particle interaction with the applied magnetic field, and $\vec{L}$ operator of infinitesimal rotation. The response coefficients of the medium are set by the conventional microrheology relations [16]:

$$\zeta_\sigma = 6\eta_\sigma V, \quad K = 6GV, \quad \sigma = N, M, \tag{2}$$

The random forces in equations (1) are defined via the fluctuation-dissipation theorem and delta-correlated (white noise):

$$\langle y_{i\sigma}(t) y_{j\sigma'}(t+\tau)\rangle = 2T\zeta_\sigma \delta_{ij}\delta_{\sigma\sigma'}\,\delta(\tau) \quad \sigma = N,\, M; \tag{3}$$

in this formula and further, temperature is taken in energy units, i.e., $k_B = 1$.

IOP Conf. Series: Materials Science and Engineering **581** (2019) 012001 doi:10.1088/1757-899X/581/1/012001

In the set (1) phase variable $\vec{Q}$ is the retarded mechanical response of the medium. In physical sense, vector $\vec{Q}$ is the torque exerted on the particle on the part of the dynamic quasi-mesh the nodes of which has reference lifetime $\tau_M = \zeta_M/K = \eta_M/G$. As mentioned, in the Jeffreys fluid there are two dissipative mechanisms working in parallel. One of them is linear (Newtonian) viscous friction that accounts for the particle interaction with the low-molecular component of the medium. Another one is the retarded interaction of the Maxwell viscoelastic type that resembles the hindered motion of the particle through the polymer environment. Note that in the limit $\tau_M \to \infty$, the Jeffreys model tends asymptotically to the case of a real gel whose mesh is not physical but chemical. Indeed, upon that limiting transition, the torque $\vec{Q}$ simplifies to: $\dot{\vec{Q}} = -K\vec{\Omega}$ pointing out that the response of the gel mesh to mechanical perturbation has only elastic component [22]. In this particular case and for planar rotation of the particle the set (1) reduces to a 1D Kelvin model considered in [22].

The kinetic equation for the particle orientation distribution function $W(\vec{e},t)$ is obtained from the set (1) via a standard procedure, see [21, 23, 24], for example:

$$\frac{\partial W}{\partial t} = \left[\frac{1}{\zeta_N}\left(\hat{\vec{L}} W \hat{\vec{L}} + \frac{\partial}{\partial \vec{Q}} W \frac{\partial}{\partial \vec{Q}} - \hat{\vec{L}} W \frac{\partial}{\partial \vec{Q}} - \frac{\partial}{\partial \vec{Q}} W \hat{\vec{L}}\right) + \frac{1}{\zeta_M}\frac{\partial}{\partial \vec{Q}} W \frac{\partial}{\partial \vec{Q}}\right] \times$$
$$\times \left(U(\vec{e}) + \frac{1}{2}KQ^2 + T\ln W\right); \tag{4}$$

here and further we use nondimensional phase variable $\tilde{\vec{Q}} = \vec{Q}/K$ but omit tilde for the sake of brevity. The equilibrium solution of Eq. (4) is rendered by the extended Boltzmann distribution

$$W_0\left(\vec{e}, \vec{Q}\right) \propto \exp\left\{-\frac{1}{T}\left[U(\vec{e}) + \frac{1}{2}KQ^2\right]\right\}. \tag{5}$$

Note an essential feature of the system under study: in equilibrium – both in the absence and in the presence of a constant magnetic field – distribution (5) does not contain cross terms, i.e., the phase variables are statistically independent.

The orientation-dependent potential of a particle with a “frozen-in” magnetic moment is

$$U = -\mu H(\vec{e}\,\vec{h}), \quad \vec{h} = \vec{k} + \varepsilon(t)\vec{p}, \quad h^2 = 1,\, p^2 = 1. \tag{6}$$

In Eq. (6) we assume that the magnetic field comprises a constant (bias) part $H\vec{k}$ directed along Oz and an AC part $\vec{H}_1 = H\varepsilon(t)\vec{p}$ with $|\varepsilon(t)| \ll 1$. The constant component endows the system with uniaxial anisotropy, so that the response to a weak probing field is characterized by a susceptibility tensor

$$\chi_{ik} = n\mu\partial\langle e_i\rangle/\partial H_{1k}, \tag{7}$$

where n is particle number concentration.

We evaluate susceptibility (7) by way of averaging the stochastic equations (1); with the potential U this set takes the form

$$\frac{\partial}{\partial t}\vec{e} = \gamma_H(\vec{e}\times(\vec{h}\times\vec{e})) + \gamma_N(\vec{Q}\times\vec{e}) + (\vec{u}\times\vec{e}), \quad \gamma_N = K/\zeta_N, \ \gamma_H = \mu H/\zeta_N, \tag{8}$$
$$\frac{\partial}{\partial t}\vec{Q} + \tilde{\gamma}\vec{Q} = -\gamma_H(\vec{e}\times\vec{h}) + \tilde{u}, \ \vec{u} = \vec{y}_N(t)/\zeta_N, \ \tilde{\vec{u}} = \vec{y}_M/\zeta_M - \vec{u}; \ \tilde{\gamma} = \gamma_N + \gamma_M = \gamma_N\,(1 + 1/q);$$

where $q = \eta_M/\eta_N$ is the viscoelasticity parameter.

With the explicit form of potential, the equilibrium distribution function (5) takes the form

$$W_0\left(\vec{e}, \vec{Q}\right) = \frac{1}{Z_\xi Z_Q}\exp\left[\xi\left(\vec{e}\,\vec{k}\right) - \frac{1}{2}\beta Q^2\right], \quad Z_\xi = 4\pi\sinh\xi, \ Z_Q = (2\pi/\beta)^{3/2}. \tag{9}$$

IOP Conf. Series: Materials Science and Engineering **581** (2019) 012001 doi:10.1088/1757-899X/581/1/012001

Here two nondimensional parameters are introduced: $\xi = \mu H/T$ for the field amplitude, and $\beta = K/T$ for inverse temperature,

The statistical moments of function (9) are expressed as

$$c_n = \left\langle \left(\vec{e}\,\vec{k}\right)^n \right\rangle_0 = \frac{2\pi}{Z_\xi}\int_{-1}^{1} dx\, x^n \exp(\xi x) = \frac{e^\xi + (-1)^{n+1}e^{-\xi}}{e^\xi - e^{-\xi}} - \frac{nC_{n-1}}{\xi};\ \ c_0 = 1. \tag{10}$$

As the system at $\varepsilon = 0$ is uniaxially anisotropic with $\vec{k}$ being the axis direction, one may introduce for it an orthogonal set of projection tensors

$$P_{ij}^{\|} = k_i k_j,\ \ P_{ij}^{\perp} = \delta_{ij} - k_i k_j,\ \ P_{ij}^{(\alpha)} P_{jk}^{(\beta)} = P_{ik}^{(\alpha)} \delta_{\alpha\beta},\ \ \alpha = \|,\ \perp, \tag{11}$$

in terms of which all the matrix operations reduce to simple scalar ones:

$$A_{ij} = A_{\|} P_{ij}^{\|} + A_{\perp} P_{ij}^{\perp},\ \ A_{ij}^{-1} = A_{\|}^{-1} P_{ij}^{\|} + A_{\perp}^{-1} P_{ij}^{\perp},\ \ A_{ij} B_{jk} = A_{\|} B_{\|} P_{ij}^{\|} + A_{\perp} B_{\perp} P_{ij}^{\perp}. \tag{12}$$

As the distribution of phase variable Q remains isotropic independently of the field presence, all its statistical averages of odd order vanish, whereas for the even-order ones we have

$$\left\langle (\vec{Q})^{2n} \right\rangle_0 = \frac{1}{Z_Q}\int d\vec{Q}\,(\vec{Q})^{2n} \exp(-\beta Q^2/2) = \tilde{T}^n (2n+1)!!,\ \ \left\langle (Q_j)^{2n} \right\rangle_0 = \tilde{T}^n (2n-1)!!,\ j = x,\ y,\ z. \tag{13}$$

The probing magnetic field perturbs the equilibrium, and the distribution function W becomes time-dependent. The linear response of the system under study in the absence of bias field was studied in Ref. [21]. It turns out, however, that the functional basis developed for that case is usable as well when a constant field is on. Indeed, from the set (8) it follows that the statistical moments are linked to each other only due to the retarded friction. For instance, to solve equation for the first basis function which is of zero order with respect to $\vec{Q}$, the first-order function $(\vec{Q}\times\vec{e})$ is required. Taking, in turn, the equation for the latter, one finds that it involves the function of second order in $\vec{Q}$ and so forth *ad infinum.* A thorough analysis reveals that the whole set of basis functions splits in three mutually orthogonal sub-sets (sub-bases):

$$\vec{M}_{2n+1} = \Psi_n^{(2)}(Q^2)\vec{M},\ \ \ \vec{K}_{2n} = \Psi_n^{(0)}(Q^2)\,\vec{e},\ \ \ \vec{L}_{2n+2} = \Psi_n^{(4)}(Q^2)\left[(\vec{Q}\,\vec{e})\vec{Q} - (1/3)Q^2\vec{e}\right]. \tag{14}$$

Inside each sub-base, the functions are orthogonalized stepwise alongAs szeen, the Gram-Schmidt procedure [25]. Requiring that the non-equilibrium distribution function is normalized, one finds the first function of the sub-basis K in the form $\vec{K}_0 = \delta\vec{e} = \vec{e} - \langle\vec{e}\rangle_0$. Scalar factors $\Psi_n^{(p)}(Q^2)$ which define the sub-basis functions of higher order are polynomials of argument Q^2. The power of $\Psi_n^{(p)}$ in this argument is indicated by the subscript (i.e., n) whereas the superscript (i.e., p) is the power of the "weight" factor needed to make those polynomials mutually orthogonal:

$$\left\langle Q^p \Psi_n^{(p)} \Psi_m^{(p)} \right\rangle_0 = \left\langle \left(\Psi_n^{(p)}\right)^2 Q^p \right\rangle_0 \delta_{nm}, \tag{15}$$

as it follows from definitions (14) of the basis functions. The details of constructing the full basis and the technique to work with it are partially given in [21] and in detail presented in [26].

In terms of the above-described functional bases, the non-equilibrium solution of kinetic equation (4) takes the form

$$W(\vec{e},\vec{Q},t) = W_0(\vec{e},\vec{Q})\left\{1 + \sum_{n=0,1,2,\ldots}\left[\vec{a}_n(t)\vec{K}_{2n} + \vec{b}_n(t)\vec{M}_{2n+1} + \vec{c}_n(t)\vec{L}_{2n+2}\right]\right\}. \tag{16}$$

IOP Conf. Series: Materials Science and Engineering **581** (2019) 012001 doi:10.1088/1757-899X/581/1/012001

Averaging the first basis function with distribution (16), one finds the explicit form of the first coefficient in that expansion:

$$a_i^{(0)}(t) = \left[\frac{P_{ij}^{\|}}{c_2 - c_1^2} + \frac{2P_{ij}^{\perp}}{1 - c_2} \right] \langle \delta e_j \rangle \,. \tag{17}$$

By progressive application of that procedure, one is able to obtain any number of expansion coefficients in Eq. (16)., i.e., to evaluate the non-equilibrium distribution function with arbitrary accuracy. As soon as $W(\vec{e}, \vec{Q}, t)$ is found, it could be used for deriving the observable magnetic moment, i.e., $\langle \vec{e} \rangle$. A plausible way to do that is to expand phase variable $\vec{e}$ with respect to the above-introduced set of basis functions $\vec{K}_j$, $\vec{M}_j$ and $\vec{L}_j$. Then the non-equilibrium average $\langle \vec{e} \rangle$ comes out as a series of equilibrium averages of those functions, whereas the coefficients of this series are rendered by the solution of an appropriate matrix equation of order N that couples all the coefficients. This treatment is done taking equation (4) as a starting point. Skipping this very cumbersome work, in the next section we present the main results of this treatment.

3. Dynamic magnetic susceptibility

To the corresponding matrix equation a harmonic probing field $\varepsilon(t) H \vec{p} = \varepsilon H \exp(-i\omega t)\vec{p}$ is substituted. Linearization with respect to ε yields the response at frequency ω, the equations for the longitudinal and normal projections of the dynamic magnetic moment with respect to axis $\vec{k}$ have the same general form

$$R_{ij}^{(\alpha)} X_j^{(\alpha)} = Y_i^{(\alpha)}, \quad \alpha = \|, \perp \,. \tag{18}$$

The exciting force and response vectors In Eq. (18) are defined as

$$\begin{aligned} \left(X_j^{(\alpha)}\right)^T &= \left(\langle \delta \vec{e}, \langle \vec{M}_1 \rangle, \langle \vec{K}_2 \rangle, \langle \vec{L}_2 \rangle, \langle \vec{M}_3 \rangle, \langle \vec{K}_4 \rangle, \langle \vec{L}_4 \rangle, \ldots \right)^{(\alpha)}, \\ \left(Y_j^{(\alpha)}\right)^T &= \varepsilon \xi f^{(\alpha)} (1, -1, 0, 0, 0, \ldots); \quad f^{\|} = \tfrac{1}{2}(1 - c_2), \; f^{\perp} = \tfrac{1}{4}(1 + c_2). \end{aligned} \tag{19}$$

where superscript T denotes matrix transposition. The initial fragment of matrix $\hat{R}$ (8-th order is taken as an example) is

$$\begin{pmatrix} g_0(\vec{e}) & -\frac{1}{2}\beta & 0 & 0 & 0 & 0 & 0 & 0 \\ -\gamma(\vec{M}, \vec{e}) & g_1(\vec{M}) & \frac{1}{3}\beta & -\frac{1}{2}\beta & 0 & 0 & 0 & 0 \\ 0 & \gamma(\vec{K}, \vec{M}) & g_2(\vec{K}) & 0 & -\frac{1}{2}\beta & 0 & 0 & 0 \\ 0 & -\frac{5}{6}\gamma(\vec{L}, \vec{M}) & 0 & g_2(\vec{L}) & \frac{1}{6}\beta & 0 & 0 & 0 \\ 0 & 0 & -\frac{5}{3}\gamma(\vec{M}, \vec{K}) & \gamma(\vec{M}, \vec{L}) & g_2(\vec{M}) & \frac{1}{3}\beta & -\frac{1}{2}\beta & 0 \\ 0 & 0 & 0 & 0 & 2\gamma(\vec{K}, \vec{M}) & g_4(\vec{K}) & 0 & -\frac{1}{2}\beta \\ 0 & 0 & 0 & 0 & -\frac{7}{6}\gamma(\vec{L}, \vec{M}) & 0 & g_4(\vec{L}) & \frac{1}{6}\beta \\ 0 & 0 & 0 & 0 & 0 & -\frac{7}{3}\gamma(\vec{M}, \vec{K}) & 2\gamma(\vec{M}, \vec{L}) & g_5(\vec{M}) \end{pmatrix} \cdot \tag{20}$$

The projection index that was omitted for brevity in (20), is recovered right below. Thus specified expressions for the diagonal elements of $\hat{R}$:

$$g_m^{(\alpha)}(\vec{B}) = \frac{m\beta}{2} \left(1 + \frac{1}{q}\right) + \gamma^{(\alpha)}(\vec{B}) - i\omega\tau_D, \tag{21}$$

IOP Conf. Series: Materials Science and Engineering **581** (2019) 012001 doi:10.1088/1757-899X/581/1/012001

where $\tau_D = \zeta_N/2T$ is the Debye rotary relaxation time and the $\gamma^{(\alpha)}$ terms are

$$\gamma^{\|}(\vec{e}) = \frac{1-c_2}{2(c_2-c_1^2)},\ \gamma^{\perp}(\vec{e}) = \gamma^{\|}(\vec{M}) = \gamma^{\perp}(\vec{K}) = \frac{1+c_2}{2(1-c_2)},\ \gamma^{\perp}(\vec{M}) = \frac{3-c_2}{2(1+c_2)},$$
$$\gamma^{\|}(\vec{K}) = \frac{1-c_2}{2c_2},\ \gamma^{\perp}(\vec{L}) = \frac{13+c_2}{2(7-c_2)},\ \gamma^{\|}(\vec{L}) = \frac{7-c_2}{2(3+c_2)}.$$

Those for the off-diagonal elements also depend on the strength of bias field, they are

$$\gamma^{\|}\left(\vec{M},\vec{e}\right) \equiv \gamma^{\|}(\vec{e}) = \frac{1-c_2}{2(c_2-c_1^2)},\ \gamma^{\|}\left(\vec{K},\vec{M}\right) = \frac{1}{\gamma^{\|}\left(\vec{M},\vec{K}\right)} = \frac{2c_2}{1-c_2},\ \gamma^{\|}\left(\vec{L},\vec{M}\right) = \frac{1}{\gamma^{\|}\left(\vec{M},\vec{L}\right)} = \frac{3+c_2}{5(1-c_2)},$$
$$\gamma^{\perp}\left(\vec{M},\vec{e}\right) \equiv \gamma^{\perp}(\vec{e}) = \frac{1+c_2}{2(c_2-c_1^2)},\ \gamma^{\perp}\left(\vec{K},\vec{M}\right) = \frac{1}{\gamma^{\perp}\left(\vec{M},\vec{K}\right)} = \frac{2(1-c_2)}{1+c_2},\ \gamma^{\perp}\left(\vec{L},\vec{M}\right) = \frac{1}{\gamma^{\perp}\left(\vec{M},\vec{L}\right)} = \frac{7-c_2}{5(1+c_2)}.$$

In general, the field-induced anisotropy does not affect the way of linking of statistical moments, it remains the same as in the isotropic case [21]. However, the matrix elements below the main diagonal are different as they become bias-field dependent.

To solve Eq. (18) by backward sweep method (see [21, 27, 28], for example), we cast it in the form of recurrence relation

$$C^-_{n-2}X_{n-2} + B^-_{n-1}X_{n-1} + A_nX_n + B^+_{n+1}X_{n-1} + C^+_{n+2}X_{n-2} = Y_n. \tag{22}$$

From the above-given matrix fragment (20), it follows that the coefficients of this equation could be presented as respective components of vectors

$$A = \left(g_0(\vec{e}), g_1(\vec{M}), g_2(\vec{K}), g_2\vec{L}, g_3(\vec{M}), g_4(\vec{K}), g_4(\vec{L}), g_5(\vec{M}), g_6(\vec{K}), g_6(\vec{L}), g_6(\vec{M}), \ldots\right),$$
$$B^- = \left(0, -\gamma(\vec{M},\vec{e}), \gamma(\vec{K},\vec{M}), 0, \gamma(\vec{M},\vec{L}), 2\gamma(\vec{K},\vec{M}), \gamma(\vec{M},\vec{L}), 3\gamma(\vec{K},\vec{M}), 0, 3\gamma(\vec{M},\vec{L}), \ldots\right),$$
$$C^- = -\tfrac{1}{6}\left(0,0,0,5\gamma(\vec{L},\vec{M}), 10\gamma(\vec{M},\vec{K}), 0, 7\gamma(\vec{L},\vec{M}), 14\gamma(\vec{M},\vec{K}), 0, 9\gamma(\vec{L},\vec{M}), 18\gamma(\vec{M},\vec{K}), \ldots\right),$$
$$B^+ = \tfrac{1}{6}\beta\left(-3,2,0,1,2,0,1,2,0,1,2,\ldots\right),\quad C^+ = -\tfrac{1}{2}\beta\left(0,1,1,0,1,1,0,1,1,0,1,\ldots\right),$$
$$Y = f\left(1,-1,0,0,0,0,0,0,0,0,0,\ldots\right). \tag{23}$$

Upon substitution an expression with two sweep coefficients and a free term – $X_n = S_nX_{n-1} + P_nX_{n-2} + a_n$ – the five-term recurrence formula (22) reduces to a three-term one. After that, by comparison with (22) one arrives at the sweep equations

$$S_n = -\left[B^-_n + \left(B^+_n + C^+_nS_{n+2}\right)P_{n+1}\right]/Z_n,\quad Z_n = A_n + S_{n+1}\left(B^+_n + C^+_nS_{n+2}\right) + C^+_nP_{n+2},$$
$$P_n = -C^-_n/Z_n,\quad a_n = \left[Y_n - \left(B^+_n + C^+_nS_{n+2}\right)a_{n-1} - C^+_na_{n-2}\right]/Z_n. \tag{24}$$

To derive an expression for the magnetic susceptibility, we combine the first line of (18) and the sweep relation for the second response component: $g_0X_1 - \frac{1}{2}\beta X_2 = Y_1$ and $X_2 = S_2X_1 + a_2$. Sweep coefficient S_2 is obtained from recurrence relations (24) under uniform initial conditions $S_{N+1} = P_{N+1} = a_{N+1} = 0$, i.e., assuming that a statistical moment is the smaller the higher its order. Then varying the largest order of the moment that is taken into account gives a convenient way to control the precision of the performed calculations. In our work $N = 10$ ensured very good accuracy of the results.

With explicit expressions for the coefficients of Eq. equations (22), one finds, first, $S_2 = \gamma(\vec{e})/Z_2$, $a_2 = -Y_1/Z_2$ and, finally, the non-equilibrium magnetic moment X_1. Using definition (7) and restoring all the temporally omitted notations, one arrives at the susceptibility expression

$$\chi^{(\alpha)}(\omega,\xi) = \frac{\chi^{(\alpha)}(0,\xi)}{1 - \dfrac{i\omega\tau_D}{\gamma^{(\alpha)}(\vec{e}) - \frac{1}{2}\beta S_2}},\quad \chi^{\|}(0,\xi) = \frac{\pi\mu^2}{T}\left(c_2 - c_1^2\right),\quad \chi^{\perp}(0,\xi) = \frac{\pi\mu^2}{T}\left(1 - c_2\right); \tag{25}$$

IOP Conf. Series: Materials Science and Engineering **581** (2019) 012001 doi:10.1088/1757-899X/581/1/012001

here the components of equilibrium (static) susceptibility $\chi(\alpha)(0,\xi)$ do not depend on the relaxation mechanism and have the same form as in Refs. [29, 28].

An approximate expression for $\chi^{(\alpha)}(\omega,\xi)$ could be obtained, as it is done in Refs. [30], in a much simpler way if to use only the first two statistical moments, i.e., to set $N = 2$. In the model under study, the viscoelastic properties of carrier medium enter the dynamic equations via the retarded friction. Because of that, the minimal set of phase variables, along with the observable quantity $\langle\vec{e}\rangle$, includes the torque $(\vec{Q}\times\vec{e})$. Given that, the general formula (25) reduces to [30]:

$$\tilde{\chi}^{(\alpha)}(\omega,\xi) = \frac{\chi^{(\alpha)}(\omega,\xi)}{\chi^{(\alpha)}(0,\xi)} = \frac{1}{1 - i\omega\tau^{(\alpha)}\left(1 + \dfrac{\frac{1}{2}\beta}{\gamma^{(\alpha)}(\vec{M}) + \frac{1}{2}\frac{\beta}{q} - i\omega\tau_D}\right)}, \quad \tau^{(\alpha)} = \frac{\tau_D}{\gamma^{(\alpha)}(\vec{e})}. \tag{26}$$

In this case the frequency behavior of the susceptibility is entirely determined by the effective elasticity β of the medium and viscoelasticity parameter $q = \eta_M/\eta_N$. As already mentioned, at high temperatures ($\beta \ll 1$) the viscoelasticity is weak, so that the susceptibility expression reduces to a simple Debye formula with the relaxation time τ_D.

For a Jeffreys matrix with substantial elasticity ($\beta \gg 1$) and under a relatively low bias field, the susceptibility formula (26) may be effectively presented as a sum of just two relaxation modes, slow (s) and fast (f):

$$\tilde{\chi}^{(\alpha)}(\omega,\xi) \simeq \frac{1 - 2\gamma^{(\alpha)}(\vec{e})/\beta}{1 - i\omega\tau_s} + \frac{2\gamma^{(\alpha)}(\vec{e})/\beta}{1 - i\omega\tau_f}, \tag{27}$$

$$\tau_s = \frac{\tau_D}{\gamma^{(\alpha)}(\vec{e})}\left[1 + \frac{\beta/2}{\gamma^{(\alpha)}(\vec{M}) + \beta/2q}\right], \quad \tau_f = \frac{2\tau_D}{\beta} = \frac{\zeta_N}{\beta}; \quad \beta > 2\gamma^{(\alpha)}(\vec{e}),\ \beta \gg 1.$$

We discuss the behavior of the obtained dynamic (complex) susceptibility in terms of its imaginary component $\mathrm{Im}\tilde{\chi}^{(\alpha)}$ that is responsible for the AC field energy absorption. In Fig. 2 along with the lines of $\mathrm{Im}\tilde{\chi}^{\parallel}$ calculated with the exact solution of KE (long series), we present the effective field (two-term) approximation from Ref. [30].

As seen, qualitatively, the approximate plots are similar to the exact ones. In particular, with the increase of matrix elasticity, the slow component of $\mathrm{Im}\tilde{\chi}^{\parallel}$ grows in magnitude whereas its peak goes to yet lower frequencies. Meanwhile, the fast component of $\mathrm{Im}\tilde{\chi}^{\parallel}$ decreases and moves to the high frequency end. Note that the shift of the low-frequency peak, although more pronounced in exact solution, is anyway very slow. From asymptotic analysis it follows that its position get indeed close to the limiting point $\omega\tau_s = \omega\tau_D(1+q)/\gamma^{(\alpha)}(\vec{e})$ only at $\beta \gg 2q$, i.e., $\tau_D \gg \tau_M$. For typical viscoelastic fluids where $q \sim 10^2 - 10^3$ (see [31], for example) and the more so for gels ($q \to \infty$), this limit cannot be attained.

With the growth of bias field, the relaxation rate of the slow mode increases (its cusp moves to high-frequency end) while its height becomes smaller. Meanwhile, the position of the fast-mode peak remains fixed, and just the magnitude of the peak enhances. In a strong bias field and at weak-to-moderate elasticity $\beta < 2\gamma^{(\alpha)}(\vec{e})$, approximation (28) tends to a simple expression

$$\tilde{\chi}^{(\alpha)}(\omega,\xi) = \frac{1}{1 - i\omega\tau_f^{(\alpha)}}, \quad \tau_f^{(\alpha)} \simeq \frac{\tau_D}{\gamma^{(\alpha)}(\vec{e})} \approx \begin{cases} \tau_H = \zeta_N/\mu H & \alpha = \perp, \\ \tau_H/2 & \alpha = \parallel; \end{cases} \quad \xi \gg \beta/2. \tag{28}$$

which reproduces the result for dynamic susceptibility of magnetically hard particles under strong bias field in a linearly viscous (Newtonian) matrix [29]. In this limit, only the fast relaxation mode that is forced by the imposed field is working whereas all the viscoelasticity effects are negligible.

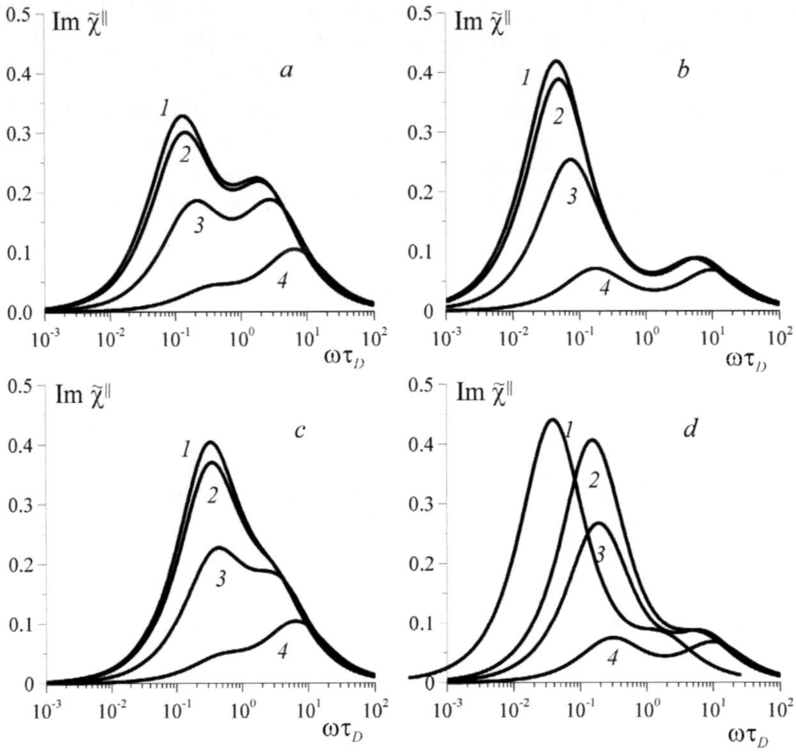

Figure 2. Normalized frequency dependences of the imaginary component of dynamic magnetic susceptibility obtained by exact solution (top row, a and b) and in mean-field approximation (bottom row, c and d) for the cases of moderate $\beta = 3$ (left column, a and c) and strong elasticity $\beta = 10$ (right column, b and d); the curves in each pane correspond to dimensionless constant fields $\xi = 0.1$ (*1*), 1 (*2*), 3 (*3*) and 10 (*4*).

4. Conclusions

Appearance of the low-frequency (slow) relaxation mode besides the customary high-frequency one, is a novel effect emerging in our model, i.e., in a system where the carrier fluid is the Jeffreys viscoelastic medium. In fact, this low-frequency mode had been already found in Ref. [30] but, as it was just a low-order (effective field) approximation, its existence could not be considered as unambigous. Now, we have demonstrated that the exact solution of the KE yields the same result. Thus, the developed high-order approximation (one may justly consider its accuracy as infinite) has shown that: (i) the slow mode really exists and (ii) the frequency dependence $\mathrm{Im}\tilde{\chi}^{\|}(\omega)$ always has a chracteristic two-peak shape.

Summing up the results, we note the following. Good general agreement between the exact and approximate plots of $\mathrm{Im}\tilde{\chi}^{\|}$ at low values of β (weak elasticity of the medium) is quite expectable because only by means of β the dynamics of the particle is affected by the retarded friction. The right-hand (high-frequency) wings of the exact and approximate lines in Fig. 2 are very similar whatever the medium elasticity, see Fig. 2. To understand that, it suffices to recall that in the model under study the most rapid relaxation process is caused by the usual Stokes mechanism (viscosity coefficient η_N) that works almost independently of the retardation one. However, if the Jeffreys media has substantial elasticity, the exact and approximate susceptibility profiles $\mathrm{Im}\tilde{\chi}^{\|}$ display notable differences. In particular, in the exact solution the leftward shift of low-frequency peak is much more pronounced than in the effective-field approximation. In other words, in complex fluid media with substantial dynamic elasticity the contributions of higher moments (they are related to the viscoelastic relaxation) are not at all negligible.

Acknowledgments
The work was done under auspices of RFBR grant 17-42-590504.

References

[1] Oswald P 2009 *Rheophysics: The Deformation and Flow of Matter* (Cambridge University Press)
[2] Pankhurst Q A, Thanh N K T, Jones S K and Dobson J 2009 *Journal of Physics D: Applied Physics* **42** 224001
[3] Veiseh O, Gunn J W and Zhang M 2010 *Advanced Drug Delivery Reviews* **62** 284–304
[4] Mody V V, Cox A, S S, Ajay S, Bevins W and Parihar H 2014 *Applied Nanoscience* **4** 385–392
[5] Ulbrich K, Holá K, Šubr V, Bakandritsos A, Tuček J and Zbořil R 2016 *Chemical Reviews* **116** 5338–5431
[6] Dutz S and Hergt R 2013 *International Journal of Hyperthermia* **29** 790–800
[7] Périgo E A, Hemery G, Sandre O, Ortega D, Garaio E, Plazaola F and Teran F J 2015 *Applied Physics Reviews* **2** 041302
[8] Suriyanto, Ng E Y K and Kumar S D 2017 *Biomedical Engineering* **16** 36
[9] El Hajj Diab D, Clerc P, Serhan N, Fourmy D and Gigoux V 2018 *Nanomaterials* **8** 468
[10] Yang C C, Yang S Y, Chieh J J, Horng H E, Hong C Y, Yang H C, Chen K H, Shih B Y, Chen T F and Chiu M J 2011 *ACS Chemical Neuroscience* **2** 500505
[11] Wiekhorst F, Steinhoff U, Eberbeck D and Trahms L 2012 *Pharmaceutical Research* **29** 1189–1202
[12] Liebl M, Wiekhorst F, Eberbeck D, Radon P, Gutkelch D, Baumgartner D, Steinhoff U and Trahms L 2015 *Biomedical Engineering* **60** 427–443
[13] Chieh J J, Wei W C, Liao S H, Chen H H, Lee n F, Lin F C, Chiang M H, Chiu M J, Horng H T and Yang S Y 2018 *Sensors* **18** 1043
[14] Nair B G, Nagaoka Y, Morimoto H, Yoshida Y, Maekawa T and Kumar D S 2010 *Nanotechnology* **21** 455102
[15] Tseng P, Judy J W and Di Carlo D 2012 *Nature Methods* **9** 1113–1119
[16] Gardel M L, Valentine M T and Weitz D A 2005 *Microrheology* (Springer, New York)
[17] J B G and Yeates D B 2007 *Journal of Chemical Physics* **127** 105106
[18] Pavlovsky L, Younger J G and Solomon M J 2013 *Soft Matter* **9** 122–132
[19] Kim J C, Seo M, Hillmyer M A and Francis L F 2013 *ACS Applied Materials and Interfaces* **5** 11877–11883
[20] Roeben E, Roeder L, Teusch S, Effertz M, Deiters U K and Schmidt A M 2013 *Colloid and Polymer Science=* **292** 2013–2023
[21] Rusakov V V, Raikher Yu L and Perzynski R 2013 *Mathematical Modelling of Natural Phenomena* **10** 1–43
[22] Rusakov V V, Raikher Yu L, Coffey W T and Kalmykov Yu P 2001 *Physical Review E* **63** 031402
[23] Coffey W T and Kalmykov Yu P 2012 *The Langevin Equation* 3rd ed (World Scientific, Singapore)
[24] Klimontovich Yu L 1995 *Statistical Theory of Open Systems* (Kluwer, Dordrecht)
[25] Korn G A and Korn T M 1961 *Mathematical Book for Scientists and Engineers* (MGraw Hill)
[26] Rusakov V V and Raikher Yu L 2019 *Colloid Journal* **81** to appear
[27] Risken H 1984 *The Fokker-Planck Equation* (Springer-Verlag)
[28] Raikher Yu L and Stepanov V I 2004 *Advances in Chemical Physics* **129** 419–588
[29] Raikher Yu L and Shliomis M I 1994 *Advances in Chemical Physics* **78** 595–751
[30] Rusakov V V and Raikher Yu L 2018 *Sensors* **18** 1661
[31] van Zanten J H and Rufener K P 2000 *Physical Review E* **62** 5389–5396

IOP Conf. Series: Materials Science and Engineering **581** (2019) 012002 doi:10.1088/1757-899X/581/1/012002

The influence of traveling magnetic field inductor asymmetric power supply on the liquid metal flow

I Sokolov[1], E Shvydkiy[1], G Losev[2], K. Bolotin[1], S. Bychkov[1]

[1] Ural Federal University, Yekaterinburg, Russia

[2] Institute of Continuous Media Mechanics of the Ural Brunch, RAS, Perm, Russia

E-mail: i.v.sokolov@urfu.ru, e.l.shvydky@urfu.ru, Losev.g@icmm.ru

Abstract. In modern times, exposure to a liquid metal by a travelling magnetic field is widely applied. There are laboratory studies on the processes of stirring and crystallization under the action of a traveling magnetic field. However, in the majority of studies it is assumed that the inductor power supply of the linear induction machine is carried out by a symmetrical three-phase system of currents with an equal phase shift, which, in some cases, is not quite correct. To approximate the model to real operating conditions, a numerical simulation of the magnetic field and the flow of liquid metal was carried out when supplied from a power source of symmetric three-phase voltage. The distortion of magnetic field, which, in turn, causes an nonuniform distribution of forces and the flow of a liquid metal, is shown. Evaluation of asymmetrical effect on the liquid metal flow was carried out by means of finite element method. That effect is caused by different coefficients of mutual coils induction of the linear induction machine, which is confirmed by experimental data.

1. Introduction

Application of magnetic fields in a modern metallurgy play an important role. Control of stirring and transport of melt can be contactless provide by magnetic field influence. That method many decades approves his advantages but from another hand it demands additional costs for research. Numerical simulation is a good tool for investigators due to opacity, high temperature and aggressiveness of liquid metals. Many developers use the simplifications in an electromagnetic part of the numerical models. It can be helpful for both model developing time and computation power demands. But sometimes that simplification includes a special aspect and can influence on a numerical results, especially in the case of non-ideal inductor [1].

Alternating magnetic field influence can be divided on three cases: one phase coil, rotating and travelling magnetic field. In the last case, the travelling magnetic field (TMF) is generated by inductor or stator of linear induction machine. That devise has a several differences with compare to rotating machines. TMF inductors have a longitudinal and transverse end effect. Besides of forces distribution that leads to unsymmetry resistance in coils and hence to differences in electrical current values. Some theoretical works consist of ideal TMF wave and do not take into account a real inductor [2]. But usually authors set the equal current density value in all three phases and shift between them 120° [3, 4]. For some types of windings it can be different in reality [5, 6].

In present work we suggest to compare ideal (symmetrical) and real (asymmetrical) coil current cases for TMF stirring application. The study is based on the installation of a travelling

IOP Conf. Series: Materials Science and Engineering **581** (2019) 012002 doi:10.1088/1757-899X/581/1/012002

magnetic field stirring shown in Fig. 1 presented in [7–9].

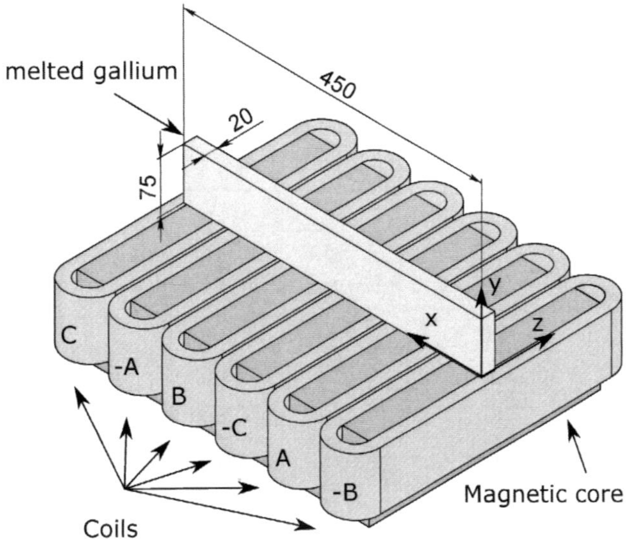

Figure 1. TMF stirring installation appearance.

2. Methods

2.1. Numerical EM and CFD model

The 3D numerical calculations were carried out by the finite element method using the Comsol Multiphysics software. At the first stage for electromagnetic calculations we used the Maxwell equations, which were formulated in terms of the magnetic vector potential in frequency domain.

$$\nabla(\nabla \times \mathbf{A}) = \begin{cases} j\mu\omega\gamma \; - \; for \; non-conducting \; domains \\ j\mu\omega\gamma + \mathbf{J}_{ext} \; - \; for \; conducting \; domains \end{cases} \tag{1}$$

$$\mathbf{E} = -j\mu\omega\mathbf{A} \tag{2}$$

where $\mathbf{A}$ is the magnetic vector potential, μ is the magnetic permeability, ω is the angular frequency, γ is the electrical conductivity, $\mathbf{J_{ext}}$ is the external current density, $\mathbf{E}$ is the electric field.

As was mentioned above the study divided in two cases. The first case is a usual uniform distribution of current density in the inductor coils. It means that external current is specified directly by the current density.

For the second case external current is specified indirectly by calculated voltage taking into account equivalent circuit and cascade coil connection. Calculation is provided according to Kirchhoffs circuit laws for equivalent circuit (Fig. 3).

$$\mathbf{E_{coil}} = \int \mathbf{E} \, dL \tag{3}$$

$$R_{coil} = \int \frac{N\mathbf{L}}{\sigma S \Sigma} \, d\Sigma \tag{4}$$

where R_{coil}, N, L, σ, S and Σ are the total resistance of the coil, the number of turns, the coil length, the wire bulk conductivity, the wire cross-section area and the total cross-sectional area of the coil domain respectively. Integration was provided over coil conductors length.

IOP Conf. Series: Materials Science and Engineering **581** (2019) 012002 doi:10.1088/1757-899X/581/1/012002

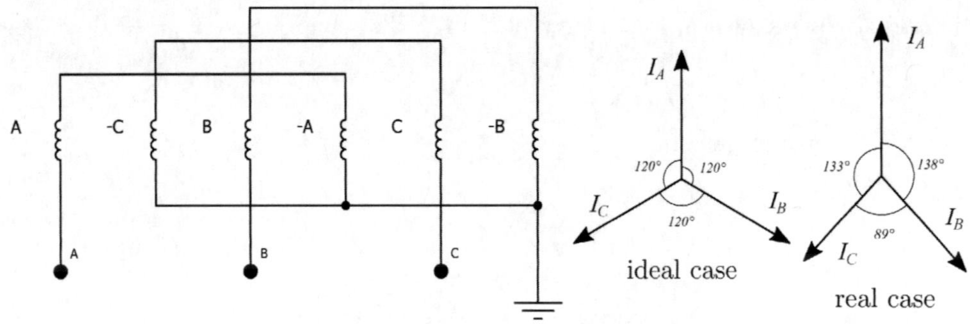

Figure 2. Windings connection scheme and coil current vectors

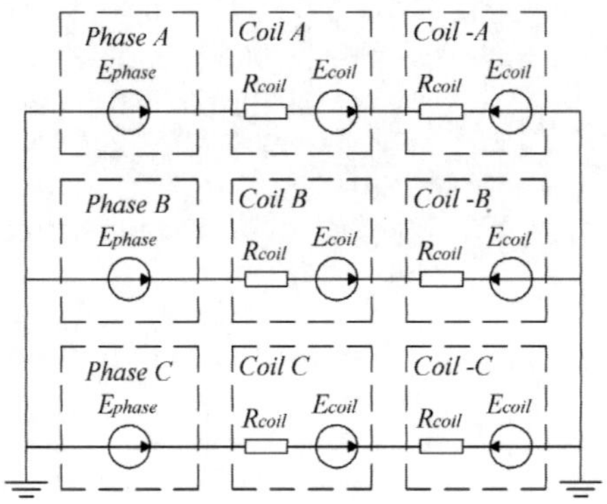

Figure 3. Equivalent circuit for asymmetry calculation

The calculations of the magnetic field, induced current and Lorenz force distribution were made with the use of the Comsol Magnetic Fields interface. Under the assumption of low magnetic Reynolds number ($Re_m = 0.065$), the induced current density was calculated by noninductive formulation [3, 10, 11].

Further the Lorenz force density are used in CFD calculations. For evaluation of turbulent flow dynamics the Large Eddy Simulation (LES) turbulence model was chosen. That approach allows to resolve the large unsteady eddies by the mesh, whereas the effect of smaller eddies at the mesh element scale or bellow are approximated. That method is demands a more computing power but from another hand can give more precision result of recirculated flow.

2.2. Experimental verification

We suggest to validate the numerical results by comparison with experimental data. The validation is carried out by magnetic flux density in the air gap between inductor and liquid metal. Magnetic flux was measured in the y and x directions. The second parameter is the time-averaged velocity profile measured by Ultrasound Doppler Velocimetry. The sensor was placed on the wall of the cell at hight 29.5 mm from the vessel bottom. The detail experimental setup and measured systems description can be found in [9, 12].

3. Results

At the first step the magnetic flux density is obtained. Cut surface in a middle of liquid metal domain is shown in Fig. 4. The two cases of symmetrical and asymmetrical inductor power supply are compared. As was expected the first case gives the uniform distribution with a local peaks close ot magnetic core tooth. For the second "asymmetrical" case differences of

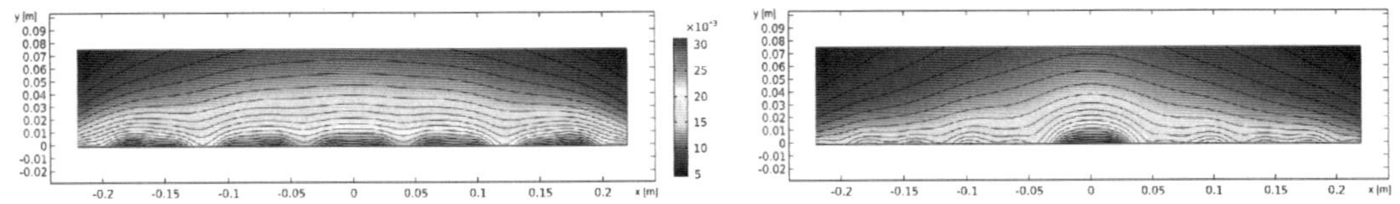

symmetrical asymmetrical

Figure 4. Magnetic flux density in the liquid metal.

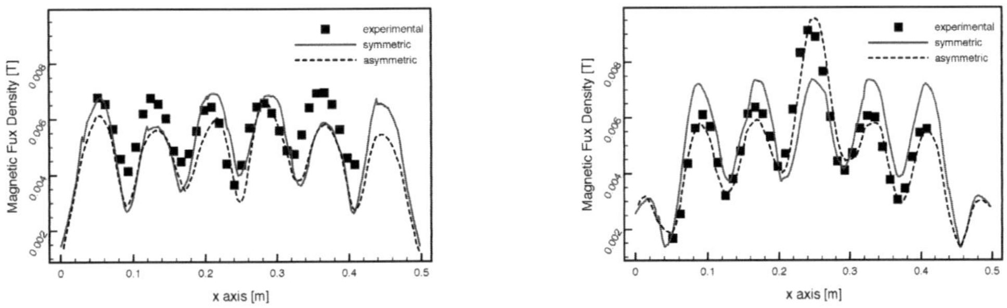

y-component x-component

Figure 5. Magnetic flux density in the air gap.

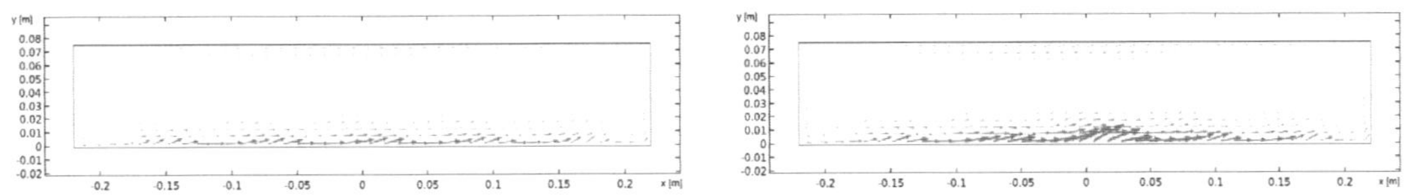

symmetrical asymmetrical

Figure 6. Lorenz force distribution in a liquid metal volume.

current magnitudes and phase shifts between them lead to the uniform distribution of magnetic flux density. In a middle of the cell the magnetic field peak is appear. That phenomena can be explained by decreasing of phase shift angle between currents I_B and I_{-A} from 120° to 89° (see Fig. 2), which is placed in one magnetic core slot.

The comparison of magnetic flux result is shown in Fig. 5. The y-component reaches maximum above magnetic core tooth and the x-component of magnetic flux has the maximum value above slots where fluxes are connected. As was expected from Fig. 4 wee can observe the peak of magnetic flux above the middle slot. Asymmetrical approach has a good agreement with the experimental data. On the other hand symmetrical approach gives the uniform distribution of magnetic flux density.

The product of magnetic flux density and current density is the Lorenz force. Fig. 6 represents the magnitude and direction of electromagnetic forces in melt. The forces are directed along x-axis. Consequently from the magnetic flux result Lorenz forces also has a difference between two cases. In the symmetrical case forces are distributed uniformly mostly in the bottom area of the melt. On the other hand asymmetrical case gave also peak on the middle of the cell and, moreover, x-component magnetic field generated the deviation of forces vectors from x to y component.

The next step of calculation is the fluid dynamic computation under the Lorenz forces (Fig. 7). Due to the fact that forces are applied mostly to the bottom region of the fluid layer and directed along x-axis the primary vortexes are generated at the bottom surface. It should be underlined

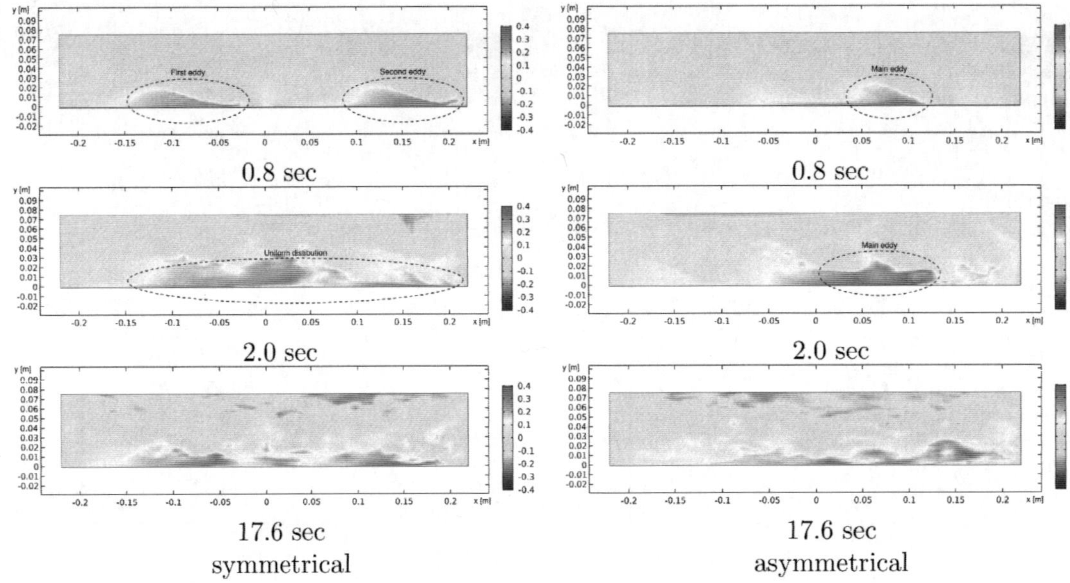

Figure 7. X-component of velocity dynamics under symmetrical and unsymmetrical TMF cases.

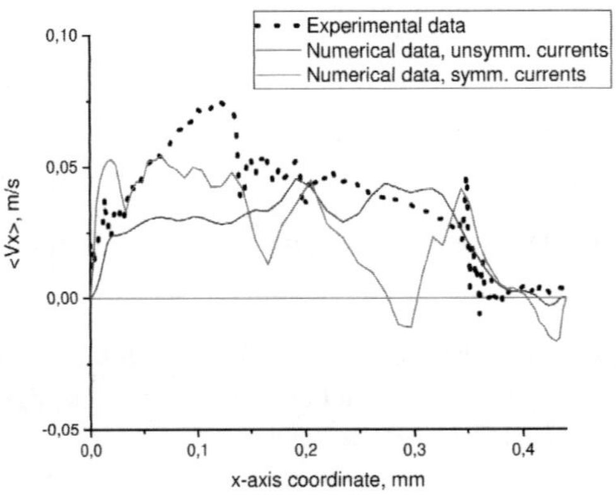

Figure 8. Comparison of mean velocity profile with UDV experimental data.

that in two cases we have the different flow pattern. In the symmetrical case the two vortexes are appear due to uniform forces distribution. In contrast, the asymmetrical case leads to one main vortex in the middle of the cell. That phenomena is explained by the magnetic field peak mentioned above. Further the flow patterns stay different. In the asymmetrical case main vortex starts generates in the middle zone. Under that influence the liquid starting under clockwise rotation. As it was expected the flow is highly unstable and one can conclude that LES approach can be useful for EM stirring calculation.

Comparison of numerical and experimental obtained x-component of time-averaged velocity profiles presented in Fig. 8. The both cases have a good agreement, but don't show near the wall eddy. On the rest part of the curve the asymmetrical case has a better agreement with experimental data.

4. Conclusions

In the presented work, a numerical simulation of a linear induction machine with a liquid metal secondary element by finite element method was carried out. Two cases of current excitation in TMF inductors coils are considered. For taking into account coil current asymmetry the equivalent circuit technique was suggested.

Considering asymmetry of TMF inductor coil currents it was shown that the asymmetry has a certain influence on simulation result. On the one hand it affects on the magnetic flux distribution, which allows to achieve well agreement with experimental data (Fig. 5). On the other hand this results in non-uniform Lorenz forces distribution and influences on the fluid flow dynamics. The magnetic flux density increase at the middle part of the inductor generates one large-scale vortex instead of uniform flow velocity distribution. Comparison of the proposed method with UDV measurements demonstrates more accurate determination of the velocity characteristics.

Considering the results obtained in the previous studies of TMF stirrers [4, 7, 13, 14] it can be mentioned that the proposed approach of currents determination allows one to increase the accuracy of model. Therefore the usage of the equivalent circuit method is justified in the case of pronounced asymmetry. So it can be concluded that the goal of the work has been achieved. The chosen method provides good comparison with the experimental data and can be widely used in the field of applied magnetohydrodynamics.

The next step in the frame of the global project is model of TMF low-frequency modulations influence on the rising MHD flows and stirring efficiency. It allows to predict the effect of the TMF and forced convection action during solidification process, the solidification rate and front shape. Concerning the developed method, it could be important to estimate the effect of other phase shifts of the LIM power supply caused by asymmetric compensation, as well as investigate other constructions of LIM for subject of asymmetry.

Acknowledgments

The work of Institute of Continuous Media Mechanics team is supporting by the RFBR grant 17-48-590539_r_a and the work of Ural Federal University team is supporting by Act 211 Government of the Russian Federation, contract 02.A03.21.0006.

References

[1] Azulay A, Mikhailovich B, Levy A and Yakhot A 2018 *Physics of Fluids* **30** 075104
[2] Stiller J, Koal K, Nagel W E, Pal J and Cramer A 2013 *The European Physical Journal Special Topics* **220** 111–122 ISSN 1951-6401
[3] Denisov S, Dolgikh V, Khripchenko S, Kolesnichenko I and Nikulin I 2014 **50** 407–422
[4] Avnaim M, Mikhailovich B, Azulay A and Levy A 2018 *International Journal of Heat and Fluid Flow* **69** 23 – 32 ISSN 0142-727X
[5] Yamamura S 1972 *Theory of Linear Induction Motors* Halsted Press book (Wiley) ISBN 9780470970904
[6] Sarapulov F N, Frizen V E, Shvydkiy E L and Smol'yanov I A 2018 *Russian Electrical Engineering* **89** 270–274 ISSN 1934-8010
[7] Shvydkiy E, Kolesnichenko I, Khalilov R, Pavlinov A and Losev G 2018 *IOP Conference Series: Materials Science and Engineering* **424** 012012 ISSN 1757-899X
[8] Shvydkiy E and Kolesnichenko I 2018 3D numerical simulation of the linear induction motor, considering magnetic saturation *Proceedings of the 2018 IEEE Conference of Russian Young Researchers in Electrical and Electronic Engineering, ElConRus 2018* ISBN 9781538643396
[9] G L Losev, I V Kolesnichenko R I K 2018 Control of the metal crystallization process by the modulated traveling magnetic field Control of the metal crystallization process by the modulated traveling magnetic field
[10] Shercliff J 1962 *Theory of electromagnetic flow-measurement* (CUP Archive)
[11] Kolesnichenko I, Khalilov R, Khripchenko S and Pavlinov A 2012 *Magnetohydrodynamics* **48** 221 234
[12] Losev G, Shvydkiy E, Sokolov I, Pavlinov A and Kolesnichenko I 2019 *Magnetohydrodynamics* **55** **(in press)**

[13] Wang X, Moreau R, Etay J and Fautrelle Y 2009 *Metallurgical and Materials Transactions B* **40** 104–113 ISSN 1543-1916
[14] Zaidat K, Sari I, Boumaaza A, Abdelhakem A, Hachani L and Fautrelle Y 2018 *IOP Conference Series: Materials Science and Engineering* **424** 012052

IOP Conf. Series: Materials Science and Engineering **581** (2019) 012003 doi:10.1088/1757-899X/581/1/012003

Mesostructural origin of the field-induced pseudo-plasticity effect in a soft magnetic elastomer

O V Stolbov and Yu L Raikher

Institute of Continuous Media Mechanics, Russian Academy of Sciences, Ural Branch, 614013 Perm, Russia

E-mail: sov@icmm.ru

Abstract. A model is presented to studying the pseudo-plasticity effect in a soft magnetic elastomer: a weakly-linked polymer matrix filled with micron-size particles of a ferromagnet with virtually zero coercivity (e.g., iron carbonyl). If the filling fraction of the composite is sufficiently high, such a material displays a remarkable ability to radically change its rheological behavior: being subjected to external magnetic field, the sample loses its elasticity and responds to external mechanical load as if being made of a ductile substance with no tendency to restore its initial shape. In our simulations we show that, considering the system at the mesoscopic scale (as an assembly of particles embedded in an elastic continuum) and taking into account just the magnetostatic and elastic interactions, one is able to qualitatively correct reproduce the field-induced pseudo-plasticity effect.

1. Introduction

Soft magnetic elastomers (SMEs) are functional composite materials obtained by filling soft polymer matrices with ferromagnet microparticles. Under moderate-to-high concentrations of the ferrophase (up to 30 wt.%), SMEs acquire high sensitivity to magnetic fields still retaining a good ability for large deformations. This combination makes these materials quite attractive for applications of which one can name adaptive dampers and vibration isolators, magnetically-controlled manipulators, gripper and other actoric devices of many kinds.

Along with the well-known magnetodeformational [1] and magnetorheological properties [2], e.g., field-induced shape changes and/or modulations of the elasticity moduli and internal friction, SMEs possess some unique and less evident effects. Of those, the most fascinating is the field-induced plasticity also known as magnetic shape memory effect [3, 4, 5, 6]. At the macroscopic scale it manifests itself as a field-induced transition in result of which a SME sample, instead of being elastic (under zero or low field) transforms in a piece of putty (under moderate-to-strong field).

The origin of the effect is doubtless: it is due to the processes which undergo at the mesoscopic scale and originates from the interparticle magnetoelastic interactions. However, mesoscopic models capable of detailing the mechanism of the pseudo-plasticity in a SME are either oversimplified or missing. In this work we present a numerical model that, in our view, on the one hand, is simple enough to make the solution comprehensible but, on the other hand, reveals the physical origin of all the essential effects – magnetic pseudo-plasticity, first place – stemming from mesoscopic structuring induced in a SME by external magnetic field and mechanical load.

IOP Conf. Series: Materials Science and Engineering **581** (2019) 012003 doi:10.1088/1757-899X/581/1/012003

2. Problem statement

Mathematic formulation of the problem splits in two parts: elasticity theory and magnetostatics.

2.1. Elastic problem

Consider a SME sample that occupies a 3D region Ω and consists of an elastic matrix with N embedded solid spherical particles. To be able to subject the sample to mechanical loads, in Ω some auxiliary region Ω_h (not necessarily simply-connected) is included. It is always positioned at the outer border of Ω and does not contain any particles. By its elastic properties, Ω_h is more rigid than the main polymer domain.

The space occupied by i-th particle is denoted as $\Omega_i^{(p)}$, so that the space of the matrix is $\Omega^{(m)} = \Omega \setminus \Omega_h \setminus \Omega_1^{(p)} \setminus \Omega_2^{(p)} \ldots \setminus \Omega_N^{(p)}$. Therefore, each geometrical point inside Ω belongs to one and only one of three elastically different sets: they are either a particle (p), the soft part of the matrix (m), or the rigid part of the matrix (h). Accordingly, we introduce three indicator functions:

$$I_i^{(p)}(\boldsymbol{r}) = \begin{cases} 1, & \boldsymbol{r} \in \Omega_i^{(p)} \\ 0, & \boldsymbol{r} \notin \Omega_i^{(p)} \end{cases}, \quad I^{(m)}(\boldsymbol{r}) = \begin{cases} 1, & \boldsymbol{r} \in \Omega^{(m)} \\ 0, & \boldsymbol{r} \notin \Omega^{(m)} \end{cases}, \quad I^{(h)}(\boldsymbol{r}) = \begin{cases} 1, & \boldsymbol{r} \in \Omega^{(h)} \\ 0, & \boldsymbol{r} \notin \Omega^{(h)} \end{cases}, \tag{1}$$

where $\boldsymbol{r}$ is radius-vector of any inner point of the SME sample.

In a general case, each i-th particle of the considered sample experiences force $\boldsymbol{F}_i^{(p)}$ that is exerted on its center. When subjected to fields and loads, the SME sample is fixed by the part Γ_u of its outer surface whereas another part of the same surface, Γ_f, is subjected to a uniformly distributed mechanical load $\boldsymbol{F}_b$.

Then the set of elasticity theory equations [7] for the problem under study takes the form

$$\nabla \cdot \boldsymbol{\sigma} + \boldsymbol{f} = 0, \quad \boldsymbol{f} = \sum_{i=1}^{N} I_i^{(p)} \boldsymbol{F}_i^{(p)} / V_i,$$

$$\boldsymbol{\sigma} = -p\mathbf{g} + 2G\mathbf{e}, \quad \mathrm{tr}\,\mathbf{e} = 0, \quad G = G_m I^{(m)} + G_h I^{(h)} + \sum_{i=1}^{N} I_i^{(p)} G_p, \tag{2}$$

$$\mathbf{e} = \frac{1}{2}(\nabla \boldsymbol{u} + \nabla \boldsymbol{u}^{\mathrm{T}}), \quad \boldsymbol{n} \cdot \boldsymbol{\sigma}|_{\Gamma_f} = \boldsymbol{F}_b / S_f, \quad \boldsymbol{u}|_{\Gamma_u} = \boldsymbol{u}^*,$$

where $\mathbf{g}$ is unit tensor, $\boldsymbol{u}$ displacement vector, G_m shear modulus of the matrix, G_h shear modulus of the fixed part of the matrix, G_p shear modulus of the particle (it is much greater than G_m), S_f is the surface area of Γ_f, and λ is Lamé constant.

At the mesoscopic level, forces $\boldsymbol{F}_i^{(p)}$, initiate translational and rotational displacements of i-th particle; at the macroscopic level, the collective effect of those motions results in displacements of the sample border Γ_f. To deal with the situation, it is convenient to introduce the averaged values

$$\boldsymbol{u}_i^{(p)} = \frac{1}{V_i} \int_{\Omega_i^{(p)}} \boldsymbol{u}\, dV, \quad \boldsymbol{u}^{(b)} = \frac{1}{S_f} \int_{\Gamma_i^{(p)}} \boldsymbol{u}\, dS, \tag{3}$$

of the particle and border displacements, respectively; here V_i is the volume of i-th particle and $\Gamma_i^{(p)}$ its surface.

Defining the generalized vectors of forces $\boldsymbol{F}$ and displacements $\boldsymbol{U}$ as

$$\boldsymbol{F} = \{\boldsymbol{F}_1^{(p)}, \boldsymbol{F}_2^{(p)}, \ldots, \boldsymbol{F}_N^{(p)}, \boldsymbol{F}^{(b)}\}, \quad \boldsymbol{U} = \{\boldsymbol{u}_1^{(p)}, \boldsymbol{u}_2^{(p)}, \ldots, \boldsymbol{u}_N^{(p)}, \boldsymbol{u}^{(b)}\}, \tag{4}$$

IOP Conf. Series: Materials Science and Engineering **581** (2019) 012003 doi:10.1088/1757-899X/581/1/012003

Figure 1. A test SME sample presented as a continuum material with piece-wise shear modulus that assumes the values: $G_m = 1$, $G_p = 500$, $G_h = 250$, see the color scale.

and introducing $\mathbf{C}$, the matrix of linear elastic response of an assembly of solid particles and of the system border Γ_f, we arrive at the relation

$$\boldsymbol{U} = \mathbf{C} \cdot \boldsymbol{F}, \tag{5}$$

that couples the averages $\boldsymbol{u}_i^{(p)}$ and $\boldsymbol{u}^{(b)}$ with the exerted forces $\boldsymbol{F}_i^{(p)}$ $\boldsymbol{F}^{(b)}$:

To find matrix $\mathbf{C}$ one has to $3(N+1)$ times solve the elastic problem (2) using esys-escript (programming tool for implementing finite element method in python [8, 9]). The force vector at i-th step is defined in such a way that $F_k = 0$ if $k \neq i$, and $F_i = 1$. Then vector $\boldsymbol{U}$ is calculated that makes i-th row of matrix $\mathbf{C}$. For further use it is convenient to introduce inverse tensor $\mathbf{L} = \mathbf{C}^{-1}$ so that equation (5) transforms into

$$\boldsymbol{F} = \mathbf{L} \cdot \boldsymbol{U}. \tag{6}$$

The increment of the work generated by forces $\boldsymbol{F}_i^{(p)}$ and $\boldsymbol{F}^{(b)}$ at respective displacements $\boldsymbol{u}_i^{(p)}$ and $\boldsymbol{u}^{(b)}$ is

$$dW = \boldsymbol{F} \cdot d\boldsymbol{U} = \boldsymbol{U} \cdot \mathbf{L} \cdot d\boldsymbol{U}. \tag{7}$$

According to energy conservation law, this work equals the elastic energy stored in the matrix. After integration one gets expression for the total elastic energy in the form

$$W = U_{\mathrm{el}} = \tfrac{1}{2} \boldsymbol{U} \cdot \mathbf{L} \cdot \boldsymbol{U}. \tag{8}$$

2.2. Magnetostatic problem

A point magnetic dipole positioned in the origin of coordinate frame possesses magnetostatic potential [10]

$$\psi = \frac{\boldsymbol{\mu} \cdot \boldsymbol{r}}{r^3}. \tag{9}$$

and induces around it magnetic field

$$\boldsymbol{h}(\boldsymbol{r}, \boldsymbol{\mu}) = -\nabla\psi = -\frac{\boldsymbol{\mu}}{r^3} + 3\frac{\boldsymbol{r}}{r^5} \boldsymbol{\mu} \cdot \boldsymbol{r}. \tag{10}$$

If the dipoles are magnetically soft, i.e., are induced by external field, then the magnetic field created In an arbitrary space point $\boldsymbol{r}$ by an assembly of N such entities positioned at

IOP Conf. Series: Materials Science and Engineering **581** (2019) 012003 doi:10.1088/1757-899X/581/1/012003

$\boldsymbol{r}_j = \boldsymbol{r}_j^0 + \boldsymbol{u}_j^p$ and possessing magnetic moments $\boldsymbol{\mu}_j$ due to the presence of external magnetic field $\boldsymbol{H}_0$, writes as

$$\boldsymbol{H}(\boldsymbol{r}) = \boldsymbol{H}_0 + \sum_{j=1}^{N} \boldsymbol{h}(\boldsymbol{r} - \boldsymbol{r}_j, \boldsymbol{\mu}_j). \tag{11}$$

Let us assume that the particles are made of a ferromagnet that magnetizes linearly, i.e., with susceptibility χ independent of the field strength. In that case, the magnetization of a spherical particle is uniform and is given by expression $\boldsymbol{\mu}/V = \boldsymbol{m} = \chi \boldsymbol{H}^{\rm int}$ with $\boldsymbol{H}^{\rm int}$ being the field inside the particle. Using relation $\boldsymbol{H}^{\rm int} = \boldsymbol{H} - 4\pi \boldsymbol{m}/3$ that links vectors of internal and external fields in case of a spherical body with magnetization $\boldsymbol{m}$, one finds the internal field as

$$\boldsymbol{H}^{\rm int} = \frac{\boldsymbol{H}}{1 + 4\pi\chi/3}, \tag{12}$$

that yields for the magnetic moment of i-th particle:

$$\boldsymbol{\mu}_i = \frac{V_i}{\chi^{-1} + 4\pi/3} \left(\boldsymbol{H}_0(\boldsymbol{r}_i) + \sum_{j=1, j\neq i}^{N} \boldsymbol{h}(\boldsymbol{r}_i - \boldsymbol{r}_j, \boldsymbol{\mu}_j) \right), \quad i = \overline{1, N}. \tag{13}$$

With the use of notations $\boldsymbol{r}_{ij} = \boldsymbol{r}_i^0 - \boldsymbol{r}_j^0 + \boldsymbol{u}_i^{(p)} - \boldsymbol{u}_j^{(p)}$, $r_{ij} = |\boldsymbol{r}_{ij}|$, $\boldsymbol{H}_0(\boldsymbol{r}_i) = \boldsymbol{H}_{0i}$ this equation takes the form

$$\sum_{j=1}^{N} \left[(1 - \delta_{ij}) \left(\frac{\mathbf{g}}{r_{ij}^3} - 3\frac{\boldsymbol{r}_{ij}\boldsymbol{r}_{ij}}{r_{ij}^5} \right) + \delta_{ij}\mathbf{g}\frac{\chi^{-1} + 4\pi/3}{V_i} \right] \cdot \boldsymbol{\mu}_j = \boldsymbol{H}_{0i}, \tag{14}$$

that explicitly presents it as a set of $3N$ linear algebraic equations for the vector components of magnetic moments $\boldsymbol{\mu}_j$ of all N particles contained in the model SME sample.

To make the left-hand side of equation (14) compact, one more notation is introduced:

$$\mathbf{A}^{ij}(\boldsymbol{U}) = \begin{cases} \mathbf{g}/r_{ij}^3 - 3\boldsymbol{r}_{ij}\boldsymbol{r}_{ij}/r_{ij}^5, & i \neq j \\ \mathbf{g}(\chi^{-1} + 4\pi/3)/V_i, & i = j \end{cases}. \tag{15}$$

so that with its aid the set (14) writes as

$$\sum_{j=1}^{N} \mathbf{A}^{ij} \cdot \boldsymbol{\mu}_j = \boldsymbol{H}_{0i}. \tag{16}$$

The last expression could be presented also in a block-matrix form::

$$\begin{pmatrix} [A_{xx}^{ij}] & [A_{xy}^{ij}] & [A_{xz}^{ij}] \\ [A_{yx}^{ij}] & [A_{yy}^{ij}] & [A_{yz}^{ij}] \\ [A_{zx}^{ij}] & [A_{zy}^{ij}] & [A_{zz}^{ij}] \end{pmatrix} \cdot \begin{pmatrix} [(\mu_x)_i] \\ [(\mu_y)_i] \\ [(\mu_z)_i] \end{pmatrix} = \begin{pmatrix} [(H_{0x})_j] \\ [(H_{0y})_j] \\ [(H_{0z})_j] \end{pmatrix} \tag{17}$$

where all the matrices are defined in the same cartesian coordinate frame.

Finally, with appropriately modified notations, the matrix set (17) may be cast in the form

$$\tilde{\mathbf{A}} \cdot \tilde{\mu} = \tilde{\boldsymbol{H}}_0, \tag{18}$$

that enables one to write the magnetostatic the energy of magnetized dipole assembly as

$$U_{mag} = -\tfrac{1}{2}\sum_{j=1}^{N} \boldsymbol{\mu}_j \cdot \boldsymbol{H}_0 = -\tfrac{1}{2}\tilde{\boldsymbol{H}}_0 \cdot \tilde{\mathbf{A}}^{-1}(\boldsymbol{U}) \cdot \tilde{\boldsymbol{H}}_0. \tag{19}$$

2.3. Coupled problem

Uniting the elastic and magnetic terms, i.e., equations (8) and (19), one arrives at the total energy of the system:

$$U_{\text{tot}} = \tfrac{1}{2}\boldsymbol{U}\cdot\mathbf{C}\cdot\boldsymbol{U} - \tfrac{1}{2}\tilde{\boldsymbol{H}}_0\cdot\tilde{\mathbf{A}}^{-1}(\boldsymbol{U})\cdot\tilde{\boldsymbol{H}}_0. \tag{20}$$

Therefore, to find the equilibrium configuration one should minimize function (20) under requirement that the particles do not overlap:

$$\boldsymbol{r}_i^0 + \boldsymbol{u}_i^{(p)} - \boldsymbol{r}_j^0 + \boldsymbol{u}_j^{(p)} > R_i + R_j, \quad i \neq j.$$

To find minimum we use sequential least squares programming algorithm [11].

2.4. Test sample

As a test SME sample we take a prolate rectangular prism pointing along Oz whose dimensions are $l_x = l_y$ and $l_z/l_x = 5$. Its three lateral sides coincide with the coordinate planes xOy, xOz, and yOz. The sample is sketched in Fig. 1, it contains 66 identical spherical particles of a linearly magnetizable ferromagnet. The particles are positioned at random, their volume content is 26%. The distribution of shear modulus in the sample is the same as indicated in at Fig. 1. Boundary conditions on the displacements $\boldsymbol{u}$ are: $u_x|_{x=0} = 0$, $u_y|_{y=0} = 0$, $u_z|_{z=0} = 0$; this excludes Euler instability of the sample under compression

3. Results and discussion

In Figure 2 we present the sequence of snapshots showing configurations of the test sample in the initial state (Fig. 2a), subjected to a longitudinal field (Fig. 2b); under the same field but compressed with force F_b (Fig. 2c) and under the same field but unloaded (Fig. 2d). As seen, the applied field makes the particles to get close and form distinctive chains, see Fig. 2b. The occurring change of longitudinal dimension of the sample is insignificant. This is quite understandable since in a sample where the particle concentration is rather large, the neighboring particles, need to cover but very small distances to pull together. Mechanical pressure exerted on the upper end-wall makes the sample to reduce its longitudinal dimension; the length of chains reduces at the expense of increasing their number, see Fig. 2c. Since all the components of the test sample are assumed to be incompressible, its longitudinal shortening is accompanied by enhancement of the transverse dimensions.

The most interesting effect is evidenced in Fig. 2d. One sees that after switching off the mechanical stress, the sample does not recover the starting configuration of Fig. 2b. Therefore, the presented example demonstrates explicitly that a mechanical load, once exerted and then removed, imparts to a magnetized sample a residual strain that would exist until the field is on. Our simulations show that only after turning off the field, the sample fully recovers its initial configuration, that of Fig. 2a.

The explanation is apparent: the studied system is described by a set of non-dissipative equations. In the initial state it dwells in equilibrium (zero field, zero load), and the internal elastic stresses are completely absent. All the actions on the sample, first, magnetization and then pressure, perturb the initial equilibrium and induce internal (mesoscopic) elastic stresses. As long as the external factors are on, they maintains the elastic stresses in the matrix at a certain level that is higher when both field and load act together, or lower when it is only the field effect. Only after all the perturbing factors are eliminated, the mesoscopic stresses become free to relax and return the system to the initial state.

Therefore, the overall cycle of the sample transformation – passage from stage (a) through (b), (c), (d) and back to (a) in terms of Fig. 2 – is reversible. However, the internal cycle that starts from stage (b) of Fig. 2 through (c) and back to (d) is irreversible. Evidently, the assembly of magnetized particles, when forced to rearrange under pressure, finds a more favorable configuration: under zero mechanical load the energy $U_{\rm tot}$ of pattern (d) of Fig. 2 is lower than that of pattern (b).

The effect of residual strain should strongly depend on the magnitude of applied field $\boldsymbol{H}_0$. Indeed, in the limiting case $H_0 = 0$ the pressure cycle is reversible as there is no magnetostatic term to counteract the elastic one in the energy expression. The pressure cycles of our test sample are presented in Fig. 3 for several values of the applied field. The strain there is defined as $\varepsilon_{zz} = u_z^{\rm top}/l_z$, where $u_z^{\rm top}$ is the displacement of the sample end-wall which in the initial state is positioned at $z = 0$; the non-dimensional applied stress is characterized by the ratio $F_b/(S_f G_m)$.

As seen, at a non-zero but low field the pressure cycle remains reversible that means that the available magnetic energy gain is yet too small to compete with the elastic stresses. The irreversibility turns up at enhanced fields, where stress–strain curves become loop-shaped that is typical for plastic materials, and the residual strain attains quite high values, up to $\sim 20\%$. Therefore, the magnetic field gives the means to modulate the plasticity of the material.

Notably, the dependence of residual strain on the applied field is not monotonic.

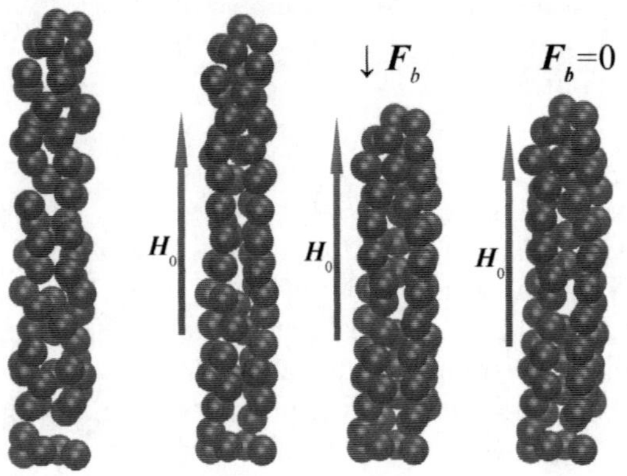

Figure 2. Distribution of the particles in the sample under: (a) – magnetic field $H_0 = 0$ and end-wall force $F_b = 0$; (b) – $H_0 = 10\sqrt{G_m}$ and $F_b = 0$; (c) – $H_0 = 10\sqrt{G_m}$, $F_b = -6\sqrt{G_m}$; (d) – $H_0 = 10\sqrt{G_m}$ and $F_b = 0$.

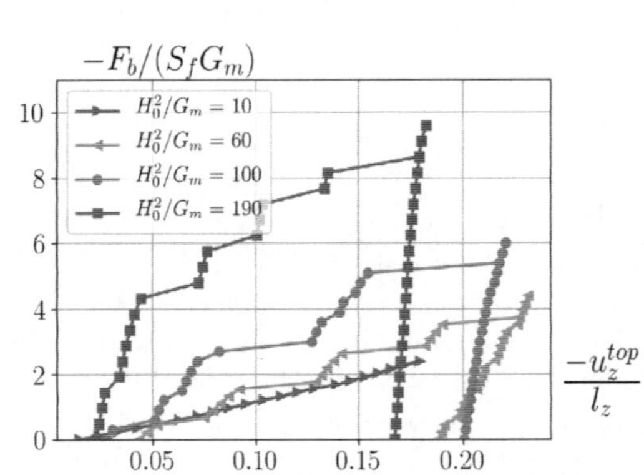

Figure 3. Stress-strain dependencies for the test sample under variation of the applied field.

4. Conclusions

A mesoscopic model of a soft magnetic elastomer that takes into account the magnetic and elastic interactions of the particles is presented. In the performed simulations, a combination of field and pressure cycles exerted on the test sample is studied. It is shown that a material with high concentration of magnetically soft microparticles, if subjected to a constant uniform magnetic field, displays a distinctive pseudo-plasticity effect in a pressure cycle. The magnitude of the effect is strongly depends on the strength of external field in a non-monotonic way. In general, the obtained model dependences agree well with the qualitative experimental evidence reported in literature.

Acknowledgments
The work was done under auspices of RBBR grants 17-41-590160 and 17-52-590504.

IOP Conf. Series: Materials Science and Engineering **581** (2019) 012003 doi:10.1088/1757-899X/581/1/012003

References

[1] Stolbov O V, Raikher Yu L and Balasoiu M 2011 *Soft Matter* **7** 8484–8487
[2] Nikitin L V, Mironova L S, Stepanov G V and Samus A N 2001 *Polymer Science, Series A* **43** 443–450
[3] Stepanov G V, Abramchuk S S, Grishin D A, Nikitin L V, Kramarenko E Yu and Khokhlov A R 2007 *Polymer* **48** 488–495
[4] Stepanov G V, Borin D Yu, Raikher Yu L, Melenev P V and Perov N S 2008 *Journal of Physics: Condensed Matter* **20** 204121
[5] Böse H 2007 *International Journal of Modern Physics B* **21** 4790–4797
[6] Nikitin L V, Stepanov G V, Mironova L S and Gorbunov A I 2004 *Journal of Magnetism and Magnetic Materials* **272** 2072–2073
[7] Landau L D and Lifshitz E M 1970 *Theory of Elasticity* (Pergamon Press)
[8] Schaa R, Gross L and Du Plessis J 2016 *Journal of Geophysics and Engineering* **13**(2) S59–S73
[9] https://launchpad.net/escript-finley
[10] Landau L D, Lifshitz E M and Pitaevskii L P 1984 *Electrodynamics of Continuous Media* (Pergamon Press)
[11] Kraft D 1994 *ACM Transactions on Mathematical Software* **20** 262–281

IOP Conf. Series: Materials Science and Engineering **581** (2019) 012004 doi:10.1088/1757-899X/581/1/012004

Inductive methods of detection the boundary of electrically conductive media in experiment

V Eltishchev, I Dimov, A Pavlinov, R Khalilov, I Kolesnichenko

Institute of Continuous Media Mechanics, Perm, Russia

E-mail: eltishchev.v@icmm.ru, dimov.i@icmm.ru, khalilov@icmm.ru

Abstract. Induction methods of localization the electrically conductive media were implemented experimentally. Local and integral techniques are considered. Relations were represented between of the local level probe (LLP) signal, its position and the frequency of alternating current. The dependences of the integral level probe (ILP) signal on the frequency of the alternating current are obtained. The methods were tested in laboratory experiments in which it was possible to determine the metal level with an error not exceeding 1 mm. The LLP is more designed for small changes in the liquid metal boundary, comparable to the size of the probe. The ILP has its advantages when used in installations in which the change in the level of the liquid metal is greater than the pitch with which the measuring coils are installed.

1. Introduction

Processes with a deformable boundary of the electrically conductive medium are often found in academic and engineering problems. There are also a number of problems in which it's necessary to control the liquid metal surface level. Such situations arise, for example, in aluminum and titanium electrolyzers, in liquid metal batteries, as well as in applications related to the metallurgy [1–3]. In these cases, fluctuations and any change in the liquid metal boundary are highly undesirable processes.

The excellent electrical conductive property of liquid metal is used as the basis for point contact, resistance and inductive types of level measuring probes. In the first type of measurement, electrodes forming a part of an electric circuit are immersed into the liquid metal, and the electrical circuit resistance is varied by the direct contact with the liquid metal [4, 5]. However, wetting is a critical factor in the use of an electrical circuit which depends upon electrode contact with the liquid metal.

High process temperatures, large masses and dimensions as well as the complexity of the technological process require a contactless measurement method [6]. The electrical conductivity of metal is considerably greater than the conductivity of the media surrounding it, and hence electromagnetic control methods are the most effective for solving this problem. These methods are used in other problems, for example in induction velocity measuring methods based on eddy currents generation. They were actively studied and used for many years. For example, velocity measurements in conductive liquids were performed by Miralles [7]. And in the last century, measurement of velocity and electroconductivity of ionized gas fluxes were made by Poberejsky [8].

Induction methods are also used in flowmeters with external placement of the generating and measuring parts, which have received basic development [9–11]. The main advantage of such systems is the possibility of placing the measuring coils into the system of thermal stabilization and cooling of the entire system, which allows them to be used even at high temperatures. Additionally, such flowmeters provide the ability to implement the liquid metal electrical conductivity measurement [12], or the registration of a gas phase in the liquid metal [13].

The task to create a reliable method of detecting the liquid metal boundary position based on inductive methods, structurally similar to the electromagnetic flowmeters faces our research team. It is not always possible to carry out investigations on real installations for various reasons, which leads to the relevance of the model laboratory studies.

In the present work, local and integral techniques for the induction determination of the liquid metal level are described, which are experimentally investigated on a duralumin cylinder. The features of the techniques realization and their comparison are given.

2. Methods

2.1. Local level probe (LLP)

The local level probe (LLP) is used as a device for local determination of the boundary between two media. It is a system of three coils placed close each other and coaxially (fig. 1a). A magnetic field is generated by the central coil, and the two outer coils are used for measuring. Measurement of the magnetic field induced by the generating coil along the probe axis shows that the measuring coils are in linear sections (fig. 1b). The LLP parameters are presented in the table 1.

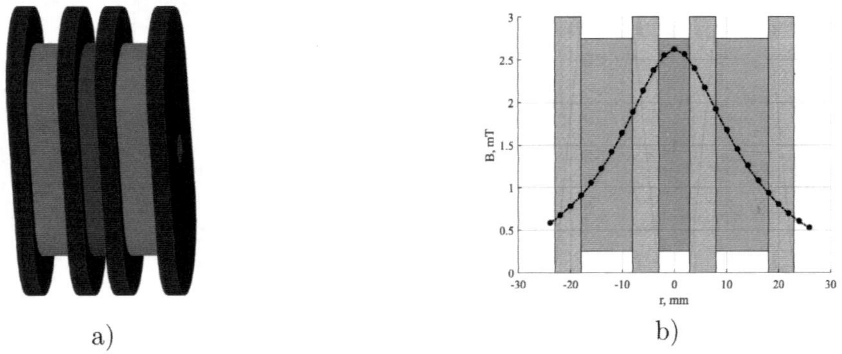

Figure 1. Local level probe: a) axonometric projection, b) cross section of the magnetic field along the central axis.

Table 1. Local level probe parameters.

	Generating coil	Measuring coil 1	Measuring coil 2
Frame width, mm	6	10	10
Frame thickness, mm	12	12	12
Frame height, mm	42	42	42
Wire section, mm	1.12	0.3	0.3
Number of turns	50	~ 1300	~ 1300
Resistance, Ohm	0.34452	44.526	43.935
Inductance, mH	0.11951	48.970	48.783

According to the law of electromagnetic induction, an alternating magnetic field created by the generating coil induces EMF in the measuring coils. If on one side of the LLP (near one of the measuring coils) there is an electrically conductive medium, then due to eddy currents arising in this medium, a counter alternating magnetic field will be created. This phenomenon underlies the skin effect. Based on the principle of superposition of magnetic fields, the potential difference in the measuring coil placed close to the electrically conductive medium is less than in the opposite one. In the absence of a conductive media, the potentials on both coils are equivalent. This phenomenon is used to determine the presence of an electrically conductive medium near the sensor.

The LLP testing is provided with the use of duralumin cylinder of the grade D16 (3) with a height of 202 mm and a diameter of 173 mm. Opposite the cylinder there is a measuring system including the LLP (2) placed on three mutually perpendicular micrometric platforms (5) (ThorLabs). With a fixed position of the measuring system, these platforms allow moving the sensor up to 50 mm in each direction in space.

An alternating current is supplied to the generating coil by the sinusoidal signal generator (GZ-109) (4). Two measuring coils are connected to the National Instruments data acquisition boards (1) (NI 9225 NI 9227). These boards allow the LabView IDE (6) records the experimental data. The values of the output voltage and current from the generator are also recorded. The coordinate system origin is placed on the edge of the duralumin cylinder (see fig. 2): the h axis is directed downward and the r axis is directed to the left.

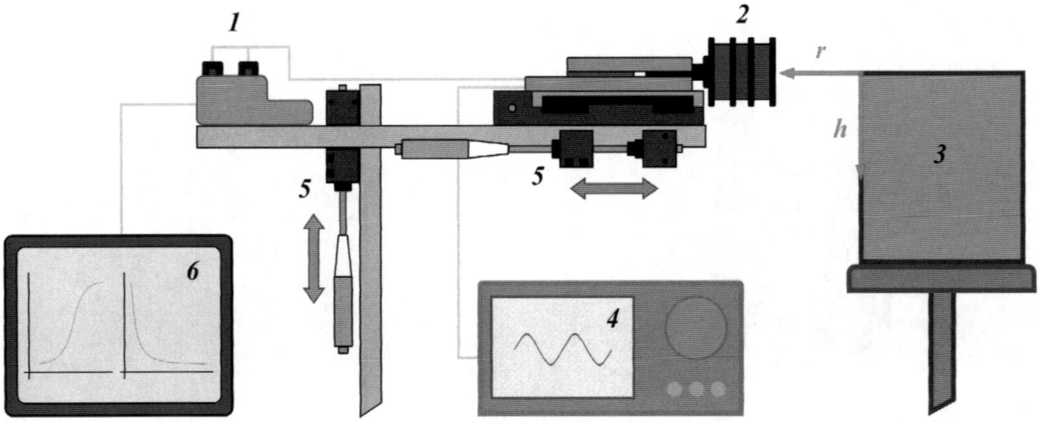

Figure 2. Laboratory setup for local level probe: 1 – data acquisition boards, 2 – local level probe, 3 – duralumin cylinder, 4 – generator, 5 – micrometric platforms, 6 – Personal Computer.

The sensitivity and characteristics of the LLP are investigated by varying the frequency of the alternating current in the generating coil, as well as by changing the vertical position of the LLP and the radial distance between it and the conducting cylinder. The parameter D_{pd} is selected as a quantitative characteristic measured in volts, which is determined by the expression:

$$D_{pd}(f,r,h) = \sqrt{\frac{\sum_{i=1}^{N} U_{2,i}^2}{N}} - \sqrt{\frac{\sum_{i=1}^{N} U_{1,i}^2}{N}} = RMS(U_2) - RMS(U_1), \quad (1)$$

where $U_{1,i}$ is the voltage at the first measuring coil (closest to the conductive medium) at the i-th time instant, $U_{2,i}$ is the voltage at the second measuring coil in i-th time instant.

2.2. *Integral level probe (ILP)*

The position of the upper boundary of the cylindrical conductive volume is also determined by using the integral level probe (ILP). In this case, a series of coils is located around the cylinder with a certain step. Each coil generates an alternating magnetic field. Each coil also resolve the resulting magnetic field by the measurements of the voltage. The voltage at the coil depends on its inductance (impedance), which depends on the position of the cylinder for fixed current. So, it is possible to calculate the impedance of each coil by measuring the voltage and current. The relationship between the impedance and the level of the coils characterizes the position of the cylinder.

The laboratory model was developed to investigate the properties of the ILP. The laboratory setup includes stainless steel wall and 38 coils. The duraluminum cylinder D16 was used as a conductive medium (fig. 3).

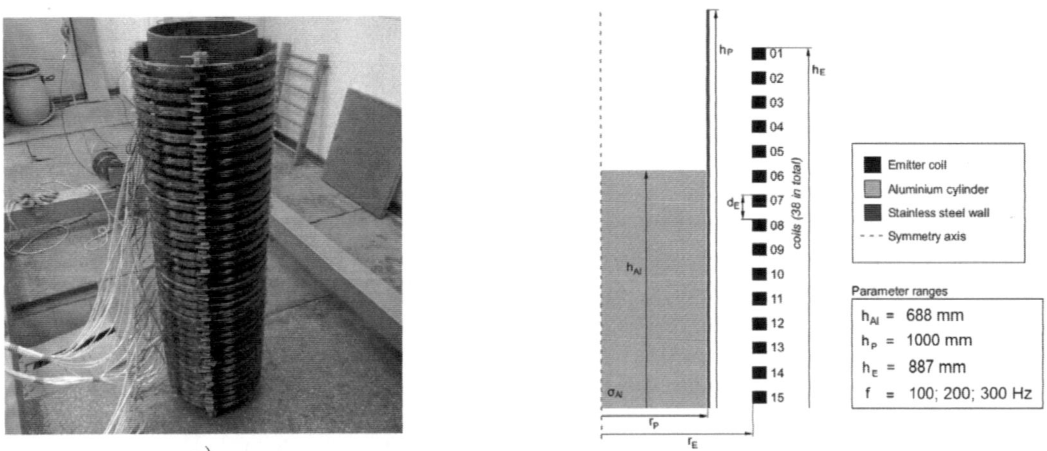

a) b)

Figure 3. Laboratory setup for integral level probe: a) real photo, b) scheme.

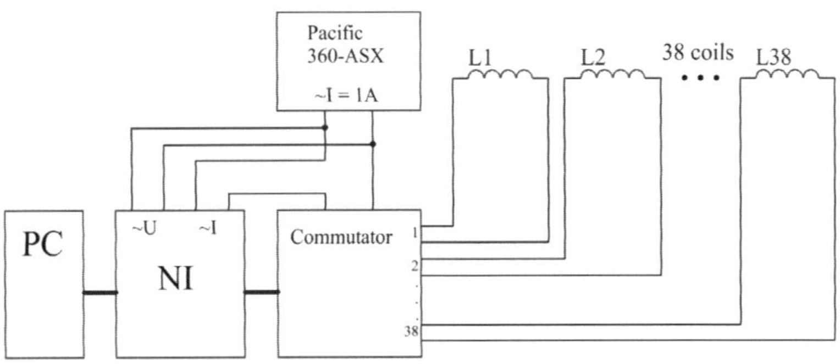

Figure 4. ILP setup schematic.

A special board (Commutator) was developed for switching the coils (fig. 4). The board is applied to reconnect the coils with the AC power source Pacific 360-ASX, which allows to change the frequency. The National Instruments boards (NI 9225 and NI 9227) are used for registration of the current and voltage on the coil. The NI 9401 board is used to communicate with the Commutator. Special software was developed in LabView IDE to automate the experiment.

The power source sets the alternating current in the coil ($I_{rms} = 1A$). During one measurement series, the coils are switched in turn. The series of experiments included the following frequencies: 100 Hz, 200 Hz, 300 Hz.

The impedance of each coil R_i is calculated from the experimental data at the selected frequency:

$$R_i = \frac{\sqrt{\frac{\sum_{j=1}^{N} U_{i,j}^2}{N}}}{\sqrt{\frac{\sum_{j=1}^{N} I_{i,j}^2}{N}}} = \frac{RMS(U_i)}{RMS(I_i)}, \quad (2)$$

where i is the coil number, $U_{i,j}$ is the voltage at the i-th coil at the j-th time instant, $I_{i,j}$ is the current in i-th coil at the j-th time instant.

Then, the plot $R(h)$ is developed, based on the obtained points R_i and h_i (height of the i-th coil). For this, a polynomial of n-th degree is calculated using the method of least squares. On the resulting plot it is possible to fix a previously known point where the boundary is located. Thus, the parameters of this relation and the real position of the boundary attached to the relation are determined. So, it is possible to analyze the measured signals and determine the unknown position of the boundary in advance using this plot. The plot of the function $R(h)$ for the new unknown position of the boundary will be shifted along the h axis relative to the initial plot. To test the sensitivity of the method to random errors $Error$, an error from 1 to 20 percent is entered into the experimental data. Then, the standard deviation δ is plotted as a function of the $Error$ in percent.

3. Results

The experimental investigation of the LLP was carried out for the different frequencies of an alternating current on the generating coil in the range [50; 500] Hz ($I_{rms} = 1.5$ A). The figure 5 shows the plots of the parameter D_{pd} as a function of height at $r = 0$ mm for various frequencies of the alternating current.

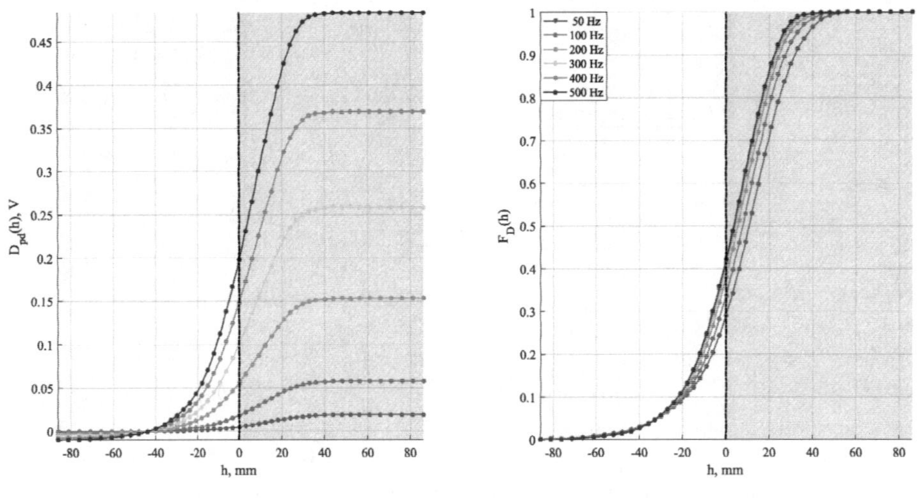

Figure 5. Relation between height at $r = 0$ mm and D_{pd} for the various frequencies of the alternating current and normalized value $F_D(h, f)$.

The normalized plots was also represented (in the conductive medium $F_D = 1$, otherwise $F_D = 0$). The black vertical dotted line shows the real boundary of the electrically conductive medium.

It should be noted that at the boundary of the electrically conductive medium, the dependence $F_D(f)$ reaches a certain constant value close to 0.5 (fig. 6).

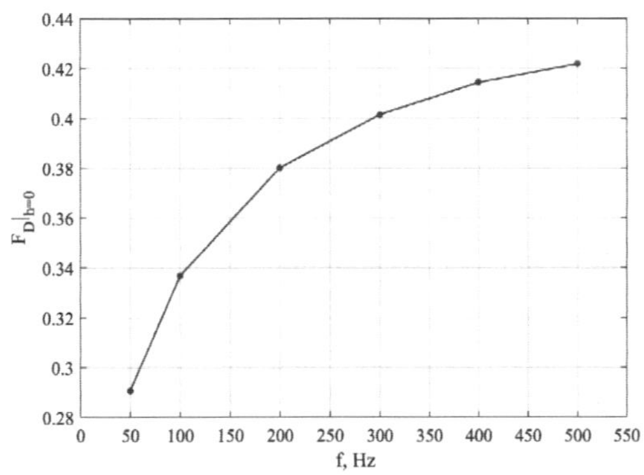

Figure 6. The calibration function F_D as a function of the frequency of the alternating current on the generating coil at $h = 0$ mm and $r = 0$ mm.

In addition, there is an exponential decay of the amplitude of the output signal of the LLP with increasing distance between the probe and the conducting cylinder. The figure 7 shows this plot for $h = 0$ mm, that is, at the boundary of the electrically conducting medium.

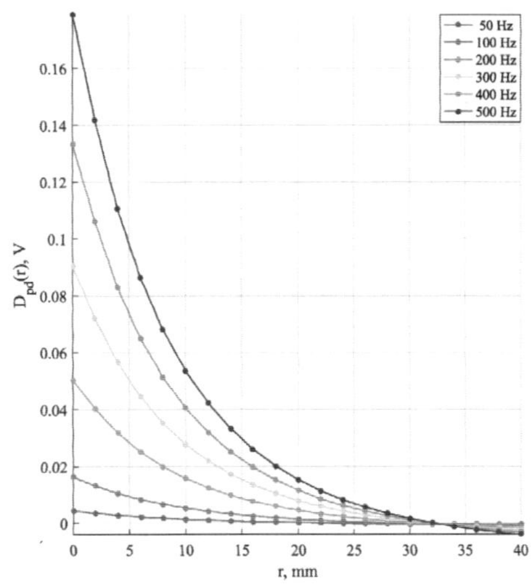

Figure 7. The parameter D_{pd} as a function of the distance at $h = 0$ mm for the different frequencies of the alternating current.

It is necessary to add that the relative error of the measured parameters does not exceed 1%.

The experimental investigation of the ILP was carried out for the following frequencies: 100 Hz, 200 Hz, 300 Hz ($I_{rms} = 1.5$ A). The figure 8 (a) shows the plots $R(h)$. The black vertical line shows the real boundary position $h = 688mm$. The figure 8 (b) represents the plot of the standard deviation as a function of the entered error in percent $\delta(Error)$ for 300 Hz.

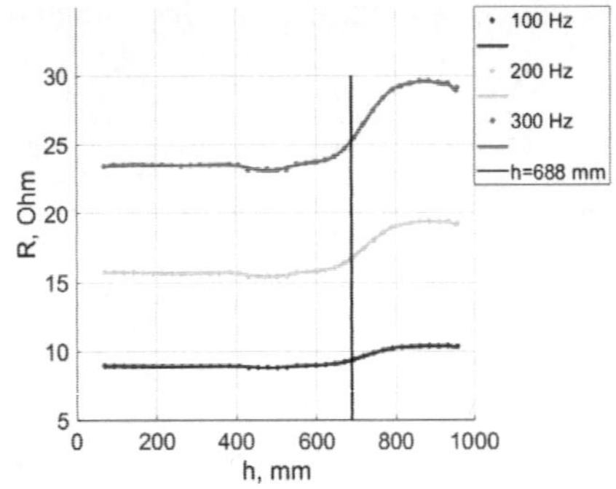

a)

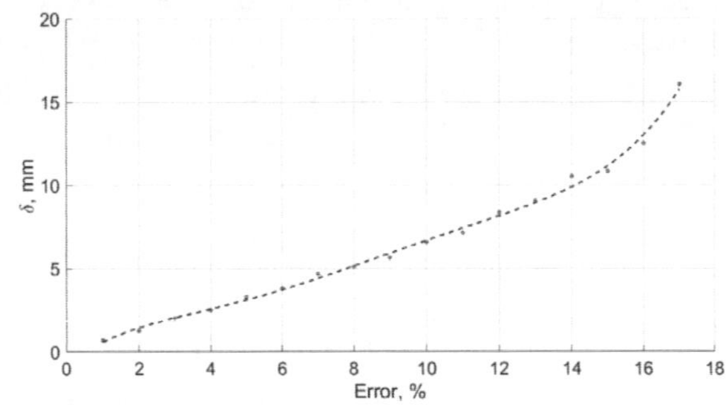

b)

Figure 8. The impedance $R(h)$ as a function of h and the standard deviation $\delta(Error)$ as a function of $Error$.

In the plots $R(h)$, crossing the boundary of the conductive medium is more clearly visible only at 300 Hz. If an additional error is entered in the experimental data, the error of the method at 100 and 200 Hz will be much higher. Therefore, for the method it is advisable to use frequencies from 300 Hz.

The accuracy of an each particular case can be estimated by plotting $\delta(Error)$.

4. Conclusions

The experimental investigation of the LLP shows the possibility of detection the boundary of an electrically conductive medium. In practice, for given values of the frequency of the alternating current and its standard deviation, the values of D_{pd} "inside" and outside the conductive medium are determined. Thus, due to the calibration function $F_D(h)$, it is possible to determine the position of the boundary of the electrically conducting medium relative to the center of the LLP. The probe position relative to the electrically conductive medium should be minimal, since the intensity of the output signal decreases exponentially with distance.

The experimental investigation of the ILP also shows the possibility of determination the level of the boundary. In practice, there are random errors in the experiment, for example, connected with the non-identity of the coils. Nevertheless, it is always possible to plot the standard deviation as a function of the entered error, estimate the accuracy of the method under various conditions and select the optimal parameters (frequency, number of coils, etc.) at which the error will be minimal.

Both methods were tested in laboratory experiments, which enabled us to estimate the measurements accuracy and the systems sensitivity to systematic and random interference. The algorithm stability analysis to random errors (Gaussian noise) and systematic errors was carried out.

In the end, it is worth adding that the LLP is more accurate in the laboratory task under consideration, which is determined by the rather precise possibility of its movement. The ILP has its advantages when used in installations in which the change in the level of the liquid metal is greater than the pitch with which the measuring coils are installed. In laboratory experiments, it was possible to determine the level of a metal with an error not exceeding 1 mm.

Acknowledgments
This work is supported by the project RSF-Helmholtz-18-41-06201.

References

[1] Weber N, Galindo V, Priede J, Stefani F and Weier T 2015 *Physics of Fluids* **27** 014103
[2] Kelley D and Weier T 2017 *arXiv:1710.03150*
[3] Krauter N, Eckert S, Gundrum T, Stefani F, Wondrak T, Frick P, Khalilov R and Teimurazov A 2018 *Metallurgical and materials transactions B*
[4] Slocomb H 1967 Liquid metal level measurement (sodium) Tech. rep. Liquid Metaf Engineering Center
[5] Singh Y, Raghuwanshi S and Kumar S 2018 *IETE Technical Review*
[6] Khalilov R, Khripchenko S, Frik P and Stepanov R 2007 *Measurement Techniques* **50** 861–866
[7] Miralles S, Verhille G, Plihon N and Pinton J F 2011 *Rev. Sci. Instrum.* **82** 095112
[8] Poberejsky L 1963 *Tech. Phys. (in Russian)* 1464–1469
[9] Forbriger J and Stefani F 2015 *Measurement science and technology* **26**
[10] Schulenberg T and Stieglitz R 2010 *Nucl. Eng. Des.* **240** 2077–2087
[11] Priede J, Buchenau D and Gerbeth G 2009 *Magnetohydrodynamics* **45** 451–458
[12] Sharma P, Kumar S S, Nashine B, Veerasamy R, Krishnakumar B, Kalyanasundaram P and Vaidyanathan G 2010 *Ann. Nucl. Energy* **37** 332–338
[13] von Weissenfluh T 1985 *Int. J. Heat Mass Transf.* **28** 1563–1574

IOP Conf. Series: Materials Science and Engineering **581** (2019) 012005 doi:10.1088/1757-899X/581/1/012005

Stirring flow of liquid metal generating by low-frequency modulated traveling magnetic field in rectangular cell

G Losev[1], A Pavlinov[1], E Shvydkiy[2], I Sokolov[2], I Kolesnichenko[1]

[1] Institute of Continuous Media Mechanics of the Ural Branch, RAS, Perm, Russia

[2] Ural Federal University, Yekaterinburg, Russia

E-mail: `Losev.g@icmm.ru`, `PAM@icmm.ru`, `E.L.Shvydky@urfu.ru`, `I.V.Sokolov@urfu.ru`, `KIV@icmm.ru`

Abstract. In the present work the influence of low frequency modulation of a travelling magnetic field (TMF) on a process of generation of electro-vortex flows in electrically conducting media are numerically and experimentally investigated. The measurements are carried out on a low melting temperature GaZnSn alloy by means of Ultrasonic Doppler Velocimetry. For numerical simulation, Comsol Multiphysics software was used. The dependencies of average and pulsating Reynolds numbers on the magnitude of electromagnetic impact and two modes of low frequency modulation are considered. A positive influence of reversed TMF modulations on the stirring process is determined. In particular the formation of a small-scale vortex structure in the main volume of liquid media.

1. Introduction

The metal production and development of alloys with specific properties is one of the great challenges of modern metallurgy. The stirring of metals during the crystallization process allows to improve the quality of ingots, homogenization of the impurity distribution and metal grain grinding [1–3].

Due to the high temperature and chemical activity of most liquid metals and salts, electromagnetic processing proved to be useful. The method of electromagnetic excitation of flows can be readily applied without contact with a conductive medium. Thereby, the contactless generation and control of flows is applicable in the field of non-ferrous and ferrous metals preparation.

In the conductive medium the alternating magnetic field excite eddy currents. The interaction of these currents with the initial magnetic field leads to the generation of an electromagnetic force. In the case of liquid medium, the force generates vortex flows. Such flows increase the intensity of heat transfer and exchange in the melts [4], smooth the crystallization front [5] and increase the energy efficiency of operations [6]. Electromagnetic stirring by rotating and travelling magnetic fields (TMF) during the crystallization of ingots is widely used [7].

In a recent time a several works are dedicated to study of influence of the TMF characteristics on a liquid metal flow and solidification process. Wang et al. [8,9] analytically and experimentally described the behavior of liquid metal flow under the reversed TMF. The parametric study is shown that flow velocity has a saturation zones and authors suggest optimal modulation period criteria. Than Hachani et al. [5, 10] experimentally approved that modulating of TMF can have a positive affect on a the solidification macrostructure and the segregation behavior. Musaeva et al. [11] experimentally shown the influence of pulsating TMF on a solidification front shape smoothing. Avnain et al. [4, 12] made a numerical and experimental study of liquid metal flow and solidification parameters under the steady applied TMF impact. All of the above mentioned works notice that modulation of TMF has a positive effect on a stirring process efficiency. Additionally the low-frequency modulation of the magnetic field can potentially reduce metal processing energy costs due to the appearance of a power supply off-period. However the question of influencies of TMF inductor parameters such as transverse longitudinal end effect, and the penetration depth of the magnetic field into the metal on a flow structure is still open.

This paper is dedicated to the experimental and numerical model of liquid metal stirring process under the external controlled travelling magnetic field (TMF). The goal of the present work is to obtain the flow characteristics under the control the stirring efficiency by low-freqency modulatation of the TMF. The geometry of the working cavity is consistent with the geometry of casting machines. The results obtained could be applied in the field of improving the quality of ingots and moldings in-process and in finished products. The verification of the proposed numerical algorithm allows it to be applied to channels and cavities of an arbitrary geometry and stirring regimes.

2. Experimental setup

The experimental setup (Fig. 2) consists of a vertical cell 1 made from plexiglass. The cell is filled with liquid GaSnZn alloy. The alloy's physical properties at the working temperature are density 6256 kg/m^3, kinematic viscosity $3 \cdot 10^{-7}$ m^2/s, and conductivity $3.56 \cdot 10^6$ Sm. The top surface of the metal was covered by a thick layer of mineral oil to prevent the oxidation of the melt. The cell measures $450 \times 20 \times 75$ mm^3. The channel was placed on a linear inductor of the TMF 2 with dimensions of 480×350 mm^2. The TMF was generated by six coils with a winding of 170 turns. The coils were fed from a three-phase programmable Pacific Smart Source 360 ASX-UPC3 3. This power supply unit made it possible to specify the shape of the output signal and to modulate the TMF. The main frequency of the current supply (and so the TMF) was 50 Hz. The modulation frequency was in the range 0.02 – 0.1 Hz.

The liquid metal flow velocity was measured by an ultrasonic Doppler velocimeter 4 (UDV) DOP 2000, Signal Processing. Four sensors 5 of the UDV were placed on the thin wall of the cell. The UDV method involves emission of ultrasound wave packets by an ultrasonic transducer, reception of reflected pulses (echoes) and calculation of the Doppler frequency shift between the emitted and received wave packets. The presence of sound-reflecting particles driven by the flow in the liquid medium is also necessary. The gallium oxide particles were used to reflect the ultrasound waves in the gallium eutectic. These particles were formed by the oxidation of gallium in the atmosphere [13].

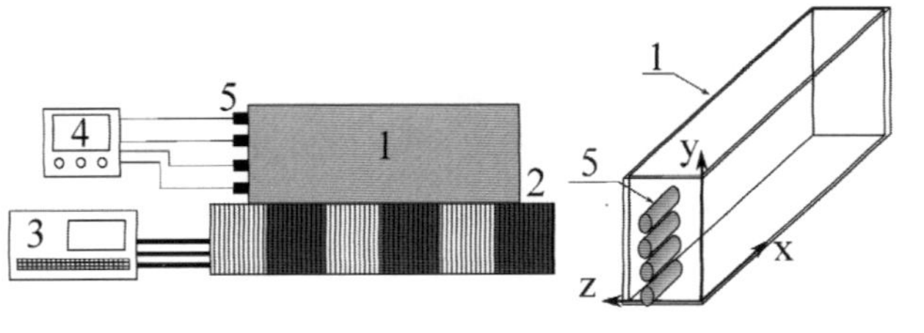

Figure 1. Experimental setup: 1. cell with liquid metal, 2. LIM, 3. power supply, 4. UDV, 5. UDV traducers.

3. Numerical model

The 3D numerical calculations were carried out by the finite element method using the Comsol Multiphysics software. At the first stage for electromagnetic calculations we used the Maxwell equations, which were formulated in terms of the magnetic vector potential. The calculations of the magnetic field, induced current and Lorenz force distribution were made with the use of the Magnetic Fields interface. Under the assumption of low magnetic Reynolds number ($Re_m = 0.065$), the induced current density was calculated by noninductive formulation [1,14,15].

The phase shift of inductor windings was 60° and the supplied current was in the range of 1 to 18 A. Analysis of magnetic saturation showed that the magnetic core saturates at 25 A and more [16], so the magnetic permeability has a constant value of $\mu = 150$. The LIM coils were modeled by a homogenized multi-turn coil, whose winding consisted of 170 turns of copper wire. The mesh for electromagnetic calculation included 152329 tetrahedral, 16246 triangular and 13200 hexahedral elements. The mesh for the cell domain was built using rectangular elements. Refinement of the mesh is performed toward the bottom layer of the domain taking into account the skin effect.

The electromagnetic force acting on a liquid metal was calculated as the product of current density and magnetic induction Then the resulting electromagnetic force field was transferred to the hydrodynamic part as a source term of the Navier-Stokes equation. Hydrodynamic flow was calculated only in the liquid metal domain which is placed on a TMF inductor. The maximum Reynolds number was $Re = 1.5 \cdot 10^4$, therefore melt flow is described by the k-ω SST turbulent model. We prescribed the slipping condition at the boundary of the top surface and the no-slip condition at the remaining walls. The computational domain contained 37500 hexahedral and 7750 quadrilateral finite elements. The maximum element size was limited to 5 mm and the mesh was refined at the domain boundaries areas.

4. Results

In the course of the work, measurements of the flow velocity profiles along the cell were made. The velocity characteristics were the main criterion for determining the quality of the mixing of the liquid metal. Fig. 2 demonstrates time-averaged velocity profiles. As can be seen, the flow has a two-vortex structure. The larger vortex occupies the central region of the cell. At the same time, a smaller vortex is formed in the near-wall region (Fig. 2a, c). Its formation is associated with the influx of fluid to the wall and the curvature of the streamlines. The rate of this vortex is slightly higher than the rate of the central vortex. With an increase in the external influence intensity, the energy of the vortexes increases, which leads to a distortion of the free metal boundary in the region of the vortexes localization.

The greatest velocity flow rate is near the bottom boundary of the cell in the region adjoining the TMF inductor (Fig. 2a). In the same area, the largest amplitude of velocity pulsations are observed. The velocity pulsation amplitude maximum is observed in the first third of the layer along its length. The pulsation magnitude varies slightly both in the height and in the length of the cell.

The imposition of low-frequency modulations on the TMF results in the restructuring of the flow pattern. The described average profiles undergo a qualitative change: the average component of the flow decreases (up to complete disappearance with small period reverse modulations), while the pulsating component tends to repeat the velocity distribution characteristic of the unmodulated force action. Fig. 3 represents the typical velocity time series in the median of the velocity profile. The imposition of low-frequency pulsations on the external electromagnetic force naturally leads to a periodicity in the generation of the flow. The frequency of the emerging hydrodynamic structures coincides with the modulation frequency of the TMF, and the magnitud of the flow velocity coincides with the magnitude of the average flow in the absence of modulations. Velocity oscillation occurs around a non-zero mean value with a periodic decrease in the velocity rate of large-scale flow to zero. In the reverse modulation mode, the velocity osciilations occur near zero mean value and the large-scale flow changes its direction with the TMF modulation frequency. High-frequency hydrodynamic pulsations are observed on the background of both average flow and large-scale low-frequency-modulated flow.

As the main characteristics for the stirring efficiency of a liquid metal, the mean Re_m and the pulsation Re_p Reynolds numbers calculated for the maximum on the average and pulsating velocity x-components were chosen. Fig. 4 represents the dependencies of Reynolds numbers on the intensity of electromagnetic field intensity and the TMF modulation period.

As the intensity of the external influence increases, the flow velocity increases nonlinearly with the release into the YsaturationY region at inductor supply currents of more than 6 A. In this case, both the average and pulsation Reynolds numbers reach the maximum (Fig. 4a, b). The results of numerical simulation correspond well with the experimental data. A difference is only observed near the upper surface of the metal due to the curvature of the free surface in the experiment, which is not taken into account in the numerical model. The best convergence of numerical and experimental data is provided at low values of the current in the coils due to the weak influence of the flow on the shape of the free surface. The maximum attainable average Reynolds number only slightly exceeds the pulsation Reynolds number maximum, which indicates a significant contribution of small-scale flows to the overall flow structure. Nevertheless, the presence of the YsaturationY region shows the ineffectiveness of controlling the mixing of the liquid metal only by a increasing in the power pumped into the system alone. This leads to

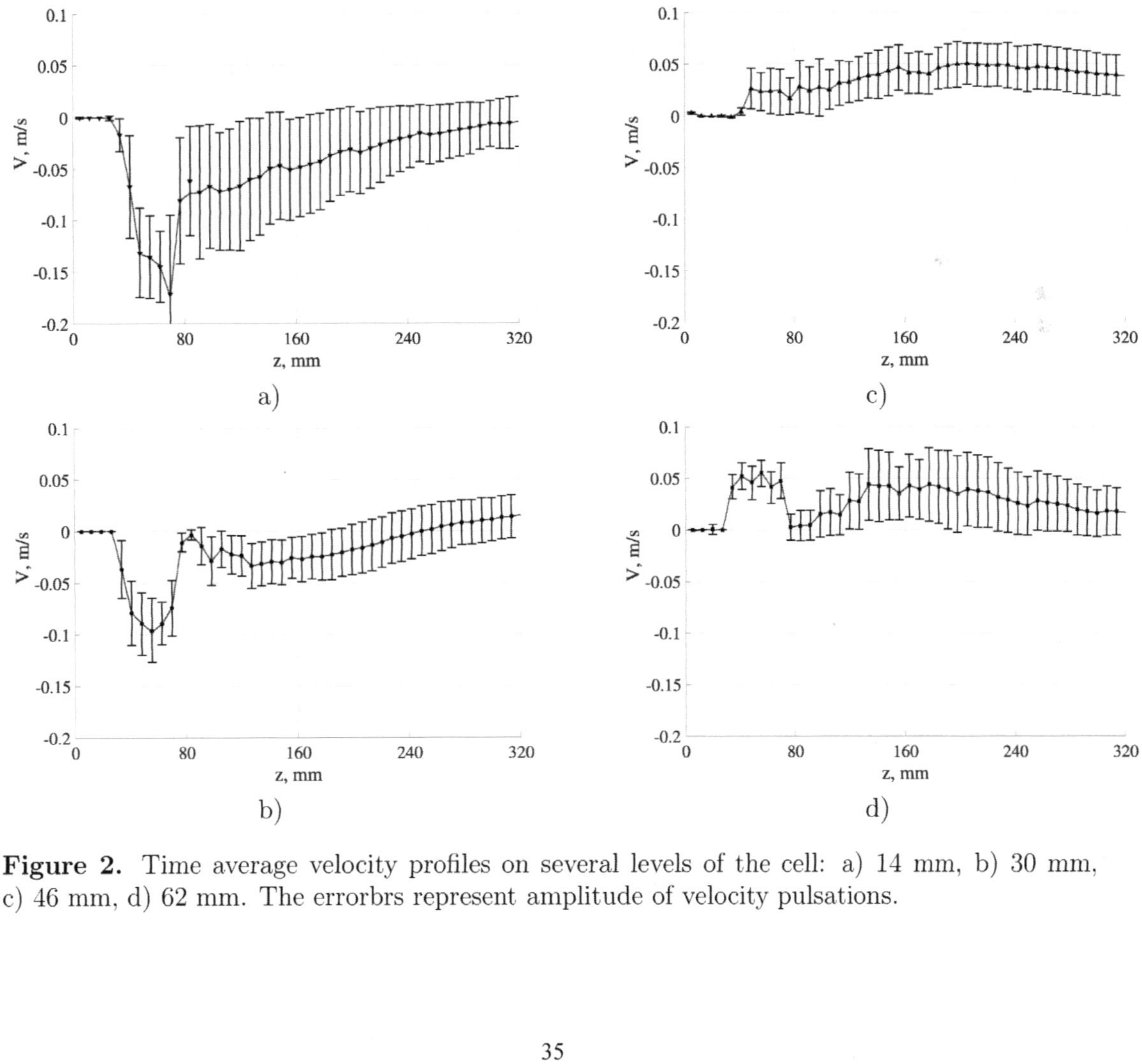

Figure 2. Time average velocity profiles on several levels of the cell: a) 14 mm, b) 30 mm, c) 46 mm, d) 62 mm. The errorbrs represent amplitude of velocity pulsations.

IOP Conf. Series: Materials Science and Engineering **581** (2019) 012005 doi:10.1088/1757-899X/581/1/012005

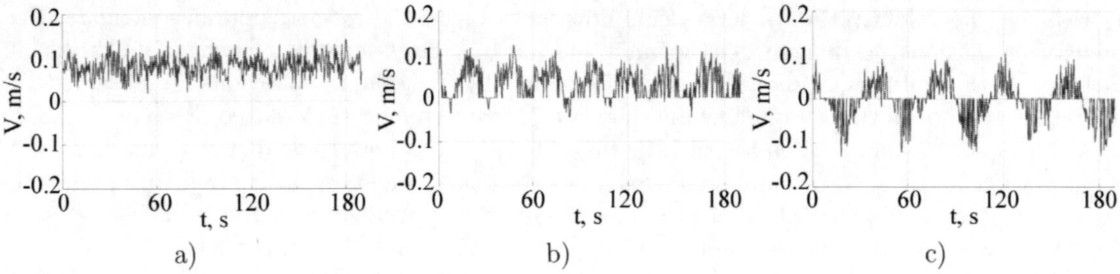

Figure 3. Velocity time series in three cases of external action: a) TMF without modulation, b) modulated TMF, c) reverse modulation TMF. Modulation period $\tau = 30$ s in both modulation cases.

the need to develop additional mechanisms for introducing small-scale hydrodynamic structures into the main flow. The modulation of electromagnetic exposure can serve as such a mechanism .

The modulated flow was created at the amplitude of the supply current of 6.0 A. Thus, the condition of minimizing energy consumption with a maximum increase in the rate of metal flow was fulfilled. In the modulation mode, the amplitude of the TMF changed according to a periodic law from zero to amplitude with a period $\tau = 5 - 50$ s. The dependences of the mean and pulsation Reynolds numbers are given in Fig. 4c, d. As the modulation period increases, the Re_m decreases. At the same time, there is an assignment increase of Re_p. The greatest intensity of hydrodynamic pulsations is observed in the region of the localization of a large-scale vortex (near the bottom of the cell and the free surface). These pulsations repeat the form of the electromagnetic effect and are a low-frequency quasi-stationary flow against the background

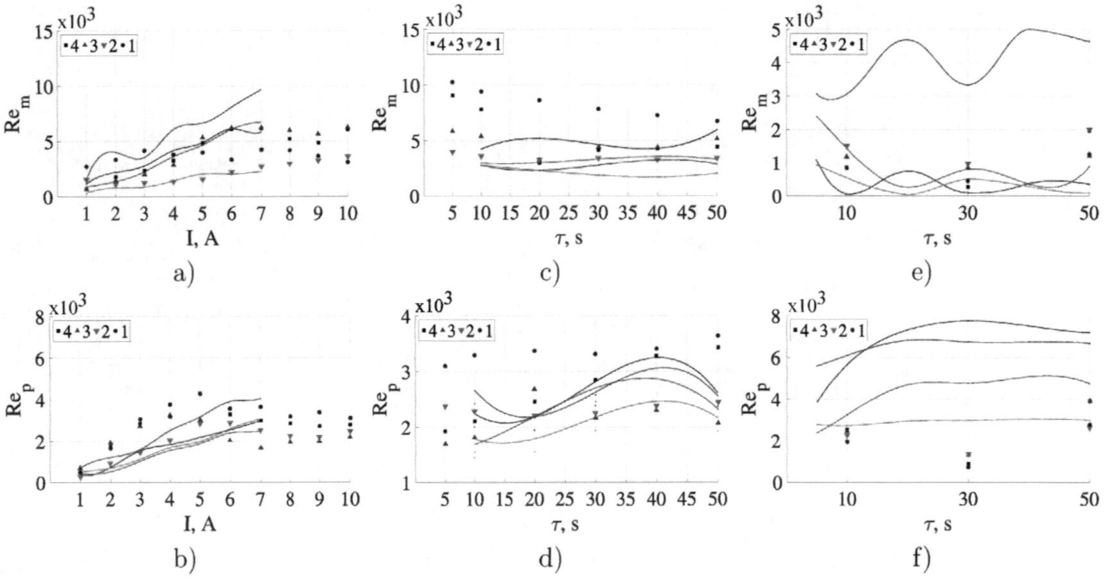

Figure 4. Time average flow velocity (a, c, e) and amplitude of velocity pulsation (b, d, f) vs supply current (a, b), period of MVT modulation (c, d) and period of reverse modulation (e, f). The line represents numerical results and the points represents experimental data. The numbers in the legend represent heights of lawer cross-sections: 1) 62 mm, 2) 46 mm, 3) 30 mm, 4) 14 mm

of which secondary small-scale structures arise.

The second considered modulation mode of the TMF we will call reverse. In this case, the magnitude of the magnetic induction varied over the period according to the sinusoidal law. In the second half-cycle, the direction of the TMF was changed. This reverse modulation generated a pronounced pulsed flow in the liquid phase. The reverse flow pattern is characterized by low values of Re_m due to the periodic change of direction of large-scale eddies. The significant difference between the experimental and numerical data in the region of the upper surface of the metal is explained by the change in the shape of the free surface with a periodic change in the direction of large-scale vortex structures rotation. However, in the main volume of the cuvette, experimental data and simulations are consistent.

The pulsating component of the reverse flow has a peculiarity as significant contribution of small-scale (in the sense of the spatial dimensions and turnover periods of the vortex) to the overall structure of the flow compared to the previously considered modes. Due to the limited resolution of the spatial and temporal scales of the vortex structures of UDV, the experimentally obtained values of Re_p are underestimated in comparison with the numerical simulation that resolves much smaller scale structures. Based on the simulation data (in the layer main volume),we can conclude about the double increasing of Re_p in the case of low-frequency reverse modulation compared with the cases of no flow modulation or amplitude modulation.

5. Conclusions

In the course of the work, the characteristics of the liquid metal vortex flow generated by the external action of TMF under the action of low-frequency modulations have been investigated. The flow characteristics and structure were found experimentally and numerically.

Under the action of steady applied TMF, the two-vortex flow appears in the metal. The intensity of the vortexes grown nonlinearly and has the «saturation» region at supply coil current more than 6 A. The formation of a large-scale vortex with an intensity independent of the magnitude of the applied electromagnetic effect could have a negative impact on the distribution of impurities and alloying additives in the volume of the metal due to the retention of impurities by the vortex. The presence of the maximum flow rate limit leads to the need to introduce new mechanisms to improve the efficiency of mixing. In this way, the control of the induction machine power supply can a solution. The obvious advantage of the method is its technical simplicity and the cheapness of modifying existing technological installations.

The low-frequency modulation of the LIM supply increases the intensity of small-scale flows generated in the volume of the mixed melt. This also has a positive effect on the heat transfer during the crystallization of castings (smoothes the crystallization front and reduces the number of grain effects). Changing the direction of TMF destroys large-scale eddies and creates multiple small-scale flows that promote efficient mixing of the medium. This LIM power mode can be achieved using the simple relay contactor circuit.

The limited capabilities of existing experimental techniques lead to the need for the development and verification of numerical models that take into account the specifics of individual tasks (including cavity geometry, mutual influence of electromagnetic fields of all installation sites, phase transformations). In this paper, an attempt to advance in the field of numerical simulation and verification of the MHD model has been made. Despite some imperfections the used model allows one to adequately describe the structure and characteristics of the flow in the main volume of the liquid metal.

Acknowledgments

The work of Institute of Continuous Media Mechanics team is supported by the RFBR grant 17-48-590539_r_a and the work of Ural Federal University team is supported by Act 211 Government of the Russian Federation, contract є 02.A03.21.0006.

IOP Conf. Series: Materials Science and Engineering **581** (2019) 012005 doi:10.1088/1757-899X/581/1/012005

References

[1] Denisov S, Dolgikh V and Khripchenko S 2014 *Magnetohydrodynamics* **4** 249 – 265

[2] Eckert S, Nikrityuk P A, Willers B, Räbiger D, Shevchenko N, Neumann-Heyme H, Travnikov V, Odenbach S, Voigt A and Eckert K 2013 *The European Physical Journal Special Topics* **220** 123–137 ISSN 1951-6401 URL `https://doi.org/10.1140/epjst/e2013-01802-7`

[3] Garrido M, Davoust L, Daudin R, Salvo L and Fautrelle Y 2018 *IOP Conference Series: Materials Science and Engineering* **424** 012001 URL

[4] Avnaim M, Mikhailovich B, Azulay A and Levy A 2018 *International Journal of Heat and Fluid Flow* **69** 23 – 32 ISSN 0142-727X URL `http://www.sciencedirect.com/science/article/pii/S0142727X17302874`

[5] Hachani L, Zaidat K and Fautrelle Y 2015 *International Journal of Heat and Mass Transfer* **85** 438 – 454 ISSN 0017-9310 URL `http://www.sciencedirect.com/science/article/pii/S0017931015001660`

[6] Fdhila R, SandJ U, Eriksson E and Yang H 2016 *ABB Review* **3**

[7] Moffatt H K 1991 *Phys. Fluids A* **3** 1336–1343

[8] Wang X, Moreau R, Etay J and Fautrelle Y 2008 *Metall. Mater. Trans. B* **40** 82 URL `https://doi.org/10.1007/s11663-008-9176-0`

[9] Wang X, Moreau R, Etay J and Fautrelle Y 2009 *Metall. Mater. Trans. B* **40** 104–113 URL `https://link.springer.com/article/10.1007/s11663-008-..`

[10] Zaidat K, Sari I, Boumaaza A, Abdelhakem A, Hachani L and Fautrelle Y 2018 *IOP Conference Series: Materials Science and Engineering* **424** 012052 URL `https://doi.org/10.1088/1757-899X/424/1/012052`

[11] Musaeva D, Baake E, Köppen A and Vontobel P 2017 *Magnetohydrodynamics* **53** 583–593 ISSN 0024998X

[12] Avnaim M, Mikhailovich B, Azulay A and Levy A 2018 *International Journal of Heat and Fluid Flow* **69** 9 – 22 ISSN 0142-727X URL `http://www.sciencedirect.com/science/article/pii/S0142727X17302886`

[13] Nowak M 2002 *Experiment in fluids* **31** 249 – 255

[14] Branover G and Cinober A 1970 *Magnetohydrodynamics of incompressible medium* (M.: Science)

[15] Kolesnichenko I, Khalilov R, Khripchenko S and Pavlinov A 2012 *Magnetohydrodynamics* **1** 221 Ч 234

[16] Shvydkiy E and Kolesnichenko I 2018 3D numerical simulation of the linear induction motor, considering magnetic saturation *2018 IEEE Conference of Russian Young Researchers in Electrical and Electronic Engineering (EIConRus)* pp 777–779

IOP Conf. Series: Materials Science and Engineering **581** (2019) 012006 doi:10.1088/1757-899X/581/1/012006

A finite element algorithm for solution of the natural vibration problem of electroelastic bodies with passive external electric circuits, interacting with a quiescent fluid

Sergey Lekomtsev, Dmitrii Oshmarin* and Natalya Sevodina

Institute of Continuous Media Mechanics of the Ural Branch of Russian Academy of Sciences, Acad. Korolev str., 1, Perm, 614013, Russian Federation

E-mail: oshmarin@icmm.ru

Abstract. In this paper we consider a mathematical statement of the problem on natural vibrations of piecewise-homogeneous electroelastic bodies with passive external electric circuits (shunting circuits) of arbitrary configuration and interacting with a quiescent fluid. The behavior of the piezoelectric body is described using the equations of electrodynamics of deformable electroelastic media in the quasi-static approximation. The motion of an ideal fluid in the case of small perturbations is considered within the framework of the acoustic approximation. Small strains in a thin plate are determined using the Reissner – Mindlin theory. The numerical solution is developed using the finite element method. The proposed algorithm is based on the approach, in which the global stiffness matrix generated with the aid of the ANSYS software package is decomposed into required constituents. The system of governing equations is constructed using the developed algorithm, which is realized in the FORTRAN language. Complex eigenvalues of the examined system are defined from the solution of the non-classic modal problem using the Mueller method. A thin plate with piezoelectric element located on the free surface of a layer of a quiescent fluid of finite size is considered as an example.

1. Introduction

Nowadays, devices made of piezoelectric materials and connected to external electric circuits are widely used to control the dynamic behavior of different structures including structures interacting with a fluid. A detailed review of studies concerning their practical applications to submerged structures is given in [1]. The authors of the paper have briefly outlined and discussed the basic results related to such topics as modal analysis, active sound and vibration control, energy harvesting and atomic force microscopes. The obtained results are used in different areas from the military and oil industries to biology and medicine.

In a number of papers the authors consider a possibility of using piezoelectric elements for sound absorption of submerged structures [2–5]. The idea of the method is in piezoelectric coating or attaching a piezoelectric patch to the acoustic surface and connecting the piezoelectric layer to a passive electric RLC-circuit. In paper [2], different kinds of shunting circuits are investigated using the proposed one-dimensional electro-acoustic model. It has been shown that the use of resistors and inductances allows achieving good absorption of sound only for a narrow

IOP Conf. Series: Materials Science and Engineering **581** (2019) 012006 doi:10.1088/1757-899X/581/1/012006

bandwidth of frequency. However, to achieve broadband reduction of sound, a series-connected RC-elements should be used [2, 3]. In papers [4, 5] in order to expand a bandwidth of controlling frequencies, a multimodal vibration damping technique is applied. For this purpose, in [4] a multi-layered piezoelectric coating is used. Each layer is connected to a passive RL-circuit that provides sound absorption at a specified frequency. A one-dimensional electro-acoustic model based on Mason equivalent circuit theory and an iteration algorithm has been developed by the authors. Among other things, the sound absorption coefficient of a four-layered specimen was determined experimentally. The results obtained from the experiment qualitatively agree with the results of the finite element simulation performed in the ANSYS software package. However the difference between the experimental results and those obtained in the framework of one-dimensional model is quite noticeable. The use of several piezoelectric elements, each of them being located at a specified site and shunted to its own circuit, is a rather wide-spread practice in the multimodal vibration damping strategy. Instead of this option, in paper [5] only one piezoelectric patch is used in combination with a blocked shunt circuit, which leads to a significant reduction in transmitted noise for a multiple modes of the panel.

Although the above mentioned papers consider the interaction with an acoustic media the models used there are formulated in terms of electric impedance with the aid of simplified (equivalent) one-dimensional equations. It means that the majority of physical processes in elastic bodies, liquid or gas media are not taken into account. When analyzing behavior of structures interacting with fluid, it is often necessary to define the response of a damped system to external harmonic or arbitrary impacts. With this in mind, it is reasonable to suggest that this can be done only in the case of an adequate description of the hydroelastic and electroelastic interactions.

A control of dynamic deformation of structures and damping of dangerous vibrations of various nature can be accomplished using either active or passive techniques. In the first case, the electric potential produced by a piezoelectric element, acting as a sensor, is further processed and amplified, and then the control action is transferred to another piezoelectric element, which serves as an actuator. In the case when the passive technique is used, it is necessary to select the parameters and configuration of the passive external electric circuits such, that could provide maximum damping of undesirable vibrations, or to perform a frequency shift, so that the resonant frequencies of the structure do not fall into the range of dangerous frequencies.

In the literature, there are a lot of studies related to active structural vibration control of thin plates [6–8] and shells [9–11] interacting with a fluid. A finite element formulation of the dynamic problem for a laminated plate, incorporating both the piezoelectric and viscoelastic layers is presented in [6]. Here, it is assumed that the structure performs harmonic vibrations in an acoustic half-space. The influence of the half-space is taken into account using the Rayleigh surface integral. The possibility of vibration control with the aid of negative velocity feedback control algorithm was demonstrated by solving a number of example problems. In paper [7], the same author offers a somewhat different algorithm of feedback between the piezoelectric sensor and the actuator. This algorithm is based on the mode shapes obtained for a structure in vacuum and for a structure interacting with a fluid. The results of experimental and theoretical studies for a partially submerged vertical cantilever plate with piezoceramic sensors and actuators are presented in [8]. The vibration control is executed by applying MIMO (Multiple-Input Multiple-Output) positive position feedback controller. The frequency response curves and the values of natural vibration frequencies obtained numerically by the Rayleigh – Ritz method have been verified and have proved to provide a qualitatively correct description of experimental data.

In the problems of passive control of structural dynamic behavior with the aid of piezoelectric elements and passive electric circuits, a special attention is paid to the analysis of the influence of external electric circuits on the damping properties of the object under study. A level of decrease in the amplitude of forced vibrations or in the rate of damping of transient processes can be

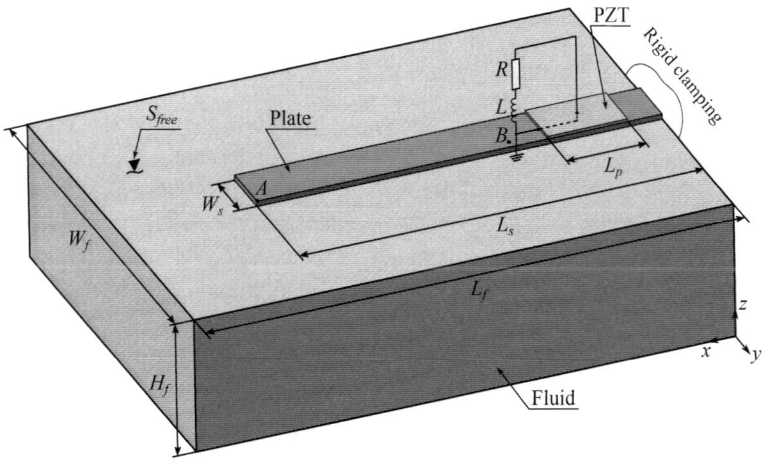

Figure 1. Computational scheme for cantilever plate located on the free surface of quiescent fluid with attached piezoelectric element and series RL-circuit.

considered as such properties. These data can be gained from the solution of the corresponding dynamic problems.

The analysis of the literature shows that comprehensive studies on the effective use of piezoelectric materials for control of dynamic behavior of complex multi-component objects and conclusions, summarizing the results of these studies, are impossible without mathematical modeling. In many papers, the finite element method is used for this purpose [12–14]. Some commercial software packages contain the relevant models and algorithms, which however are insufficient for covering all the requirements of researchers. As it was noted in [15], the modal analysis is the most efficient tool for constructing optimization algorithms due to rather low calculation costs. In the present paper, a previously developed approach [15] will be used to analyze the dynamic behavior of electroelastic structures connected to passive external electric circuits and interacting with a quiescent fluid. The distinguishing feature of this approach is solving the non-classic eigenvalue problem by the Mueller method programmatically realized in the FORTRAN language. Here the possibilities of ANSYS finite element software package are used for assembling the global matrices.

2. Mathematical statement of the problem

As an example, let us consider a structure, which is thin plate of length L_s, width W_s and thickness h_s. This plate is located on the free surface of a layer of a quiescent fluid of finite dimensions ($L_f \times W_f \times H_f$) (see Figure 1). The variational equation of motion of the combined body of volume $V = V_s + V_p$, consisting of the elastic plate of volume V_s and the attached piezoelectric element of volume V_p, is formulated in terms of the relations of Reissner – Mindlin theory, linear theory of elasticity and quasi-static Maxwell equations [15–18]. The electroded coatings applied to the top (S_{top}) and the bottom (S_{bot}) parts of the surface of piezoelectric body are assumed to be ideal conductors with negligible mass. The final governing equation in the absence of body forces and free surface charges in the matrix form can be written as

$$\int\limits_{V_p} \Big(\{\delta\varepsilon_p\}^{\mathrm{T}}[\mathbf{D}_p]\{\varepsilon_p\} - \{\delta\varepsilon_p\}^{\mathrm{T}}[\mathbf{e}]\{\mathbf{E}\} - \{\delta\mathbf{E}\}^{\mathrm{T}}[\mathbf{e}]^{\mathrm{T}}\{\varepsilon_p\} - \{\delta\mathbf{E}\}^{\mathrm{T}}[\beta]^{\mathrm{T}}\{\mathbf{E}\}\Big)\, dV + \qquad (1)$$

$$\int_{V_p} \{\delta \mathbf{u}_p\}^{\mathrm{T}} \rho_p \{\ddot{\mathbf{u}}_p\} dV + \int_{S_s} \{\delta \boldsymbol{\varepsilon}_s\}^{\mathrm{T}} [\mathbf{D}_s] \{\boldsymbol{\varepsilon}_s\} dS + \int_{V_s} \{\delta \mathbf{u}_s\}^{\mathrm{T}} \rho_s \{\ddot{\mathbf{u}}_s\} dV - \int_{S_\sigma} \{\delta \mathbf{u}_s\}^{\mathrm{T}} \{\mathbf{t}_s\} dS = 0.$$

Here, the subscripts "p" and "s" refer the corresponding variables to the piezoelectric element and the plate. The following notations are accepted: column vectors $\{\boldsymbol{\varepsilon}\}$ contain the components of strain tensors and are defined according to the well-known relations [16, 17]; $[\mathbf{D}]$ are the elastic constant matrices; $[\mathbf{e}]$, $[\boldsymbol{\beta}]$ are the matrices of piezoelectric and dielectric coefficients; $\{\mathbf{E}\}$ is the electric field intensity vector; $\{\mathbf{u}_p\}$ is the vector of piezoelectric element displacements; $\{\mathbf{u}_s\}$ is the generalized displacement vector for the plate, which includes rotation angles; ρ is the material density; $\{\mathbf{t}_s\}$ is the vector of the surface traction caused by the interaction with the fluid on the part of the plate surface S_σ.

The potentiality condition is fulfilled for an electric field

$$-\{\mathbf{E}\} = \nabla \varphi, \tag{2}$$

where φ is the electric potential.

The bottom electroded surface of the piezoelectric element S_{bot} is grounded and has a zero-value electric potential. The corresponding boundary condition can be written as $\varphi = 0$. On the electrode-free parts of the surface of piezoelectric body there are no free electric charges. When an external voltage supply is absent, the other parts of the electrode surface are considered free. In this case, the open circuit mode (o/c) is realized. On the other hand, it can be treated as having zero-value electric potential, in which case the short circuit mode is realized (s/c).

By using the electroded surfaces, the external electric circuits of an arbitrary configuration, including resistive (R), capacitive (C) or inductive (L) elements, can be attached to the surface of the system under consideration. If these circuits are not supplied by an external power, they are classed as the internal elements of the system and then the following term should be added to equation (1) [15]:

$$\sum \delta A = \sum_{p=1}^{n_L} \delta A_{L_p} + \sum_{q=1}^{n_R} \delta A_{R_q} + \sum_{r=1}^{n_C} \delta A_{C_r} = \sum_{p=1}^{n_L} \frac{1}{L_p} \iint \left(\varphi_1^{L_p} - \varphi_2^{L_p}\right) \delta\varphi \, dt dt + \tag{3}$$
$$\sum_{q=1}^{n_R} \frac{1}{R_q} \int \left(\varphi_1^{R_q} - \varphi_2^{R_q}\right) \delta\varphi \, dt + \sum_{r=1}^{n_C} C_r \left(\varphi_1^{C_r} - \varphi_2^{C_r}\right) \delta\varphi,$$

where n_L, n_R, n_C are the quantities of the inductive, resistive and capacitive elements, respectively; L_p, R_q, C_r are the values of the inductance, resistance and capacitance of the corresponding circuit elements; $\varphi_1^{cir} - \varphi_2^{cir}$ is the potential difference in the corresponding circuit element ($cir = L_p$, R_q, C_r).

The small amplitude vibrations of an ideal compressible fluid in the region V_f is described by the well-known Euler equations, continuity equation and state equation. The elimination of velocity from them gives the Helmholtz equation governing the hydrodynamic pressure p

$$\nabla^2 p = \ddot{p}/c^2, \tag{4}$$

where c is the speed of sound in the fluid.

In the proposed mathematical model, we assume that sloshing of the free surface of the fluid (S_{free}) is absent. The corresponding boundary condition is written as $p = 0$. At the interface between the liquid and rigid wall S_w, the normal velocity of the fluid is equal zero.

At the fluid-structure interface the pressure must satisfy the boundary condition

$$\frac{\partial p}{\partial n} = -\rho_f \ddot{w}, \tag{5}$$

IOP Conf. Series: Materials Science and Engineering **581** (2019) 012006 doi:10.1088/1757-899X/581/1/012006

where n denoting the normal to the fluid region, w is the normal displacement of the plate, ρ_f is the density of the fluid.

Equation (4) together with the boundary conditions and the impermeability condition (5) are converted to a weak form using the Bubnov – Galerkin method [19]. Finally, we have

$$\int\limits_{V_f} \delta p \left(\frac{1}{c^2}\ddot{p} + \nabla^2 p\right) dV + \rho_f \int\limits_{S_\sigma} \delta p \ddot{w}\, dS = 0. \tag{6}$$

To take into account the fluid-structure interaction, the traction integral in equation (1) can be expressed as

$$\int\limits_{S_\sigma} \{\delta \mathbf{u}_s\}^{\mathrm{T}}\{\mathbf{t}_s\}\, dS = \int\limits_{S_\sigma} \{\delta \mathbf{u}_s\}^{\mathrm{T}}\{\mathbf{n}_f\} p\, dS, \tag{7}$$

where $\{\mathbf{n}_f\}$ is the vector of outward normal to the fluid region. We supposed that positive pressure is defined in compression and $\{\mathbf{n}_s\} = -\{\mathbf{n}_f\}$.

Let us consider a perturbed motion of the fluid and the plate with the attached piezoelectric element defined as

$$\mathbf{U}(\mathbf{x},t) = \{\mathbf{u}_p(\mathbf{x},t),\ \varphi(\mathbf{x},t),\ \mathbf{u}_s(\mathbf{x},t),\ p(\mathbf{x},t)\} = \tilde{\mathbf{U}}(\mathbf{x})e^{i\lambda t}, \quad \mathbf{x} = \{x,\ y,\ z\}, \tag{8}$$

where $\tilde{\mathbf{U}}(\mathbf{x})$ is the function of coordinates, $\lambda = \omega + i\gamma$ the characteristic parameter, ω corresponds to the circular natural frequency of vibrations and γ is the rate of its damping and $i = \sqrt{-1}$. Taking into account the above mentioned form of the solution, instead (1), (3) and (6) we get (wavy line is dropped)

$$\int\limits_{V_p} \left(\{\delta\varepsilon_p\}^{\mathrm{T}}[\mathbf{D}_p]\{\varepsilon_p\} - \{\delta\varepsilon_p\}^{\mathrm{T}}[\mathbf{e}]\{\mathbf{E}\} - \{\delta\mathbf{E}\}^{\mathrm{T}}[\mathbf{e}]^{\mathrm{T}}\{\varepsilon_p\} - \{\delta\mathbf{E}\}^{\mathrm{T}}[\boldsymbol{\beta}]^{\mathrm{T}}\{\mathbf{E}\}\right) dV - \tag{9}$$

$$-\lambda^2 \int\limits_{V_p} \{\delta\mathbf{u}_p\}^{\mathrm{T}}\rho_p\{\ddot{\mathbf{u}}_p\}dV + \int\limits_{S_s} \{\delta\varepsilon_s\}^{\mathrm{T}}[\mathbf{D}_s]\{\varepsilon_s\}dS - \lambda^2 \int\limits_{V_s} \{\delta\mathbf{u}_s\}^{\mathrm{T}}\rho_s\{\ddot{\mathbf{u}}_s\}dV - \int\limits_{S_\sigma} \{\delta\mathbf{u}_s\}^{\mathrm{T}} p\{\mathbf{n}_f\}dS -$$

$$-\sum_{k=1}^{n_L} \frac{1}{\lambda^2 L_k}\left(\varphi_1^{L_k} - \varphi_2^{L_k}\right)\delta\varphi + \sum_{q=1}^{n_R} \frac{1}{i\lambda R_q}\left(\varphi_1^{R_q} - \varphi_2^{R_q}\right)\delta\varphi + \sum_{r=1}^{n_C} C_r\left(\varphi_1^{C_r} - \varphi_2^{C_r}\right)\delta\varphi = 0,$$

$$\int\limits_{V_f} \delta p \left(-\frac{\lambda^2}{c^2}p + \nabla^2 p\right) dV - \lambda^2 \rho_f \int\limits_{S_\sigma} \delta p w\, dS = 0. \tag{10}$$

3. Finite element formulation

Using standard procedures of the finite element method we can convert the problem of the natural vibrations of electroelastic bodies with an external circuit interacting with quiescent fluid, which is described by equations (9)–(10), to the following algebraic eigenvalue problem:

$$\left(-\lambda^2\,[\mathbf{M}] + [\mathbf{C}(\lambda)] + [\mathbf{K}]\right)\left\{\tilde{\mathbf{U}}\right\} = 0, \tag{11}$$

where

$$\mathbf{M}=\begin{bmatrix}\mathbf{M}_p & 0 & 0 & 0\\ 0 & 0 & 0 & 0\\ 0 & 0 & \mathbf{M}_s & 0\\ 0 & 0 & \rho_f\mathbf{Q}^{\mathrm{T}} & \mathbf{M}_f\end{bmatrix},\mathbf{C}(\lambda)=\begin{bmatrix}0 & 0 & 0 & 0\\ 0 & \mathbf{C}_\varphi(\lambda) & 0 & 0\\ 0 & 0 & 0 & 0\\ 0 & 0 & 0 & 0\end{bmatrix},\mathbf{K}=\begin{bmatrix}\mathbf{K}_p & \mathbf{K}_{p\varphi} & 0 & 0\\ (\mathbf{K}_{p\varphi})^{\mathrm{T}} & \mathbf{K}_\varphi+\mathbf{K}_\varphi^C & 0 & 0\\ 0 & 0 & \mathbf{K}_s & -\mathbf{Q}\\ 0 & 0 & 0 & \mathbf{K}_f\end{bmatrix},$$

$$[\mathbf{C}_\varphi(\lambda)]=-\sum_{p=1}^{n_L}\frac{1}{\lambda^2 L_p}\left[\mathbf{C}_\varphi^{L_p}\right]+\sum_{q=1}^{n_R}\frac{1}{i\lambda R_q}\left[\mathbf{C}_\varphi^{R_q}\right],\quad [\mathbf{K}_\varphi^C]=\sum_{r=1}^{n_C}C_r\left[\mathbf{K}_\varphi^{C_r}\right].$$

Here, the subscripts "f" and "φ" refer the corresponding variables to the fluid and the electric potential, $[\mathbf{Q}]$ is the fluid-structure interaction matrix [19], and all zeroes are matrices also. The matrices $\left[\mathbf{C}_\varphi^{L_p}\right]$, $\left[\mathbf{C}_\varphi^{R_q}\right]$, $\left[\mathbf{K}_\varphi^{C_r}\right]$ contain the coefficients "1" and "-1" only in positions that refer to the nodal variables of the corresponding elements of electric circuit. The remaining typical matrices of individual finite elements are determined by the known manner [19, 20]. Discretization of the fluid, piezoelectric element and plate computational domains is based on the spatial 20-node brick elements and the 8-node flat rectangular finite elements with a quadratic approximation of the unknown variables.

Equation (11) essentially differs from the generalized eigenvalue problem due to the presence of the matrix $[\mathbf{C}_\varphi(\lambda)]$, which describes the external electric circuit. The solution was performed using the capabilities of the ANSYS software package for generating the finite element matrices. We employ the program written in FORTRAN language for assembling the system of governing equations (11) using the sparse matrix technology for basic operations and computation of complex eigenvalues by the Mueller method [15].

4. Approbation of the algorithm

In this Section using the obtained results we demonstrate the reliability of computing the complex natural vibration frequencies of the examined system on the basis of the developed algorithm. In order to apply the obtained results to future studies, the numerical simulation was performed using the geometry and characteristics of the real experimental setup. The fluid volume was a rectangular prism 200 mm long, 150 mm wide and 195 mm high. In view of the specific features of the excitation device, a light neodymium magnet weighting 0.10 g was glued to the plate surface at a distance of 66 mm from the clamped edge of the plate and 3 mm away from its long side. It was modeled as a mass element, which was defined at a single node with the specified concentrated mass and rotational inertia. The following characteristics were used in the numerical simulations: plate (duraluminium alloy) — Young's modulus $E = 68.5$ GPa, Poisson's ratio $\nu = 0.3$, $\rho_s = 2714$ kg/m^3; fluid (water) — $c = 1500$ m/s, $\rho_f = 1000$ kg/m^3; and piezoelectric element was made of PZT-19 piezo-ceramics (lead zirconate-titanate), polarized along the z-axis, with the following physical and mechanical properties: $C_{11} = C_{22} = 10.9 \times 10^{10}$ Pa, $C_{13} = C_{23} = 5.4 \times 10^{10}$ Pa, $C_{12} = 6.1 \times 10^{10}$ Pa, $C_{33} = 9.3\times10^{10}$ Pa, $C_{44} = C_{55} = C_{66} = 2.4\times10^{10}$ Pa, $\beta_{13} = \beta_{23} = -4.9$ C/m^2, $\beta_{33} = 14.9$ C/m^2, $\beta_{51} = \beta_{42} = 10.6$ C/m^2, $\epsilon_{11} = \epsilon_{22} = 8.2 \times 10^{-9}$ F/m, $\epsilon_{33} = 8.4 \times 10^{-9}$ F/m, $\rho_p = 7500$ kg/m^3.

The parameters of the finite element mesh, which is used for subsequent calculations, are determined from the analysis of asymptotical convergence of the solution depending on the number of nodal unknowns. During this research the maximal relative difference between the first eleven complex eigenvalues was evaluated for different degree of discretization. The option was considered as acceptable, if the obtained results differed from the previous ones by less than 1% with a significant increase in the number of unknowns. In our example, the finite element model was defined by 50623 degrees of freedom.

A series of numerical calculations were carried out to verify the validity of the results obtained with the use of the proposed algorithm. For these computations a series RL-circuit was connected to the electrodes of piezoelectric element (Figure 1). The case when both parameters of the inductance and resistance are set negligibly small (of the order of magnitude of about 10^{-15}), is equivalent to the case when the piezoelectric element operates in the short circuit mode. And vice versa, extremely high values of both parameters of inductance and resistance (of the order of magnitude of about 10^{15}), corresponds to the case when the piezoelectric element operates in the open circuit mode. This peculiarity allows us to make a comparison between the results obtained with the help of the proposed algorithm and the results obtained with the use of the ANSYS software. To determine natural vibration frequencies in the calculations done in the ANSYS package, the open circuit and short circuit modes were realized by setting the corresponding boundary conditions for electric potential.

Table 1. Natural vibration frequencies of the plate with piezoelectric element connected to the series RL-circuit and located on the free surface of quiescent fluid.

ω	Short circuit mode		Open circuit mode	
	ANSYS	Proposed solution	ANSYS	Proposed solution
ω_1	19.9329	19.9329	19.8982	19.8982
ω_2	122.8790	122.8793	122.8008	122.8008
ω_3	346.0930	346.0931	346.0619	346.0619
ω_4	385.7038	385.7038	385.7038	385.7038
ω_5	684.8710	684.8710	684.8572	684.8572

Table 1 show the first five natural vibration frequencies for the plate with the attached piezoelectric element connected to the series RL-circuit and interacting with a quiescent fluid (Figure 1). The parameters of the electric circuit correspond to the piezoelectric element operating in the open circuit ($L = 10^{15}$ H, $R = 10^{15}$ Ω) and short circuit ($L = 10^{-15}$ H, $R = 10^{-15}$ Ω) modes. In should be noted that the eigenvalues of the system of equations (11) are purely real in both cases, so $\lambda = \omega$. The data presented in Table 1 show that the results obtained in the ANSYS and on the basis of the proposed algorithm differed by less than 0.02%.

Table 2. Complex eigenfrequencies of the plate with piezoelectric element connected to the series RL-circuit ($L = 3000$ H, $R = 2000$ Ω) and located on the free surface of quiescent fluid.

	λ_1	λ_2	λ_3	λ_4	λ_5
ω	19.2269	20.4141	122.8814	346.0932	385.7038
γ	0.0312	0.0219	0.0000	0.0000	0.0000

The solution of the modal problem was performed using the proposed algorithm with following values of parameters for the external electric circuit: $L = 3000$ H, $R = 2000$ Ω. In this case the natural vibration frequency of the external electric circuit is incorporated into initial spectrum of the system (Table 1). If the values of parameters of external electric circuit is chosen in such a way that incorporated frequency turns out to be equal to the certain mode of vibrations, then this vibration mode can be damped. The values of the first five complex natural vibration frequencies for the examined system are presented in Table 2. Verification of the solution for this case was done by comparing the real parts of the calculated eigenvalues with peaks of the frequency response curve obtained using the module of harmonic analysis of the ANSYS software

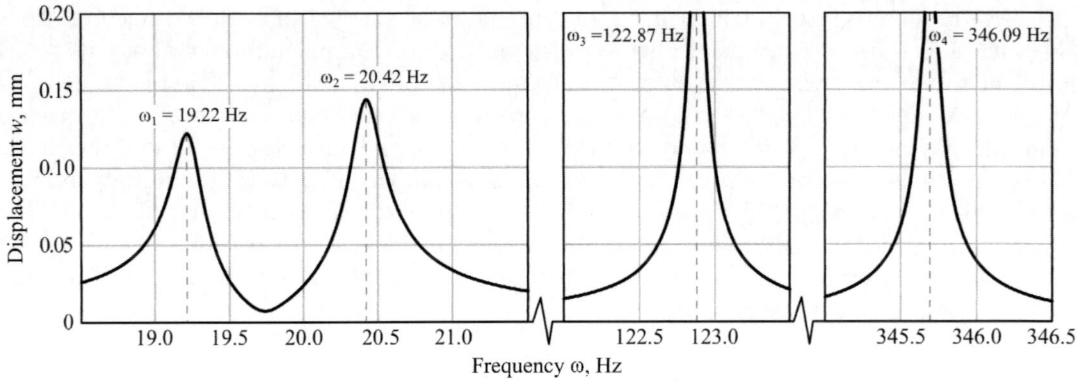

Figure 2. Frequency response curve for cantilever plate located on the free surface of quiescent fluid with attached piezoelectric element and series RL-circuit.

package. For this purpose the problem of forced harmonic vibrations of the plate with attached piezoelectric element with connected external series RL-circuit and interacting with a quiescent fluid was considered. The excitation of vibrations was realized by applying the harmonic force $\{\mathbf{F}\} = \{0, 0, -1\}$ to the point B of the structure (Figure 1). Figure 2 shows frequency response curve of the normal component w of the displacement vector of the plate for the point A. The locations of the peaks obtained in the vicinity of a resonance correspond to the values of ω presented in Table 2 with the tolerance up to 0.1% for the frequency step equal to 0.01 Hz.

It should be noted that the incorporated frequency is presented on the frequency response curve (Figure 2). The new resonance peak appears at the frequency 19.22 Hz. Herewith the first natural vibration frequency of the system changes its value from 19.9329 Hz to 20.4141 Hz (Table 2). Figure 2 demonstrates that resonance peaks ω_1 and ω_2 have finite amplitude. This fact evidences the presence of damping in the system. It corresponds to the appearance of imaginary part for these complex eigenvalues when the modal problem is solved (Table 2).

5. Conclusions

A mathematical formulation of the natural vibration problem for a piecewise-homogeneous electroelastic body connected to a shunt external electric circuit and interacting with a quiescent fluid has been proposed. The process of obtaining the problem solution was based on the finite element method. The possibilities of the ANSYS software package were used for generating global matrices. The final system of governing equations was assembled based on the developed algorithm written in FORTRAN language. The sparse matrix technology was used for basic operations. The solution of non-classic eigenvalue problem was found using the Mueller method.

A thin rectangular plate located at the free surface of the fluid with an attached piezoelectric element connected to the external series RL-circuit was considered as an illustrative example. The functionality of the developed program and the reliability of numerical results were verified by comparing the obtained eigenvalues with the ones computed with the aid of the ANSYS package for two extreme cases of the piezoelectric element operation.

A mathematical statement of the modal problem proposed in this paper has no analogues in modern commercial software packages for finite element modeling. This formulation allows us to determine natural vibration frequencies and corresponding rates of their damping regardless of the form of the exciting impact. In this regard, the problem being considered can be viewed as an efficient tool for constructing the optimization algorithms, which can be readily applied to problems of passive damping of hydroelastic vibrations.

Acknowledgments
The study was supported by the grant of the Russian Scientific Foundation (project No. 18-71-10054).

References
[1] Presas A, Luo Y, Wang Z, Valentin D and Egusquiza M 2018 *Sensors* **18** 2251
[2] Zhang J, Chang W, Varadan V and Varadan V 2001 *Smart Mater. Struct.* **10** 414
[3] Kim J and Jung Y C 2006 *J. Acoust. Soc. Am.* **120** 2017
[4] Sun Y, Li Z, Huang A and Li Q 2015 *J. Sound Vib.* **355** 19
[5] Kim J and Kim J H 2004 *J. Acoust. Soc. Am.* **116** 942
[6] Li S and Zhao D 2004 *J. Sound Vib.* **272** 109
[7] Li S 2011 *J. Sound Vib.* **330** 5545
[8] Kwak M and Yang D H 2015 *J. Fluids Struct.* **54** 848
[9] Pan X, Tso Y and Juniper R 2008 *J. Sound Vib.* **311** 224
[10] Kwak M and Yang D H 2013 *J. Sound Vib.* **332** 4775
[11] Loghmani A, Danesh M and Keshmiri M 2016 *J. Sound Vib.* **363** 345
[12] Benjeddou A 2000 *Comput. Struct.* **76** 347
[13] Mackerle J 1998 *Modelling Simul. Mater. Sci. Eng.* **6** 293
[14] Mackerle J 2003 *Modelling Simul. Mater. Sci. Eng.* **11** 707
[15] Iurlova N, Sevodina N, Oshmarin D and Iurlov M 2019 *Int. J. Smart Nano Mat.* **10** 156
[16] Reddy J N 2015 *An introduction to nonlinear finite element analysis* 2nd ed (Oxford: Oxford University Press)
[17] Washizu K 1982 *Variational methods in elasticity and plasticity* (London: Pergamon Press)
[18] 1988 *IEEE Standard on Piezoelectricity* (ANSI/IEEE Std)
[19] Zienkiewicz O C and Taylor R L 2000 *The finite element method* 5th ed vol 1,2 (Oxford: Butterworth-Heinemann)
[20] Allik H and Hughes J R 1970 *Int. J. Numer. Meth. Eng.* **2** 151

IOP Conf. Series: Materials Science and Engineering **581** (2019) 012007 doi:10.1088/1757-899X/581/1/012007

Influence of aspect ratio on heat transfer in non-uniformly heated cylindrical fluid layers

A. Evgrafova[1], **A. Sukhanovskii**[1]
[1]Institute of Continuous Media Mechanics, Academ. Korolyov, 1, Perm, 614013, Russia
E-mail: eav@icmm.ru

Abstract. In this paper experimental study of heat transfer processes in a cylindrical container non-homogeneously heated from below is presented. The heater has a circular form and was placed in the center of the vessel. The experiments were performed in the range of Rayleigh numbers from $4.4{\cdot}10^6$ to $1.2{\cdot}10^7$ and fixed Prandtl number ($Pr = 67$). Special attention was paid to the influence of the aspect ratio (the ratio of the height of the fluid layer to the diameter of the heater) on the flow structure and the intensification of heat transfer.

1. Introduction

It is known that convection is the main mechanism of heat transfer in a number of technological and geophysical processes. In thermal convection, the control parameters are the Rayleigh number Ra and the Prandtl number Pr. The intensity of heat exchange in convective systems is described by the Nusselt number Nu (the dimensionless heat flux). And the key question in heat transfer studies is to obtain and understand the dependences $Nu(Ra, Pr)$.

So far the convective heat transfer are studied mostly for Rayleigh-Benard convection when the entire top surface is cooled and the entire bottom surface is heated (vertical temperature gradient) [1-4]. Another interesting case that attracts a lot of attention is a horizontal convection when the flow is initiated by a horizontal temperature gradient [5, 6].

Convection from a localized heat source, which will be discussed in this article, is the result of both vertical and horizontal temperature gradients. The flow in such a system is a complex combination of multiscale convective structures. It consists of a large-scale advective cell, which occupies the entire cavity, and secondary flows forming over the heating region. The formation and structure of large-scale circulation, laminar or turbulent, were considered in [7-9]. The formation of secondary structures in thermal boundary layer over a localized heat source was observed in [9,10] and their dynamics was described in [11-13]. At the same time, in various studies it was shown that the structure of convective flow and their dynamics are sensitive to the heating conditions, the shape and size of the heat source, to the properties of the working fluids. This leads to the need for a systematic approach for study of heat and mass transfer processes in such systems.

Heat transfer in a horizontal fluid layer over an nonuniform heated surface was investigated in [14] for a wide range of Rayleigh and Prandtl numbers. It was shown that the decrease in the viscosity of the fluid under study leads to a significant change in the dynamics of secondary flows, but does not have a significant effect on the intensity of heat exchange. The influence of different boundary conditions (constant temperature or a constant heat flux) on heat transfer in the same configuration also was considered. The next logical step is to study the effect of aspect ratio (the ratio of the height of the liquid layer to the size of the heater) on the heat exchange processes. Results of this investigation are presented in this paper.

Despite the fact that we considered only a particular configuration we strongly believe that our results will be helpful in solving of various practical problems such as optimization of cooling systems for electronic equipment, semiconductor industry, design of nuclear reactors.

2. Experimental setup

Two cylindrical vessels were used. The first vessel (I) has diameter $D = 300$ mm (figure 1(a)). The heater is a brass cylindrical plate mounted flush with the bottom. The diameter of the plate $2R$ is 104 mm, and its thickness is 10 mm. The second vessel (II) has diameter $D = 690$ mm. The bottom of the larger vessel is the textolite plate with thickness of 20 mm. In the center of plate the copper heater of diameter $2R = 195$ mm was placed flush with the bottom. The temperature of the heaters was controlled by computer. Silicon oil PMS-5 (Pr=67 at T = 25 ^{0}C) was used as a working fluid. The surface of the fluid was always open. The depth of the layer for the vessel I was fixed $h = 30$ mm. For the vessel II a series of experiments was carried out for different values of h. The Table 1 provides the geometric parameters of the experiments for convenience of the readers, where α is the aspect ratio (the ratio of layer height to the diameter of the heater). The room temperature was kept constant by air-conditioning system, and cooling of the fluid was provided mainly by the heat exchange with a surrounding air on the free surface and some heat losses through sidewalls. The temperature of the fluid was measured by a horizontal row of 12 copperconstantan thermocouples. The location of the thermocouples is shown in figure 1(b). The coordinate origin is at the center of the bottom. The thermocouples can be moved through the whole layer depth with a step of 1 mm by a motorized translation stage. The data from the thermocouples were obtained by an Agilent 34970A data acquisition switch unit with a 16-channel multiplexer module 34902A. Temperature was measured in the central vertical cross-section above the heating area. To estimate the mean temperature of the fluid, one thermocouple measured temperature inside the fluid layer near the periphery.

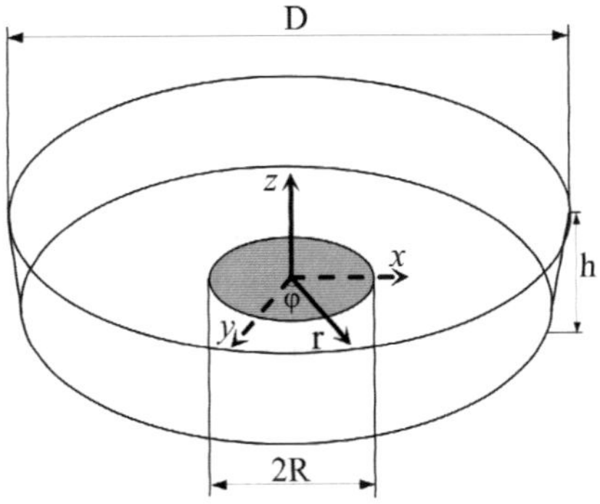

(a)

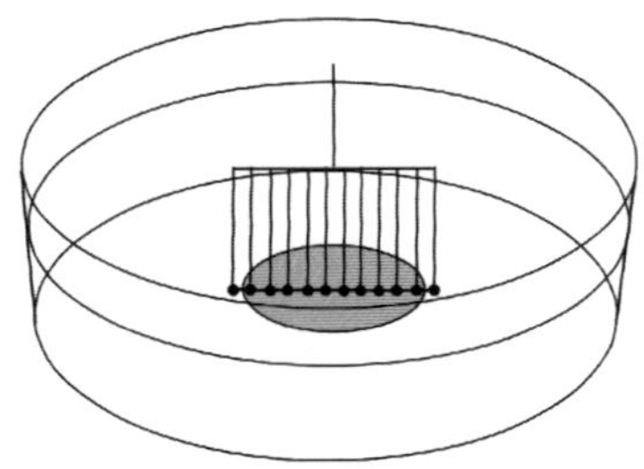

(b)

Figure 1. Experimental setup, dimensions and location of the coordinate system.

We use the set of the non-dimensional parameters which are commonly used for similar problems. These are Prandtl number Pr and Rayleigh number Ra:

$$Pr = \frac{\nu}{\kappa} \quad (1)$$

$$Ra = \frac{g\beta h^3 \Delta T}{\nu^2} \quad (2)$$

where g is the gravitational acceleration, h is the layer depth, β is the coefficient of thermal expansion, ΔT - temperature difference between temperatures of the heater and the room, ν is the coefficient of kinematic viscosity.

Heat transfer was characterized by the Nusselt number defined as

$$Nu = \frac{q}{q_\lambda} \tag{3}$$

where q_λ is heat flux due to conduction, and q is overall heat flux. The heat fluxes q and q_λ were calculated as

$$q = \frac{P}{S}; q_\lambda = \lambda \frac{T_h - T_s}{h} \tag{4}$$

where P is the power of the heater, S is the heater surface area, T_h is the temperature of the heater, and T_s is the temperature of the fluid surface. The temperature of the fluid surface T_s was received by the following way. The row of thermocouples measured the temperature of fluid above the heater consistently at z = 28, 29 and 30 mm. The time of measurements was about 1000 seconds for each height. Then, the temperature was averaged over time and space in this thin near-surface layer, and the resulting temperature (T_s) was used to calculate the heat flux q_λ. Here, we neglect heat losses through the bottom and side walls.

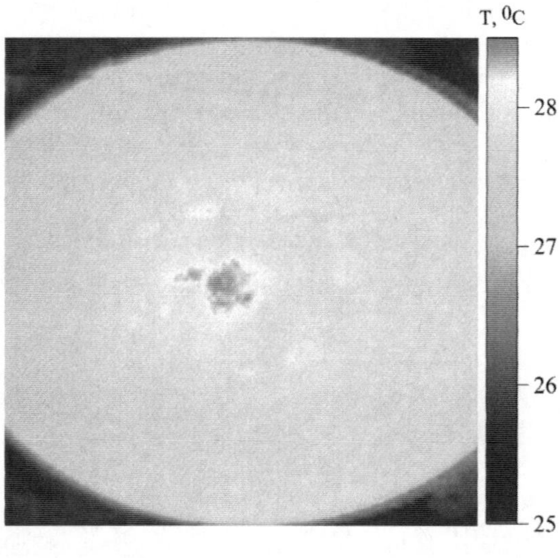

Figure 2. Thermal image of the upper surface of the fluid for $Ra = 4.2 \cdot 10^6$ (model I), $T_h = 32$ °C.

The temperature T_s is the mean value of temperature along the thermocouple row, which is placed in the central part of the vessel above the heater. Temperature fields on a fluid surface were obtained by a thermal imager Fluke Ti32 system for different regimes. Temperature sensitivity of the Fluke Ti32 is high (about 0.05 °C), but temperature measurement accuracy is ±2 °C and it needs to be calibrated for every regime. Therefore it was used only for getting information about homogeneity of temperature distribution over the entire fluid surface. On figure 2 the thermal image of free surface of fluid is shown for $Ra = 4.2 \cdot 10^6$. It is clearly seen that temperature field is not homogeneous in both radial and azimuthal directions and this the main source of error in T_s measurements. However according to our estimations, the inaccuracy of T_s measurements does not exceed 5 % and the one for Nu calculation is less than 7 % (taking into account the heat leakage through the sidewalls).

Table 1.

		h, mm	D, mm	2R, mm	$\alpha = h/2R$
I	1	30	300	104	0.3
II	2	30	690	100	0.3
	3	40	690	100	0.4
	4	50	690	100	0.5
	5	60	690	100	0.6

3. Results

Detailed description of the basic flow structure can be found in [12] but brief description of the general structure of the large-scale flow is necessary for a better understanding of the results. The heat flux in the central part of the bottom is a source of the intensive upward motion above the heater. Warm fluid cools at the free surface and moves toward the periphery where the cooled fluid moves downward along the side wall. Large-scale advective flow occupies the whole vessel (figure 3). Along with the main updraft in the center there are less intensive but pronounced upgoing convective flows close to the periphery of the heater.

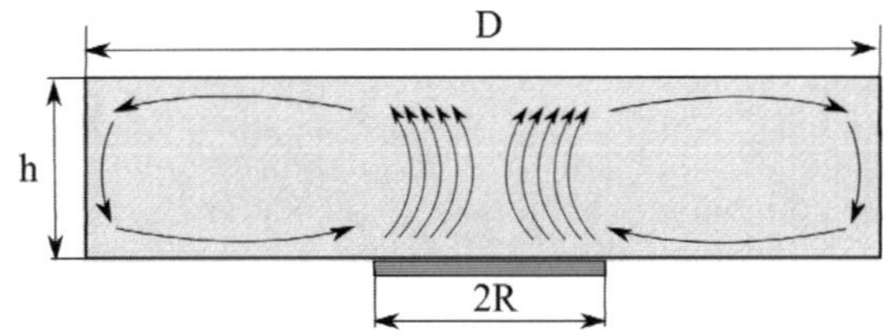

Figure 3. Large-scale circulation.

The convergent flow in the lower part of the layer leads to the formation of a boundary layer with potentially unstable temperature stratification above the heater. Examples of the experimental temperature profiles for different values of Ra and aspect ratio are presented in figure 4. In this boundary layer secondary flows appear. The structure and dynamics of secondary flows are strongly dependent on heating intensity. Weak heating regimes are characterized by the appearance of ring-like rolls. An increase of the heat flux leads to more complex convective patterns - the superposition of a spiral and a system of radial rolls on the periphery. For the more detailed information about dynamics and structure of secondary flows we refer the reader to [12].

One of the most remarkable results obtained in [12] is that ring-like rolls and spirals appear periodically. Dependence of characteristic frequency of secondary structures formation F on Rayleigh number for fixed Pr and different aspect ratio is presented in figure 5. The best fitting for all data is the power law ($F \propto Ra^{0.72}$) but there is significant quantitative difference between measurements for different values of layer depth (aspect ratios). The formation of secondary structures occurs in an unstable boundary layer over the heated area. Its thickness δ is determined by the properties of the working fluid, the value of the aspect ratio and the heating intensity. Changing the depth of the fluid layer for the fixed Ra leads to an increase in the thickness of the boundary layer and to the reduction temperature gradient. Thus, for fixed values of Prandtl and Rayleigh numbers, the increase of aspect ratio leads to a decrease in the frequency of secondary structures formation. It is also interesting that remoteness of lateral boundaries have no substantial impact on secondary flows formation.

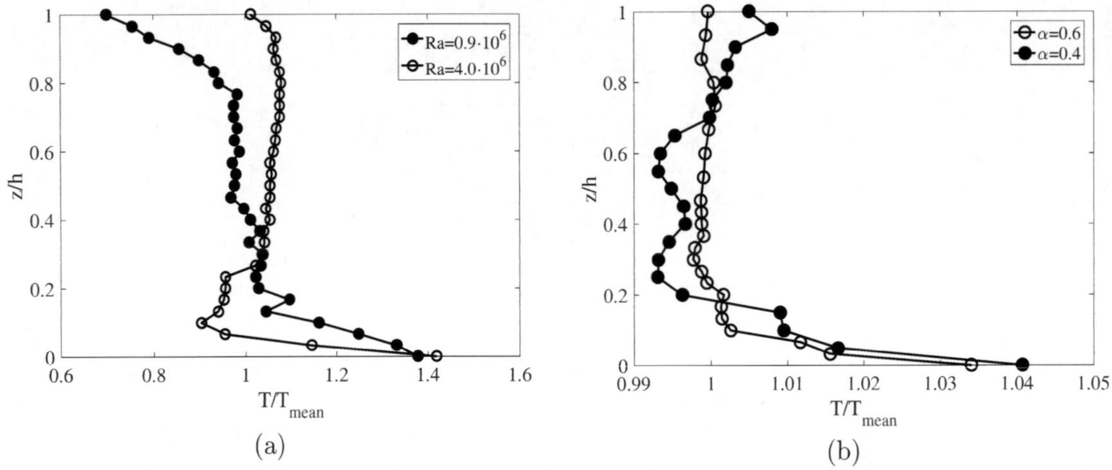

Figure 4. a - Temperature profiles at r = 3 cm for α=0.3(model I);b - Temperature profiles at r = 3 cm for $Ra = 1.3 \cdot 10^7$ (model II)

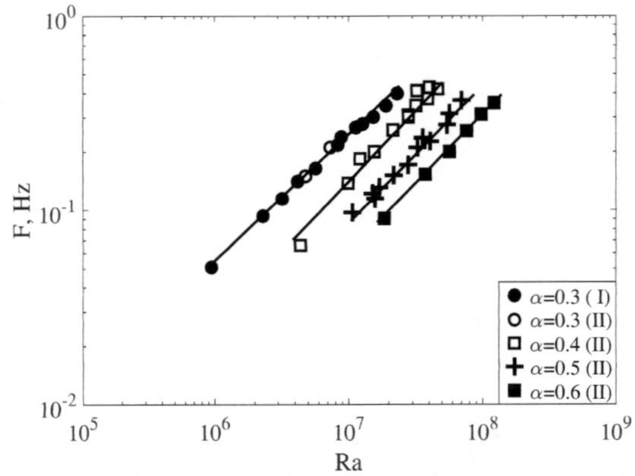

Figure 5. Characteristic frequence F versus Rayleigh number

In figure 6(a) the dependence of the Nusselt number on the Rayleigh number is presented in log-log scale. One can see that the heat transfer does not depend on the aspect ratio and increases with increasing Rayleigh number($Nu \propto Ra^{0.28}$). This exponent value defines heat transfer for quite different flow configurations [14]. By definition the Nusselt number (Nu) is the ratio of convective to conductive heat transfer across (normal to) the boundary. In other words, the Nusselt number is proportional to the ratio of the thickness of the fluid layer to the thickness of the temperature boundary layer. Our results show that for the described configuration the Nusselt number for the fixed Ra does not depend on the aspect ratio. This means that the thickness of the boundary layer varies linearly with the fluid layer depth. The characteristic time between formation of two successive plumes can be estimated as follows $\tau \sim \delta^2/\chi$ [15]. Thus using relation $\delta \sim h$ we proposed a new parameter $f = F \cdot h^2/\chi$, which is the dimensionless frequency of secondary structures formation. The dependence of f on the Rayleigh number is shown in figure 6 (b). We see that all experimental points lie on the same curve, which proves our assumptions.

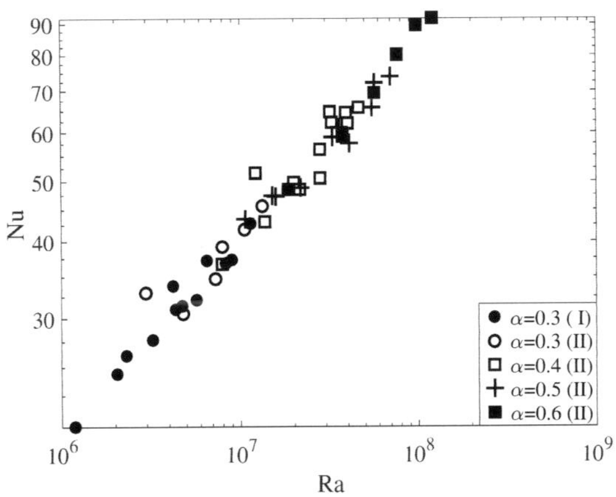

(a)

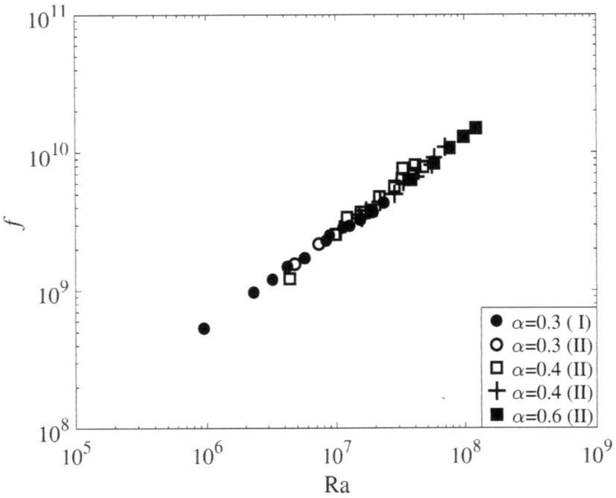

(b)

Figure 6. Nusselt number (a) and dimensionless characteristic frequency F (b) versus Rayleigh number for different aspect ratio

4. Conclusions

Heat transfer from the localized heat source in a cylindrical layer was studied experimentally for different values of aspect ratio. The flow in such a system consists of large-scale advective circulation and secondary structures appeared in the thermal boundary layer above the heated area. Our temperature measurements showed that the aspect ratio do not cause any substantial changes in the dependence of heat transfer on the Rayleigh number. Based on this result we proposed a new parameter that characterizes the formation of secondary structures.

References

[1] Ahlers G et al. 2009 *Rev.Modern Phys.***81(2)**. 503
[2] Chilla F and Schumacher J 2012 *Eur. Phys. J. E***35(7)**. 58
[3] Goldstein R et al 1990 *J. Fluid Mech.***213**. 111-126
[4] Siggia ED 1994 *Annu. Rev. Fluid Mech.***26(1)**. 137-168
[5] Hughes GO and Griffiths RW 2008 *Annu. Rev. Fluid Mech.***40**. 185-208
[6] Shishkina O et al 2016 *Geophys. Res. Lett.***43(3)**. 1219-1225
[7] Torrance KE 1979 *J. Fluid Mech.***95**. 477-495
[8] Sezai I and Mohamad A 2000 *Int. J. Heat Mass Transf.***43**. 2257-2266
[9] Boubnov B and Van Heijst G 1994 *Exp Fluids.***16(3-4)**. 155-164
[10] Lu J et al 1997 *J. Ap. Met.* **36**. 1377
[11] Bogatyrev GP 1990 *Soviet J. Exp. Theoret. Phys. Lett.***51**. 630
[12] Sukhanovskii A et al 2016 *Physica D.***316**. 23-33
[13] Torrance KE 1969 *J. Fluid Mech***36**. 21-31
[14] Evgrafova A and Sukhanovskii A 2019 *Int. J. Heat Mass Transf.***135**. 761-768
[15] Cioni S et al. 1997 *J. Fluid Mech.***335**. 111-140

IOP Conf. Series: Materials Science and Engineering **581** (2019) 012008 doi:10.1088/1757-899X/581/1/012008

Stabilization of direct numerical simulation for finite truncations of circular cumulant expansions

I V Tyulkina[1,2], D S Goldobin[1,2] and A Pikovsky[3,4]

[1]Institute of Continuous Media Mechanics UB RAS, 1 Akademika Koroleva street, Perm 614013, Russia
[2]Department of Theoretical Physics, Perm State University, 15 Bukireva street, Perm 614990, Russia
[3]Institute for Physics and Astronomy, University of Potsdam, 24/25 Karl-Liebknecht-Strasse, Potsdam 14476, Germany
[4]Department of Control Theory, Nizhny Novgorod State University, 23 Gagarin Avenue, Nizhny Novgorod 606950, Russia

E-mail: irinatiulkina95@gmail.com, Denis.Goldobin@gmail.com, pikovsky@uni-potsdam.de

Abstract. We study a numerical instability of direct simulations with truncated equation chains for the "circular cumulant" representation and two approaches to its suppression. The approaches are tested for a chimera-bearing hierarchical population of coupled oscillators. The stabilization techniques can be efficiently applied without significant effect on the natural system dynamics within a finite vicinity of the Ott–Antonsen manifold for direct numerical simulations with up to 20 cumulants; with increasing deviation from the Ott–Antonsen manifold the stabilization becomes more problematic.

1. Introduction

In studies of collective phenomena, especially interesting are the cases, where a network of elements exhibits nontrivial dynamics, while the inherent dynamics of elements is simple and coupling between elements or/and an external driving of them are small. This is the case of populations of self-sustained periodic oscillators with a weak mutual coupling or/and subject to a weak common forcing. For description of such system a phase reduction proved [1] to be applicable and the paradigmatic Kuramoto model [2] was found to be relevant for many physical, chemical and biological systems [3, 4]. The examples include both continuous distributed parameter systems and networks of discrete elements.

Later on, Kuramoto system as well as a broad class of paradigmatic phase systems of form

$$\dot{\varphi}_k = \Omega(t) + \mathrm{Im}(2h(t)\,e^{-i\varphi_k})\,, \qquad k = 1, ..., N\,, \tag{1}$$

where $\Omega(t)$ and $h(t)$ are arbitrary real and complex-valued functions of time, were found to be partially integrable. Watanabe–Strogatz [5, 6, 7, 8] and Ott–Antonsen theories [12] were developed for such systems. While WS theory allowed studying the dynamics of finite populations [9, 10, 11] and made a solid foundation for OA theory, the latter one, constructed for the thermodynamic limit of large populations, and especially its extension to the case of nonidentical elements [13, 14, 15], found broad applications and helped to elucidate many sophisticated collective phenomena [16, 17, 18, 19, 20, 21, 22, 23, 24, 25, 26, 27].

Simultaneously, the peculiarity of the systems of type (1) necessitated construction of a perturbation theory for the OA approach, since real system can be close to (1) but never perfectly possess such a form. In [28], a formalism of "circular cumulants" formally corresponding to the Kuramoto–Daido order parameters $a_n = \langle (e^{i\varphi})^n \rangle = N^{-1}\sum_{k=1}^{N} e^{in\varphi_k}$ was suggested. Specifically, with a moment-generating function $F(\zeta) = \sum_{j=0}^{\infty} a_j \zeta^j / j!$, one can introduce circular cumulants $\varkappa_j$ via power series of the generating function $\zeta \frac{\partial}{\partial \zeta} \ln F(\zeta) = \sum_{j=1}^{\infty} \varkappa_j \zeta^j$. For instance, the first three cumulants are: $\varkappa_1 = a_1$, $\varkappa_2 = a_2 - a_1^2$, $\varkappa_3 = (a_3 - 3a_2 a_1 + 2a_1^3)/2$. This formalism turned out to be fruitful for generalization of the Ott–Antonsen theory [28, 29, 30].

In particular, an important case of violation of the applicability conditions of OA theory is the one of population on phase elements

$$\dot{\varphi}_k = \Omega_k + \mathrm{Im}(2h(t)\, e^{-i\varphi_k}) + \sigma \xi_k(t), \qquad k = 1, ..., N \tag{2}$$

with natural frequencies Ω_k, which are distributed around the mean value Ω_0 according to the Lorentzian distribution of half-width γ, and individual white gaussian noises of strength σ: $\langle \xi_k(t) \rangle = 0$, $\langle \xi_k(t) \xi_m(t') \rangle = 2\delta_{km}\delta(t - t')$. The collective dynamics of population (2) for large N is governed by the following system of equations for circular cumulants [28]:

$$\dot{\varkappa}_j = j(i\Omega_0 - \gamma)\varkappa_j + h\delta_{1j} - h^*(j^2 \varkappa_{j+1} + j \sum_{m=1}^{j} \varkappa_{j-m+1}\varkappa_m) - \sigma^2 (j^2 \varkappa_j + j \sum_{m=1}^{j-1} \varkappa_{j-m}\varkappa_m), \tag{3}$$

where $\delta_{1j} = 1$ for $j = 1$ and $\delta_{1j} = 0$ for $j \neq 1$. For $\sigma = 0$, equation system (3) admits a particular solution $\varkappa_1 = a_1$, $\varkappa_{j \geq 2} = 0$; this case reproduces the OA theory.

While equations (3) are useful for analytical studies and for construction of perturbation theories [28, 29, 30], direct numerical simulations with them require truncations of the infinite system. Truncations with more than the first two cumulants kept were revealed to be able to cause a numerical instability. This issue turned out to be especially important for chimera-bearing systems [7, 16, 19], where neutrally stable or weakly stable degrees of freedom appear to be rather snesitive to the numeric instabilities. On the other hand, chimera regimes are in focus of many research works (e.g., [31, 32]) also due to similarities with turbulence intermittency in fluid dynamics [33, 34].

In this paper we suggest two approaches to suppression of the numerical instability of direct simulations with truncated chains (3) and test them for a chimera-bering system [7, 16] described in section 2. In section 3, the stabilization by means of a dissipation term for the high-order cumulants is studied. In section 4, we consider the approach based on the suppression of terms causing instability. In section 5, the findings are summarized.

2. Hierarchical population of coupled oscillators

Abrams et al. [16] studied the collective dynamics of an ensemble of hierarchically coupled oscillators, composed by two subpopulations of equal size N; the strength of a Kuramoto–Sakaguchi coupling between oscillators [2, 35, 36] within each subpopulation, $(1+A)/2$, is larger than the coupling strength $(1-A)/2$ between subpopulations:

$$\begin{aligned} \dot{\varphi}_k &= \Omega + \frac{1+A}{2N}\sum_{j=1}^{N} \sin(\varphi_j - \varphi_k - \alpha) + \frac{1-A}{2N}\sum_{j=1}^{N} \sin(\psi_j - \varphi_k - \alpha) + \sigma\xi_k(t)\,, \\ \dot{\psi}_k &= \Omega + \frac{1+A}{2N}\sum_{j=1}^{N} \sin(\psi_j - \psi_k - \alpha) + \frac{1-A}{2N}\sum_{j=1}^{N} \sin(\varphi_j - \psi_k - \alpha) + \sigma\eta_k(t)\,, \end{aligned} \tag{4}$$

where α is the coupling phase shift, σ is the intrinsic noise strength, ξ_k and η_k are independent white Gaussian noises. For vanishing noise, the system satisfies the conditions of Watanabe–Strogatz and Ott–Antonsen theories [5, 6, 12], which provided a substantial progress in understanding its dynamics [7, 16]. The chimera states, where one of subpopulations is fully synchronized $\psi_1 = \psi_2 = ... = \Psi$, while the other one is in a partial synchrony state, were attracting the main interest of researchers [16].

However, the intrinsic noise violates these properties, and the circular cumulant approach proved to be useful for the description of the noisy system [28, 29]. In the thermodynamic limit of large N, one can introduce the circular cumulant description, with cumulants $\varkappa_j$ for subpopulation $\{\varphi_k\}$ and κ_j for $\{\psi_k\}$ (cf. equation (3)):

$$\dot{\varkappa}_j = j(i\Omega_0 - \gamma)\varkappa_j + h_\varphi\delta_{1j} - h_\varphi^*(j^2\varkappa_{j+1} + j\sum_{m=1}^{j}\varkappa_{j-m+1}\varkappa_m) - \sigma^2(j^2\varkappa_j + j\sum_{m=1}^{j-1}\varkappa_{j-m}\varkappa_m), \quad (5)$$

$$\dot{\kappa}_j = j(i\Omega_0 - \gamma)\kappa_j + h_\psi\delta_{1j} - h_\psi^*(j^2\kappa_{j+1} + j\sum_{m=1}^{j}\kappa_{j-m+1}\kappa_m) - \sigma^2(j^2\kappa_j + j\sum_{m=1}^{j-1}\kappa_{j-m}\kappa_m), \quad (6)$$

where $h_\varphi = 0.25[(1+A)\varkappa_1 + (1-A)\kappa_1]e^{-i\alpha}$ and $h_\psi = 0.25[(1+A)\kappa_1 + (1-A)\varkappa_1]e^{-i\alpha}$. In the classical "Abrams' system" [16] the oscillator natural frequencies are identical: $\gamma = 0$. In this paper we use the Abrams' system (4) as a model for testing the approaches to stabilizing numerical simulation with truncated equation chains (5), (6).

3. Introducing auxiliary dissipation

From numerical viewpoint, eqs. (3) is a system of ODEs, which, if properly truncated, can be solved with the Runge-Kutta method. However, for some systems, truncation of the system can lead to a numerical instability. In this section we introduce additional dissipation-like terms into the original equations,

$$\begin{aligned}\dot{\varkappa}_j = j(i\Omega_0 - \gamma)\varkappa_j + h\delta_{j1} - h^*\left(j^2\varkappa_{j+1} + j\textstyle\sum_{m=1}^{j}\varkappa_{j-m+1}\varkappa_m\right)\\ - \sigma^2\left(j^2\varkappa_j + j\textstyle\sum_{m=1}^{j-1}\varkappa_{j-m}\varkappa_m\right) - G(\gamma+\sigma^2)(j-1)(j-2)2^j\varkappa_j\,, \quad (7)\end{aligned}$$

and study their effect on the numerical stability of calculations. The choice of such a form of the stabilizing term is guided by the following reasons:

- The G-term introduces additional dissipation into the system dynamics. Similar to natural dissipation terms, it depends on parameters γ and σ, but is rather strong for large j.
- Since the stabilizing term must make a stronger impact on the higher-order cumulants, i.e. grow with j much faster than the $\sigma^2 j^2\varkappa_j$-term, we adopt the j-dependence of the form $\propto j^2 s^j\varkappa_j$. As simulations with only the first or two first cumulants are always stable, it is reasonable to amend the j^2-multiplier by changing it to $(j-1)(j-2)$. The numerical simulations revealed that the choice of $s = 2$ is optimal.

We perform numerical simulations with the following initial conditions: the second subpopulation is synchronized $\kappa_j = \delta_{1j}$ and the elements of the first subpopulation form a two-bunch distribution, which is a superposition of two wrapped Cauchy distributions with complex order parameters b_1 and b_2. In terms of moments $a_j = \langle e^{ij\varphi}\rangle$, the latter means

$$a_j = q_1(b_1)^j + q_2(b_2)^j,$$

where q_1 and $q_2 = 1 - q_1$ are the fractions of elements in the first and second bunch, respectively. Terms $(b_1)^j$ and $(b_2)^j$ represent the probability density distributions of φ corresponding to OA

IOP Conf. Series: Materials Science and Engineering **581** (2019) 012008 doi:10.1088/1757-899X/581/1/012008

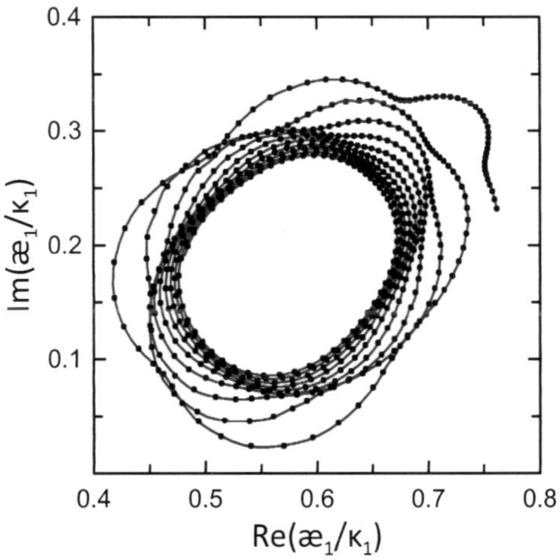

Figure 1. Trajectory of system (5)–(6) for $A = 0.3$, $\alpha = \pi/2 - 0.15$, $\gamma = 0.001$, $b_{1,2} = (0.8 \pm 0.05)e^{i(0.3 \pm 0.1)}$. Black dots are the result of "exact" integration, blue curve is the result of numerical integration of equations (7); first 15 cumulants are used.

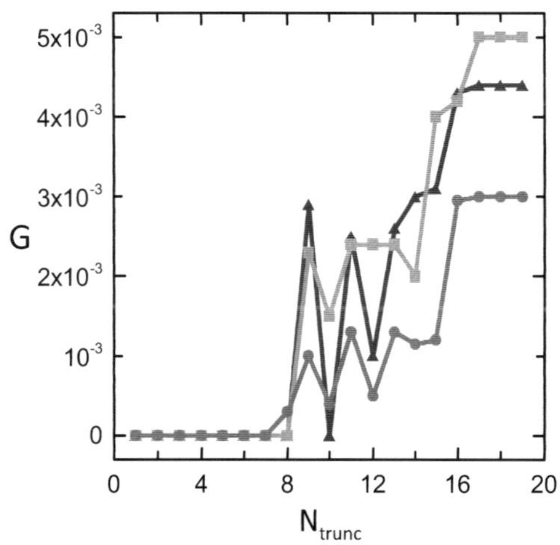

Figure 2. The critical value of the coefficient G versus the number of cumulants N_{trunc} is plotted for $\gamma = 0.001$. Blue triangles: $\sigma = 0$, $q_1 = q_2 = 0.5$; red circles: $\sigma = 0.01$, $q_1 = q_2 = 0.5$; green squares: $\sigma = 0$, $q_1 = 0.4$, $q_2 = 0.6$.

solutions with complex order parameters b_1 and b_2, respectively. The distance between b_1 and b_2 is a convenient parameter for controlling the proximity to the OA manifold; the case $b_1 = b_2$ yields an OA solution.

Let us first consider the case of equipartition between two bunches: $q_1 = q_2 = 0.5$. In this case, simulations can be performed accurately, due to the fact that all odd cumulants, except the first one, turn to zero [30]. Indeed, one can perform simulations for $\varkappa_1$ and $\varkappa_{2n}$ with terms $\varkappa_{2n+1} = 0$ wiped-out from equations for $n \geq 1$; the numerical calculations are stable for any order of truncation in this case. The trajectory of the system is shown in figure 1.

Figure 2 provides the results which can guide the choice of the values of coefficient G for numerical simulations. In the graph, one can see the dependence of the critical value G, above which the numerical calculations are stable, on the number of cumulants kept in a truncated chain, N_{trunc}. The choice of value of G, moderately exceeding the critical value, allows achieving compromise: the simulations are free of numerical instability, while the system trajectory is practically unaffected by the stabilizing term.

In figure 3(a) we demonstrate the explosion of numerical simulation due to numerical instability below the stabilization threshold; one can see that the blow-up is driven by the highest-order cumulant, the equation for evolution of which is affected by truncation of the cumulant chain. Above the threshold, in figure 3(b), the evolution is regular, fast fluctuations of cumulants decay with time, and the system forms a hierarchy of smallness for cumulants as predicted theoretically [28, 29].

Stabilization can be achieved with a minor distortion of the original system dynamics for not more than 20 cumulants and a non-large distance between the two bunches in the unsynchronized subpopulation. When using more than 20 cumulants, the value of G required to stabilize the numerical calculations grows very quickly with the number of cumulants: the stabilizing term becomes significantly distorting the 'true' dynamics of the original system.

Let us consider the case of noise strength $\sigma = 0.01$. In the unsynchronized subpopulation,

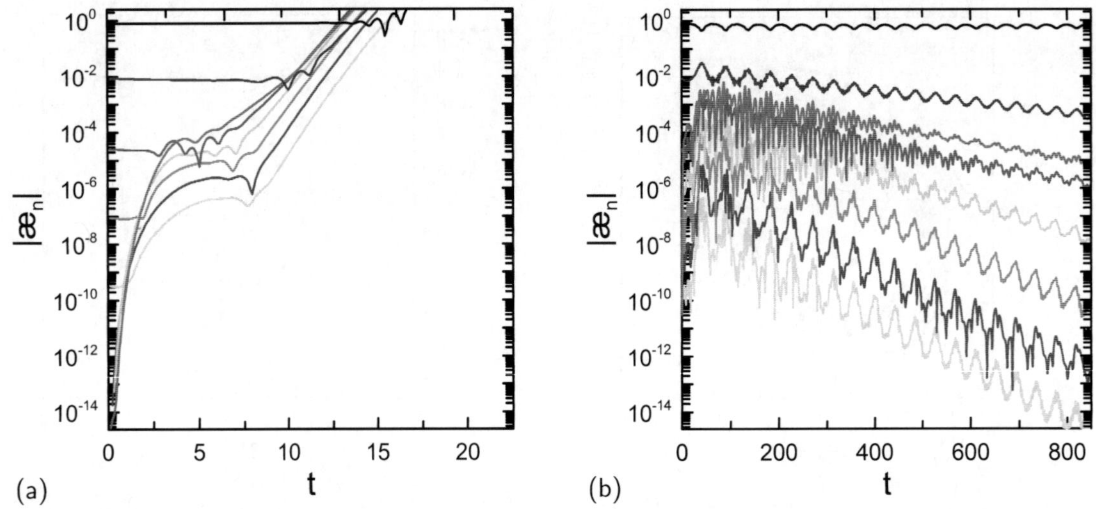

Figure 3. Dependencies of cumulants $\varkappa_n$ on time t (from top to bottom: from $\varkappa_1$ to $\varkappa_8$). (a) Parameter G is chosen below the stability threshold of the method (see figure 2). (b) Parameter G is above the stability threshold of the method.

the oscillators are equally distributed between two bunches, i.e. $q_1 = q_2 = 0.5$. In figure 2 one can see that the critical value of G for the noisy case (red circles) is lower than for the noise-free case (blue triangles). In figure 4(a) the example of system trajectory is plotted. One can notice that the trajectory for a marginal value of G, although being significantly disturbed by fluctuations, does not diverge with the trajectory calculated for a larger value of G, for which the computational fluctuations are suppressed.

Testing with the equipartition case $q_1 = q_2$ at $\sigma = 0$ is important, since in this case the integration of the modified equation system can be accurately performed without a numerical

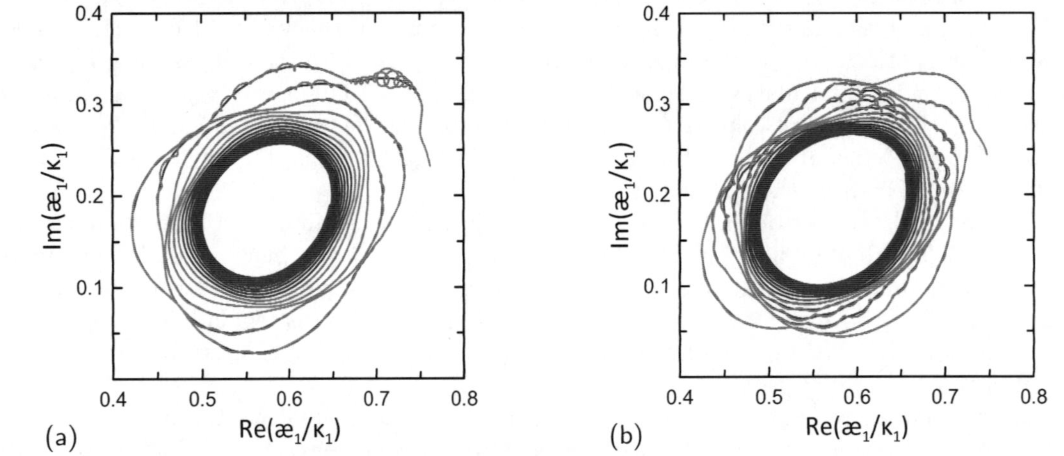

Figure 4. Trajectory of the system is plotted for (a) $q_1 = q_2 = 0.5$, $\sigma = 0.01$ and (b) $q_1 = 0.4$, $q_2 = 0.6$, $\sigma = 0$. The blue 'smooth' curves: $G = 0.01$; the red 'curly' curves: G at the stability boundary. Calculations with first 15 cumulants.

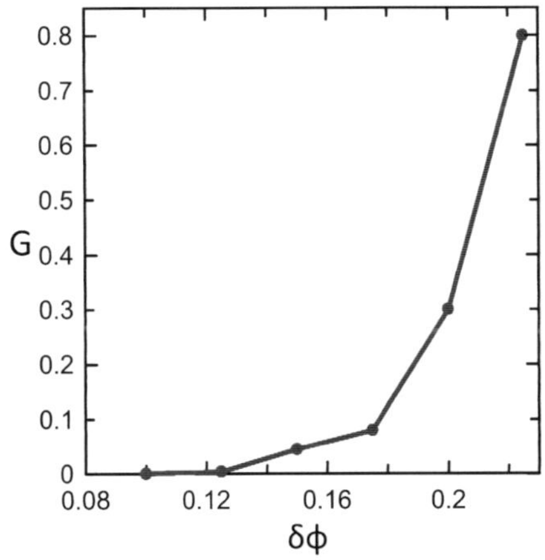

Figure 5. The critical value of the coefficient G versus the phase distance $\delta\phi$ for $A = 0.3$, $\alpha = \pi/2 - 0.15$, $\delta R = 0.05$, $R_0 = 0.8$, $\Phi_0 = 0.3$, $\gamma = 0.001$. Here $N_{\text{trunc}} = 12$ cumulants are used for calculations; $\sigma = 0.01$ and the distribution of oscillators between two bunches is unequal: $q_1 = 0.4$ and $q_2 = 0.6$.

instability; one can check the results of numerical integration with stabilization terms against the numerical solutions for the unaltered system. However, the specificity of the equipartition solutions can also effect their numerical stability properties. Hence, an additional treatment for the case of unequal partition between two bunches is required. The numerical results for $\sigma = 0$, $q_1 = 0.4$ and $q_2 = 0.6$ are presented in figures 2 (green squares) and 4(b). In figure 2 one can see that the characteristic critical values of G for equal and unequal partitions between two bunches are close. For a noisy system, smaller values of the coefficient G are required, since noise makes a stabilizing effect.

From the provided results on the stability boundary of the numerical simulations, one can conclude that, for the considered initial conditions, the stabilizing term is not required if less than 8 cumulants are used. For a simulation with larger number of cumulants, a stabilizing term is required.

The efficiency of stabilization procedure depends on initial conditions. For initial conditions, the distance between the complex order parameters

$$b_{1,2} = (R_0 \pm \delta R) e^{i(\Phi_0 \pm \delta\phi)}$$

in amplitude and phase (δR, $\delta\phi$) of two bunches is required to be non-large, otherwise the suppression of the numerical instability becomes impossible (simulation with two first cumulants is always stable). For these situations, the required value of the coefficient G becomes too large and affects the natural dynamics of the system. In figure 5, the dependence of the critical value of G on the phase difference $\delta\phi$ is plotted for a fixed distance in amplitude between two bunches.

4. Suppression of the terms responsible for instability

An alternative approach to stabilization is the suppression of the terms which are primarily responsible for numerical instability in the original system. We modify one of the terms of the original equation by an additional exponential factor $e^{-\lambda(j-1)}$:

$$\begin{aligned} \dot{\varkappa}_j = j(i\Omega_0 - \gamma)\varkappa_j + h\delta_{j1} - h^* \Big(j^2 \varkappa_{j+1} e^{-\lambda(j-1)} \\ + j \textstyle\sum_{m=1}^{j} \varkappa_{j-m+1}\varkappa_m \Big) - \sigma^2 \Big(j^2 \varkappa_j + j \textstyle\sum_{m=1}^{j-1} \varkappa_{j-m}\varkappa_m \Big). \end{aligned} \quad (8)$$

This modification is guided by the following reason:

- The numerical instability is related to the $\varkappa_{j+1}$-term in $\dot{\varkappa}_j$. We modify it in such a way that

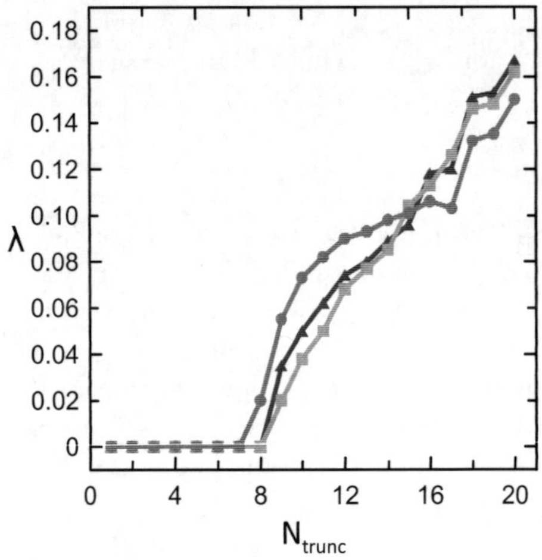

Figure 6. The critical value of the exponential coefficient λ versus the number of cumulants N_{trunc} for $\gamma = 0.001$. Blue triangles: $\sigma = 0$, $q_1 = q_2 = 0.5$; red circles: $\sigma = 0.01$, $q_1 = q_2 = 0.5$; green squares: $\sigma = 0$, $q_1 = 0.4$, $q_2 = 0.6$.

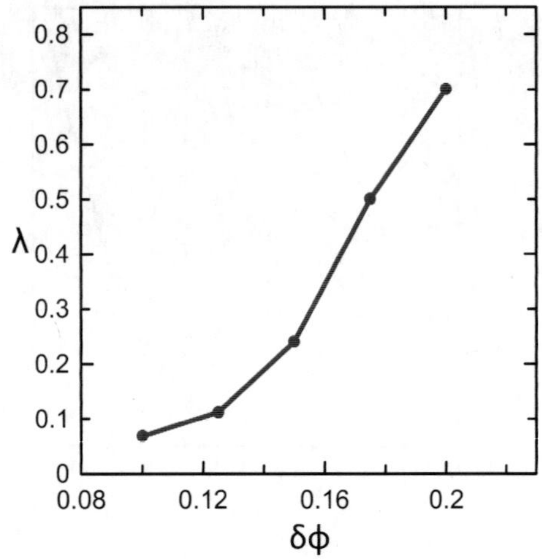

Figure 7. The critical value of the exponential coefficient λ versus the phase difference $\delta\phi$ for $A = 0.3$, $\alpha = \pi/2 - 0.15$, $\delta R = 0.05$, $R_0 = 0.8$, $\Phi_0 = 0.3$, $\gamma = 0.001$, $\sigma = 0.01$, unequal partition between two bunches, $q_1 = 0.4$, $q_2 = 0.6$; $N_{\text{trunc}} = 12$.

for large j it vanishes, while for small j it is nearly unchanged, that is the modification does not affect the accuracy of calculations significantly.

We apply this stabilization procedure to the same examples as in the previous section. The second subpopulation is nearly perfectly synchronized, in the first subpopulation oscillators are partitioned between two bunches. The cases to be considered:
1) Oscillators are equally distributed between two bunches, i.e. $q_1 = q_2 = 0.5$; no noise, $\sigma = 0$.
2) Oscillators are equally distributed between two bunches, i.e. $q_1 = q_2 = 0.5$, but noise is added to the system $\sigma = 0.01$.
3) Oscillators are unequally distributed between two bunches, $q_1 = 0.4$ and $q_2 = 0.6$; $\sigma = 0$.

The minimal exponential coefficient λ required for stabilization as a function of the number of cumulants N_{trunc} is presented in figure 6. Initial conditions are the same as in the previous section (see the caption for figure 1).

We also calculate the dependence of the coefficient λ on the phase difference $\delta\phi$ between two bunches. In figure 7, one can see such a dependence for the case of unequal partition of oscillators between two bunches, $q_1 = 0.4$ and $q_2 = 0.6$.

Similarly to the previous stabilization method, this method can be implemented without affecting the natural system dynamics for direct numerical simulation with no more than 20 cumulants. The strength of the required stabilizing correction depends on the proximity to the Ott–Antonsen manifold; too far away from the OA manifold (for a large distance between two bunches), numerical simulations with large number of cumulants become impossible.

5. Conclusion

We have considered two approaches to suppress numerical instability of direct simulations of truncated systems of circular cumulant equations (3). With the first approach, an additional dissipation term for high-order cumulants is introduced (see equation 7). With the second approach, the the coupling term term $\propto \varkappa_{j+1}$, responsible for instability of truncated chains, is exponentially suppressed for high-order cumulants (see equation 8). The stability domains have been constructed for these stabilization methods; the dependence of stability on the initial conditions has been studied as well.

One can conclude, that the stabilization can be practically achieved without significant effect on the natural system dynamics for no more than 20 cumulants. Stabilization efficiently works in a small but finite vicinity of the Ott–Antonsen manifold. For a larger deviation from the OA manifold, direct numerical simulations with fewer number of cumulants can be stabilized.

Acknowledgments
The work has been financially supported by the Russian Science Foundation (Grant no. 19-42-04120).

References

[1] Winfree A T 1967 *J. Theor. Biol.* **16** 15–42
[2] Kuramoto Y 1975 *International Symposium on Mathematical Problems in Theoretical Physics* (*Lecture Notes Phys.* vol 39) ed. H Araki (New York: Springer) pp 420–2
[3] Kuramoto Y 2003 *Chemical Oscillations, Waves and Turbulence* (New York: Dover)
[4] Acebrón J A, Bonilla L L, Vicente C J P, Ritort F and Spigler R 2005 *Rev. Mod. Phys.* **77** 137–85
[5] Watanabe S and Strogatz S H 1993 *Phys. Rev. Lett.* **70** 2391–4
[6] Watanabe S and Strogat S H 1994 *Phys.* D **74** 197–253
[7] Pikovsky A and Rosenblum M 2008 *Phys. Rev. Lett.* **101** 264103
[8] Marvel S A, Mirollo R E and Strogatz S H 2009 *Chaos* **19** 043104
[9] Marvel S A and Strogatz S H 2009 *Chaos* **19** 013132
[10] Zaks M A and Tomov P 2016 *Phys. Rev.* E **93** 020201(R)
[11] Chen B, Engelbrecht J R and Mirollo R 2017 *J. Phys.* A **50** 355101
[12] Ott E and Antonsen T M 2008 *Chaos* **18** 037113
[13] Ott E and Antonsen T M 2009 *Chaos* **19** 023117
[14] Mirollo R E 2012 *Chaos* **22** 043118
[15] Pietras B and Daffertshofer A 2016 *Chaos* **26** 103101
[16] Abrams D M, Mirollo R, Strogatz S H and Wiley D A 2008 *Phys. Rev. Lett.* **101** 084103
[17] Martens E A, Barreto E, Strogatz S H, Ott E, So P and Antonsen T M 2009 *Phys. Rev.* E **79** 026204
[18] So P and Barreto E 2011 *Chaos* **21** 033127
[19] Laing C R 2009 *Phys.* D **238** 1569–88
[20] Nagai K H and Kori H 2010 *Phys. Rev.* E **81** 065202
[21] Laing C R 2012 *Chaos* **22** 043104
[22] Laing C R 2014 *Phys. Rev.* E **90** 010901
[23] Luke T B, Barreto E and So P 2014 *Front. Comput. Neurosci.* **8** 145
[24] Pazó D and Montbrió E 2014 *Phys. Rev.* X **4** 011009
[25] Montbrió E, Pazó D and Roxin A 2015 *Phys. Rev.* X **5** 021028
[26] Pimenova A V, Goldobin D S, Rosenblum M and Pikovsky A 2016 *Sci. Rep.* **6** 38518
[27] Hannay K M, Forger D B and Booth V 2018 *Sci. Adv.* **4** e1701047
[28] Tyulkina I V, Goldobin D S, Klimenko L S and Pikovsky A 2018 *Phys. Rev. Lett.* **120** 264101
[29] Goldobin D S, Tyulkina I V, Klimenko L S and Pikovsky A 2018 *Chaos* **28** 101101
[30] Tyulkina I V, Goldobin D S, Klimenko L S and Pikovsky A 2019 *Radiophys. Quantum Electron.* **61** 640–9
[31] Omel'chenko O E, Maistrenko Y L and Tass P A 2008 *Phys. Rev. Lett.* **100** 044105
[32] Omelchenko I, Maistrenko Y, Hovel P and Schöll E 2011 *Phys. Rev. Lett.* **106** 234102
[33] Barkley D and Tuckerman L S 2005 *Phys. Rev. Lett.* **94** 014502
[34] Moxey D and Barkley D 2010 *PNAS* **107** 8091–6
[35] Sakaguchi H and Kuramoto Y 1986 *Prog. Theor. Phys.* **76** 576–81
[36] Sakaguchi H, Shinomoto Sh and Kuramoto Y 1988 *Prog. Theor. Phys.* **79** 1069–79

IOP Conf. Series: Materials Science and Engineering **581** (2019) 012009 doi:10.1088/1757-899X/581/1/012009

Numerical study of electro-vortex flow in long cylinder with localized current supply

S Mandrykin, V Ozernykh, I Kolesnichenko

Institute of Continuous Media Mechanics, Perm, Russia

E-mail: msd@icmm.ru

Abstract. Numerical simulations of the electrovortex flow (EVF) of the GaSnZn alloy in a closed cylindrical volume are performed. The electric current is applied with two localized opposed electrodes on the cylinder side wall. It is shown that in the entire considered range of electric current values from 50 to 1000 A the EVF arises in a cylinder, in a form of the two-toroidal flow. The flow intensity increases with increasing current values and its maximum intensity is 12 cm/s (at 1000 A). Even at the lowest electric current value of 50 A, a non-stationary behavior of EVF is observed. However, the two-toroidal flow pattern is present in all considered regimes. It is shown that with increasing current the velocity pulsations in the near-electrode region have a frequency up to 0.26 Hz.

1. Introduction

Electrovortex flows (EVF) arise due to the interaction between the electric current in conducting medium and the magnetic field, induced by the current itself, in case of inhomogeneous electric current distribution in the medium [1]. EVFs are found in many industrial applications. For example, in titanium arc furnaces [2, 3], which are used in modern industrial metallurgy.

In dependence on the problem configuration, EVF can be stationary, or not. A number of academic tasks are known, where the possibility of obtaining stationary EVF is show. EVFs are well-studied in canonical configurations, such as hemispheres [4] and flat layers [5, 6]. In most industrial configurations EVFs appear in non-stationary form. The instability of such flows is shown to be either positive (for example, better mixing of a liquid metal) or negative factor. Liquid metal batteries [7, 8] — a promising power storage unit, in which the arising non-stationary EVF may result in device malfunction, are good example of undesired EVFs. At the same time, as mentioned before, EVFs can help in better liquid metal stirring, improving crystalline structure of obtained ingot. Thus, it is of interest to study EVF in close to industrial ones configurations, with the aim of investigating the possibilities of controlling their intensity. An additional parameter of the task is the method of electric current supply to the conducting medium. Of particular interest is the localized current application, since it is the method used in described above industrial applications, and also in aluminum reduction cells [9] (including the multi-electrode current supply [5]).

In this work, the EVF in a closed cylindrical volume, in the case of inconsistency of the cavity geometry and topology of the arising vortices, is studied numerically. A localized electric current application with two opposed electrodes on the cylinder side wall is considered.

IOP Conf. Series: Materials Science and Engineering **581** (2019) 012009 doi:10.1088/1757-899X/581/1/012009

2. Methods

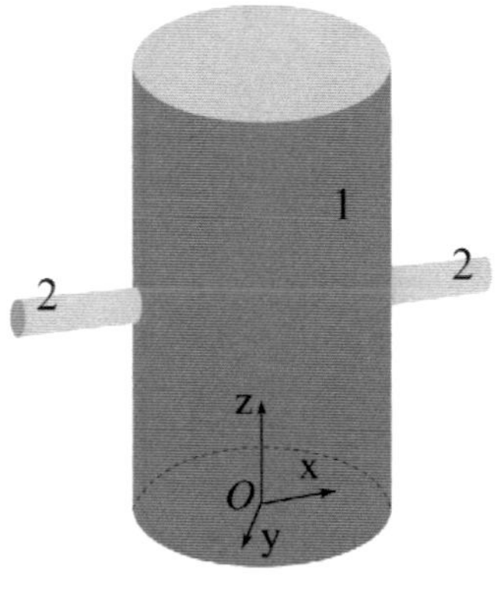

Figure 1. Computational domain scheme

The electrovortex flow of a gallium alloy GaSnZn arising in a cylindrical volume is considered. The problem is solved numerically. The solution of the electromagnetic part is carried out in the ANSYS Emag software, and the hydrodynamic one — in ANSYS Fluent. The computational domain is a cylinder 1 with a diameter of $D = 0.068$ m and a height of $L = 0.2$ m (see Fig. 1). The Cartesian coordinate system is used in calculations and in the presentation of results. Its origin is located at the bottom of the cylinder, the y axis is parallel to the line connecting the eletrodes 2, and the z axis coincides with its vertical symmerty axis. The electrodes are located oppositely on the side wall at $z = 0.1$ m and have a diameter of 0.02 m. The current values were varied from 50 to 1000 A. The cylinder is filled with GaSnZn gallium alloy with following material parameters: kinematic viscosity $\nu = 3.1 \cdot 10^{-7}$ m^2/s, density $\rho = 6265$ kg/m^3, electric conductivity $\sigma = 3.56 \cdot 10^6$ Sm. An unstructured mesh with tetrahedral elements is used whose linead size does not exceed 0.002 m. To resolve the boundary layer, computational nodes distribution density is increased as it approaches the side wallsd of the cylinder. The problem is considered to be isothermal. No-slip conditions are set on all boundaries for the velocity. All simulations began from a state of zero velocity.

For additional data processing, wavelet analysis [10, 11] was used. This type of analysis was selected due to the appearance of non-harmonic oscillations. The wavelet transform translates the one-dimensional time series $f(t)$ into a two-dimensional frequency-time plane (ν, t), according to

$$W(\nu, t) = \sqrt{\nu} \int f(\tau)\, \psi(\nu, (\tau - t)) \mathrm{d}\tau.$$

The Morlet wavelet was selected as an analyzing wavelet — one of the most popular wavelets:

$$\psi(t) = \exp\left[-t^2/2\sigma^2\right] \exp\left[i2\pi t\right],$$

which better suits for signals of complex nature. Here σ is an adjustable parameter. By changing its value one can obtain optimal resolutions of time t and frequency ν. Low σ values provide better time resolution, while large ones can help achieving better frequency resolution of wavelet. However, an uncontrolled increase in σ is undesirable, since it leads to attenuation of the analyzing function. The commonly adopted value of σ is 1 [12], while the limit $\sigma \to \infty$ corresponds to the Fourier transform.

The map $W(\nu, t)$ has an easily interpreted structure only in case of the simple data analysis. But in the case of the real data, wavelet map interpretation can be much more difficult. The wavelet transform integral spectrum is one of the simplest its characteristic:

$$\mathrm{M}(\nu) = \int |W(\nu,t)|^2 \mathrm{d}t,$$

which is calculated as a convolution of $W(\nu,t)$ along the time axis. The integral spectrum is a smoothed Fourier transform. Increasing the σ parameter reduces the degree of spectrum smoothing.

3. Results

The three-dimensional velocity fields of gallium alloy in the cylinder, as well as statistical characteristics of the process, are obtained. The current values were varied in the range from 50 to 1000 A, which corresponds to the MHD interaction parameter $S = \mu_0 I^2/(\rho\nu^2)$ values in range $S = (0.52 \div 208) \cdot 10^7$. In an experimental study [13, 14] the lower current value was determined by the resolution of the Doppler ultrasound velocimeter [15], and the upper one — by the technological properties of the setup. In numerical simulations such limitations are not imposed, which made it possible to expand the range of considered regimes. The intensity of emerging EVF is moderate — the average flow rate does not exceed 12 cm/s for the highest current value of 1000 A, which corresponds to the Reynolds number determined through the diameter ($\mathrm{Re} = \langle V\rangle D/\nu$) in the range of $438 \div 26761$.

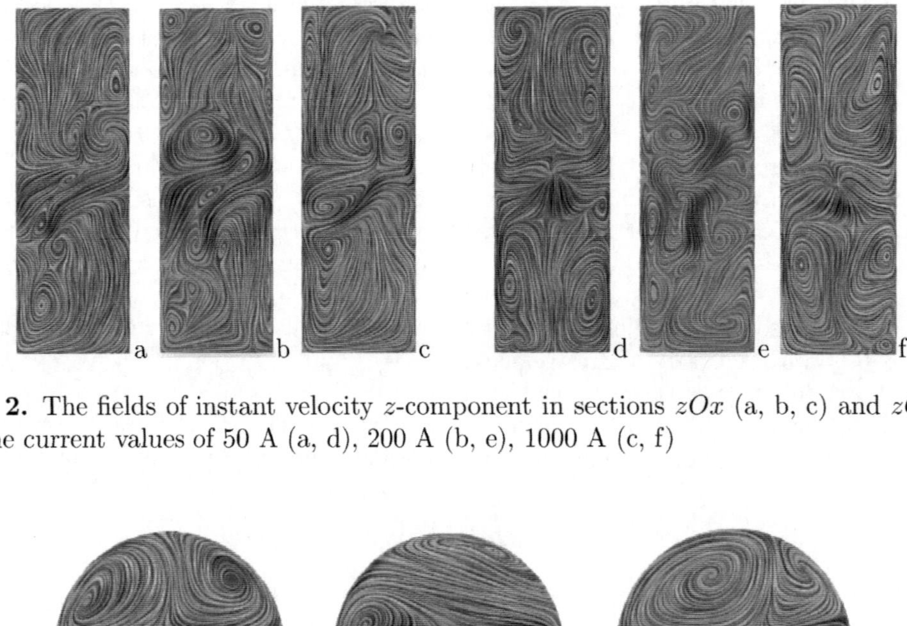

Figure 2. The fields of instant velocity z-component in sections zOx (a, b, c) and zOy (d, e, f) for the current values of 50 A (a, d), 200 A (b, e), 1000 A (c, f)

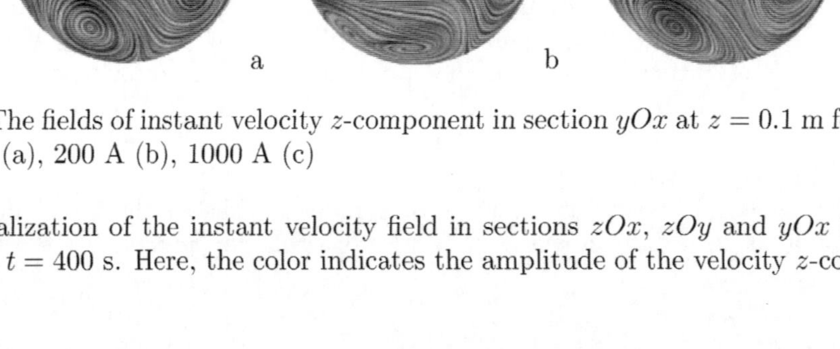

Figure 3. The fields of instant velocity z-component in section yOx at $z = 0.1$ m for the current values 50 A (a), 200 A (b), 1000 A (c)

The visualization of the instant velocity field in sections zOx, zOy and yOx is pictured in Figs. 2, 3 at $t = 400$ s. Here, the color indicates the amplitude of the velocity z-component and the velocity vector field interpretation in current section is shown with lines. The figure shows that the flow intensity is the highest in areas close to the electrodes and that it is noticeably lower in other parts of the cylinder. Also, one can clearly see the distorted two-toroidal flow pattern. The observed asymmetry stands for non-stationarity of the EVF, even for the lowest electric current value of 50 A.

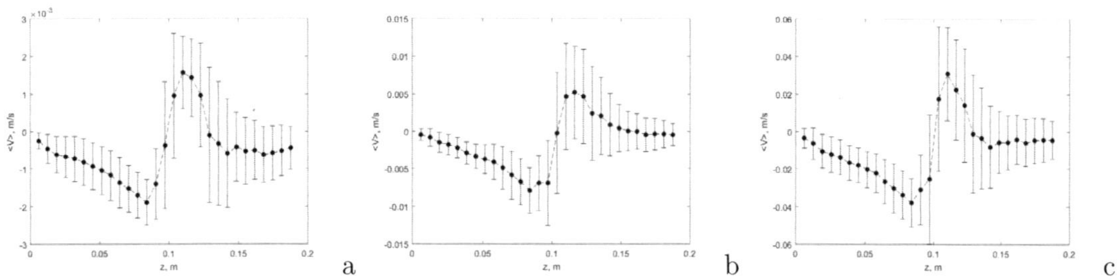

Figure 4. Liquid metal velocity profiles along the line $(-0.025, 0, z)$ m for the current values a — 50 A, b — 200 A, c — 1000 A

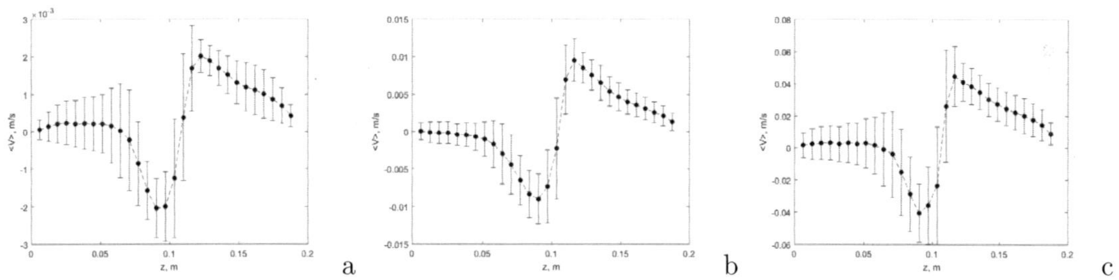

Figure 5. Liquid metal velocity profiles along the line $(0.025, 0, z)$ m for the current values a — 50 A, b — 200 A, c — 1000 A

Figs. 4, 5 presents the liquid metal velocity time-averaged profiles in a cylinder, obtained along the lines parallel to the cylinder vertical symmetry axis. Namely, $(0.025, 0, z)$ m and $(-0.025, 0, z)$ m. Here, a negative velocity value means the flow towards the bottom cylinder face, and a positive value — towards the top one. Thus, the presented profiles indicate the presence of a large-scale two-toroidal vortex structure in volume. The velocity pulsations depicted by vertical lines have an amplitude of up to 6 cm/s, which indicates a strong oscillatory behavior of EVF. It can be seen that the velocity profiles are slightly asymmetric with respect to the center of the vertical axis of the cylinder, and this asymmetry differs for profiles along the lines $(0.025, 0, z)$ m and $(-0.025, 0, z)$ m, which indicates the distortion of the two-toroidal flow pattern.

To analyze the statistical characteristics of the process, the spatiotemporal maps of the vertical velocity component were produced (see Fig. 6). As in the Fig. 10, one can see here that the highest flow amplitude is observed in areas near the electrodes. The change in the velocity amplitude in time indicated that the process is non-stationary. Also, an increase in electric current value leads to an increase in both oscillations intensity and frequency.

Next, points with $z = 0.10$ m were selected on the spatiotemporal maps, that correspond to the areas with the highest velocity values, for obtaining oscillation characteristics. For them, the wavelet spectrograms were obtained (see Figs. 7, 8). Here $a = f_s/\nu$, where $f_s = 0.1$. The Fig. 7a

IOP Conf. Series: Materials Science and Engineering **581** (2019) 012009 doi:10.1088/1757-899X/581/1/012009

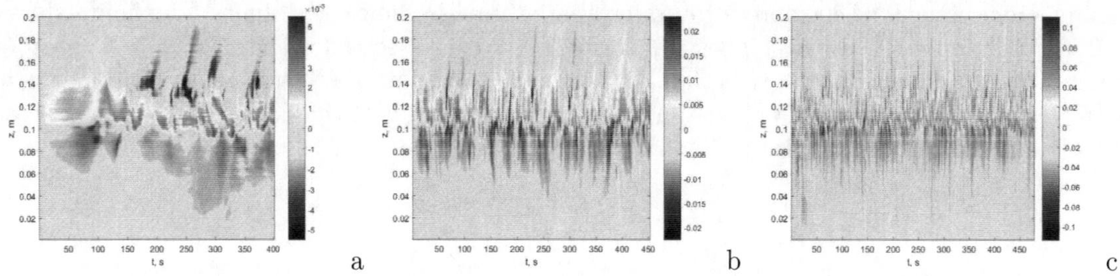

Figure 6. Spatiotemporal maps of the liquid metal flow along the line $(-0.025, 0, z)$ m for the current values a — 50 A, b — 200 A, c — 1000 A

shows that even at the electric current value of 50 A it is possible to resolve oscillations with frequency approx 0.069 Hz, which can be also seen from Fig. 6. With subsequent increase in the electric current oscillation frequency increases to 0.075 Hz (see Fig. 7b). This observed peak remains at higher current values, while at 490 A additional peaks are present, corresponding to 0.147 and 0.222 Hz. Further increase in the electring current results in moderate oscillation frequency growth, while the peaks positions on wavelet spectrograms change. This can indicate a traversing of large-scale vortices with time.

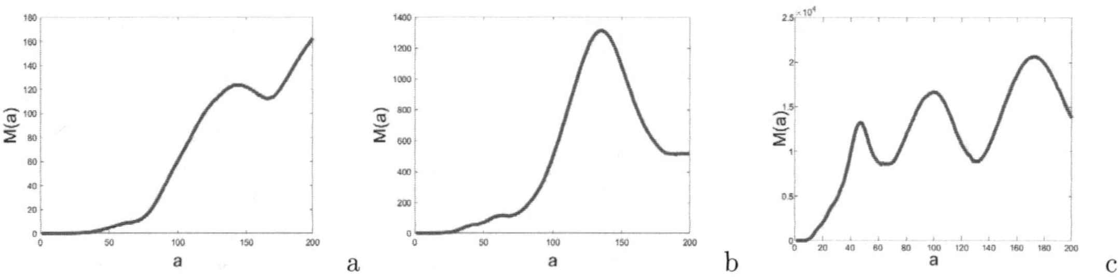

Figure 7. Wavelet spectrograms for the flow velocity at $(0.025, 0, 0.10)$ m and current values a — 50 A, b — 100 A, c — 490 A

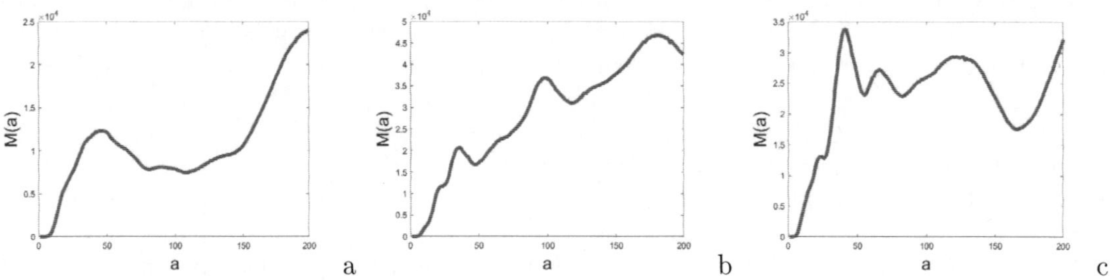

Figure 8. Wavelet spectrograms for the flow velocity at $(0.025, 0, 0.10)$ m and current values a — 700 A, b — 900 A, c — 1000 A

The dependencies of obtained in the wavelet analysis frequencies on the electric current value are pictured in Fig. 9. The figure shows that low-frequency oscillations are present at all electric current values. From 200 A and above, additional high-frequency oscillations appear. While the number of dominant frequencies increases with increase in electric current, the overall change in those frequencies values is rather small.

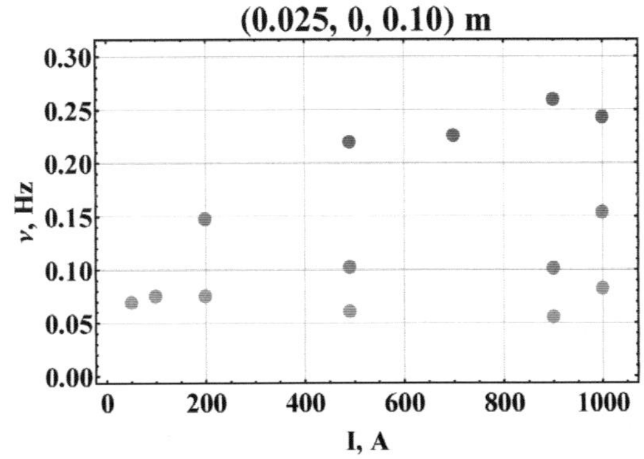

Figure 9. Oscillation frequency in point (0.025, 0, 0.10) m versus electric current value

An estimation of the maximum mean flow velocity shows an increase in the EVF intensity with an increase in the electric current magnitude (see Fig. 10). In the considered range of current values, this dependence is best fitted with quadratic trend. The figure shows that at the highest considered electric current value of 1000 A the maximum mean flow velocity reaches 12 cm/s.

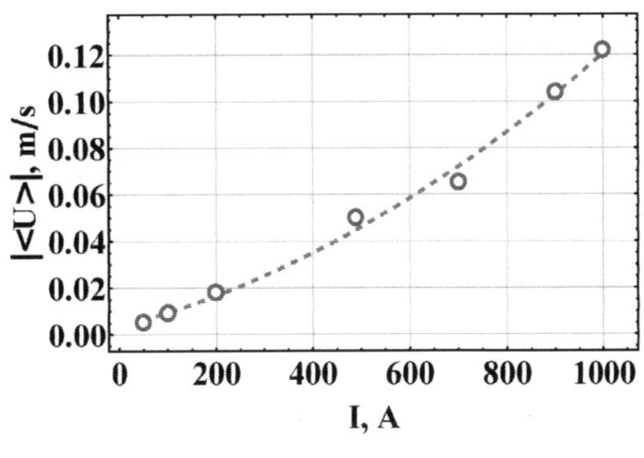

Figure 10. Maximum mean flow velocity versus the electric current value

4. Conclusions

The EVF in a closed cylindrical volume, generated by a direct electric current supply with two localized opposite electrodes on the side wall, is studied numerically. For all electric current values a large-scale two-toroidal vortex flow is formed in the volume, supplemented with several small-scale vortices. The flow is non-stationary even at the lowest current of 50 A. With increasing current, the velocity pulsations amplitude and the flow asymmetry increase, as well as the number of secondary small-scale vortices. Maximum EVF intensity is 12 cm/s, at 1000 A (which is the highest electric current value in this work). The distorted two-toroidal flow pattern remains at all studied regimes. At all electric current values the presence of a dominant velocity pulsations frequency, up to 0.26 Hz, is observed.

IOP Conf. Series: Materials Science and Engineering **581** (2019) 012009 doi:10.1088/1757-899X/581/1/012009

Acknowledgments

The study was supported by the Government of Perm Region (Program for the support of Scientific Schools of Perm Region, grant C-26/788).

References

[1] Bojarevics V, Freibergs Y, Shilova E and Shcherbinin E 1989 *Electrically induced vortical flows* (Kluwer Academic Publishers, Dordrecht.)
[2] Rabiger D, Zhang Y, Galindo V, Franke S, Willers B and Eckert S 2014 *Acta Materialia* **79** 327–338
[3] Kazak O 2013 *Metallurgical and Materials Transactions B* **44** 1243–1250 ISSN 1543-1916
[4] Vinogradov D A, Teplyakov I O, Ivochkin Y P and Klementeva I B 2017 *Journal of Physics: Conference Series.* **899** 082006
[5] Pedchenko A, Molokov S, Priede J, Lukyanov A and Thomas P J 2009 *EPL (Europhysics Letters)* **88** 24001
[6] Khripchenko S Y 1991 *Magnetohydrodynamics* **27** 77–83
[7] Weber N, Galindo V, Stefani F and Weier T 2014 *Journal of Power Sources* **265** 166–173
[8] Weber N, Galindo V, Priede J, Stefani F and Weier T 2015 *Physics of Fluids* **27** 014103
[9] Sneyd A D and Wang A 1996 *Magnetohydrodynamics* **32** 487–493
[10] Holschneider M 1999 *Wavelets: An Analysis Tool* (*Oxford Mathematical Monographs* no Paperback: pages Publisher: (July 29, 1999) Language: English) (Oxford University Press) ISBN 0198505213
[11] Mallat S 2008 *A wavelet tour of signal processing, third edition: the sparse way* 3rd ed (New York: Academic Press)
[12] Kolesnichenko I, Pavlinov A, Golbraikh E, Frick P, Kapusta A and Mikhailovich B 2015 *Experiments in Fluids* **56**
[13] Mandrykin S D, Kolesnichenko I V, Losev G L and Frick P G 2018 *Bulletin of Perm University. Physics* **40** 20–27
[14] Mandrykin S, Kolesnichenko I and Frick P 2019 (in press) *Magnetohydrodynamics* **55** 3–11
[15] Kolesnichenko I, Khripchenko S, Buchenau D and Gerbeth G 2005 *Magnetohydrodynamics* **41** 39–51

IOP Conf. Series: Materials Science and Engineering **581** (2019) 012010 doi:10.1088/1757-899X/581/1/012010

Influence of length of partitions on the generation of transit flow in MHD-channel

Kolesnichenko I, Dolgikh V, Pavlinov A, Khalilov R
Institute of Continuous Media Mechanics of the Ural Branch RAS, Perm, Russia
E-mail: kiv@icmm.ru, dolgikh@icmm.ru

Abstract. The process of the transit flow generation in the magnetohydrodynamic channel of the pump is investigated. Its design provides for the variation of the flow-rate of the liquid metal without changing the current in the channel. The pressure in the pump channel creates by an electromagnetic force through the interaction of a longitudinal electric current with its own magnetic field. C-shaped ferromagnetic cores are put on the channel. The current generates its own magnetic field which magnetizes the cores. The total magnetic induction vector of the current's magnetic field and the field of the cores is orthogonal to the channel plane. The arising electromagnetic force generates a transit flow through the channel which is caused by the presence of flow partitions in it. The pump is tested on the gallium circuit. The pressure drop at the locked mode and variable flow rate mode for different values of channel overlapping by C-cores and their different numbers has been measured in the experiment. The pressure drop — flow-rate characteristics of the channel are determined at different values of overlap the channel.

1. Introduction

The use of various magnetohydrodynamic (MHD) devices in the liquid metals transport systems has been carried out for a long time. A significant development in this area was observed in the 60s and associated with progress in the field of nuclear energy due to the cooling of a nuclear reactor in which liquid sodium has been used as a coolant. At about the same time, the intensive application of MHD technology in metallurgy has began. The transportation of liquid metal using a ladle to the site of casting is most often used in production. This process is difficult to mechanization and automatation and so involves a considerable amount of hard and dangerous operations. It is should be noticed that when the metal is poured the ingots of relatively poor quality are obtained due to the content of non-metallic inclusions. It is also very difficult task to regulate the consumption of the melt. Besides the liquid metal can ignite in contact with atmosphere air. The typical example are liquid sodium and magnesium.

Currently, the safer and better way is to transport liquid metal using various electromagnetic pumps [1–6]. The development of industry needs the emergence of new type of pumps or the improvement of existing ones. Such a pump should be cheap to production and simple in its design. Thus, the study of MHD processes related to various operations in the foundry and the metallurgy is very important.

One of the promising areas is the use of the MHD channel with a ferromagnetic core as the main structural element. Consider the scheme of a conduction MHD channel using a C-shaped ferromagnetic core (Fig.1a) witch was taken from [7]. When the electric current **j** passes through the channel the ferromagnetic core is magnetized by the currents own magnetic field. The induction of **B** is creating in the gap. The interaction of the initial current and magnetic field generates an electromagnetic force $\mathbf{f}^{em} = \mathbf{j} \times \mathbf{B}$, which causes the flow of a liquid metal.

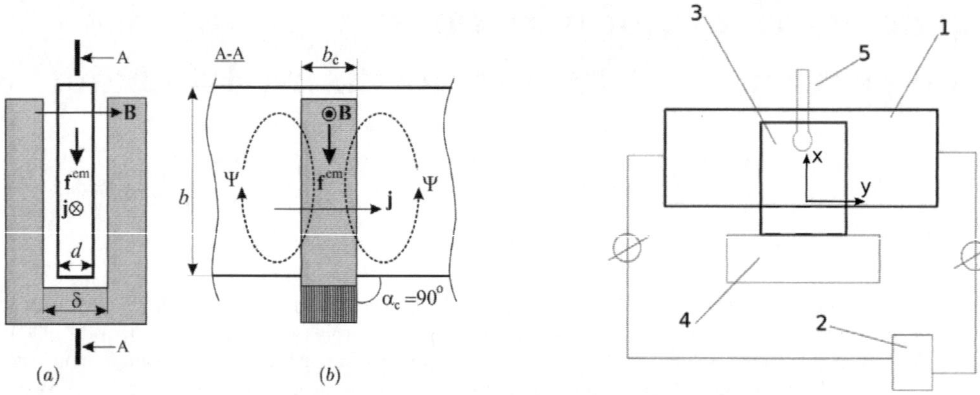

Figure 1. The sketch of the MHD channel with a ferromagnetic core (left) and the sketch of the experiment to determine the characteristics of the core and the generated electromagnetic force (right).

In this paper, the process of generating a transit flow in the magnetohydrodynamic channel of the pump is studied. The design of the pump provides the varying the flow-rate Q of liquid metal without changing the current I in the channel. A simple and reliable solution in this case is the mechanical movement of the cores which regulates the degree of channel overlap by ferromagnetic C-cores. Another advantage of this type of pump is the absence of windings. It allows to operate at high temperatures. The pressure drop P in the pump channel creates by electromagnetic force. The total magnetic field induction vector is orthogonal to the channel plane. It consists of current's magnetic field in the nonmagnetic gap of the cores and the field of the C-cores. The arising electromagnetic force generates the transit flow through the channel which is caused by the presence of partitions in it.

The features and advantages of these channels are as follows. The current is supplied to the channel from an external source. Therefore, there are no copper windings in the channel. So, the pump can be used at high temperatures. The ferromagnetic C-cores are magnetized by their own electromagnetic field. This significantly increases the electromagnetic force and allows to govern its direction.

In [3,6–8] the approach of MHD channel to control the flow of liquid metal has been developed. For example, there were the transportation of metal between areas of metallurgical production. The works have been made in the laboratory of ICMM UB of RAS (Perm, Russia). In the above studies, the performance of the MHD channel was regulated by the magnitude of the current. Therefore, every characteristics are defined under the fixed position of the cores. The influence of the C-core position on the characteristics of the channels has not been studied.

There are tasks where it is necessary to fix the value of the current strength and at the same time change the flow-rate of the pump. An example is the electromagnetic purification of liquid metal in the MHD channel with transit flow [9,10]. One of the solution is to change the flow-rate by mechanically moving ferromagnetic cores. The purpose of this work is to study the influence of the ferromagnetic cores position and the geometry of the channel on the characteristics of the MHD pump.

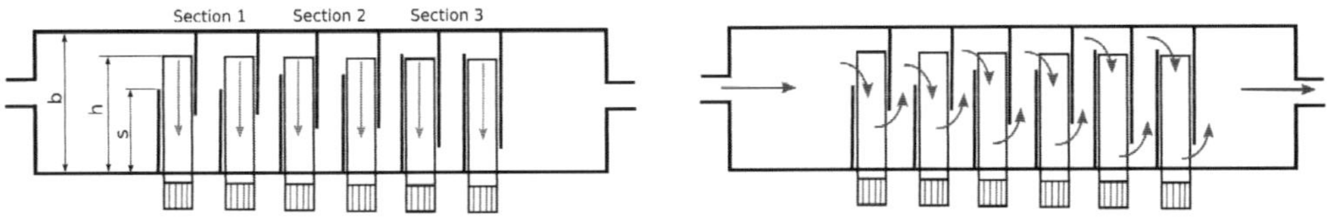

(a) (b)

Figure 2. Sketches of the pump channel: (a) red arrows are electromagnetic force vectors, b) blue arrows are liquid metal flow direction.

2. Methods

At the first stage the experimentally determination of the characteristics of the C-core and the electromagnetic force (fig. 1b) has been performed. In this experiment, it was necessary to obtain characteristics of the C-core, without using it in a real channel with a liquid metal. The channel modelled by the plate 1 to which the C-core 3 was placed. The electrodes were connected to the tails of the plate. When the current form the current source 2 Heiden was turned the C-core was drawn into the plate (or vice versa). In this case, the mass change before and after the current was switched on was measured by the electronic scales 4. Thus, the dependence of the electromagnetic force on the supply current was measured at different positions of the C-core relative to the plate. The Hall sensor 5 of the Lakeshore 421 Gaussmeter was measured the components of the magnetic field in two directions depending on the current.

Figure 3. Photo of the pump with C-cores in testing in gallium loop.

Further studies were carried out to the MHD channel (Fig. 2,3). In this paper the conduction type MHD channel is studied. An electromagnetic force is created in the pump channel by means of a conduction mechanism: electric current is supplied to the channel inlet and outlet from an external source. The flat channel of the pump having width b is made of stainless steel. The fluid partitions with length s are placed inside the channel (Fig. 2a). They are staggered and partially overlap the channel. There are small holes for exit the air during the channel filled with the liquid metal. The channel may be completely empty after the drain of liquid metal. The C-shaped ferromagnetic cores are placed on the channel. The overlap length is h (Fig. 2a). The current generates its own magnetic field which pass into the cores. They are magnetized and the magnetic induction closes in the gap. Its vector is orthogonal to the current density vector. The arising electromagnetic force generates the flow through the channel (Fig. 2b), which is caused by the presence of partitions. The main advantage of the presented MHD pump is the ability to control the flow of liquid metal without changing the current in the channel. Sometimes it is necessary to fix a certain value of the electric current in the channel but to change the metal flow-rate. The best and reliable solution in this case is the mechanical movement of the C-cores to adjust the degree of channel overlap h.

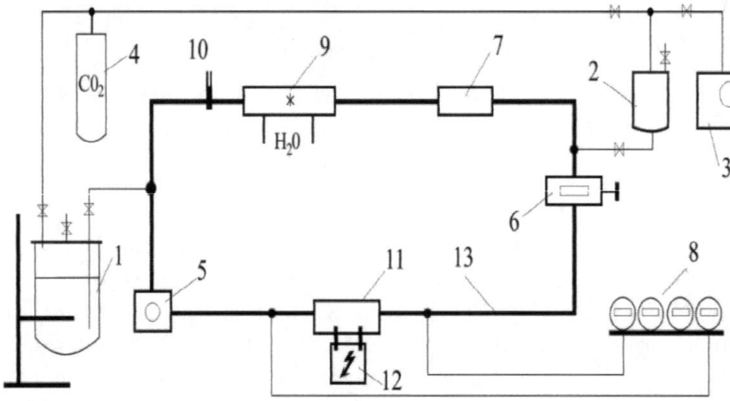

Figure 4. Sketch of the gallium circuit.

The working regimes of the pump was studied with the use of the gallium circuit of the ICMM UB RAS (Perm, Russia). The tubes having the internal diameter of 20 mm filled with a gallium alloy (Ga87.5%+Sn10.5%+Zn2%). The circuit equipment allows to change the hydraulic resistance of the circuit and to measure the pressure drop P developed by the pump, the flow-rate Q of gallium alloy and the current I flowing through the pump channel. In the experiment the PQ-characteristics for different values of channel overlapping by C-cores s and their different numbers N were measured. The circuit includes storage tank 1 for the alloy (Fig. 4), expansion tank 2, vacuum system 3, system with inert gas 4 and filter 5. The flow is regulated by a valve 6 and the flow-rate Q is measured by electromagnetic flow meter 7. The pressure drop P is determined by the pressure gauges 8. The circuit is equipped with a water cooling system 9 and thermocouples 10. The pump 11 is connected to the pipeline 13 circuit with flanges. All measured data was transmitted to the computer using the NI DAQ and processed in the LabView system. The measurement error does not exceed 5%.

3. Results

The electromagnetic force generated by the MHD channel quadratically depends on the current (fig. 5a) passing through the channel. This indicates the absence of saturation of the ferromagnetic C-core. The magnetic field quickly disperses from the edges of the C-core (fig. 6). The greatest magnitude of the force is concentrated in the tail of the C-core in gap.

In the main part of the study the PQ-characteristics were obtained for various configurations of the MHD channel. Each of the PQ-characteristic is the dependence of the pressure drop between the channel entrance and its output on the flow-rate of the liquid metal flowing through the channel. The loss of pressure due to hydraulic resistance is significantly lower than the developed pressure.

In the every studied configuration there is a specific value of the overlap s by the ferromagnetic C-core. If s is less than critical the pump gives insignificant productivity. The productivity is the smallest for s = 60 mm (7a). The using only one C-core makes it impossible to precise control the characteristics of the pump output.

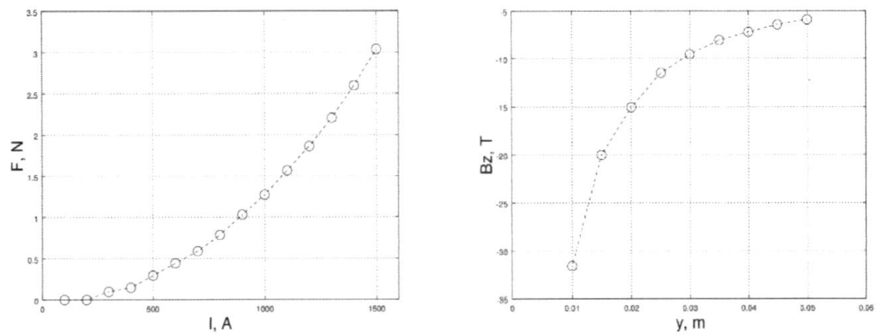

(a) (b)

Figure 5. The dependencies of the electromagnetic force on the supply current (a) and the z-component of the magnetic field induction on the x-coordinate.

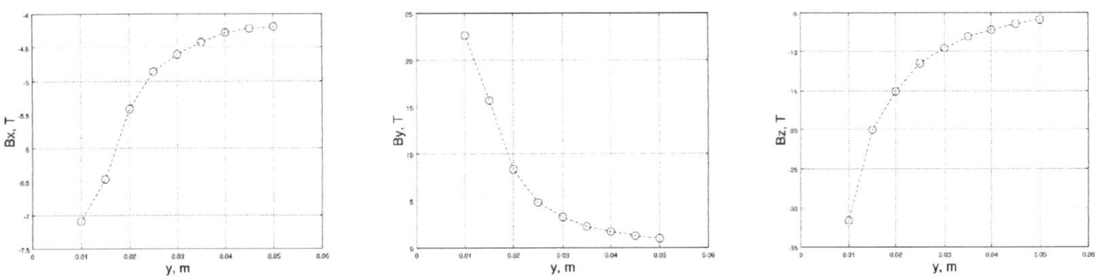

(a) (b) (c)

Figure 6. The dependencies of the magnetic field components on the y-coordinate: (a) x-component, (b) y-component, (c) z-component.

When the two cores are used in the same section, the result changes significantly (7b). For three upper positions of the C-cores the PQ-characteristics of the pump are close to each other and for two lower positions the PQ-characteristics are also close but in a different region. This feature of the systems behaviour does not allow to recommend this configuration for smoothing flow-rate control. The analysis showed that the least drastic change in the MHD pump output is for channel overlap $s = 70$ mm.

Figure 8 represents the PQ-characteristics of the pump for different current strengths at fixed position of one C-core having maximum overlap of the channel. At small values of the electric current the pump flow-rate is very weak. For the two lower current values the characteristics are close. The increase in current leads to increase in output which however stops after reaching the penultimate current value. This indicates the asymptotic evolution of the system to a certain state. After that the increase in current leads to insignificant increase in output. Thus there is no need to use the greatest values of the current strength to achieve the greatest output.

The PQ-characteristics of most of the channel configurations are close to linear. It indicates a small value of hydraulic resistance. At the same time the nonlinear segment is observed (fig. 8b) for the case of using all C-cores at a high flow-rates. The large flow-rate leads to increase the influence of gallium circuit hydraulic resistance on the output. However there is a clear quantitative difference between PQ-characteristics for different currents in the case of using all the C-cores in the pump.

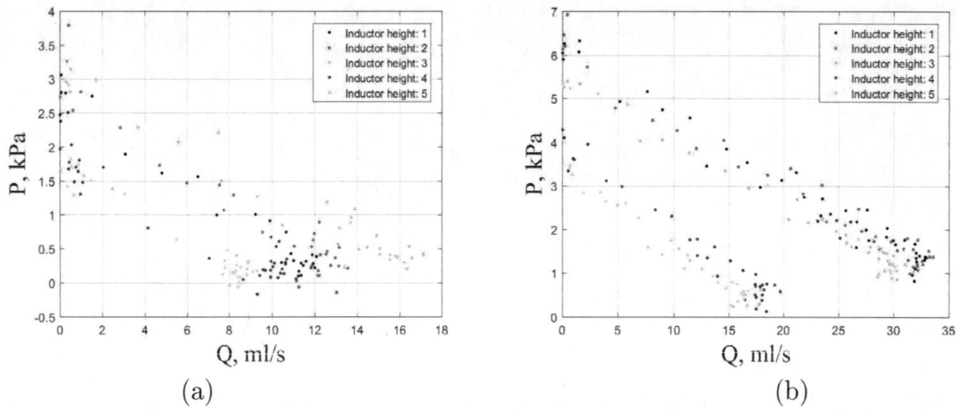

Figure 7. *PQ*-characteristics for the various positions of the ferromagnetic cores: (a) one core in the first section, (b) two cores in the first section.

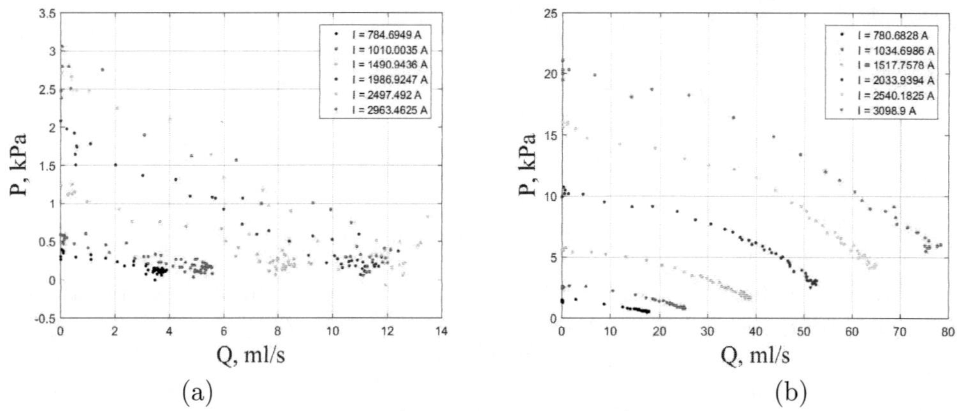

Figure 8. *PQ*-characteristics for the various positions of the ferromagnetic cores: (a) one core in the first section, (b) all six cores in all sections.

4. Conclusions

In the present study, several channels with C-shaped cores have been investigated. The electromagnetic force quadratically depends on the current passing through the channel. It indicates the absence of saturation of the ferromagnetic C-core. The magnetic field quickly disperses from the edges of the C-core. The maximum of the magnetic field is in the gap between the poles at a distance of 10 mm from its tail. The output of the pump is insignificant when the value of C-core overlap length h is smaller than some value for all configurations of the channel. This value is the smallest for partitions size $s = 60$ mm. The most smooth change in the output of the pump occurs for partitions size $s = 70$ mm.

The PQ-characteristics is flat and close to the linear for the most channel geometries. It indicates a small value of hydraulic resistance of the channel. The results of th study will be used to verify the mathematical models of these devices which has been developed in [3,6,8,11]. The MHD pumps of the output capability mentioned above can be used in metallurgy for pumping and pouring of molten metals.

Acknowledgments
This work was supported by the RFBR grant 17-48-590539_r_a.

References
[1] Verte L 1965 *Electromagnetic transport of liquid metal* (Metallurgia, Moscow) (in Russian)
[2] Lielpeter Y 1969 *Liquid metal induction MHD machines* (Zinatne, Riga) (in Russian)
[3] Khripchenko S, Kolesnichenko I, Dolgikh V and Denisov S 2008 *Magnetohydrodynamics* **44** 303–313
[4] Denisov S, Dolgikh V, Khalilov R, Kolesnichenko I and Khripchenko S 2013 *Magnetohydrodynamics* **49** 223–229
[5] Khalilov R and Kolesnichenko I 2015 *Magnetohydrodynamics* **51** 95–103
[6] Denisov S, Dolgikh V, Khripchenko S and Kolesnichenko I 2016 *Magnetohydrodynamics* **52** 25–33
[7] Dolgikh V and Kolesnichenko I 2015 *Magnetohydrodynamics* **51** 709–720
[8] Kolesnichenko I and Khripchenko S 2002 *Magnetohydrodynamics* **38** 391–398
[9] Kolesnichenko I, Sukhanovsky A, Khripchenko S and Gladkov I 2008 Impact on inclusions in a shallow mhd channel *Proc. of 7-th International Conference on Fundamental and Applied MHD. Giens (France).* pp 179–183
[10] Kolesnichenko I 2013 *Magnetohydrodynamics* **49** 217–222
[11] Denisov S, Dolgikh V, Khalilov R, Kolesnichenko I and Khripchenko S 2012 *Magnetohydrodynamics* **48** 197–202

Electromagnetic flowmeter for wide-temperature range intensive liquid metal flows

A Pavlinov[1]**, R Khalilov**[1]**, A Mamykin**[1]**, I Kolesnichenko**[1,2]
[1] Institute of Continuous Media Mechanics, Perm, Russia
[2] Perm National Research Polytechnic University, Perm, Russia

E-mail: pam@icmm.ru, Khalilov@icmm.ru, mad@icmm.ru, kiv@icmm.ru

Abstract.
The electromagnetic velocity sensor of Ricou and Vives with permanent magnet is well known. The probe is difficult to use at industrial conditions like metallurgy or nuclear power plants close to the Curie temperature. This article describes a reliable technique to provide controlled magnetic field distribution at the measurement area, generated by induction coils and concentrated by magnetic cores. The effect of magnetic field displacement, as well as weight and size of the assembled device, were minimized using numerical calculation of the field distribution for the case of a conducting media of different conductivity moving with different velocities. Calibration experiments showed high accuracy of measurements with low sensitivity to external electromagnetic noise.

1. Introduction

There are a large number of liquid metal velocity sensors designs. Each of them has it's own advantages and disadvantages. Potential probes [1, 2] and flowmeters [3] are widespread. Their advantages are the simplicity of implementation and the wide possibilities of the velocity pulsations characteristics measurement. Disadvantages are the need for a good electrical contact with liquid metal and the vulnerability of permanent magnets to high temperatures. Electromagnetic flowmeters of sodium loop ICMM UB RAS are made using NdFeB permanent magnets with magnetic cores which create strong localized magnetic field of about 0.3 T. Electromagnets must be used for higher temperatures [4]. There are techniques based on the measurement of cross-correlations between the signals of thermocouples placed along the flow [5, 6]. The disadvantage is the need to immerse the thermocouples into the flow. It is also possible to note modern methods based on measuring the Lorentz force [7] and the induced motion [8], which are in the stage of research and experimental application.

The flow of an electrically conductive medium in area with the applied magnetic field leads to the generation of an electromotive force (EMF) on the channel walls, which is proportional, in the first approximation, to the average flow velocity [9]. The potential difference $\Delta\phi$ induced between the two electrodes, which are spaced by Δl, is governed by Ohms law:

$$\vec{j} = \sigma(\vec{E} + \vec{u} \times \vec{B}_m) \approx \sigma(-\frac{\Delta\phi}{\Delta l}\vec{e}_p + \vec{u} \times \vec{B}_m) \ , \qquad (1)$$

where $\vec{j}$ is the electric current density, $\vec{u}$ the fluid velocity, $\vec{e}_p$ the unit vector between the tips of the wires, and $\vec{B}_m$ the applied magnetic field. Under the assumption of absence of induced and external electric currents for orthogonally oriented $\vec{B}_m$, $\vec{e}_p$ and $\vec{u}$, the equation 1 reduces to the simple linear relation $u = \Delta\phi/(B_m \Delta l)$.

Construction of flowmeters based on this principle requires consideration of a number of complicating factors. Firstly, in the case of large magnetic fields, on the one hand, the sensitivity of the sensor increases, and on the other hand, the Hartmann number $Ha = \frac{\sigma L^2 B^2}{\rho\nu}$ increases: the flow profile changes significantly and the braking effect appears. Secondly, in the case of high conductivity of a liquid medium and high speeds, the magnetic Reynolds number $Re_m = \mu\mu_0\sigma LV$ increases: magnetic field displacement from the measuring volume appears and, therefore, the nonlinearity of the characteristic comes out.Thirdly, the displacement effect is dependent on the electrical conductivity which varies with temperature. Also, if the channel of the flowmeter is made of an electrically conductive material, then the EMF induced in the fluid is shunted by a wall through which the electrical currents are closed. The shunting factor is expressed in terms of the wall σ_w and liquid σ_l conductivities ratio and the ratio of the external D_{out} and internal D_{in} diameters as follows:

$$K_{sh} = E_{wall}/E_{free} = \frac{2D_{in}/D_{out}}{1 + D_{in}^2/D_{out}^2 + \frac{\sigma_w}{\sigma_l}(1 - D_{in}^2/D_{out}^2)} \quad (2)$$

2. Methods

2.1. Flowmeter design

In order to provide safe temperature conditions for electronic circuits the flowmeter was separated on two units connected by signal and power cables.

The flowmeter's head unit with a channel of nominal diameter $DN = 89\ mm$ is shown on Fig.1(a). The head unit consists of a channel 1 made of stainless-steel AISI 321 and inductor, creating a magnetic field in the channel. The inductor consists of a magnetic core 2 and induction coils 3 connected electrically in series. All elements are rigidly fixed on the frame. To prevent overheating of the induction coils, the magnetic core is enclosed in a heat insulator. The temperature is controlled by a set of thermocouples in several regions: the temperature of the liquid metal in the measurement volume, the temperature of the magnetic core, the temperature of the induction coils, and the surface temperature of the heat insulating casing.

In the required range of temperatures and ratios of the electrical conductivities the shunting coefficient (see eq.2) of this channel is of the order of $K_{sh} \approx 6 \div 9\%$ and its temperature change is automatically taken into account by the algorithm given below.

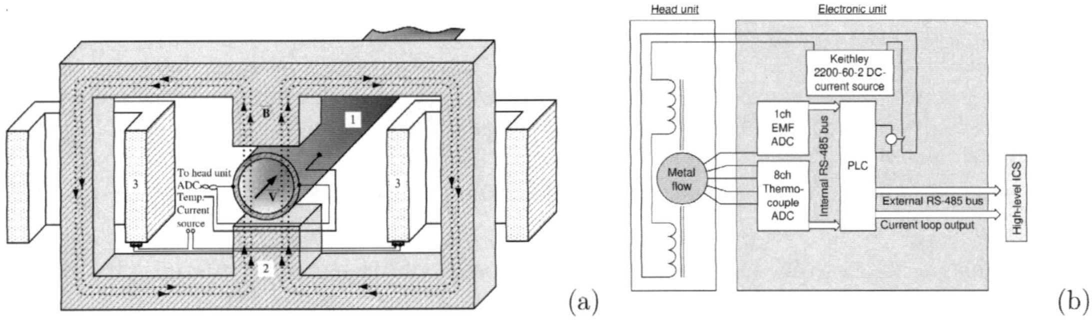

Figure 1. Sketch of the flowmeter's head unit (on the left) and block diagram of internal connections (on the right).

The electronic unit (Fig.1(b)) provides the following functions:

- inductor stabilized DC power supply,

IOP Conf. Series: Materials Science and Engineering **581** (2019) 012011 doi:10.1088/1757-899X/581/1/012011

- amplification of thermocouples and EMF signals,
- measured signals to the flow rate conversion and the thermal regime of the head unit monitoring to prevent its overheating,
- signals coding and transmitting to the upper level control system over the communication line.

The flowmeter has:

- analog DC current 4..20 mA output interface, proportional to the value of measured flowrate in range $0 \div 100\%$ for the load resistance $R_{cl} \leq 500\ \Omega$;
- digital interface RS-485 which allows to acquire the measured and transformed values by a Modbus-RT protocol;
- digital indication of the measured flow rate on the display.

2.2. Magnetic field displacement problem solving

The method of taking into account the nonlinearity of the flowmeter characteristics due to the magnetic field displacement dependence on the conductivity (temperature) of the medium [10] implies a complication of both the calibration procedure and the measurement technique itself.

Calibration technique. In the process of calibration the volumes of liquid metal are pumped through the flowmeter with various combinations of average flow rate and temperature. Each combination gives values of $E_i[V]$ (measured EMF value), $T_i[^oC]$ (temperature around the measuring volume) and $Q_i[m^3/h]$ (reference value of the average flow rate during the operation, obtained by the tanks level sensors signals analysis). Subsequent regression analysis of E_i, T_i, Q_i arrays gives the values of the A, C, β calibration coefficients (procedure is shown below). The values of ρ_0, ρ_1, ρ_2 coefficients are taken from the liquid metal electrical resistance temperature dependence approximation $\rho(T) = \rho_0 + \rho_1 \cdot T + \rho_2 \cdot T^2$ from reference books. T_0 - minimal operating temperature.

- The initial value of $\beta \approx -8 \cdot 10^{-5}$ is set.
- $B_i = 1 + \beta(T_i - T_0)$ is calculated.
- The transition to the new variables: $X_i = Q_i/\rho(T_i)$, $Y_i = \frac{E_i}{Q_i \cdot B_i}$
- The least squares method gives a linear approximation $Y_i = A - C \cdot X_i$ (A, C coefficients and standard deviation ϵ)
- Actions are repeated with a different β values to minimize the ϵ.

The final values of A, C, β together with used values of $\rho_0, \rho_1, \rho_2, T_0$ are fixed in the flowmeter read-only memory, calibration procedure is over.

Measurement technique. During normal operation of a calibrated device, the nonlinearity is taken into account as follows:

- The instantaneous values of E (average compensated EMF with the turned off magnetic field EMF substracted) and T (average temperature) are transferred in digital form from the ADC output to the PLC input.
- For the instant temperature the values of $\rho(T)$ (instant electrical resistance) and $B(T) = 1 + \beta(T - T_0)$ (nonlinear correction) are calculated.
- Auxiliary coefficients $Q_0 = \frac{A \cdot \rho}{2C}$ and $E_0 = \frac{A \cdot B}{2} \cdot Q_0$ are calculated.
- The resulting value of the instantaneous flowrate is $Q = Q_0 \cdot (1 - \sqrt{1 - \frac{E}{E_0}})$.

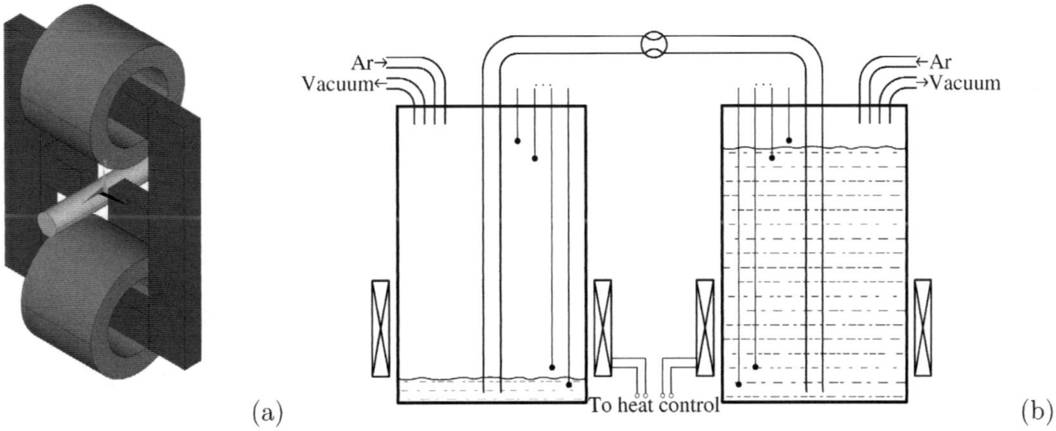

Figure 2. Model domains configuration for numerical simulation (on the left) and sketch of the calibration facility (on the right).

2.3. Calibration facility

The flowmeter calibration is carried out by volumetric-time method on ICMM UB RAS sodium facility (see Fig. 2(b)). Liquid metal flow through the channel of the flowmeter is provided by pressure-difference between two tanks (300 l each). One of the tanks is filled with sodium and overpressured by argon ($0.02 \div 0.10$ MPa). The other one is empty and vacuumed. Each tank has a system of contact level gauges, whose signals are recorded by a high-speed data acquisition system and used to determine the reference flowrate value. The flowmeter polling is provided synchronously. Both tanks are placed in a thermocase with electric heaters supplied by thermal control devices. This allows to maintain the sodium temperature in the required range throughout the calibration procedure.

3. Results

Calculations of magnetic field distribution into the measurement volume were done for different configurations of magnetic cores and different combinations of liquid media conductivity and flowrate. Electrodynamics was simulated by ANSYS Emag software. Magnetic cores and coils of different size and geometry were considered to produce a strong localized magnetic field in the orthogonal cross-section of a stainless-steel pipe filled with moving or resting volume of conducting media (see Fig. 2(a)). The magnetic field profiles along the pipe's diameter for stationary and moving cases are shown on Fig. 3. Minimum value corresponds to the center of the pipe. In the case of moving media it's absolute value decreases while it's position moves towards the direction of flow. Increasing the size of the core in direction of the flow suppresses the displacement effect, but on the other hand it dramatically increases the core weight.

It is clearer to analyze dependency of magnetic field in the center of the channel on mean velocity. The obtained quadratic-like dependency is shown on Fig. 4(a). These points were calculated for the final configuration of the core and show less than 10% decrease in magnetic field that can be resolved by the proposed calibration and measurement techniques. The assembled inductor was tested experimentally and shown magnetic field distribution coincident with numerically obtained results for stationary case.

Calibration experiments were performed on ICMM UB RAS sodium facility at different reference values of flowrate and temperatures. An example of calibration data points is shown on Fig. 4(b). The value of registered EMF-signals is large enough and is consistent with analytical

IOP Conf. Series: Materials Science and Engineering **581** (2019) 012011 doi:10.1088/1757-899X/581/1/012011

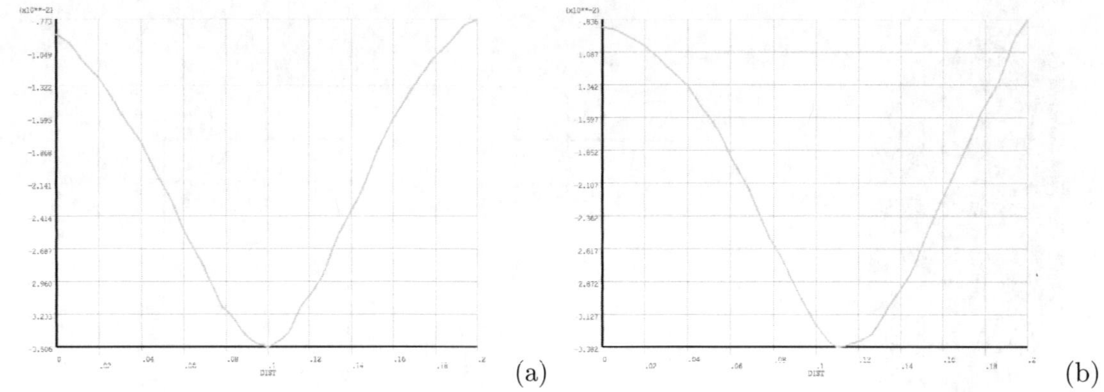

Figure 3. Magnetic field profiles obtained for stationary case (on the left) and for liquid sodium flow of mean velocity $V = 1\ m/s$ and temperature $T = 170^oC$.

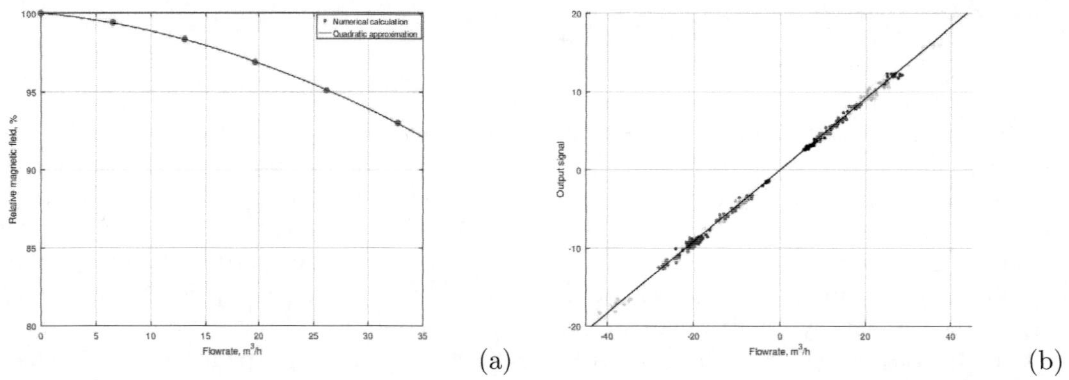

Figure 4. Relative magnetic field displacement at the center of the measurement volume obtained by numerical simulation of electrodynamic statement of the problem: sodium, $T = 170^oC$ (on the left), calibration data points experimentally obtained at $T = 200^oC$: points of different colors correspond to different series, the solid black line is a linear fit (on the right).

and numerical estimates. At the beginning and at the end of each pumping operation there is a non-stationary flow regime. The corresponding intervals can be automatically excluded by error filtering during experimental data processing. Due to integrated industrial frequency noise filtration the close spaced apparatuses like actuators and magnetohydrodynamical pump did not have a significant impact on measurements. The obtained characteristic is strongly symmetrical. The standard deviation of the calculated flowmeter's characteristic from experimental data did not exceed 3%.

4. Conclusions

Electric potential difference probe using globally applied magnetic field has been developed and manufactured to measure mean velocity ($Q \leq 40\ m^3/h$) in a high-temperature ($T \leq 450^oC$) flow of liquid metal. Magnetic field displacement numerical calculation allowed to determine the core configuration which provides acceptable magnetic field localization in the measurement volume for the whole required range of temperatures and flow rates. This reduced weight and size of the device essentially. Long-time tests showed the efficiency of the inductor thermal

IOP Conf. Series: Materials Science and Engineering **581** (2019) 012011 doi:10.1088/1757-899X/581/1/012011

protection. Calibration experiments confirmed weak magnetic field displacement at ultimate flow rates. This will reduce the weight and size of the inductor in the next modification of the device. Series of tests were carried out, which showed high accuracy of measurements with low sensitivity to external electromagnetic noise. The design of the core allows to install the larger cross-section channels to reduce hydraulic resistance with the aim of measuring higher flowrates (up to $\approx 200\ m^3/h$). This flowmeter, because it has the property of finding the magnitude and direction of velocity, allows the measurement of mass transfer in the heat exchangers of nuclear reactors using alkaline metals as coolant fluids. It also makes possible the dosage of molten metal during different technological processes in metallurgy.

Acknowledges

This work was supported by the RFBR grant 17-48-590539_r_a.

References

[1] von Weissenfluh T 1985 *Int. J. Heat Mass Transf.* **28** 1563–1574
[2] Ricou R and Vives C 1982 *Int. J. Heat Mass Transf.* **25** 1579–1588
[3] Shercliff J A 1962 *Theory of electromagnetic flow measurement* (UK: Cambridge University Press)
[4] Hayashi H, Becker A and Evans J 1999 *Metallurgical and Materials Transactions B* **30** 623–630
[5] Belyaev I A, Razuvanov N G, Sviridov V G and Zagorsky V S 2015 *Magnetohydrodynamics* **51** 673–684
[6] Frick P, Khalilov R, Kolesnichenko I, Mamykin A, Pakholkov V, Pavlinov A and Rogozhkin S 2015 *EPL* **109** 14002
[7] Thess A, Votyakov E and Kolesnikov Y 2006 *Phys. Rev. Lett.* **96** 164501(4)
[8] Priede J, Buchenau D and Gerbeth G 2009 *Magnetohydrodynamics* **45** 451–458
[9] Cramer A, Varshney K, Gundrum T and Gerbeth G 2006 *Flow Measurement and Instrumentation* **17** 1–11
[10] Adamovskii L, Golovanov V, Inkin Y and Kozubov E 1987 *Atomnaya Energiya* **62** 120–122

Multidisciplinary Approach to the Design of Superconducting Electrical Machines

K Kovalev[1], V Penkin[1], N Ivanov[1, a], N Kosheleva[2], G Serovaev[2]

[1] Moscow aviation institute (National research university)
[2] Institute of Continuous Media Mechanics UB RAS, Perm, Russia

E-mail: [a] n.s.ivanov88@gmail.com

Abstract. Compared to traditional machines, superconducting electrical machines have significant advantages, i.e: improved weight and dimensions, higher efficiency, lower values of inductive parameters, reduced noise level, greater stability when operating in electrical networks. Development of such machines requires the solution of a whole series of scientific and engineering problems related to many fields of science and technology. Along with the solution of traditional problems of the usual "warm" electrical engineering, one has to solve the problems associated with cooling of superconducting windings and the stability of their superconducting state under the action of external heat leakage, centrifugal and ponderomotive forces, as well as vibrations. The design of cryogenic machines should provide for the compensation of thermal compression of structural elements and windings during cooling process. This paper presents some interdisciplinary tasks, the solution of which will allow to reach a new level in the field of superconducting electrical machines and to create samples of electric motors and generators, surpassing all existing analogues in specific parameters.

1. Introduction

Compared to traditional machines, superconducting electrical machines have significant advantages, i.e: improved weight and dimensions, higher efficiency, lower values of inductive parameters, reduced noise level, greater stability when operating in electrical networks [1]. The development of superconducting electromechanical transducers in the Russian Federation and abroad has been underway since the early 1960s. For the first time for the needs of the navy in 1966, the British company International Research & Development manufactured a superconducting unipolar motor with a capacity of 37.5 kW and a rotational speed of about 2000 rpm [2, 3]. In the United States, Westinghouse and General Electric have developed a number of general industrial and aviation generators [2,4,5]. In Russia in aggregate design office Yakor together with MAI, an onboard aviation synchronous generator based on low-temperature superconductors (LTS) with a capacity of 780 kVA [1] was designed and manufactured. In the USA, a 36.5 MW marine superconducting synchronous electric motor has been created. In Germany, a 4 MW shipboard HTS electric motor and a 4 MVA HTS generator were developed [6]. The design schemes have a number of drawbacks, such as a bulky cryogenic supply system, the presence of rotating cryostats with HTS windings, and, as a rule, low rotational speed (for HTS devices).

Until now, superconducting machines have predominantly experimental status. There are single cases of their use in industry [1,7,8]. This is due, firstly, the high cost of their production and operation. And secondly, it is not yet possible to achieve the required level of specific (kW/kg) and volumetric (kW/m^3) power. Superconducting electrical machines are complex technical devices, the creation of which requires the solution of a whole series of scientific and engineering problems related to many fields of science and technology. Along with the solution of traditional problems of the usual "warm" electrical engineering, one has to solve the problems associated with cooling of superconducting windings and the stability of their superconducting state under the action of external heat leakage, centrifugal and ponderomotive forces, as well as vibrations. The design of cryogenic machines should provide for the compensation of thermal compression of structural elements and windings during cooling process.

This paper presents some interdisciplinary tasks, the solution of which will allow to reach a new level in the field of superconducting electrical machines and to create samples of electric motors and generators, surpassing all existing analogues in specific parameters.

2. Features of modern superconducting materials

There are two types of wires used in the manufacturing of electrical machine windings: composite wires based on low-temperature superconducting (LTS) alloys and intermetallic compounds and tapes based on high-temperature superconducting (HTS) ceramics. The latter, in turn, are divided into tapes of the first (1G) and second generation (2G). The tapes of the first generation are made of $Bi_2Sr_2Ca_2Cu_3O_{10}$ bismuth ceramics and represent a multilayer structure in a silver matrix, which leads to their high cost compared to LTS wires. The tapes of the second generation are also a multi-layer structure, but the current-carrying layer with a thickness of 1 μm is unique and made of $ReBa_2Cu_3O_x$. ceramics. Here Re is one of the rare earth metals, usually yttrium or gadolinium.

Currently, a large number of works on the creation of SC machines involve the use of HTS tapes and wires from the MgB_2 binary compound [9–11]. Despite the similar dimensions of the first and second generations of HTS tapes, the minimum bending radius of the second-generation tapes is several times smaller. This means that a pole winding made of this tape will provide a large fill factor. It is also seen that the current density of second-generation HTS tapes in the presence of an external magnetic field is several times higher [12,13].

Currently, HTS tapes from different manufacturers have similar properties. For example, at temperatures above 65 K, all tapes have a rather low critical current, but at a temperature of 20 K, the difference becomes noticeable (see Figure 1a) [13]. In addition, the critical current depends not only on temperature, but also on an external magnetic field. Figure 1b shows the dependences for the temperature of 77 K [13]. It is seen that as the field increases, the current decreases 10 times. In an electric machine, the characteristic value of magnetic induction in the region of HTS windings without a ferromagnetic core can be 1 T at liquid nitrogen temperatures (Figure 2) or more with decreasing temperature. Therefore, to create electric machines with high values of specific and volumetric power it is necessary to choose the right HTS tape.

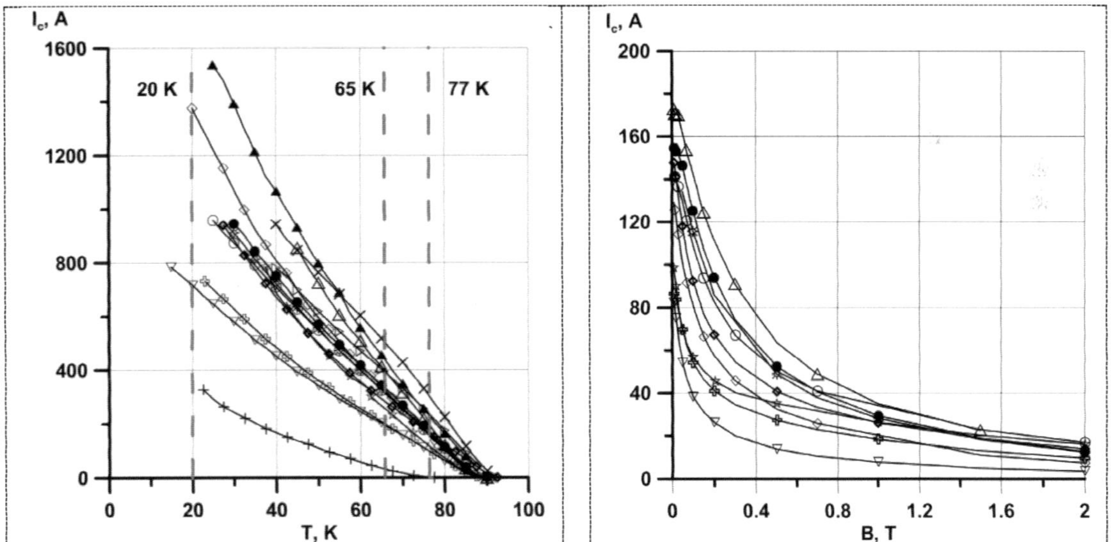

a) b)

Figure 1. Critical current of HTS vs: a) - temperature; b) - external field at 77K.

The mechanical stresses arising in the HTS tape during its operation in the composition of the windings of electrical machines also reduce its critical current. In addition, the disadvantage of tapes compared with round wires is associated with a critical magnetic field for a superconductor depending on the orientation of the field relative to the tape surface (Figure 3).

One of the main problems in the production and use of HTS tapes is related to the fact that any thermomechanical and mechanical stresses lead to the destruction of the superconducting layer [14] and, consequently, to a decrease in the critical current in HTS tapes [15,16]. In [17] experimental studies of the destruction process of HTS tapes under tension of the tape in the transverse direction are given. Samples were tested in the form of strips with various widths, which were cut from an YBCO tape. In the tension in the transverse direction, the stress of the sample was measured. It should be noted that the voltage was constant until the beginning of the critical current drop. In [18] the necessity in the research of mechanical properties of HTS tapes in order to reduce the possibility of destruction of the superconducting layer under different operating loads was shown. In [19], it is demonstrated how the superconducting layer is destroyed after bending the HTS tapes. Photos from a microscope are presented, where you can see the destroyed superconducting layer in the HTS tape. In subsequent papers [20] solutions are presented, as well as the results of numerical modeling of this problem. The problem of reducing the critical current in a HTS tape during bending is devoted to work that offers various options for tapes laying in a cable. For example, in [21] the results of a finite-element parametric study of a cable with HTS tapes under the torsion are presented. HTS tapes were stacked in a multi-level way while modeling. These results were verified experimentally. However, in order to achieve a reliable quantitative level of prediction using FEM, it is necessary to optimize the cable model, as well as to provide additional extensive experiments to simplify the cable configuration. In [22] the effect of the influence of the configuration of torsional chirality on the electromechanical behavior of multilayer superconducting tapes is studied. Also, the authors give recommendations on the methods of laying and selection of HTS for the optimal configuration of the cable.

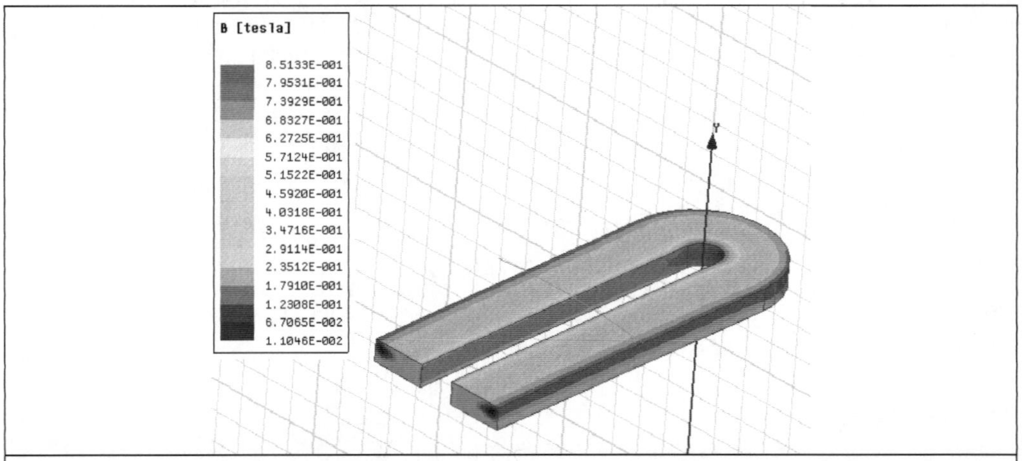

Figure 2. An example of the distribution of magnetic induction in the region of HTS coils.

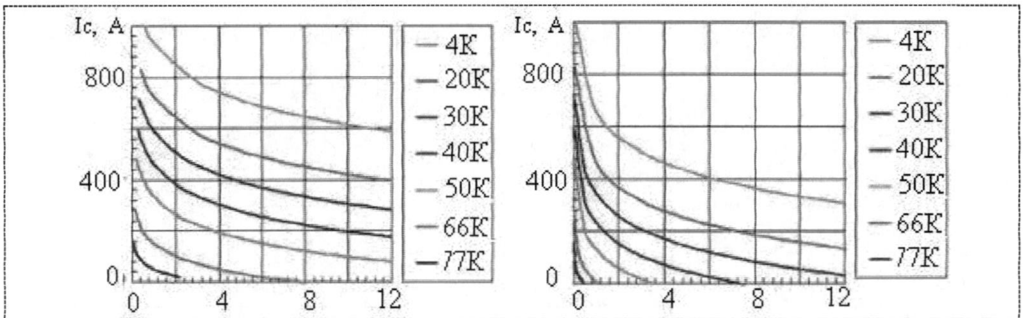

Figure 3. Critical current vs magnetic field dependencies for BSCCO wires from Sumitomo Electric: a) - parallel magnetic field, b) – perpendicular magnetic field.

Thus, constructions containing HTS require the development of numerical models that could take into account the following features: the specific heterogeneous geometry of this material, the ultralow thickness of some layers, the nonlinear behavior of the constituent material under different operating loads. Since the development of an analytical solution seems to be time consuming, optimal numerical models should be created and used.

3. Determination of the stress-strain state of HTS coils

In contrast to the usual magnetic systems, in superconducting ones the permissible deformations and stresses are determined not only by the corresponding strength and by stiffness limits, but also depend on a number of factors related to the stability of the superconducting state. When working in the windings of electric machines, the coils from HTS absorb centrifugal forces during rotation of the rotor, as well as electromagnetic forces arising due to the presence of strong magnetic fields. In addition, low operating temperatures invoke thermal stresses in the structural elements of the machine, especially when it is thermally cycled. It is experimentally established, that the effect of degradation arise, which means that the critical current in the superconducting winding turns out to be lower than the current of the short sample. It may be due to mechanical stresses in the SC and friction on the contact surfaces adjacent to the bandage and the frame, and between the layers [23]. The considered features of the mechanical behavior of SC systems indicate the need for a comprehensive analysis of the stress-strain state of the system. The following requirements can be formulated for the main elements of HTS electrical machines. The design of the rotor must meet the requirements of mechanical strength in the temperature range of 300-20 K and vibratory requirements and also provide access to the elements of the HTS windings for their diagnosis, repair, or replacement. The design of HTS windings must meet the requirements of mechanical strength under the action of centrifugal and electrodynamic forces, allow for element-by-element testing of coils, contain the minimum possible number of junctions and ensure their placement in the zone of the minimum magnetic field during intensive cooling. It also mandatory to eliminate degradation of the SC by organizing effective cooling of each conductor, and the reliable mechanical fixation. The system of diagnostics of winding parameters should have a minimum error, have working temperatures in the range of 300-20 K and provide an indication of the thermal and electrical state of the machine's structural elements and their stress-strain state.

As mentioned above, modern HTS tapes have a composite structure. In some cases, the coils are compounded to provide mechanical strength and increase thermal conductivity. However, the properties of such compounded modules at low temperatures are not fully understood. If non-vacuum compounding is used, unfilled zones may be present (region II in Figure 4). In addition, it is necessary to take into account the presence of an insulation layer, which can be up to 20% of the thickness of an HTS tape. At the same time, the properties of the materials forming the coil differ significantly. It should also be noted that the coil is wound on a frame (area I in Figure 4), which can have a rather complex shape. In this regard, and to reduce weight, these structural elements can be made of plastics using 3D printing technology. Thus, the HTS coil is a composite structure of 3-4 different materials. It is obvious that such a structure is inhomogeneous, and its properties are anisotropic.

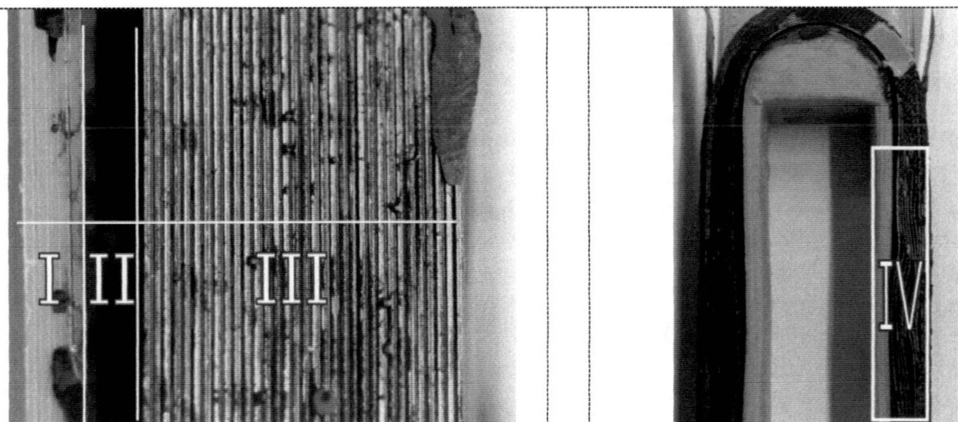

Figure 4. Example of HTS coil: I is the frame area; II is the coil separation region (not filled with compound tape), III is the uniform winding region of the HTS tape, IV is deformation of the tape layers.

IOP Conf. Series: Materials Science and Engineering **581** (2019) 012012 doi:10.1088/1757-899X/581/1/012012

In addition to the homogeneous zone there is an area, where the deformation of the tape layers occurs, the connection between the layers and the frame is broken (region IV in Figure 4). The presence of such defects can lead to a decrease in the operating current. This will especially affect the operation of the coil on alternating current, when under the action of electromagnetic forces uncoupled coils start to move. As a result, there is an increase in losses, and it can lead to a coil failure.

A non-uniform temperature field, internal stresses and thermoelastic deformations will occur in the operational process under the temperature influence and mechanical loads in the HTS coil. To estimate these parameters, it is necessary to solve the associated dynamic problem of thermoelasticity, the main relations of which are given below:

$$\sigma_{ij,j} + F_i = \rho\,\ddot{}\quad (1)$$

$$\sigma_{ij} = C_{ijkl}\varepsilon_{kl} - \beta_{ij}\Delta T \quad (2)$$

$$\varepsilon_{ij} = \frac{1}{2}(u_{i,j} + u_{j,i}) \quad (3)$$

$$\frac{\partial T}{\partial t} - \frac{k_{ij}}{c_p\rho}\frac{\partial^2 T}{\partial x_i \partial x_j} = w \quad (4)$$

where σ_{ij} - the stress tensor components; ε_{ij} - the strain tensor components; C_{ijkl} - the stiffness tensor components; F_i - the volume force vector components; ρ - material density; u_i - the displacement vector components; β_{ij} - the thermoelastic coefficients tensor components; k_{ij} - the heat conductivity tensor components; c_p - the specific heat capacity; w - the power density of heat sources.

Thus, it is necessary under certain initial and boundary conditions to find the 6 components of the stress tensor σ_{ij} , 6 strain tensor components ε_{ij} , the displacement vector components u_i and temperature T, which should satisfy the equations of motion (1), the relations between stresses and strains (2), the relations between strains and displacements (3) and the heat equation (4).

The calculation of stresses arising in composite structures is a complex scientific task requiring the use of numerical calculation methods. In particular, various software packages of the finite element analysis can be used, such as ANSYS, ABAQUS, COMSOL Multiphysics, etc., which allow to carry out calculations of coupled problems of thermoelasticity, thermoelectroelasticity. Finite element modeling allows to study objects with a complex geometric configuration, to take into account materials with anisotropic properties and demonstrating both linear and non-linear behavior. In addition, modern numerical approaches allow to model various processes of destruction, including delamination between the components of the composite structure.

The determination of the stress-strain state of HTS tapes in the coil is an important task. Its solution is one of the necessary steps on the way to solving problems of increasing the reliability of machines, reducing the overall dimensions of the windings, increasing the filling factor of the cross section of the HTS winding with a tape and also increasing the operating current. All this will lead to an increase in the electromagnetic loads of the machine and an increase in its specific (kW/kg) and volumetric power (kW/m^3).

4. Determination of the state of HTS coils

The second problem to be solved is the determination of the state of the HTS coils. First of all, it refers to the temperature of the individual layers. When designing HTS electric machines, thermal calculation is important. It is known that traditional electric machines can operate with a certain excess of a given initial temperature. In this case, the working temperature lies in the range in which the working capacity of the machine is maintained. When it comes to SC machines, the problem of thermal calculation is complicated by the need to maintain the temperature of the SC windings. A temperature jump of 1 degree can lead to a decrease in the current-carrying capacity of HTS, and in the case when the working temperature is close to critical (77 K), exceeding it can lead to an avalanche-like transition of the SC

windings to a normal state with a further increase in heat generation and an emergency situation.
Determination of the temperature of HTS windings can be carried out in several ways. The greatest accuracy is ensured by the installation of temperature sensors on various structural elements of the HTS machine, including the windings. When using composite multilayer coils of HTS, the temperature of the various layers may differ. In this case, the most dangerous can be both internal turns and medium ones, which is determined by the design of the coils. Accurate determination of the temperature of the various turns in the coil will make it possible to develop recommendations for intensifying cooling and, accordingly, increasing the operating current of the windings.
The greatest effect is achieved by controlling both the temperature of the windings and the stresses arising in them. Fiber optic sensors can be used for this purpose.
Fiber optic sensors (FOS) based on the Bragg grating have several advantages over other sensors. In addition to the fact that FOS can measure various physical and mechanical quantities, they are small and can be embedded in almost any part / structure without any special consequences or glued to the surface of the part. It is easy and accessible to produce FOS. A single fiber can contain several sensors, which makes it possible to obtain data along the entire length of the part. Under harsh conditions FOS have advantages in sensitivity compared to other sensors. FOS are not affected by electromagnetic noise. They are safe when using flammable liquids and pressure. They can be used to monitor the parts in real time, including at operational loads. Also FOS can be easily integrated with other equipment for remote control of parts or objects.
In addition, FOS have a large range of operating temperatures depending on the outer coating. A lot of scientific literature is devoted to the study of the FOS behavior at cryogenic temperatures. For example, the experimental characteristics of FBG sensors with three different wavelengths in the range from 123 K to 273 K were checked in [24] The FBG sensors demonstrated a non-linear temperature response, which can be approximated by a third-order polynomial. Under isothermal conditions, FOS showed a linear relationship with changes in the strains. In the study was shown a stable sensitivity of FBG to strains and the dependence of its sensitivity on temperature. The strains arising in the FBG increased with decreasing temperature in a cryogenic medium. In [25] was provided a very useful tool for determining the minimum thickness of a metal coating (FOS coating) required to achieve maximum sensitivity of FOS based on the Bragg grating. Materials studies with different properties allow to choose the best coating depending on the purpose of measurement (for example, the temperature range), which make it possible to optimize different sensors for each measured temperature range. In [26] sensors in a zinc and glass coating were investigated. The conducted studies show the stability of the readings and the sensitivity of the sensors in different temperature ranges. Also the comparison with the sensors in a standard polyimide coating is shown. Recommendations on the use of FOS and their calibration are given. In [27] the strain characteristics of a FOS with silicon dioxide coating in a dynamic experiment from 30 to 273 K was investigated. Also the tests on static tension in a medium with liquid nitrogen were performed. Based on the analysis of experimental data, a formula for the strain calibration coefficient of FBG with silicon dioxide coating in the temperature range of 30–273 K was obtained and proposed.
However, despite the abundance of scientific research, the main disadvantage of fiber-optic sensors based on the Bragg grating is a significant decrease in sensitivity with decreasing temperature, mainly due to the critical decrease in the thermo-optical fiber coefficient and very low thermal expansion coefficient (CTE) of fused silica at cryogenic temperatures. Thus, it is important to increase the sensitivity to temperature, as well as to choose the right coating in order to possess a higher CTE, but in the same time does not detract from the sensitivity of fiber-optic sensors based on the Bragg grating. Therefore, it is important to develop models that would take into account the particularities of this challenging issue: temperature, optimum location of sensors, issues of gluing, input and output of sensors, etc.

5. Rational solutions for the design of modern HTS machines

Currently, various papers contain design schemes of HTS electrical machines without a steel ferromagnetic core [4,11]. The ability to abandon iron elements and replace them with non-magnetic ones, such as titanium, aluminum, or various non-metals, turns out to be very attractive for reducing the mass of the machine. In this case, the increased magnetic resistance is compensated by the growth of the EMF of the excitation winding due to the high current in the HTS winding. In this case, the use of

constructions made of a combination of different materials may be relevant.
Figure 5 shows an example of the of field of winding pole of a synchronous electric machine. Traditionally, the pole core and the tip are performed in one piece. In the framework of a weight reduction task, an alternative design may be proposed (Figure 5b). In this case, the metal plate plays a role of a supporting structure that provides rigidity and strength of the coil and the pole, and the non-metallic pole tip is designed to ensure the streamlining of the rotor during its rotation in a gas or liquid medium.

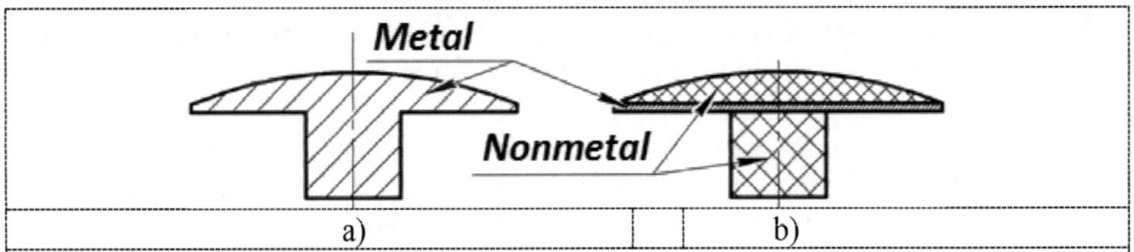

Figure 5. The design of the field winding pole: a) - traditional, b) – composite.

Another example is the use of a polyimide composite material for constructive machine parts, for example, a housing and bearing shields. In this case, not only the mass can be reduced, but also the heat influx from the environment into the cryogenic volume. These composites were developed at Salyut Design Bureau and are polyimide films with one or two sided fluoroplastic coating [28]. They have low gas permeability, increased strength, high chemical resistance, which ensures their performance at cryogenic temperatures. Another important advantage is the ability to weld these composite products and steel parts [28]. These materials are used as current-insulating and heat-insulating parts of current leads intended for the electrical connection of the current-carrying elements of the HTS power electrical cable with the cable network elements [29]. The use of such solutions in the design of electrical machines can reduce the mass of structural elements by 15-20%, as well as reduce heat gain from the external environment.

6. Conclusions

This paper presents some interdisciplinary tasks, the solution of which will allow us to reach a new level in the field of superconducting electrical machines and to create models of motors and generators, surpassing all analogs existing today in specific parameters. The following is a brief list of these tasks:

1. Development of analytical and numerical models of structures containing HTS, that take into account the following features: the specific heterogeneous geometry of this material, the ultralow thickness of some layers, and the nonlinear behavior of the constituent material under various operational loads.
2. Determination of the stress-strain state of HTS tapes in the composition of the coil, which is a composite structure with anisotropic properties.
3. Determination of state of superconducting coils, especially temperature and mechanical stress.
4. Determination of the possibility of strain registration as well as the appearance and development of damage in the HTS coil using FOSS under various operating loads.
5. Development of new design solutions based on composite materials and additive technologies.

The experience accumulated at the MAI and the Institute of Continuous Media Mechanics of the Ural Branch of the Russian Academy of Sciences allows us to develop solutions to the problems posed and search for new SC technologies.

7. Acknowledgments

This work was supported by Russian Science Foundation (project No. 17-19-01269)

References

[1] Penkin V and Kovalev K 2018 *Synchronous electrical machines with composite and bulk superconductors for transport systems* (Moscow: MAI publisher)
[2] Bertinov A 1982 *Special electric machines* (Moscow: Energoizdat)

[3] Appleton A D, Potts R and Grand B J C 1980 Assessment of Potential Performance of a 45 MW Marine Superconducting Motor *Canada, Transp. Canada, Int. Res. Dev. Co. Ltd.* 1–15

[4] Haran K S, Kalsi S, Arndt T, Karmaker H, Badcock R, Buckley B, Haugan T, Izumi M, Loder D, Bray J W, Masson P and Stautner E W 2017 High power density superconducting rotating machines - Development status and technology roadmap *Supercond. Sci. Technol.* **30**

[5] Barnes P N, Sumption M D and Rhoads G L 2005 Review of high power density superconducting generators: Present state and prospects for incorporating YBCO windings *Cryogenics (Guildf).* **45** 670–86

[6] Qiang L 2015 Superconducting Electric Machines *NIST/DOE Workshop on Enabling Technologies for Next Generation of Electric Machines. NIST/Gaithersburg*

[7] Glebov I, Danilevich Y and Shahtarin V 1981 *Superconducting turbogenerators* (Leningrad)

[8] Chubraeva L 1991 *Nonconventional generators* (Leningrad: Nauka)

[9] Qu T, Li Y, Song P, Hao C, Wu Q, Gu C, Zhu J and Han Z 2018 Design Study of a 10-kW Fully Superconducting Synchronous Generator *IEEE Trans. Appl. Supercond.* **28** 2–6

[10] Magnusson N, Eliassen J C, Abrahamsen A B, Helleso S M, Runde M, Nysveen A, Moslatt L E, Bjerkli J and King P 2018 Fabrication of a Scaled MgB2 Racetrack Demonstrator Pole for a 10-MW Direct-Drive Wind Turbine Generator *IEEE Trans. Appl. Supercond.* **28**

[11] Dezhin D, Ivanov N, Kovalev K, Kobzeva I and Semenihin V 2018 System Approach of Usability of HTS Electrical Machines in Future Electric Aircraft *IEEE Trans. Appl. Supercond.*

[12] Holzapfel B 2017 Status of Industrial Coated Conductor Production and Properties Conductor Application *EUCAS 2017, Geneva*

[13] Anon High-temperature superconducting wire critical current database *http://htsdb.wimbush.eu/*

[14] Laan D C van der, Ekin J W, Clickner C C and Stauffer T C 2007 Delamination strength of YBCO coated conductors under transverse tensile stress *Supercond. Sci. Technol.* **20** 765–70

[15] van der Laan D C and Ekin J W 2008 Dependence of the critical current of YBa2Cu3O7−δ coated conductors on in-plane bending *Supercond. Sci. Technol.* **21** 115002

[16] van der Laan D C, Ekin J W, Douglas J F, Clickner C C, Stauffer T C and Goodrich L F 2010 Effect of strain, magnetic field and field angle on the critical current density of Y Ba 2 Cu 3 O 7−δ coated conductors *Supercond. Sci. Technol.* **23** 072001

[17] Gorospe A, Nisay A and Shin H-S 2014 Delamination behaviour in differently copper laminated REBCO coated conductor tapes under transverse loading *Phys. C Supercond. its Appl.* **504** 47–52

[18] Hwanjun Jeong, Heecheol Park, Seokho Kim, Minwon Park, In-Keun Yu, Sangjin Lee, Taejun Park, Kideok Sim, Hong-Soo Ha, Sang-Su Oh and Seung-Hyun Moon 2012 De-Lamination Characteristics of Coated Conductor for Conduction Cooled HTS Coil *IEEE Trans. Appl. Supercond.* **22** 7700804–7700804

[19] Bruzek C E, Allais A, Morice S, Theune C-F, Petit S, Mikolajczyk M, Dechoux N, Jimenez C, Sarigiannidou E, Porcar L, Soubeyroux J, Odier P and Waeckerle T 2012 New HTS 2G Round Wires *IEEE Trans. Appl. Supercond.* **22** 5800204–5800204

[20] Kosheleva N, Shahrour I, Bruzek C E, Vega G and Kolmogorov G 2016 Development of a Simplified Numerical Model for the Design of 2G High-Temperature Superconductors *IEEE Trans. Appl. Supercond.* **26** 1–5

[21] Anvar V A, Ilin K, Yagotintsev K A, Monachan B, Ashok K B, Kortman B A, Pellen B, Haugan T J, Weiss J D, van der Laan D C, Thomas R J, Prakash M J, Hossain M S A and Nijhuis A 2018 Bending of CORC ® cables and wires: finite element parametric study and experimental validation *Supercond. Sci. Technol.* **31** 115006

[22] Ta W, Liu Y, Wang K, Liu L and Gao Y 2019 Effect of the twisting chirality configuration on the electromechanical behavior of multilayer superconducting tapes *Phys. Lett. A* **383** 949–56

[23] Alievskiy B 1993 *Superconducting electrical machines ang magnetic systems* (Moscow: MAI publisher)

[24] Guo Z-S, Feng J and Wang H 2012 Cryogenic temperature characteristics of the fiber Bragg grating sensors *Cryogenics (Guildf).* **52** 457–60

[25] Vendittozzi C, Felli F and Lupi C 2018 Modeling FBG sensors sensitivity from cryogenic temperatures to room temperature as a function of metal coating thickness *Opt. Fiber Technol.*

42 84 91

[26] Swinehart P R, Maklad M, Courts S S, Weisend J G, Barclay J, Breon S, Demko J, DiPirro M, Kelley J P, Kittel P, Klebaner A, Zeller A, Zagarola M, Van Sciver S, Rowe A, Pfotenhauer J, Peterson T and Lock J 2008 CRYOGENIC FIBER OPTIC SENSORS BASED ON FIBER BRAGG GRATINGS *AIP Conference Proceedings* vol 985 (AIP) pp 940–6

[27] Li L, Lv D, Yang M, Xiong L, Luo J and Tan L 2018 Strain characteristics of the silica-based fiber Bragg gratings for 30–273 K *Cryogenics (Guildf).* **92** 93–7

[28] Anon 2006 *Technical developments of OKB-23 - KB "Salut"* (Moscow)

[29] Kostyuk V V., Blagov E V., Antyukhov I V., Firsov V P, Vysotsky V S, Nosov A A, Fetisov S S, Zanegin S Y, Svalov G G, Rachuk V S and Katorgin B I 2015 Cryogenic design and test results of 30-m flexible hybrid energy transfer line with liquid hydrogen and superconducting MgB2 cable *Cryogenics (Guildf).*

IOP Conf. Series: Materials Science and Engineering **581** (2019) 012013 doi:10.1088/1757-899X/581/1/012013

On the Optimal Location of Several Peizoelectric Elements on the Structure Surface

Dmitrii A Oshmarin, Maksim A Iurlov, Natalia V Sevodina, Nataliia A Iurlova [a]

Institute of Continuous Media Mechanics UB RAS, Russia

E-mail: [a] yurlova@icmm.ru

Abstract. In this paper, based on the solution of the natural vibration problem an approach for determining the best options for the arrangement of piezoelectric elements on the surface of a structure allowing an essential improvement of its dynamic characteristics is proposed. The condition for a maximum value of the generalized electromechanical coupling coefficient, which characterizes the fraction of the converted mechanical energy of vibrations into the electrical energy, is adopted as an optimality criterion. The performance of the proposed approach is demonstrated by solving an example problem of locating two piezoelectric elements on the surface of a spatial structure providing damping of its vibrations at the selected frequencies. The obtained results are substantiated by the frequency response plots of the electric potential, obtained for different options of placement of piezoelectric elements on the surface of a structure under forced steady-state vibrations.

1. Introduction

Nowadays, damping of structure vibrations by converting the energy of mechanical vibrations into the electrical energy with the aid of attached piezoelectric elements, which, as the structure elements, are connected to the electrode surfaces of external electrical circuits, is an intensively developed area of research in mechanics.

To make reliable progress in this field, it is essential to find such technological solutions, in which piezoelectric elements will serve most effectively the ultimate purpose of their application in smart structures (vibration damping, control of the structure shape under the action of various external factors, flaw detection, etc.). The efficiency of piezoelectric elements depends not only on the control parameters (corresponding to the schemes of external electrical circuits), but also on the location of piezoelectric elements that act as sensors and actuators. At the same time, as practice shows, with reference to a specific purpose of using smart technology, the optimal locations of sensors / actuators can be different. This is reflected in the selection of conditions for the optimal fitting of structural elements, as well as in the algorithms for the numerical implementation of the problem under consideration.

Conventionally,devices for passive vibration damping are located at points where the deformation energy in the structure is maximal [1]. As a rule, these locations are determined from the analysis of the deformation behavior of the structure. Nevertheless, the question of choosing the location of the piezoelectric element attracts attention of many researchers, whose studies are targeted at the development of both the optimization methods and the criteria for the optimal location of piezoelectric elements [1-5]. As noted in [6–8], this is also due to the fact that the location of sensor at points, where they fail to generate a significant electric potential, leads even to a decrease in the rate of vibration damping.

Deeper insight into the state of art in this area can be obtained from the review article [9], which summarizes the approaches presented in 109 papers for identification of the optimal location of sensors or actuators in various structures. The condition for the optimal location of a piezoelectric element can be specified based on the maximization of forces, moments or strains transferred by the piezoelectric element acting as an actuator, minimization of changes in the dynamic characteristics of

IOP Conf. Series: Materials Science and Engineering **581** (2019) 012013 doi:10.1088/1757-899X/581/1/012013

the original structure, as well as on the need of obtaining the desired dynamic characteristics of the original structure, drastic reducing the control function and maximal damping of vibrations of the initial structure, etc.

Nevertheless, as it was noted in [9], in the previous studies no consideration was given to spatial thin-walled structures. Moreover, despite the differences in the construction of optimization algorithms and approaches, the methods proposed in [5-8] are linked together by the fact that they do not consider the electric potential generated on the surface of a piezoelectric element, although it is a key factor in damping the structure vibrations with piezoelectric elements and external electrical circuits.

An attempt to calculate the value of the electric potential generated on the surface of piezoelectric element with the aim to determine its optimal location was made in [1] using a finite element analysis. However, this study examined the possibility of damping only the first mode of vibrations, which in its self is quite simple. The study of the electromechanical behavior of the piezoelectric element attached to the structure under more complicated vibration modes is of much greater interest to researchers from the viewpoint of its use as a sensor or actuator (the value of the electric potential generated on its surface).

In work [10], using a beam as an example, the authors showed that one piezoelectric element would suffice only for control in cases of simple loading, when the deflection curve is convex or concave along the entire beam length. In more complicated situations it is necessary to use several piezoelectric elements to induce oppositely directed electric fields. It was demonstrated that under complex loading conditions the optimal effect can be produced by a greater number of piezoelectric elements.

Thus, the purpose of this work is to develop an algorithm that will allow us to determine the optimal location of several piezoelectric elements providing the maximum value of the relative electric potential for spatial structures of arbitrary geometry subject to deformation at the specified vibration frequencies.

Here we restrict ourselves to the problem of locating two piezoelectric elements on the surface of a spatial structure such as to ensure the possibility of damping the specified vibration modes.

2. The object of study

The efficiency of using piezoelectric element in various structures is generally determined, along with other factors, by the value of the electric potential generated on their electrode surfaces during the dynamic processes. Therefore, a search for such variants of piezoelectric element location on the structure surface, at which the electric potential can reach the maximum possible value, is the problem that needs to be solved in each specific case of using piezoelectric elements as a tool for controlling the dynamic behavior of the object under consideration.

The solution of this problem involves the analysis of the dependence of the electromechanical coupling coefficient (EMCC) on the location of the piezoelectric element and the mode of vibrations, to which it corresponds.

The concept of a generalized electromechanical coupling coefficient (EMCC) was introduced in [11], and its physical meaning was considered in detail in [12]. This coefficient characterizes the fraction of the mechanical energy of the *i-th* vibration mode converted into the electrical energy for each individual piezoelectric element and is determined by the formula:

$$K_i = \sqrt{\frac{\left(\omega_{o/c}^2\right)_i - \left(\omega_{s/c}^2\right)_i}{\left(\omega_{s/c}^2\right)_i}} \tag{1}$$

where $\left(\omega_{o/c}\right)_i$, $\left(\omega_{s/c}\right)_i$ are the *i*-th natural vibration frequencies of the structure with the attached piezoelectric element in the open circuit mode (o/c) and short circuit(s/c) mode. The o/c mode is realized in the case when one of the electrode surfaces of the piezoelectric element is grounded (the value of electric potential is equal to zero) while the other electrode surface is free of loads. In the s/c mode the zero-valued electric potential is prescribed at both electrode surfaces. The mathematical statement of the problem of natural vibrations of electro-viscoelastic bodies and the algorithm for its numerical realization are given in [13].

In [14, 15], the authors have revealed a correlation between the value of the electric potential generated on the electrode surfaces of piezoelectric elements during dynamic processes, the amount of the strain energy in the zone of the piezoelectric element location, and the value of electromechanical coupling coefficient. This relationship was derived by solving an example problem, in which an optimal location of a piezoelectric element is searched for to provide the most effective damping of the specified vibration modes.

However, the best placement for the piezoelectric element in spatial structures cannot always be chosen based on the values of the electric potential and the strain energy in the zone of location of the piezoelectric element, which were obtained by solving the natural vibration problem. These values are of relative nature and consider meaningful only within the limits of a single simulation made to estimate the efficiency of the piezoelectric element at different modes. Moreover, it is not always correct to compare the values of the electric potential obtained by solving the problem of natural vibrationsfor different positions of the piezoelectric element, its various dimensions, physical and mechanical characteristics of the material, etc. At the same time, K_i is an absolute quantity, which allows us to carry out a qualitative comparison of the obtained results.

As an optimality criterion for the location of the piezoelectric element, it is reasonable to use the condition of maximum value of the electromechanical coupling coefficient (EMCC). The generalized electromechanical coupling coefficient demonstrated its high efficiency in determining the optimal location of a single piezoelectric element [14] and, therefore, it seems attractive to apply this approach to the case of several piezoelectric elements. From a practical point of view, this is necessary in the case of locating several piezoelectric elements on the structure surface. In doing so, it is desirable to ensure their independent operation, that is, to minimize their reciprocal influence, which necessitates additional investigations.

The attachment of the piezoelectric element to the structure surface changes the spectrum of natural frequencies and vibration modes interacting with each other. Accordingly, each subsequently attached piezoelectric element causes a change in the already existing vibration mode and, the deformation of the previously located piezoelectric elements. This leads to a change in the patterns of distribution of the initially obtained values of K_i. Thus, the zones of optimal location of the second and each successive piezoelectric element may vary at different frequencies. It therefore seems incorrect to estimate the efficiency of location of several piezoelectric elements in the structure based solely on the analysis of the distribution patterns of K_i obtained for one piezoelectric element. On the other hand, the arrangement of successive piezoelectric elements also influences the expected performance of the previously located piezoelectric elements. Hence, the method proposed in [14], when extended to the case of locating several piezoelectric elements, must be modified in order to allow evaluating the cross impact of piezoelectric elements on the efficiency of their performance.

The key idea of the method of determining the optimal locations of piezoelectric elements, which is based on the condition for a maximum value of K_i, can be formulated as follows. After finding an optimal location for a single piezoelectric element ensuring the most effective damping of vibrations at the specified frequency, new piezoelectric elements are added one by one to the structure, each being designed to damp vibrations at other frequencies, taking into account the effect of previously located piezoelectric elements.

In this case, along with fixing the value of the electromechanical coupling coefficient of the new piezoelectric element, it is necessary to monitor its variation in the previous piezoelectric elements, the location of which is fixed.

3. The placement of piezoelectric element on the shell surface according to the proposed approach. Numerical example

Let us demonstrate the performance of the proposed method using as an example a spatial structure in the form of a thin-walled shell schematically represented in Fig.1.The geometrical dimensions of the shell are as follows: $L = 290$ mm, $B = 150$ mm, $H = 120$ mm, $t = 0.26$ mm. The shell is made of an isotropic elastic material with the following physical and mechanical characteristics: $E = 6.5 \cdot 10^{10}$ Pa, $\nu = 0.28$, $\rho = 2900$ kg/m^3. The edges of the shell are clamped along their entire length.

It is necessary to select the locations of a few piezoelectric elements such that each of the elements can generate the maximum possible electric potential at the specified vibration frequency.

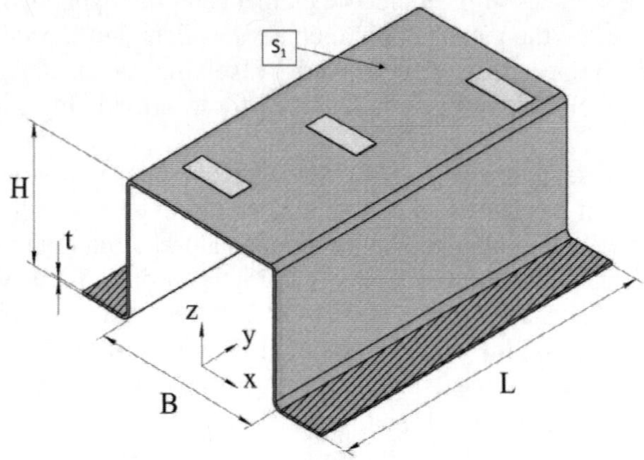

Figure 1. Computational model of the structure

To illustrate the performance of this approach, we consider the problem of locating two piezoelectric elements in the segment S1 of the structure surface (the upper horizontal surface of the shell with the coordinate $z=H$, 290x120 mm in size, see Fig. 1). The values of the first 5 natural frequencies of the shell are 13.63, 41.82, 47.20, 67.35, 81.63 (Hz), respectively.

For further investigations, the second and third natural vibration frequencies (41.82 and 47.20 Hz) were chosen as closest to one another as possible and, as a result, showing the maximum effect of reciprocal influence. Accordingly, the first piezoelectric element must generate the maximum electric potential at the second natural vibration frequency, and the second piezoelectric element - at the third natural vibration frequency.

The piezoelectric elements are made of PZT-19 piezoelectric ceramics, with upper and lower electrode surfaces. The lower electrode surface is grounded (the electric potential there is equal to zero), and the material is polarized in the z-axis direction. The dimensions of piezoelectric elements are as follows: 50x20x0.36 mm. Piezoceramics has the following physico-mechanical properties:

$C_{11} = C_{22} = 10.9 \cdot 10^{10}$ Pa, $C_{13} = C_{23} = 5.4 \cdot 10^{10}$ Pa, $C_{12} = 6.1 \cdot 10^{10}$ Pa, $C_{33} = 9.3 \cdot 10^{10}$ Pa, $C_{44} = C_{55} = C_{66} = 2.4 \cdot 10^{10}$ Pa, $\beta_{13} = \beta_{23} = -4.9$ C/m^2, $\beta_{33} = 14.9$ C/m^2, $\beta_{51} = \beta_{42} = 10.6$ C/m^2, $э_{11} = э_{22} = 8.2 \cdot 10^{-9}$ F/m, $э_{33} = 8.4 \cdot 10^{-9}$ F/m, $\rho = 7500$ kg/m^3.

3.1. Determination of the optimal location of the first piezoelectric element

The determination of the optimal location of the piezoelectric element on the surface S_1 to obtain the maximum possible electric potential at the second natural frequency of vibrations was carried out by the direct search method. A series of calculations was carried out for different options of location of piezoelectric element, which were determined by the coordinates of its center of mass.

For convenience of presentation of the obtained results, the zone of clamp and lateral surfaces of the shell were excluded, and the results are presented in the form of a flat picture only for a part of the surface S_1.The x- coordinate is plotted on the horizontal axis, and the y - coordinate is plotted on the vertical axis. The axis of symmetry of the shell is aligned along the y-axis.

Figure 2 shows the patterns of distribution of the EMCC values at the second natural frequency of vibrations in the case of vertical (Fig. 2a) and horizontal (Fig. 2b) orientation. The obtained results demonstrate that the values of EMCC obtained at the horizontal orientation of the piezoelectric element are 1.5 times higher than the values obtained at the vertical orientation. Therefore, further calculations were made on the assumption of horizontal orientation of the first piezoelectric element.

The analysis of the distribution patterns of the electromechanical coupling coefficient shown in Fig. 2b, allowed us to determine the coordinates of the center of mass of the piezoelectric element, which are optimal for obtaining the maximal value of the EMCC at vibration with the frequency corresponding to the natural vibration frequency of the second mode (K_2=0.037): x=0 mm, y=145 mm). For such arrangement of the piezoelectric element, the value of the EMCC obtained at the frequency corresponding to the natural vibration frequency of the third mode is small (K_3=0.002). It means that the above mentioned position of the piezoelectric element at this frequency is not optimal.

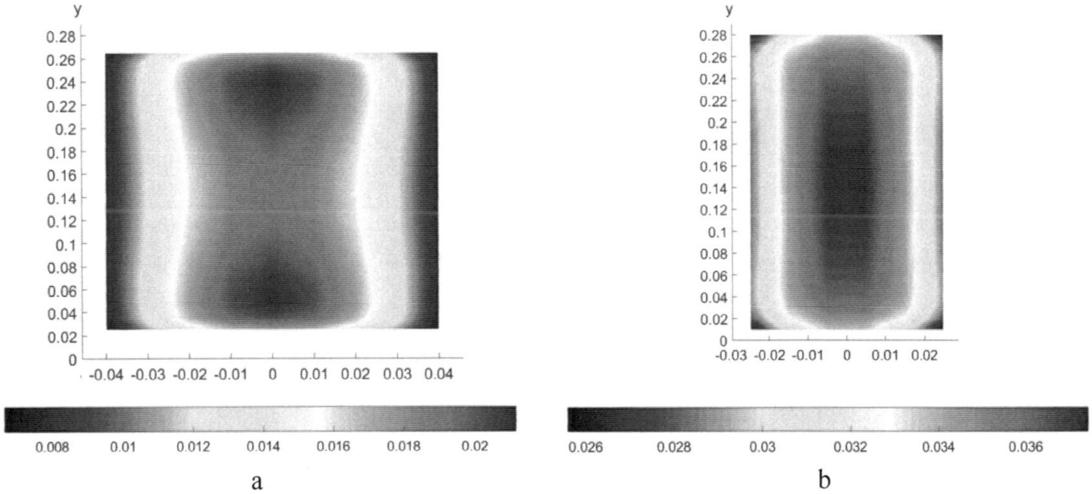

Figure 2. Patterns of distribution of the electromechanical coupling coefficient (EMCC) at the second frequency in the case of vertical (a) and horizontal (b) arrangement of the piezoelectric element.

Consider the problem of forced steady-state vibrations of the shell under the impact of the force $F_Z = 1$ N, applied to the surface S_l of the structure at the point with coordinates x = 0 mm, y = 0 mm (Fig. 3), for three options of the piezoelectric element location. The coordinates of the center of mass of the piezoelectric element in the optimal option of location (x=0 mm, y =145 mm) were determined by the method described above, whereas for the intermediate (x=20 mm, y=210 mm) and non-optimal locations (x=25 mm, y=280 mm a) they were chosen arbitrarily (table 1).

Figure 4 shows the amplitude- frequency response plots of the electric potential generated by the piezoelectric element at different locations. It can be seen from the plots that when the location of the piezoelectric element is optimal for its effective performance at the second natural frequency, the electric potential is maximum and equal to 120.1 V. In this case, the EMCC also takes the maximum value.

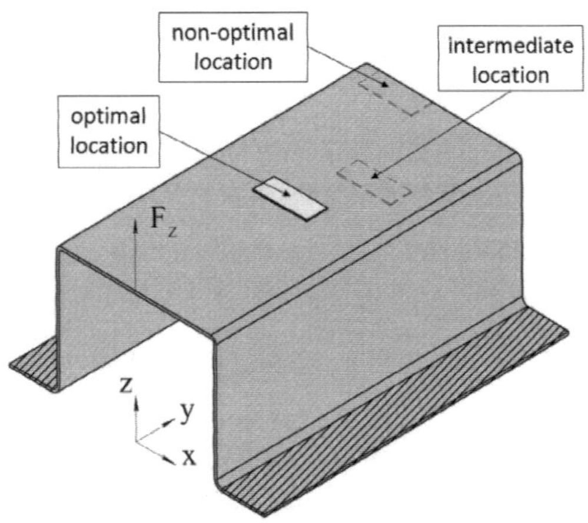

Figure 3. Computational model for the problem of forced steady-state vibrations.

It should also be noted that in the intermediate option of piezoelectric element location, a sufficiently high level of the electric potential (~ 100 V) is generated on its surface at the frequencies corresponding to the second and third natural vibration frequencies, which indicates its ability to work equally effectively at two frequencies.

Table 1. The values of the second natural frequency of vibrations and the electromechanical coupling coefficient depending on the location of the first piezoelectric element

	Options of piezoelectric element location		
	Optimal	Intermediate	Nonoptimal
Coordinates of the center of mass of piezoelectric element, (mm)	x=0 y=145	x=20 y=210	x= – 25 y=280
Second natural frequency of the structure with piezoelectric element $f_{o/c}$ (Hz)	42.17	41.95	42.03
Electromechanical coupling coefficient K_2	0.037	0.032	0.025

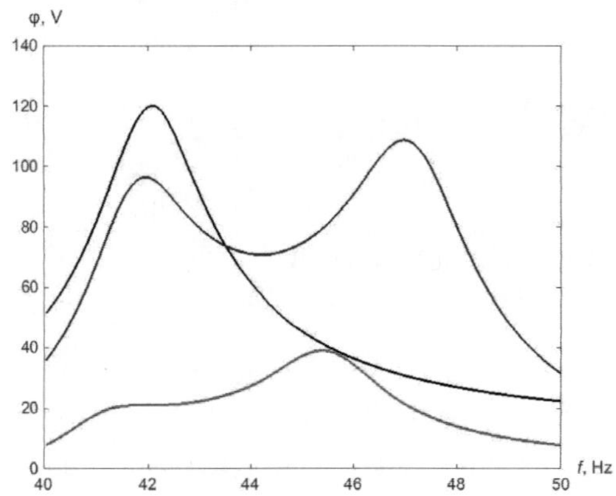

Figure 4. The amplitude-frequency response plots of the electric potential generated by the piezoelectric element at different locations: optimal (black line), intermediate (blue line), non-optimal (red line).

3.2. The effect of electrical boundary conditions on the value of the coefficient of the electromechanical coupling coefficient

As it was noted in [3], the attachment of the piezoelectric element to an external electric circuit can produce changes in its stiffness properties in the range determined by the two limiting cases: short circuit (s / c) and open circuit (o / c) condition. The changes in the stiffness properties give rise to the changes in the natural vibration frequencies and vibration modes of the structure. In this regard, it is important to know what effect these variations can exert on the electromechanical coupling coefficient.

Table 2. Natural vibration frequencies and electromechanical coupling coefficients for the first option of location of the second piezoelectric element with the coordinates of the center of mass: x = 15 mm, y = 70 mm

Number of frequency	$f_{o/c}$ (Hz)	$f_{s/c}$ (Hz)	K_i
EMCC for the first piezoelectric element, the second piezoelectric element is in the o/c mode			
2	41.89801	41.87122	0.035777
3	47.18087	47.18073	0.002464
EMCC for the first piezoelectric element, the second piezoelectric element is in the s/c mode			
2	41.87619	41.84944	0.035761
3	47.16129	47.16114	0.002525
EMCC for the second piezoelectric element, the first piezoelectric element is in the o/c mode			
2	41.89801	41.87619	0.032285
3	47.18087	47.16129	0.028821
EMCC for the second piezoelectric element, the first piezoelectric element is in the s/c mode			
2	41.87122	41.84944	0.032267
3	47.18073	47.16114	0.028826

Table 3. Natural vibration frequencies and electromechanical coupling coefficients for the second option of location of the second piezoelectric element with the coordinates of the center of mass: x = 15 mm, y = 110 mm

Number of frequency	$f_{o/c}$ (Hz)	$f_{s/c}$ (Hz)	K_i
EMCC for the first piezoelectric element, the second piezoelectric element is in the o/c mode			
2	41.80375	41.77910	0.034357
3	47.48214	47.48213	0.000358
EMCC for the first piezoelectric element, the second piezoelectric element is in the s/c mode			
2	41.78501	41.76023	0.034455
3	47.47708	47.47707	0.000375
EMCC for the second piezoelectric element, the first piezoelectric element is in the o/c mode			
2	41.80375	41.78501	0.029953
3	47.48214	47.47708	0.014610
EMCC for the second piezoelectric element, the first piezoelectric element is in the s/c mode			
2	41.77910	41.76023	0.030066
3	47.48213	47.47707	0.014600

To estimate the influence of the electric boundary conditions on the value of the EMCC, a series of computational experiments were carried out, in which the EMCC was determined for the first piezoelectric element at the time when the second piezoelectric element was operating in the open- and short circuit modes and in the reverse order. In this case, the first piezoelectric element was held in its optimal position, whereas for the second piezoelectric element we considered two options of piezoelectric element location differing in the coordinates of the center of mass of the element.

In the first case, x = 15 mm, y = 220 mm, which means that it was located at a distance from the first piezoelectric element. In the second case x = 15 mm, y = 180 mm, which means that it was located in the immediate vicinity of the first piezoelectric element. The results obtained are presented in tables 2 and 3

From the results given in Tables 2 and 3, it follows that a change in the boundary conditions with respect to the electrical component has inessential effect both on the natural vibration frequencies and on the electromechanical coupling coefficient (a change is less than 2%). Thus, if there are several piezoelectric elements in the structure, the electromechanical coupling coefficient for each piezoelectric element is determined by formula (1) regardless of the boundary conditions for the electrical component specified on the other piezoelectric elements. Therefore, in all subsequent calculations, the evaluation of the EMCC for one piezoelectric element is carried out under the condition of short-circuit operation specified for the other piezoelectric elements (zero potential on both electrode surfaces).

3.3. Determination of the optimal location of the second piezoelectric element

Now, let us choose for the second piezoelectric element such position that the generation of the maximum electric potential on its surface occurs at the third natural vibration frequency. In this case, it is essential that the electric potential of the first piezoelectric element remains maximum (or close to the maximum) at the second frequency.

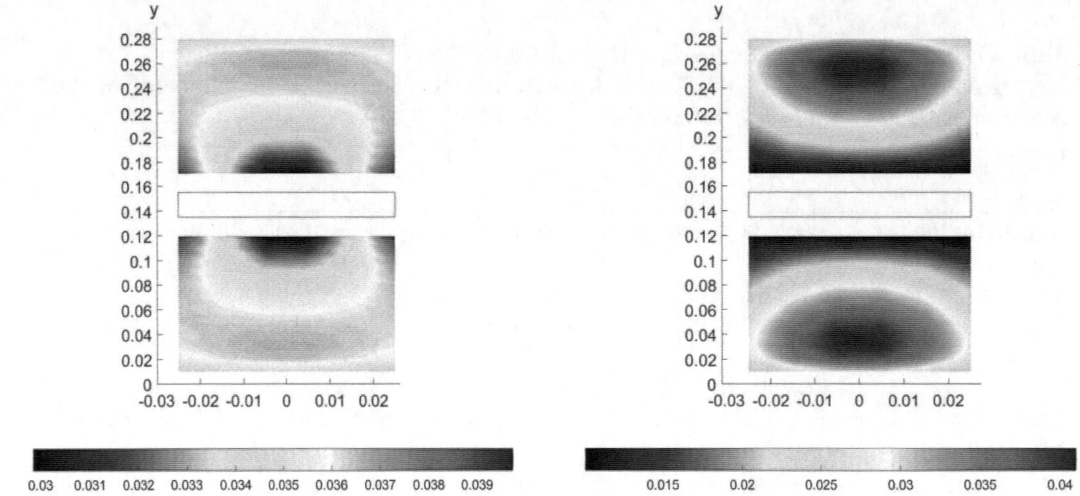

a b

Figure 5. Patterns of the distribution of electromechanical coupling coefficient for the first piezoelectric element at the second natural vibration frequency (a) and for the second piezoelectric element at the third natural vibration frequency (b) at different locations of the second piezoelectric element.

Using the direct search method described above, we obtained a series of the distribution partens of the electromechanical coupling coefficient for the first and second piezoelectric elements according to the location of the second piezoelectric element. Figure 5 shows the distribution patterns of the electromechanical coupling coefficient for the first piezoelectric element at the second natural vibration frequency (a) and for the second piezoelectric element at the third natural vibration frequency (b). Each point on these graphs corresponds to the position of the center of mass of the second piezoelectric element. The white rectangle shows the location of the first piezoelectric element.

As is evident from the above figures, a change in the location of the second piezoelectric element significantly affects the value of EMCC of the first piezoelectric element (the maximum change in its value is about 20% compared to the value obtained for a single piezoelectric element, which is equal to 0.037).

Table 4. The values of the second and third natural frequencies of vibrations and the corresponding electromechanical coupling coefficients depending on the location of the second piezoelectric element

	Options of piezoelectric element location		
	Optimal	Intermediate	Nonoptimal
Coordinates of the center of mass of the piezoelectric element, (mm)	x=0 y=260	x=0 y=210	x= – 25 y=170
Second natural frequency of the structure with the attached piezoelectric element, $f_{o/c}$ (Hz)	42.72	42.28	42.35
Third natural frequency of the structure with the attached piezoelectric element, $f_{o/c}$ (Hz)	46.55	47.14	47.08
Electromechanical coupling coefficient K_2	0.037	0.035	0.039
Electromechanical coupling coefficient K_3	0.041	0.027	0.011

As it follows from Fig.5a, the second piezoelectric element can be located in such a way that the EMCC of the first piezoelectric element will increase. The obtained results demonstrate that the patterns of distribution of the EMCC values for a single piezoelectric element cannot be used as a criterion for locating several piezoelectric elements.

Figure 5.b shows that, for the third natural vibration frequency the optimal location of the second piezoelectric element is such that the center of its mass coincides with the points with coordinates x = 0 mm, y = 30 mm, or x = 0, y = 260 mm. In this case, for the second piezoelectric element, the value of the electromechanical coupling coefficient is $K_3 = 0.041$, whereas the value of the EMCC for the first piezoelectric element at the second frequency is $K_2 = 0.037$ (the value of the EMCC without the second piezoelectric element), which indicates that the reciprocal influence of the piezoelectric elements at this particular locations is minimal. With this location of the second piezoelectric element, its electromechanical coupling coefficient at the frequency of the second vibration mode is $K_2 = 0.035$ and is close in the magnitude to the EMCC of the first peizoelectric element.

Based on the results obtained, it can be concluded that the locations found for both piezoelectric elements are indeed optimal from the viewpoint of the maximum value of the electromechanical coupling coefficient, characterizing the effective operation of the piezoelectric element.

Consider the problem of forced steady-state vibrations of the shell under the impact of the force F_Z =1 N applied to the surface S_l of the structure at the point with coordinates x = 0 mm, y = 0 mm (Fig.6).The location of the first piezoelectric element is considered to be fixed, and the center of mass of the second piezoelectric element will be located as follows: at the point with coordinates x = 0 mm, y = 260 mm (the optimal location), x = 0 mm, y = 210 mm (intermediate location) and x = 25 mm, y = 170 mm (non-optimal location) (table 4).

Figure 7 shows the amplitude- frequency response plots of the electric potential measured on the first (Fig. 7a) and second (Fig. 7b) piezoelectric elements for the three options of location of the second piezoelectric element. The curves presented in Fig. 7b show that the electric potential of the second piezoelectric element under the vibrations with a frequency corresponding to the third natural vibration frequency reaches a maximum value of 192 V when its center of mass is located at the point with the coordinates corresponding to the optimal location (see Fig. 7).

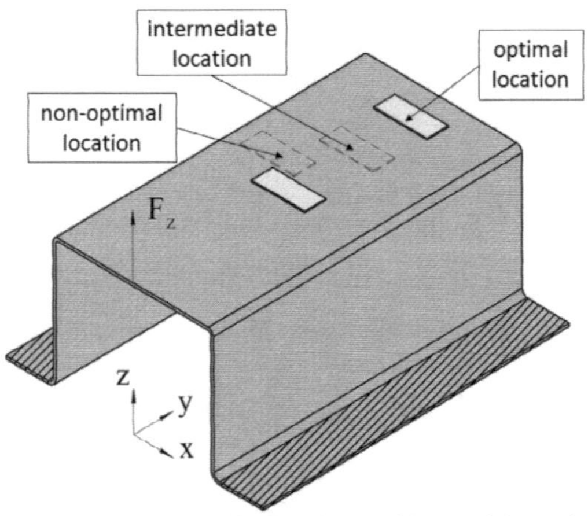

Figure 6. Computational model for the problem of forced steady-state vibrations.

However in this case, the electric potential of the first piezoelectric element is reduced to 113.5 V (Fig. 7a), which is also reflected in the value of the electromechanical coupling coefficient.

When the second piezoelectric element occupies a non-optimal location (see Fig. 7), the value of the electric potential generated by the first piezoelectric element at the frequency of the structure vibrations, corresponding to the natural vibration frequency of the second mode, increases reaching 128.4 V, which is also reflected in the value of the electromechanical coupling coefficient.

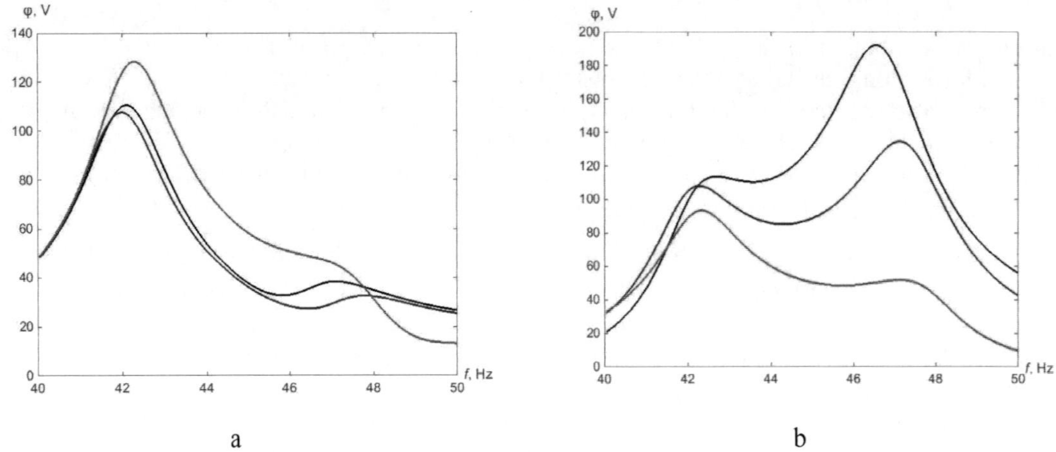

Figure 7. The amplitude- frequency response plots of the electric potential generated by the first (a) and second (b) piezoelectric elements depending on location of the second piezoelectric element: optimal location x = 0 mm, y = 260 mm (black line), intermediate location x = 0 mm, y = 210 mm (blue line), non-optimal location x = 25 mm, y = 170 mm (red line).

4. Conclusions

The method, which is generally used to determine the location of a single piezoelectric element on the surface of the structure which will ensure the maximum possible value of the electric potential generated on its electrode surface under vibration at the specified frequency, has been extended to the case of several piezoelectric elements.

A numerical investigation has shown that the electrical boundary conditions specified for other piezoelectric elements do not affect the value of the electromechanical coupling coefficient of the examined piezoelectric element.

It has been found that in the case of using several piezoelectric elements, they can be so arranged that their efficiency can be essentially improved. It is also possible to find such options of location, at which the piezoelectric elements can generate a sufficient electrical potential at several vibration frequencies, which will allow control over the dynamic characteristics of the structure at these frequencies.

Acknowledgements
The work is supported by the RFBR (projects No 18-31-00080-mol_a).

References

[1] Venna S, Lin Y-J 2013 An Effective Approach for Optimal PZT Vibration Absorber Placement on Composite Structures *Modern Mechanical Engineering* **No 3** 21-26.
[2] Han J Y, Lee I 1999 Optimal placement of piezoelectric Sensors and actuators for vibration control of a composite plate using genetic algorithms *Smart Materials and Structures* **8, No 2** 257-267.
[3] Lammering R, Jia J, Rogers C A 1994 Optimal Placement of Piezoelectric Actuators in Adaptive Truss Structures *Journal of Sound and Vibration* **171, No 1** 67-85.
[4] Barboni R, Mannini A, Fantini E, Gaudenzi P 2000 Optimal placement of PZT actuators for the control of beam dynamics *Smart Mater. Struct.* **No 9** 110–120.
[5] Onoda Y, Li J, Minesugi K 2002 Simultaneous optimization of piezoelectric actuator placement and feedback for vibration suppression *Acta Astronautica* **50, No 6** 335–341.
[6] Holman R E, Spencer S M, Austin E M, Johnson C D (1997) Passive damping technology demonstration *SPIE* **3045, No 1** 51-59.
[7] Wu S, Turner T L, Rizzi S A 2010 Piezoelectric shunt vibration damping of F-iS panel under high acoustic excitation *SPIE* **3989, No 2** 276-287.
[8] Hollkamp J J, Gordon R W 1996 An experimental comparison of piezoelectric and constrained layer damping *Smart Mater. Struct.* **5** 715–722.

[9] Gupta V, Sharma M, Thakur N 2010 A Technical Review Optimization Criteria for Optimal Placement of Piezoelectric Sensors and Actuators on a Smart Structure *J. Intelligent Material Systems and Structures* **21** 1227-1243.

[10] Bruch Jr J C, Sloss J M, Adali S, Sadek I S 2000 Optimal piezo-actuator locations/lengths and applied voltage for shape control of beams *Smart Mater. Struct.* **9** 205–211.

[11] Hagood N W., von Flotow A 1991 Damping of structural vibrations with piezoelectric materials and passive electrical networks *Journal of Sound and Vibration* **146, No 2** p. 243-268.

[12] Trindade M A, Benjeddou A 2009 Effective electromechanical coupling coefficients of piezoelectric adaptive structures: critical evaluation and optimization *Mechanics of Advanced Materials and Structures* **16, No 3** 210-223.

[13] Matveenko, V.P., Oshmarin, D.A., Sevodina, N.V. and Iurlova, N.A. (2016). Natural vibration problem for electroviscoelstic body with external electric circuits and finite-element relations for its numerical implementation *Computational Continuum Mechanics* **9, No** 4 476-485.

[14] Iurlova N A, Matveenko V P, Oshmarin D A, Sevodina N V, Yurlov M A. 2016 Layout optimization of piezoelectric elements with external electric circuits in smart constructions based on solution of the natural vibrations problem *ECCOMAS Congress 2016 VII European Congress on Computational Methods in Applied Sciences and Engineering Proceedings*, (*Crete Island, Greece, 5–10 June 2016)* **1** 1920-1929.

[15] Sevodina N V, Yurlova N A, Oshmarin D A 2015 The optimal placement of the piezoelectric element in a structure based on the solution of the problem of natural vibrations *Solid State Phenomena* **243** 67-74.

IOP Conf. Series: Materials Science and Engineering **581** (2019) 012014 doi:10.1088/1757-899X/581/1/012014

An Algorithm for Identifying Texture Components in the Framework of Statistical Crystal Plasticity Models

K V Ostapovich [a], P V Trusov and A Yu Yanz

Department of Mathematical Modelling of Systems and Processes, Perm National Research Polytechnic University, 29 Komsomolsky prospekt, Perm 614990, Russian Federation

E-mail: [a] ostkirvad@gmail.com

Abstract. An approach based on involving clustering techniques to identify texture components for a given polycrystalline aggregate is formulated. A heuristic iterative clustering algorithm operating only with a sample of lattice orientations and the concept of closeness between them is proposed for this task. The mentioned closeness concept is introduced via a special pseudometric distance, which is induced by the natural Riemann metrics and takes the lattice symmetry into consideration. The developed procedure allows representing an inhomogeneous orientation distribution by localized clusters in the space of orientations and thus can be used to reduce the dimension of orientation distribution data in statistical crystal plasticity models. Two examples of such an application of the clustering procedure are provided.

1. Introduction

Polycrystalline metals and alloys consist of anisotropic elements with lattice structure, i.e., crystallites, the orientations of which have a significant influence on macroscopic properties of such materials. During severe inelastic deformation, the crystallites can rotate leading to inhomogeneity in the orientation distribution, i.e. the formation of a crystallographic texture. The analysis of the texture plays an important role in material science problems related to the study and description of the interdependence between the evolution of the microstructure and changes in macroscopic properties of polycrystals.

Quantitative description of texture formation is generally based on the methods of multilevel modeling and explicitly or implicitly refers to the orientation distribution density in a polycrystalline representative volume element or its statistical model, i.e., the polycrystalline aggregate. This density (or the corresponding measure) determines volume fractions of a crystal lattice in various orientational configurations. For its numerical representation in most applications, certain reduction techniques are being involved. Some approaches [1–3] include the Fourier expansion with respect to the symmetrized generalized spherical harmonics; the proper orthogonal decomposition and the snapshot method, which is its simplified analogue; the texture components method; and omnifarious techniques based on the explicit discretization of the orientational space.

The mentioned methods and their modifications are in widespread use in texture analysis, in particular, for the purpose of calculating orientation distribution densities by the inversion of diffraction pole figures. In [4–11], these approaches are applied to formulating reduced-order models of texture formation during deformation. Such models are of significant interest in polycrystalline material design. Relevant inverse boundary value problems in this field require solving repeatedly the equations for microstructure evolution during various thermo-mechanical processes. Due to high resource-intensity of these problems, the possibility of compressing orientation distribution data is quite important. One of the ways to reduce such information in statistical polycrystalline models, which is developed in this work, is based on identifying pronounced texture components by using clustering techniques [12–13]. Here, an emphasis is placed on the formulation of a heuristic algorithm appropriate for this task.

Clusters are generally treated as disjoint subsets of similar (close) elements. The procedure described below allows distinguishing orientation clusters for a given polycrystalline aggregate by considering only its sample of orientations. In fact, these clusters can be considered as regions in the space of orientations, in which the distribution density is higher.

2. A procedure for clustering orientations

One should note that almost all clustering schemes in one way or another refers to the concept of closeness between data elements. In the current case, these points are the orientations of a crystal lattice, the geometric symmetry of which needs to be taken into account. Such a problem is solved in this work by introducing a special pseudometric distance for comparing orientations. Unlike a metric one, a pseudometric distance does not assume the non-degeneracy on unequal arguments. Moreover, it should be equal to zero if and only if the orientations being compared are geometrically indistinguishable from each other.

Consider a single-phased polycrystalline aggregate of M crystallites with the same volume fractions. Their orientations can be defined by the elements, $o^{(m)}, m \in \overline{1,M}$, of the special orthogonal group, O^+. Let $\mathrm{S}^+ \subset \mathrm{O}^+$ be the rotational symmetry group of the given lattice type, so any $o \in \mathrm{O}^+$ is geometrically indistinguishable from so for all $s \in \mathrm{S}^+$, i.e., symmetrically equivalent to them. The pseudometric distance, $d[\cdot,\cdot]$, between arbitrary $o_1, o_2 \in \mathrm{O}^+$ is introduced as

$$d[o_1,o_2] = \min_{s\in \mathrm{S}^+} d_o[o_1, so_2], \tag{1}$$

where $d_o[\cdot,\cdot]$ is some metric distance. As for this distance, it is convenient to use the one [] induced by the natural Riemann metrics. In such a case, by adopting the quaternion parametrization, one can write:

$$\begin{aligned} d[o_1,o_2] = 2\min_{s\in \mathrm{S}^+} \arccos\big|&\mathrm{Re}\,s\left(\mathrm{Re}\,o_1\,\mathrm{Re}\,o_2 + \mathbf{Im}\,o_1 \cdot \mathbf{Im}\,o_2\right) \\ &-\mathbf{Im}\,s \cdot \left(\mathrm{Re}\,o_1 \mathbf{Im}\,o_2 - \mathrm{Re}\,o_2 \mathbf{Im}\,o_1 + \mathbf{Im}\,o_1 \times \mathbf{Im}\,o_2\right)\big|. \end{aligned} \tag{2}$$

Here, $\mathrm{Re}\,o$ and $\mathbf{Im}\,o$ denote, respectively, the real and the vector (imaginary) parts of the quaternion corresponding to $o \in \mathrm{O}^+$. This distance determines the minimal angle of rotations from one orientation-argument to any symmetrically equivalent one of another.

Due to high complexity of the space being considered, a clustering approach based exclusively on the comparison of the sample orientations seems to be quite performant. In such a framework, it is reasonable to introduce the misorientation matrix, $\{D_{mn}\}$ (of $M \times M$ size):

$$D_{mn} = d\left[o^{(m)}, o^{(n)}\right]. \tag{3}$$

This is a symmetric matrix with zero main diagonal, which contains almost all information required for applying abstract clustering methods of a rather wide class.

2.1. Extracting a layer

Generally, the exact amount of clusters for a given texture is not known a priori and the possibility to distinguish them is quite suppositional. In order to obtain an adequate partition, one should associate the locations of clusters with the regions of higher orientation distribution density. Appropriate techniques are based on grouping close orientations (or maybe their neighborhoods) into distinct subsets. However, such methods are usually sensitive to the "noise" created by rarefied orientations. To overcome this flaw, it is proposed to successively exclude from the sample under consideration the layers of the orientations that have the highest density in their neighborhoods. Each of the layers obtained in such a way is supposed to be further independently clustered.

Let $M_{l\varepsilon}^{(m)}$ be the number of the orientations that left in the sample after excluding the l-th layer and belongs to the pseudo-metrical ε-neighborhood of $o^{(m)}$. In the framework of the approach being described, one constructs a layer by choosing a lower bound, $\underline{M}_{l\varepsilon}$, and grouping the orientations with $M_{l\varepsilon}^{(m)} \geq \underline{M}_{l\varepsilon}$.

2.2. Clustering by reachability

Orientations of a layer are considered as the vertexes of a connectivity graph. The connectivity criterion is defined as the closeness between orientations or their neighborhoods. In the simplest case, $o^{(m)}$ and $o^{(n)}$ belongs to an edge if and only if $D_{mn} \leq \varepsilon$. The maximal connected subgraphs of the introduced graph are treated as the initial clusters.

The selection of appropriate values for $\underline{M}_{l\varepsilon}$ is restricted by the variations of distribution density in the ε-neighborhoods and hence depends on the texture under consideration. To define them, a histogram analysis for the distributions of $\underline{M}_{l\varepsilon}$ over sample orientations at different layers can be engaged. Here, it seems reasonable to choose ε in such a manner that the amount of orientations in almost each ε-neighborhood is statistically sufficient for introducing their local density (with the exception of relatively few neighborhoods containing rarefied orientations). In the case of the uniform distribution, the fraction, μ_0, of orientations within an ε-neighborhood under the condition that there are no pairs of symmetrically equivalent elements (which can be always achieved by choosing not too large values of ε) is given by

$$\mu_0 = \frac{S}{\pi}(\varepsilon - \sin\varepsilon), \tag{4}$$

where S is the cardinality of S^+.

2.3. Medoid-based reclustering

Clustering by reachability does not provide an efficient control of the forms, sizes and amount for the clusters being obtained. Meanwhile, in order to simplify the reduced representation of a texture, the possibility to distinguish well enough localized clusters with pronounced centers is of a significant interest. Due to this, it seems reasonable to introduce an additional procedure for identifying the concentration points of orientations in clusters and then, if necessary, performing a repartition around them. This task is resolved in this work by applying the so-called medoid-based reclustering.

The medoid of a set is understood as its element such that the mean pseudometric distance from this element to other elements in the set is minimal. Medoid-based reclustering on the l-th layer with the fixed number, K_l, of clusters being formed can be considered as a heuristic procedure to solve the following discrete optimization problem. It is required to find the indexes, m_{lk}, $k \in \overline{1, K_l}$, of orientation-medoids and the discrete function, $c_l[\cdot]$, which maps orientation indexes onto corresponding cluster indexes, such that the objective,

$$J_l^{K_l} = \sum_{k=1}^{K} \sum_{m \in c_l^{-1}[k]} D_{mm_{lk}}, \tag{5}$$

reaches its minimum value.

The minimization strategy is realized by alternating between determining medoids of current clusters and reassigning each orientation to the cluster with the closest medoid. One should note that, in general, such an approach provides only a sub-optimal (i.e., unimprovable by any further iterations) solution. Other optimization techniques [14] can be also involved to solve this problem.

2.4. Splitting clusters

To get a partition such that each cluster is well enough localized in the sense that, e.g., the mean pseudometric deviation of its orientations from the medoid does not exceed a critical value, $\bar{\delta}^{avg}$, the described procedure is iterated. Herewith at each step, the least localized cluster is splitted by the pair of its most distant orientations which are treated as new medoids.

The scheme of the clustering procedure described above is shown in Figure 1. This operation sequence is repeated until there are no orientations left unassigned to any cluster. The orientation clusters obtained by the proposed algorithm can be treated as texture components. Their local characteristics give a reduced representation of the orientation distribution in a polycrystalline aggregate.

More specifically, the orientation distribution measure, $F[\cdot]$, of the given polycrystalline aggregate is formally defined for any Borel subset, $\mathrm{O} \in \mathrm{O}^+$, as

$$F[\mathrm{O}] = \frac{1}{M}\sum_{m=1}^{M} o^{(m)} \Delta_{o^{(m)}}[\mathrm{O}]. \tag{6}$$

Here, $\Delta_o[\cdot]$ denotes the Dirac measure concentrated in $o \in \mathrm{O}^+$, i.e., $\Delta_o[\mathrm{O}] = 1$ if and only if $o \in \mathrm{O}$ and $\Delta_o[\mathrm{O}] = 0$ otherwise. Further, let L be the amount of layers, $K_l, l \in \overline{1,L}$ be the amounts of clusters at the layers, $M_{lk}, l \in \overline{1,L}, k \in \overline{1,K_l}$ be the cardinalities of the clusters and $m_{lk}, l \in \overline{1,L}, k \in \overline{1,K_l}$ be the indexes of the corresponding medoids. Then one can propose the simplest variant for a reduced orientation distribution measure, $\tilde{F}_{\bar{L}}[\cdot]$, in the form:

$$\tilde{F}_{\bar{L}}[\mathrm{O}] = \frac{1}{M}\sum_{l=1}^{\bar{L}}\sum_{k=1}^{K_l} M_{lk} \Delta_{o^{(m_{lk})}}[\mathrm{O}], \tag{7}$$

where $\bar{L}$ is the amount of the layers taken into consideration.

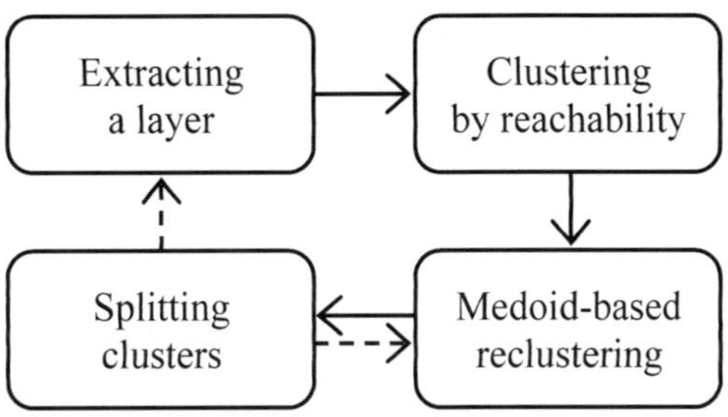

Figure 1. The scheme of the clustering procedure. Dashed arrows point at conditioned stages.

3. Some examples

In this section, examples of applying the clustering procedure to two numerically obtained textures are presented. The two-scale elasto-visco-plasticity model [15] is used in simulations. Kinematic loads of a copper (the face-centered cubic lattice) polycrystalline aggregate are realized in a simple shear and an one-axial tension tests with a constant velocity gradient. The aggregate is consisted of $M = 1000$ crystallites with initially uniformly randomly distributed orientations. The clustering parameters are set as $\varepsilon = \bar{\delta}^{avg} = 0.158$ ($\mu_0 = 0.005$).

To represent input and output data of the procedure in each test, direct and reduced (characterizing cluster cardinalities) pole figures are plotted for the $\{100\}$ family of planes. One should note that these figures are used only for the purpose of visualization since the formulated algorithm operates directly with orientations, not the projections of the corresponding directions. To specify characteristic axes of loads, a stationary (laboratory) coordinate system with the biorthogonal basis vectors, $\mathbf{l}_i, i \in \overline{1,3}$, is defined.

3.1. Simple shear

In this case, the shear along $\mathbf{l}_2$ in the plane perpendicular to $\mathbf{l}_3$ is simulated. The sample of obtained orientations is depicted in figure 2a. Partitioning into $L = 7$ layers is used for clustering. However, the orientations of the last layer (13.6% of the total amount) are highly rarefied so the procedure distinguishes a large number of degenerate "clusters" consisting of only one element. This layer is excluded from consideration. The medoids of the first 6 clustered layers are shown in figure 2b.

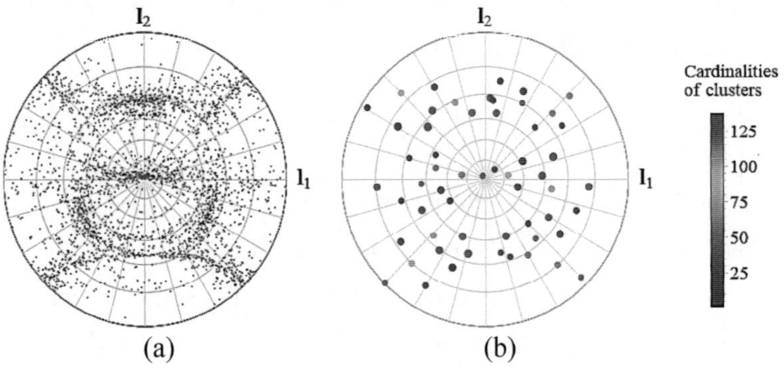

Figure 2. Pole figures for the simple shear texture: the direct pole figure for the aggregate orientations (a) and the reduced pole figure for the identified cluster medoids (b)

The adequacy of the resulted medoid-based representation is verified for aggregating elastic properties. The Voigt averages of the elasticity tensors are evaluated with the full orientation distribution measure (6) and with its reduced analogue (7) for $\overline{L} = 6$. The difference between the corresponding responses defined by the anisotropic Hooke's law is estimated to be less than 10%.

3.2. One-axial tension

In this test, the tension along $\mathbf{l}_3$ is simulated. The sample of obtained orientations is presented in figure 3a. Clustering with $L = 6$ layers is performed in this example. By reasoning as for the previous case, the last layer (8.1% of the total amount) is omitted. The medoids of the first 5 clustered layers are given in figure 3b.

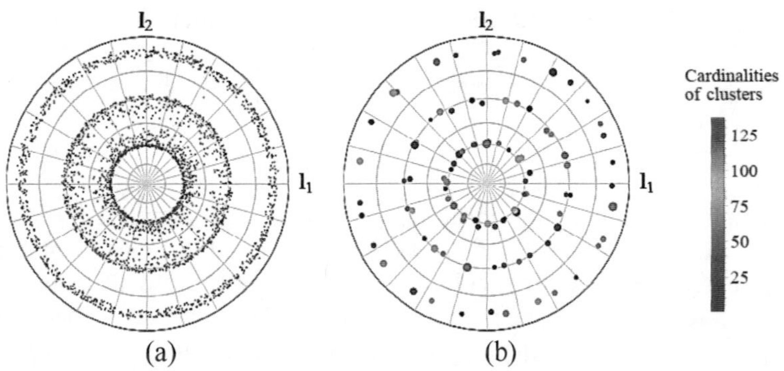

Figure 3. Pole figures for the one-axial tension texture: the direct pole figure for the aggregate orientations (a) and the reduced pole figure for the identified cluster medoids (b)

The verification of the adequacy of this medoid-based representation is carried out in the same way as before. The use of the reduced measure (7) with $\overline{L}=5$ instead of the full measure (6) leads to an error in the response, which is less than 5%.

4. Conclusion

To sum up, this work is devoted to the formulation of the heuristic iterative algorithm for identifying texture components by analysing the sample of lattice orientations with the help of clustering techniques. The proposed procedure consists of the following stages: extracting a layer of orientations; clustering the layer by reachability (in the sense of the transitive closure of the closeness criterion being adopted for orientations); medoid-based reclustering of the layer; and splitting weakly localized clusters. The mentioned methods operate with the concept of closeness between orientations. To formalize it, the pseudometric distance that is invariant to the lattice symmetry transformations of one argument is introduced on the basis of the natural Riemann metrics. The developed clustering approach can be used to reduce the dimension of orientation distribution data in statistical crystal plasticity models. Its application is considered in the examples of the numerically simulated simple shear and one-axial tension textures. The results demonstrate the satisfactory adequacy of the reduced orientation distribution measures corresponding to the medoid-based representations of the textures.

Acknowledgments
The work was supported by the Russian Federation Ministry of Science and Higher Education (the basic part of the PNRPU state assignment, No. 9.7434.2017/8.9), Russian Foundation for Basic Research (grant No. 17-01-00379-a) and the RF President Grant No. MK-2970.2019.1.

References

[1] Chateigner D 2013 *Combined Analysis* (New York: John Wiley & Sons) chapter 4 pp 95–190
[2] Holmes P, Lumley J L, Berkooz G and Rowley C W 2012 *Turbulence, Coherent Structures, Dynamical Systems and Symmetry* (Cambridge: Cambridge University Press) chapter 3 pp 68–105
[3] Kumar A and Dawson P R 1996 *Comput. Methods Appl. Mech.* Eng. **130**(3–4) 227–46
[4] Adams B L, Henrie A, Henrie B, Lyon M, Kalidindi S R and Garmestani 2001 *J. Mech. Phys. Solids* **49**(8) 1639–63
[5] Kalidindi S R, Houskamp J R, Lyons M and Adams B L 2004 *Int. J. Plast.* **20**(8–9) 1561–75
[6] Kalidindi S R, Houskamp J R, Proust G and Duvvuru H 2005 *Mater. Sci. Forum* **495-497** 23–30
[7] Brahme A, Winning M and Raabe D 2009 *Comput. Mater. Sci.* **46** 800–4
[8] Sundararaghavan V and Zabaras N 2005 *Acta Mater.* **53**(4) 1015–27
[9] Sundararaghavan V and Zabaras N 2009 *Stat. Anal. Data Min.* **1**(5) 306–21
[10] Acharjee S and Zabaras N 2003 *Acta Mater.* **51**(18) 5627–46
[11] Ganapathysubramanian S and Zabaras N 2004 *Comput. Methods Appl. Mech. Eng.* **193**(45–47) 5017–34
[12] Ostapovich K V and Trusov P V *Proc. 12th Int. Conf. on Mechanics, Resource and Diagnostics of Materials and Structures* (Ekaterinburg, RF, 21–25 May 2018) pp 271–72.
[13] Mokrova S M, Petrov R P and Milich V N 2016 *Vestn. Udmurt. Univ. Mat. Mekh. Komp. Nauk* **26**(3) 336–44
[14] Hastie T, Tibshirani R and Friedman J 2009 *The Elements of Statistical Learning: Data Mining, Inference, and Prediction* (New York: Springer-Verlag)
[15] Trusov P V, Shveykin A I and Yanz A Yu 2016 *Phys. Mesomech.* **19**(2) 47–65

IOP Conf. Series: Materials Science and Engineering **581** (2019) 012015 doi:10.1088/1757-899X/581/1/012015

Nonlinear Effects in the Behavior and Fracture of Composite Materials

E V Lomakin[1], B N Fedulov[2, a] and A N Fedorenko[2]

[1] Lomonosov Moscow State University, GSP-1, Leninskie Gory, Moscow, Russia, 119991

[2] CDM2, Skolkovo Institute of Science and Technology, Bolshoy Boulevard 30, bld. 1 Moscow, Russia 121205

E-mail: [a] fedulov.b@mail.ru

Abstract. This paper presents an approach to the formulation of models characterizing nonlinear deformation and damage of composite materials. The research includes the consideration as the simplest model as more complex ones. In this research, a number of material relations to build up a theory for failure prediction in laminated composites are presented. The considered approach deals with the degradation and failure of the material using a phenomenological introduction of damage parameters explicitly influencing on the composite material stiffness. The part of this work is dedicated to the extension of formulated approach to capture and take into account the complex effects such as initial nonlinear shear deformation properties of laminated composites and the influence of high strain rate on the strength characteristics.

1. Introduction

Currently there is no well-established and universally accepted approach to strength modelling of composite materials. The failure analysis of heterogeneous materials plays a major role in engineering applications. For areas where the use of lightweight and high strength structures is essential for product design, as in aerospace industry, the capability to predict failure in composite materials is vital for realization of full potential of a material. Restricting one's attention to phenomenological models only, there can be found a number of works and researches demonstrating essentially different approaches and interpretations of mechanical behavior of heterogeneous materials. Some of them focus on the study of failure criteria for unidirectional composites [1–9], while others study response of stress concentrators in particular lay-ups [10 –13], one can also find works studying nonlinear relations between stress and strain components [14 –22]. There are a lot of review publications discussing these methods [23 –27]. In this research, the different steps necessary to build a theory for failure analysis are discussed, which then are implemented in the simplest form, using standard composite characteristics, as an example. The part of the research describes the way to extend simple failure model to the formulation of more complex models taking into account different effects such as shear nonlinearity of deformation properties of materials and strain rate dependence of their strength. It gives the possibility to understand the necessity of further refinement of approaches and models for the characterization of different effects in the behavior of composite materials and what experimental studies ought to be conducted to improve the models.

2. Damage parameters

Considering the simplest way of introducing damage characteristics, only two parameters, ψ_1 and ψ_2, are chosen, where first parameter corresponds to fiber failure and the second one to matrix failure:

$$\begin{cases} \psi_1 = 0\, fiber\ failure, \psi_1 = 1\ initial\ value \\ \psi_2 = 0\, matrix\ failure, \psi_2 = 1\ initial\ value \end{cases} \tag{1}$$

The modified constitutive relations with damage parameters can be formulated as follows [28, 29]:

$$\begin{Bmatrix} \varepsilon_{11} \\ \varepsilon_{22} \\ \varepsilon_{33} \\ \gamma_{12} \\ \gamma_{13} \\ \gamma_{23} \end{Bmatrix} = \begin{bmatrix} \frac{1}{\psi_1 E_{11}} & -\frac{\psi_2 \nu_{21}}{E_{22}} & -\frac{\psi_2 \nu_{31}}{E_{33}} & 0 & 0 & 0 \\ -\frac{\psi_2 \nu_{12}}{E_{11}} & \frac{1}{\psi_2 E_{22}} & -\frac{\psi_2 \nu_{32}}{E_{33}} & 0 & 0 & 0 \\ -\frac{\psi_2 \nu_{13}}{E_{11}} & -\frac{\psi_2 \nu_{23}}{E_{22}} & \frac{1}{\psi_2 E_{33}} & 0 & 0 & 0 \\ 0 & 0 & 0 & \frac{1}{\psi_2 G_{12}} & 0 & 0 \\ 0 & 0 & 0 & 0 & \frac{1}{\psi_2 G_{13}} & 0 \\ 0 & 0 & 0 & 0 & 0 & \frac{1}{\psi_2 G_{23}} \end{bmatrix} \begin{Bmatrix} \sigma_{11} \\ \sigma_{22} \\ \sigma_{33} \\ \sigma_{12} \\ \sigma_{13} \\ \sigma_{23} \end{Bmatrix} \tag{2}$$

At initial state, the damage parameters are equal one $\{\psi_1, \psi_2\} = \{1, 1\}$, and constitutive equations (2) coincide with classical linear elastic one.

The way of evaluation of damage parameters depends on the failure criteria. Considering the simplest one based on the maximum stresses, the following conditions can be formulated:

$$X_c \le \sigma_{11} \le X_t, Y_c \le \sigma_{22} \le Y_t, |\sigma_{12}| \le S \tag{3}$$

where:

X_c – compression failure stress in fiber direction
X_t – tension failure stress in fiber direction
Y_c – compression failure stress in transversal direction
Y_t – tension failure stress in transversal direction
S – in-plane shear failure stress

Once the failure criterion is chosen, we can formulate rule for modifying damage parameters using following equations:

$$\begin{cases} \psi_1 = 1, if\ X_c \le \sigma_{11}^{el} \le X_t, or\ \psi_1 = 0\ if\ not \\ \psi_2 = \min(\psi_2^{22}, \psi_2^{12}, \psi_2^{13}, \psi_2^{23}, \psi_2^{33}), \end{cases}$$

where

ψ_2^{22} – solution of equation $\sigma_{22}^{el} = Y_t$, if $\sigma_{22}^{el} > Y_t$,
ψ_2^{22} – solution of equation $\sigma_{22}^{el} = Y_c$, if $\sigma_{22}^{el} < Y_c$,
$else\ \psi_2^{22} = 1$
$\psi_2^{12} = S/|\sigma_{12}^{el}|$ if $|\sigma_{12}^{el}| > S$, else $\psi_2^{12} = 1$,   (4)
$\psi_2^{13} = S/|\sigma_{13}^{el}|$ if $|\sigma_{13}^{el}| > S$, else $\psi_2^{13} = 1$,
$\psi_2^{23} = S/|\sigma_{23}^{el}|$ if $|\sigma_{23}^{el}$ $> S$, else $\psi_2^{23} = 1$,
ψ_2^{33} – solution of equation $\sigma_{33}^{el} = Y_t$, if $\sigma_{33}^{el} > Y_t$,
ψ_2^{33} – solution of equation $\sigma_{33}^{el} = Y_c$, if $\sigma_{33}^{el} < Y_c$,
$else\ \psi_2^{33} = 1,$

where σ_{ij}^{el} – stress tensor components, defined by equations (4) before the damage parameters are updated.

In this case, for the determination of the value of parameter ψ_2 we have the condition $\sigma_{22}^{el}(\psi_2^{22}) = Y_t$, which can be represented in the following form:

$$\begin{aligned}&\left[E_{22}\nu_{13}E_{33}\psi_1{}^2\left((-\nu_{13}\varepsilon_{22}+\varepsilon_{33}\nu_{12})E_{22}+\nu_{23}(2Y_C\nu_{12}\psi_1+E_{11}\varepsilon_{11})\right)\right]{\psi_2^{22}}^3\\&+\left[\psi_1\left(\nu_{12}(Y_C\nu_{12}\psi_1+E_{11}\varepsilon_{11})E_{22}{}^2+E_{33}(Y_C\nu_{13}{}^2\psi_1+E_{11}\varepsilon_{33}\nu_{23})E_{22}\right.\right.\\&\left.\left.+E_{11}E_{33}Y_C\nu_{23}{}^2\psi_1\right)\right]{\psi_2^{22}}^2+\left[E_{11}E_{22}{}^2\varepsilon_{22}\right]\psi_2^{22}-E_{11}E_{22}Y_C=0\end{aligned}\tag{5}$$

The minimum positive real root less than one must be chosen in the solution of this cubic equation. If this requirement is not satisfied then $\psi_2^{22}=1$. Typical loading curves for the equations (4) are shown on figure 1.

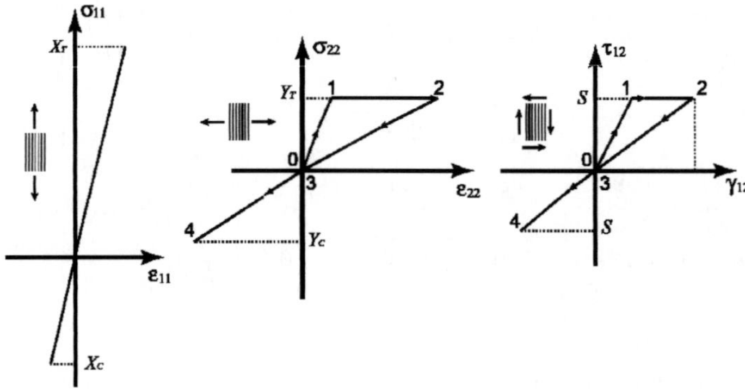

Figure 1. Typical loading curves.

3. Nonlinear elasticity

To characterize the effect of different possible response of the composite material to the type of loading, the formalization of stress state is required for the developing of the mathematical model. By means of $\sigma=\sigma_{ii}/3$ which is the hydrostatic stress component, and $\sigma_0=\sqrt{3/2S_{ij}S_{ij}}$ which is the effective stress, where $S_{ij}=\sigma_{ij}-\sigma\delta_{ij}$ is a deviator stress, it is possible to formulate the parameter $\xi=\sigma/\sigma_0$. This parameter ξ is a good candidate for proposed formalization due to its mechanical sense, invariant nature and scalar simplicity. This introduced parameter can be found in the literature under the name of the stress triaxiality.

3.1. Plane shear formalization

The parameter, which can be used to describe the influence of the degree of shear loading on material properties can be formalized as a shear stress component in the case of plane shear loading and formulated in the invariant form as

$$Q=D_{ij}\sigma_{ij},$$

where

$$D_{ij}=\begin{bmatrix}0&1/2&0\\1/2&0&0\\0&0&0\end{bmatrix}.\tag{6}$$

Thus, it obviously gives in the chosen coordinate system, $Q=\sigma_{12}$.

3.2. Constitutive equations

It is possible to show that the introduction of parameters of stress state ξ and degree of shear loading Q into coefficients of elastic potential

$$\Phi=A_{ijkl}(\xi,\,Q)\sigma_{ij}\sigma_{kl},\tag{7}$$

gives a robust set of constitutive equations that can capture different effects of elastic nonlinearities exhibited by composite materials. In case of 2D plane stress problem, the constitutive equations have the form:

IOP Conf. Series: Materials Science and Engineering **581** (2019) 012015 doi:10.1088/1757-899X/581/1/012015

$$\varepsilon_{11} = A_{1111}(\xi)\sigma_{11} + A_{1122}(\xi)\sigma_{22} + \left[\left(\frac{1}{3\xi} + \frac{3}{2}\xi\right)\sigma - \frac{3}{2}\xi\sigma_{11}\right]\Phi_1\sigma_0^{-2},$$
$$\varepsilon_{22} = A_{1122}(\xi)\sigma_{11} + A_{2222}(\xi)\sigma_{22} + \left[\left(\frac{1}{3\xi} + \frac{3}{2}\xi\right)\sigma - \frac{3}{2}\xi\sigma_{22}\right]\Phi_1\sigma_0^{-2},$$
$$\varepsilon_{12} = B(\xi, Q) - \frac{3}{2}\xi\Phi_1\sigma_0^{-2}, \qquad (8)$$

where

$$\Phi_1 = \frac{1}{2}[A'_{1111}(\xi)\sigma_{11}^2 + A'_{2222}(\xi)\sigma_{22}^2 + 2A'_{1122}(\xi)\sigma_{11}\sigma_{22} + A'_{1212}(\xi, Q)\sigma_{12}^2],$$

here ' – denotes $\frac{d}{d\xi}$.

In spite of the complex form of equation, in the case of proportional loading, the stress state parameter ξ becomes constant, and all equations remain nonlinearity only due to shear parameter Q. Moreover, for uniaxial tension and uniaxial compression loadings, when $\xi=\pm 1/3$, the first two equations become similar to classical ones without any nonlinear components. This gives possibility for relatively easy validation of required parameters of the model.

3.3. Example and test correlation

Let us consider as an example, the simplest approach with linear dependency on parameter ξ of compliance coefficients:

$$A_{1111}(\xi) = a_{11}^0 + c_{11}\xi,$$
$$A_{2222}(\xi) = a_{22}^0 + c_{22}\xi,$$
$$A_{1122}(\xi) = a_{12}^0 + c_{12}\xi, \qquad (9)$$

and the case when coefficient for pure shear compliance has no influence of stress traxiality:

$$B(\xi, Q) = \Gamma(Q).$$

The last component can be found as $\Gamma(\sigma_{12}) = \varepsilon_{12}/\sigma_{12}$ from the pure in-plane shear test. While coefficients a_{11}^0, a_{12}^0 и c_{11}, c_{12} can be determined from uniaxial tests. The pictures in Fig. 2 show the possibilities of this model by correspondence between the theoretical dependencies and the results of tests of composite material on the base of fiberglass fabric and polyether matrix for tension, compression and shear conditions ~~tests~~.

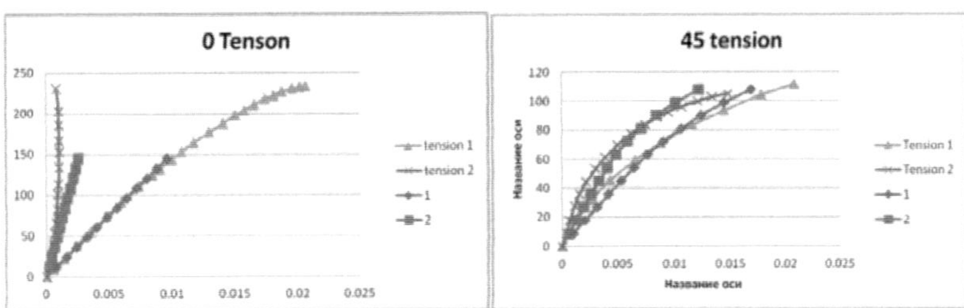

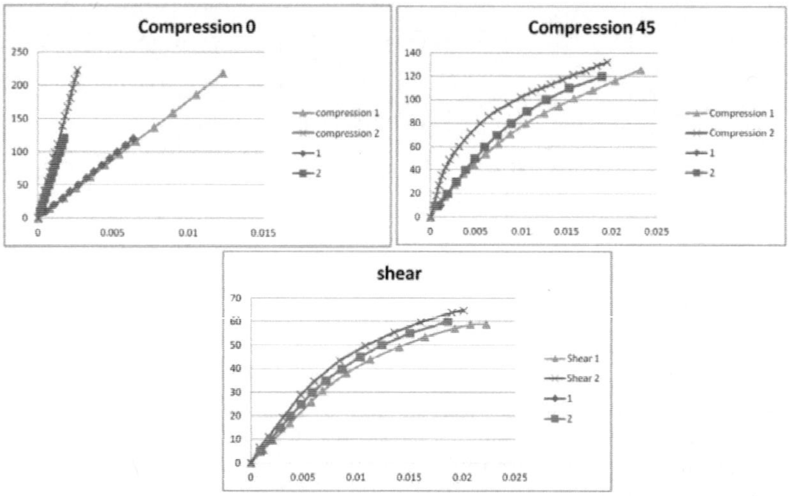

Figure 2. Test correlation for fiberglass fabric composite; 1,2 –predicted loading diagrams.

4. Influence on damage

More over such kind of effects have the influence on the damage distribution in strength analysis. The pictures in Figs. 3-5 show the difference in matrix damage distribution for results based on linear elastic model and for nonlinear one presented in this research. Modeling was realized for the compression of graphite fibers and epoxy matrix composite specimen with open-hole and ±45 layup.

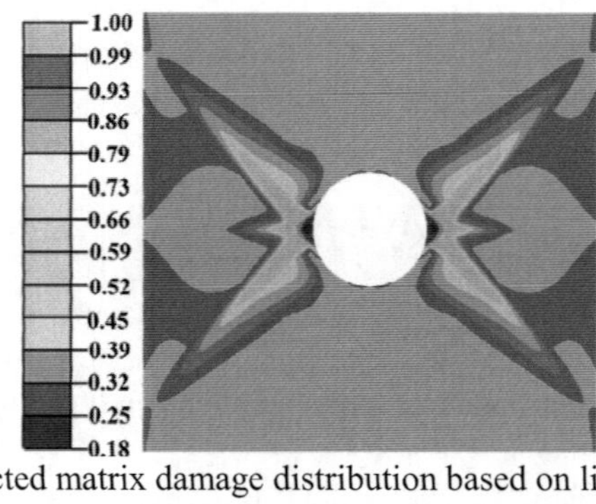

Figure 3. Predicted matrix damage distribution based on linear elastic model

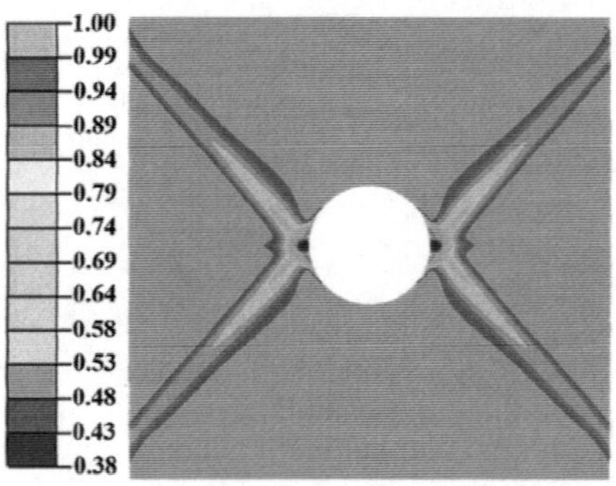

Figure 4. Predicted matrix damage distribution based on nonlinear elastic model

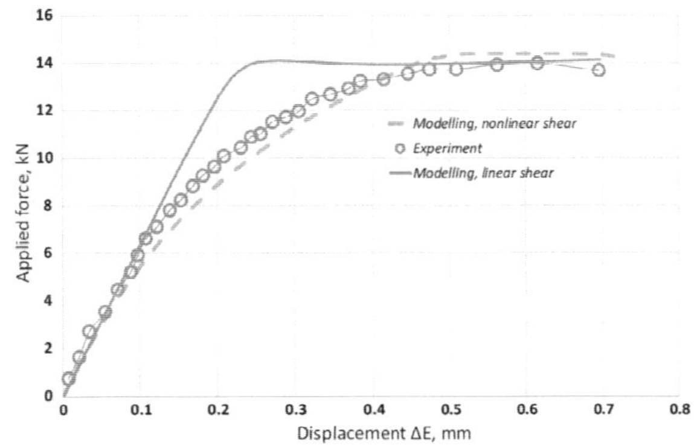

Figure 5. Predicted and experimentally obtained loading diagrams

5. Conclusions

An example of failure modeling approach utilizing only standard strength characteristics is demonstrated. It can be seen that proposed failure modeling approach based on formulated assumptions and utilizing only minimum experimental data gives a good correlation with the results of biaxial loading experiments. The approach has a block form and can be modified at all key steps such as elastic relations or first ply failure criterion.

6. Acknowledgement

This research was supported by the Russian Foundation for Basic Research (Grants no. 18-31-20026 and 17-51-52001).

References

[1] Hashin Z. Failure Criteria for Unidirectional Fiber Composites. J Appl Mech 1980;47 (2):329–334.

[2] Puck A., Schürmann H. Failure analysis of FRP laminates by means of physically based phenomenological models. Compos. Sci. Technol 1998;58(7):1045–1067.

[3] Reddy J. N., Pandey A. K. A first-ply failure analysis of composite laminates. Computers & Structures 1987; 25(3):371–393.

[4] Rotem A. The Rotem failure criterion: theory and practice. Compos Sci Technol 2002;62 (12–13):1663–1671.

[5] Tsai S. W., Wu E. M. A general theory of strength for anisotropic materials. J Compos Mater 1971;5 (1):58–80.

[6] Hashin Z., Rotem A. A fatigue failure criterion for fiber reinforced materials. J Compos Mater 1973;7 (4):448–464.

[7] Y. Wan, B. Sun, B. Gu, Multi-scale structure modeling of damage behaviors of 3D orthogonal woven composite materials subject to quasi-static and high strain rate compressions, Mechanics of Materials 2016 94: 1–25.

[8] P. Maimı´a, P.P. Camanhob, J.A. Mayugoa, A. Turona, Matrix cracking and delamination in laminated composites. Part I: Ply constitutive law, first ply failure and onset of delamination, Mechanics of Materials, 2011 43 (4): 169–185.

[9] R. M. Christensen, Compressive failure of composites using a matrix-controlled failure criterion with the kink band mechanism, Mechanics of Materials, 2000 32(9): 505–509.

[10] McCarthy C. T., McCarthy M. A., Lawlor V. P. Progressive damage analysis of multi-bolt composite joints with variable bolt-hole clearances. Composites Part B 2005;36 (4):290–305.

[11] Camanho P. P., Matthews F. L., A progressive damage model for mechanically fastened joints in composite laminates. J Compos Mater 1999;33 (24):2248–2280.

[12] Moure M. M., Otero F., García-Castillo S. K., Sánchez-Sáez S., Barbero E., Barbero E.J. Damage evolution in open-hole laminated composite plates subjected to in-plane loads, Composite Structures, 2015; (133): 1048-1057.

[13] Iarve E. V., Pagano N. J. Singular full-field stresses in composite laminates with open holes, International Journal of Solids and Structures, 2001; 38 (1, 5): 1-28
[14] Bogetti T. A., Hoppel C. P. R., Harik V. M., Newill J. F., Burns B. P. Predicting the nonlinear response and progressive failure of composite laminates. Compos Sci Technol 2004; 64(3–4):329–342.
[15] Soden P. D., Hinton M. J., Kaddour A. S. Lamina properties, lay-up configurations and loading conditions for a range of fibre-reinforced composite laminates. Compos Sci Technol 1998;58(7):1011–1022.
[16] Lomakin, E. V. Constitutive Models of Mechanical Behavior of Media with Stress State Dependent Material Properties, Mechanics of Generalized Continua. Advanced Structured Materials, 2011; (7): 339-350.
[17] Lomakin E.V., Fedulov B.N., Nonlinear anisotropic elasticity for laminate composites, Meccanica (2015) 50: 1527.
[18] Petit P. H., Waddoups M. E., A Method of Predicting the Nonlinear Behavior of Laminated Composites. J. Comp Mat 1969;(3) 2.
[19] Chamis, C. C. and Sinclar, J. H. Ten-Deg Off-Axis Shear Properties in Fiber Composites. Experimental Mechanics. 1977; 339-346.
[20] Hahn H. T., Tsai S.W. Nonlinear Elastic Behavior of Unidirectional Composite Laminae Journal of Composite Materials January. 1973 (7): 102-118.
[21] Lomakin E. V., Fedulov B. N. Nonlinear anisotropic elasticity for laminate composites, Meccanica 2015 (50):1527–1535.
[22] Hashin Z. Analysis of Composite Materials—A Survey. J Appl Mech 1983;50(3):481–505.
[23] Hinton M. J., Kaddour A. S., Soden P. D. A comparison of the predictive capabilities of current failure theories for composite laminates, judged against experimental evidence. Compos Sci Technol 2002;62(2–13):1725–1797.
[24] Icardi U., Locatto S., Longo A., Assessment of Recent Theories for Predicting Failures of Composite Laminates. Appl Mech Rev 2007;60 (2):76–86.
[25] Thom H. A review of the biaxial strength of fibre-reinforced plastics. Composites Part A: Appl Sci Manuf 1998;29(8):869–886.
[26] Garnich M. R., Venkata M. K. Akula. Review of degradation models for progressive failure analysis of fiber reinforced polymer composites. Appl Mech Rev 2009;62(1):010801-1–010801-33.
[27] Fedulov B. N., Fedorenko A. N., Kantor M. M. and Lomakin E. V.. Failure analysis of laminated composites based on degradation parameters. Meccanica, 2017;53:359–372.
[28] Fedulov B. N., Fedorenko A.N., Safonov A.A. and Lomakin E.V., Nonlinear shear behavior and failure of composite materials under plane stress conditions. Acta Mechanica. 2017 Jun 1;228(6):2033-40.

IOP Conf. Series: Materials Science and Engineering **581** (2019) 012016 doi:10.1088/1757-899X/581/1/012016

Experimental Setup for Studying Thermosolutal Convection in Moist Air

S A Somov[1,2] and A S Ivanov[1,2]

[1]Institute of Continuous Media Mechanics Ural Branch of Russian Academy of Sciences the affiliate of Perm Federal Scientific Research Center, 1 Korolev str., Perm, 614000, Russia;
[2] Federal state budgetary educational institution of higher education Perm State National Research University, 15 Bukirev str., 614990, Perm, Russia.

E-mail: somov.s@icmm.ru

Abstract. The paper is concerned with designing an experimental scheme for studying convection of moist air under conditions closely approximating the atmospheric conditions. Detailed numerical estimates, necessary conditions and dimensionless gas-dynamic parameters characteristic of the problem of dry and moist air convection are presented and analyzed. The distinctive features of convection of dry and wet air at constant pressure caused by changes in the concentration of water vapor as a result of its evaporation (condensation) are discussed. The effect of water vapor concentration on the convective instability of moist air is analyzed in the context of solutal convection, characterized by the Rayleigh concentration number. The obtained numerical estimates are used to demonstrate the possibility of differentiating and comparing between the processes of convective heat and mass transfer in humid and dry air under laboratory conditions. In future, this will allow researchers to obtain experimental data on the influence of the water-vapor phase transition on the intensity of convective motion in moist air. The structure and detailed technical characteristics of the constructed experimental setup designed for studying convection in moist air in the temperature range from 10 to 30 C are described.

1.Introduction

Convection in a gaseous environment (air) is an internal motion caused by a disturbance of mechanical equilibrium in the system aimed at recovering the equilibrium state. The processes of complex convective heat and mass transfer observed in the atmosphere of the Earth require a physical and mathematical description in terms of gas dynamic models, which must be subjected to experimental verification. The study of convection in the atmosphere is of great importance for the development and improvement of methods for weather and climate forecasting. Convection has a considerable dynamic effect on large-scale atmospheric phenomena [1,2]. This problem has been the focus of a large number of studies [4–8], the results of which make it possible to relate various types of atmospheric circulation with changes in the weather.

In the water-water vapor system, the first-order phase transition plays an important role in convection of the atmospheric moist air [3], but in most cases this phenomenon is considered in terms of thermodynamics, i.e. the computation of the latent heat of phase transition as a result of condensation / evaporation and the evaluation of the effect of the associated released / absorbed heat on the behavior of the system. On the other hand, the Rayleigh numbers (thermal) Ra_T for moist and dry air in the atmospheric temperature range (up to 50 C) are found to be close - the observation, which the researchers take into use when performing the experiments and calculations for the parameters of dry air, ignoring its humidity. Such a simplified approach is based on the fact that in the specified temperature range the volume fraction of water vapor in moist air does not exceed 15% and has proved to be valid only in the absence of phase transition. We consider it inappropriate to study heat and mass transfer in moist air solely in the framework of a thermal convection model including the elements of thermodynamics to take into account the caloric effects associated with the release / absorption of the latent heat of phase transition. We believe that this problem should also be studied in

IOP Conf. Series: Materials Science and Engineering **581** (2019) 012016 doi:10.1088/1757-899X/581/1/012016

the context of solutal convection, since heating or cooling of moist air changes the mass percent composition of the dry air-water vapor mixture due to evaporation and condensation of moisture, respectively.

It should be noted that the laboratory study of thermosolutal convection and the construction of a model of the atmosphere seems to be insolvable problem, since the atmosphere is a substantially stratified system with the characteristic size of ten kilometers and the characteristic velocity of the air mass flow of ten meters per second. It is impossible to build a laboratory model for the description of the process which would satisfy all criteria for similarity of the real Earth's atmosphere. Therefore, our study will concentrate only on a single aspect of the problem - the influence of the water-vapor phase transition on the intensity of convection in moist air.

The aim of this study is to design an experiment to investigate convection of moist air in conditions closely approximating the atmospheric conditions based on the analysis and numerical estimates of the dimensionless gas-dynamic parameters characterizing convection in dry and moist air. The distinguishing features of convection of dry and moist air caused by changes in the concentration of water vapor as a result of its evaporation / condensation are discussed.

2.Constitutive parameters

Planning of an experiment based on the approximate calculations is the most essential part of experimentalist's work. In order to conduct an experiment correctly, it is necessary to choose similarity criteria (dimensionless quantities), which form the framework for constructing a laboratory model of the examined physical system ensuring maximum possible accuracy in the description of the main features of processes occurring in the atmosphere. In practice, qualitative estimates often turn to be inaccurate in the order of magnitude leading to incorrect result interpretation, which can significantly affect the final version of the model. Therefore, the researchers tried to obtain a more accurate numerical estimate of free convection [9–11] characterized by a dimensionless Rayleigh number. Since in the problem under study, heating and concentration variation in a mixture of two gases (dry air and water vapor) occur simultaneously, it is necessary to use two independent Rayleigh numbers, which take into account both of these mechanisms, namely the thermal Ra_T and solutal Ra_C Rayleigh numbers. Their sum corresponds to the effective Rayleigh number Ra_E:

$$Ra_T = \frac{gC_p\rho^2 d^3 \Delta T}{\kappa\eta T_0},\ Ra_C = \frac{gd^3 \Delta\rho_c}{\eta D},\ Ra_E = Ra_T + Ra_C, \tag{1}$$

where d is the characteristic size of the cavity with moist air, ΔT is the temperature difference, T_0 –is the average temperature value in terms of the cavity volume, κ is the thermal conductivity, C_p is the specific heat of the gas mixture, η is the dynamic viscosity, D is the kinematic diffusion coefficient for the gas mixture, ρ is the average gas density, $\Delta\rho_c$ is the change in density caused by the change in the concentration composition (volume percent composition) of the mixture. In a non-isothermal system, local values of thermophysical parameters can vary significantly, therefore, a functional relation between the dimensionless quantities depends on which temperature and concentration is chosen as a governing factor. In the case when the medium-averaged temperature is taken as the governing factor, the temperature dependence of the parameters of a liquid or gas has inessential effect on the experimental results. The same observation applies to the concentration dependence. Therefore, the formulas for Ra_T and Ra_D, involve the average values of temperature and concentration.

To find correct values of Ra_T, Ra_C, it is necessary to use the exact values of the thermophysical parameters of gas available in the reference literature [12]. The dependence of the thermophysical properties of gases on temperature and pressure has been well studied. For water vapor there exist few experimental skeleton tables [12], but in recent studies it is common practice to use the well-known exact interpolation formulas, rather than tabular values. In what follows, we will also use semi-empirical equations, which take into account the dependence of the parameters C_p, κ, η, D, ρ on pressure, temperature and concentration pattern of the gaseous mixture of dry air and water vapor.

For pure gases and gas mixtures in the temperature range from 270 to 400 K and at pressures from 10^{-3} to 10, the Mendeleev-Clapeyron equation is fulfilled and the use of equations for an ideal gas has often proved its worth. In this case, the parameters of the state of the air and water vapor conditions (far from the phase transition) calculated within the framework of the classical theory of an ideal gas is accurate within 1-2%. With this in mind, for the density of moist air ρ it is reasonable to use the formula, which takes into account the variation in the partial pressure of water vapor P_1 and of dry air P_2 with temperature:

$$\rho = \frac{P_1 M_1 + P_2 M_2}{k_b T Na}, \tag{2}$$

where M_1 = 0.018 kg / mol and M_2 = 0.029 kg / mol is the molar mass of water vapor and dry air, respectively, k_b = 1.38 10^{-23} J / K is the Boltzmann constant, T is the gas mixture temperature (in degrees K), Na = 6.02 10^{23} mol^{-1} is the Avogadro number. The sum of partial pressures P_1 and P_2 is equal to the atmospheric pressure P_0, which is known from the experimental conditions:

$$P_1 + P_2 = P_0. \tag{2.1}$$

The partial pressure of saturated water vapor P_1 is a well-studied function of temperature [12, 13]:

$$\log_{10}\left(\frac{P_1}{P_0}\right) = 10.79586\left(1 - \frac{273.16}{T}\right) - 5.02808 \cdot \log_{10}\left(\frac{T}{273.16}\right) + 1.50474 \cdot 10^{-4}\left(1 - 10^{-8.29692\left(\frac{T}{273.16} - 1\right)}\right) + $$
$$+ 0.42873 \cdot 10^{-3}\left(10^{4.76955\left(1 - \frac{273.16}{T}\right)} - 1\right) - 2.2195983, \tag{3}$$

where the value of $[P_1]$ is measured in mbar. Thus, knowing the temperature of moist air T at the atmospheric pressure P_0, one can calculate the corresponding partial pressure of water vapor from (3), the partial pressure of dry air from (2.1) and the desired density from (2).

The kinematic diffusion coefficient $D_{1,2}$ of a binary mixture was calculated using the Hirschfelder formula [18]:

$$D_{1,2} = \frac{BT^{3/2}}{P_0 \sigma_{1,2}^2 \Omega_{1,2}^{(1,1)}} \left(\frac{M_1 + M_2}{M_1 M_2}\right)^{1/2}, \tag{4}$$

$$B = 0.00214 - 0.000492\left(\frac{M_1 + M_2}{M_1 M_2}\right)^{1/2}, \tag{5}$$

$$\Omega^{(1,1)} = 1.075\left(\frac{k_b T}{\varepsilon_{1,2}}\right)^{-0.01615} + 2\left(\frac{10 k_b T}{\varepsilon_{1,2}}\right)^{-0.74 \cdot \log\left(\frac{10 k_b T}{\varepsilon_{1,2}}\right)}, \tag{6}$$

where B is the refined Wilke coefficient [19], $[D_{1,2}]$ is measured in cm^2 / s, the mixture pressure $[p]$ is measured in atm, $\varepsilon_{1,2}$ and $\sigma_{1,2}$ are constant forces in the Lennard-Jones potential, describing the molecular interaction in gas. The constants of interactions $\varepsilon_{1,2}$ and $\sigma_{1,2}$ were calculated by the method [14], using the tabulated values for the components of moist air: ε_1 / k_b = 380 K, σ_1 = 0.265 nm and ε_2 / k_b = 97.0 K, σ_2 = 0.3617 nm for molecules of water and dry air, respectively. The constant forces $\varepsilon_{1,2}$ and $\sigma_{1,2}$ were calculated using the formulas for a mixture of polar (water vapor) and non-polar (dry air) gases:

$$\sigma_{1,2} = 0.5(\sigma_1 + \sigma_2)\, \xi^{-1/6} = 0.3103 \text{ nm},$$
$$\varepsilon_{1,2} / k_b = (\varepsilon_1 \varepsilon_2)^{0.5} \xi^2 / k_b = 215.8 \text{ K}, \tag{6.1}$$
$$\xi = 1 + 2^{-0.5}(\alpha_2^*)(t_1^*)(\varepsilon_1/\varepsilon_2)^{0.5} = 1.0603,$$

where t_1^* = 1.2 is the dimensionless Stockmayer parameter for a water molecule, and $\alpha_2^* = \alpha_2 / \sigma_2^3 = 0.03592$ is the dimensionless polarizability of an air molecule. The parameter t_1^* characterizes the difference in the behavior of polar and non-polar substances. The relationship $D_{1,2}(p, T)$ calculated by formula (4) is described by the fitting function with accuracy of less than 0.5%:

$$D_{1,2} = (p^{-1})(0.2177 + 1.5562 \cdot 10^{-3}\, T + 2.2813 \cdot 10^{-6}\, T^2).$$

The approximate calculations were carried out on the assumption that the coefficient $D_{1,2}$ does not depend on the relative concentration of the components, since the ratio of water vapor to dry air at any temperature realized in the experiment does not exceed 1/8. The validity of this statement can be demonstrated by comparing our results with the known data for other gas mixtures. Thus, according to experimental data [14] for a mixture of H_2 and CO_2, a change in the molar ratio from 3/1 to 1/1 leads to an increase in $D_{1,2}$ by only 1.8%. It is obvious that in our case the dependence of $D_{1,2}$ on the percentage ratio of the mixture components will be an order of magnitude less.

To calculate the dynamic viscosity, heat capacity, and thermal conductivity, we have developed semi-empirical formulas by approximating the data from the skeleton tables of the National Standard Reference Data System NSRDS 125-88 [20] for moist air. The calculation of these thermophysical parameters in the temperature range of 270-400 K and pressures of 10^{-3} – 10^1 is independent of pressure at a relative humidity of φ = 100% (the calculation error does not exceed 1%):

$$\eta = -2.45614 \cdot 10^{-15} t^5 + 2.56583 \cdot 10^{-13} t^4 - 1.32329 \cdot 10^3 t^3 + 2.17221 \cdot 10^{-10} t^2 +$$

$$+4.59113\cdot10^{-8}t+1.72539\cdot10^{-5}, \quad (7)$$

$$C_p = 1785.60629\text{-}372.25007\,t+68.79781\,t^2\text{-}6.57854\,t^3+0.36963\,t^4\text{-}0.012978\,t^5+$$

$$+0.00029\,t^6\text{-}4.22899\cdot10^{-6}t^7, \quad (8)$$

$$\kappa = -3.77888\cdot10^{-12}t^5+5.97899\cdot10^{-10}t^4-3.80594\cdot10^{-8}t^3+1.06486\cdot10^{6}t^2+$$

$$+5.88323\cdot10^{-5}t+0.02435, \quad (9)$$

The temperature scale $[t]$ is Celsius degree (C), viscosity $[\eta]$ – Pa·s, heat capacity $[C_p]$ – J / (kg·K), thermal conductivity $[\kappa]$ – W / (m·K).

The known thermophysical parameters of moist air (2-9) allow us to calculate the convective factor k_c of the Rayleigh number (1)

$$Ra_E = k_c d^3 \Delta T. \quad (9.1)$$

It has been known that the onset of convection corresponds to the critical value of the Rayleigh number $Ra^* = 1700 \pm 50$ [17], which must be substituted into (9.1) in order to find the main parameters of the experimental setup - the characteristic size of the experimental cell d and the temperature difference ΔT between the cooler and the heater. The order of magnitude of the quantities entering into (9-1) at $Ra_E = Ra^*$ is

$$d \sim 10^{-2}\ \text{m}, \quad \Delta T \sim 1\ \text{K}.$$

It is obvious that the characteristic scale of d and ΔT makes it difficult to experimentally study convection in moist air and would necessitate searching for a compromise solution. Due to the cubic dependence $Ra_E \sim d^3$, even a slight (desirable) increase in the size of the experimental cell d leads to a sharp increase in Ra_E, while a hypothetical decrease in the temperature difference ΔT is limited by the ability of the experimental equipment to control and maintain constant temperatures at the cooling and heating units with acceptable accuracy (~0.1 K). To determine the optimal compromise values of d and ΔT, we constructed a table (see Table 1) and a graph (Fig. 1).

The final combination of the values for the parameters d and ΔT will be determined after analyzing the potentials of the proposed experimental technique in the next section of the paper.

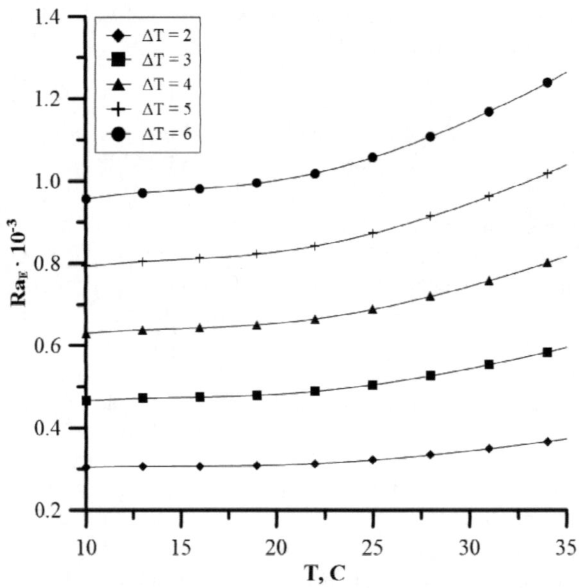

Fig. 1. The plot of the effective Rayleigh number versus temperature for cavity height d = 1 cm. The average temperature T_0 varies in the range from 15 to 30 C, since exactly in this temperature range it is easier to maintain a stable temperature difference ΔT.

Table 1. Desired cell size for the onset of convection

	d, mm				
T_0, C	ΔT=2 C	ΔT=3 C	ΔT=4 C	ΔT=5 C	ΔT=6 C
15	17.7	15.3	13.8	12.8	12.0
20	17.6	15.2	13.7	12.7	11.9
25	17.4	15.0	13.5	12.5	11.7
30	17.0	14.6	13.2	12.2	11.4

3. Experimental setup

Let us select a suitable experimental technique for the observation and study of free convection in moist air. As far as we know, the most widely used standard methods for studying the phenomena of heat and mass transfer in transparent media are as follows: the shadow method, the interferometric method using the Fabry-Perrot scheme, and holographic interferometry [15, 16]. The use of classical shadow and interference methods is complicated by insufficient sensitivity at small temperature differences ΔT (which is desirable) and low sensitivity due to a small coherence length of the laser radiation (for the most widely used He-Ne laser, the coherence length is approximately 20 cm). For these reasons, we chose the method of holographic interferometry. This method allowed us to visualize the distribution of the heat and concentration fields in the objects of complicated configuration in real time. A significant advantage of the holographic interferometer is the lack of requirements to the coherence length of the laser radiation and the quality of optical components (lenses and mirrors).

To study convection in moist air, we have developed and constructed an experimental setup shown in Fig. 2.

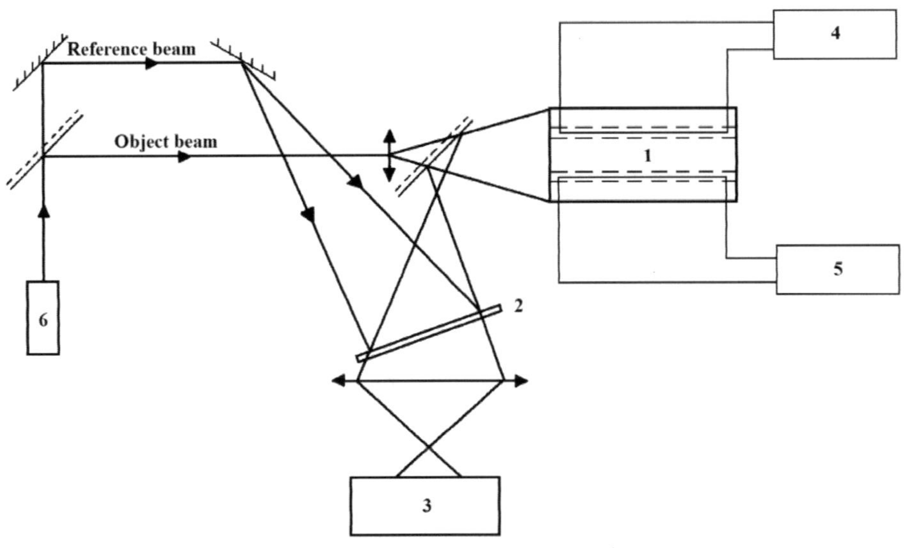

Fig. 2. Schematic representation of the experimental setup: 1 - measuring cell filled with moist air; 2 - holographic plate; 3 - video camera; 4 - heat exchanger and cooler thermostat, 5 - heat exchanger and heater thermostat 6 - He-Ne laser

Fig. 2 elucidates the operation of the setup. The source of coherent radiation in the setup is a He-Ne laser with a wavelength of 632.8 nm. Using translucent mirror the laser beam is split into two beams (reference and object beams). The reference beam after it has been reflected several times from the system of mirrors falls on a holographic photographic plate, where it interferes with the object beam, which passes twice through the measuring cell with moist air. The interferometric pattern obtained on a holographic plate is recorded by a video camera for further image processing and interpretation of experimental observations.

However, the acquisition of useful information on convection in moist air is possible only in the case when the experimental setup has sufficient sensitivity. The sensitivity and resolution of the installation (Fig. 2) ultimately depends on whether the test medium (inhomogeneously heated moist air) filling the measuring cell is able to change significantly the difference in the optical path length (the optical path difference) between the object beam and the reference beam. The greater the path difference of the beam, the more interference fringes can be observed - the higher the resolution of the setup, and eventually, the more detailed picture of the gas-dynamic flow can be obtained. It has been found that to characterize the processes occurring in the object under study, it is necessary to obtain at least five interference fringes (i.e. ten alternations of dark and light bands).

So, the next step is to obtain numerical estimates, which will allow us to determine what dimensions the measuring cell should have in order to ensure the required resolution of the setup with the known ΔT. To this end, we consider the dependence of the refractive index of light on two parameters: temperature and concentration of water vapor:

$$n(T,C) = n_0 + \frac{dn}{dT}(T - T_0) + \frac{dn}{dC}(C - C_0), \tag{10}$$

where n_0 is the refraction index at $T = T_0$ and $C = C_0$ is the concentration of water vapor at $T_0 = 20$ C. The relation between the refraction index of the mixture of gases and its density, and, consequently, its concentration is determined by the Lorentz-Lorentz equation [16]

$$\frac{n^2 - 1}{n^2 + 2} = \sum_i R_i \rho_i, \tag{11}$$

where R_i is the specific refraction, ρ_i is the specific density. For moist air equation (11) takes the following form:

$$\frac{n^2 - 1}{n^2 + 2} = R_1\rho_1 + R_2\rho_2, \tag{12}$$

in which R_1 and R_2 are the specific refraction of dry air and water vapor, ρ_1 and ρ_2 – are the specific densities of these gases, respectively.

To find the constant refractive index R_1 and R_2, we transform the expression (11):

$$R_i = \frac{n^2 - 1}{(n^2 + 2)\rho_i}, \tag{13}$$

The first component can be calculated by the formula:

$$(n_1 - 1)\cdot 10^8 = 8340.78 + \left(\frac{2405640}{\left(130 - \frac{1}{\lambda_{vac}^2}\right)}\right) + \left(\frac{15994}{(38.9 - \frac{1}{\lambda_{vac}^2})}\right), \tag{14}$$

$$\rho\left(\frac{g}{m^3}\right) = 348.328\,\frac{P}{T}\left[1 + P\left(57.90\cdot 10^{-8} - \frac{0.94581\cdot 10^{-8}}{T} + \frac{0.25844}{T^2}\right)\right] \quad , \tag{15}$$

At the gas temperature of 150 C, the pressure of 1013.25 mbar and $\lambda_{vac} = 632.8$ nm, the specific refraction of dry air is $1.48 \cdot 10^{-7}$. The second component is calculated by the formula [12,13]:

$$(n_1 - 1)\cdot 10^8 = 295.235 + 2.6422\frac{1}{\lambda_{vac}^2} - 0.032380\frac{1}{\lambda_{vac}^4} + 0.004028\frac{1}{\lambda_{vac}^6}, \tag{16}$$

$$\rho\left(\frac{g}{m^3}\right) = 216.582\,\frac{P}{T}\left[1 + P^2(1 + (3.7\cdot 10^{-4}))\left(-2.37321\cdot 10^{-3} + \frac{2.23366}{T} - \frac{710.792}{T^2} + \frac{7.75141\cdot 10^4}{T^3}\right)\right], \tag{17}$$

At the temperature of 200 C, pressure of 13.33 mbar, and $\lambda_{vac} = 468$ nm, the specific refraction of water vapor is $1.99 \cdot 10^{-7}$. The refraction index at arbitrary temperature and concentration of molecules in the gas mixture is corrected according to the Mendeleev-Clapeyron equation and (2). Table 2 shows the values of dn/dT and dn/dC in the temperature range from 15 to 30 C.

Table 2. Temperature and concentration contributions to the refraction index depending on temperature

t, C	$dn/dT \cdot 10^7$	$dn/dC \cdot 10^7$
10	9.8	5.9
15	9.4	5.8
20	9.1	5.7
25	8.8	5.6
30	8.5	5.5

As is evident from Table 2, the derivatives of the refraction index with respect to concentration and temperature have the same order of magnitude. Hence, when calculating the resulting value of $n(T)$, in equation (10) it is necessary to take into account both these terms, since the contributions of both terms to the final sensitivity of the interferometer are equivalent. Therefore, the cell length L, which is necessary for obtaining 5 interference fringes (as mentioned earlier) is calculated based on the relationship between the optical path difference L Δn and the number of wavelengths N by the formula

$$L = \frac{N \cdot \lambda}{2\Delta n} \approx 0.32 \text{ m}, \tag{18}$$

where Δn is a change in the refraction index of moist air caused by the temperature difference ΔT between the cooler and the heater, λ is the He-Ne laser wavelength of 632.8 nm, $N = 5$ is the required number of interference fringes. The coefficient 2 in the denominator (18) implies that the laser beam passes the cell twice. Here, we again face the problem of finding a compromise solution associated with increasing ΔT (in order to increase the number of interference fringes N and reduce the length of the installation L) and decreasing the height of the cell d (9-1). With the aim of determining the optimal compromise values for L, ΔT and d we have constructed a table (see Table 3).

Table 3. Required cell length L for obtaining $N = 5$ interference fringes

	L, cm				
T, C	$\Delta T = 2$ C	$\Delta T = 3$ C	$\Delta T = 4$ C	$\Delta T = 5$ C	$\Delta T = 6$ C
15	53.1	31.8	22.7	17.7	14.5
20	54.1	32.4	23.2	18.0	14.7
25	54.8	32.9	23.5	12.5	14.9
30	55.3	33.2	23.7	18.4	15.1

Summarizing the data of tables 1 and 3, we may conclude that for the experimental study of convection in moist air it is necessary that the experimental setup should have the following optimal parameters: maintained temperature difference $\Delta T = 3$ K, cavity height $d = 15$ mm, cavity length $L = 32$ cm.

These parameters were used to develop a computer model of the experimental setup with the aid of three-dimensional design program Kompas-3D. The laboratory model was constructed according to its computer version.

In addition to the visual study of convection using holograms, the experimental setup allows taking measurements of heat and mass transfer in the experimental cell by means of embedded thermocouples (not shown in Fig. 3).

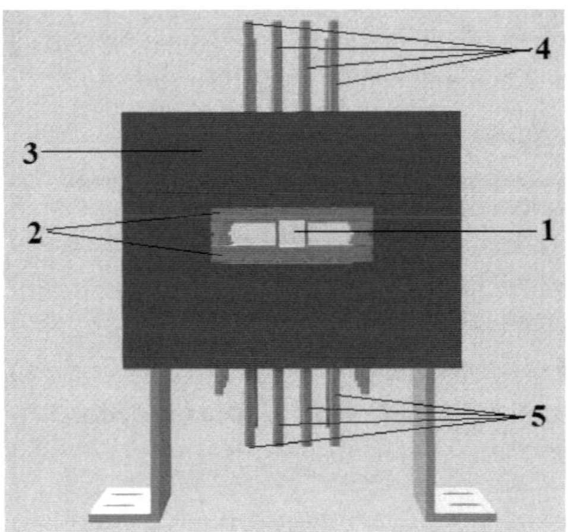

Fig. 3 Model of the experimental setup for studying convection in moist and dry air:
1) convective cell, 2) heat exchangers, 3) setup casing, 4) heat transfer pipes of the cooler, 5) heat transfer pipes of the heater

4.Conclusion

A review of the specialized literature devoted to the numerical simulation of the thermophysical properties of moist air has been made. The numerical estimates of the effective Rayleigh number characterizing the thermal convection of moist air have been obtained. The numerical estimates substantiate the possibility of distinguishing and comparing the convective heat and mass transfer in wet and dry air under laboratory conditions. The numerical calculations have been performed to determine the optimal technical parameters of the laboratory experimental setup designed for studying convection by the method of holographic interferometry in moist air in the temperature range from 10 to 30 C.

References

[1] Lambaerts J, Lapeyre G and Zeitlin V 2011 *J. Atmos. Sci.* **68** 1234
[2] Yano J, Geleyn J F, Köhler M, Mironov D, Quaas J, Soares P M M, Phillips V T J, Plant R S, Deluca A and Marquet P 2015 Basic concepts for convection parameterization in weather forecast and climate models *J. Atmosphere* **6** 88
[3] Shmerlin B Y and Kalashnik M V 2013 *Phys. Uspekhi* **56** 473
[4] Chandrasekhar S 1961 *Hydrodynamic and Hydromagnetic Stability* (UK: Oxford)
[5] Martynenko O G, Khramtsov P P 2005 *GemanyFree-Convective Heat Transfer* (Berlin: Springer)
[6] Ogura Y 1963 *Meteorol. Monogr.* **5** 65
[7] Arakawa A and Jung J H 2011 *Atmos. Res.* **102** 263
[8] Derbyshire S H, Beau I, Bechtold P, Grandpeix, J Y, Piriou J M, Redelsperger J L and Soares P M 2004 *Q. J. R. Meteorol. Soc.* **130** 3055
[9] Alekseev V V and Gusev A M 1983 *Sov. Phys. Uspekhi* **26** 906
[10] Rayleigh L 1916 *Philos. Mag.* **32** 529
[11] Ginsburg T I, Guliayev V T and Zhalkhovskaya LV 1987 *Dynamic models of the free atmosphere* (Novosibirsk: Nauka) p 290
[12] Vukalovich M P 1955 *Thermodynamic properties of water and water vapor* vol 5 (Moskow: Mashinostroenie) p 92
[13] Goff J A and Gratch S 1946 *Trans. Am. Soc. Heat. Vent. Eng* **52** 95
[14] Bretschnider S 1966 *Properties of fluids and gases* (Leningrad: Himiy) p 536
[15] Hauf V and Grigul U 1973 *Optical methods in heat transfer* (Moscow: Mir) p 242
[16] Ioffe B 1983 *Refractometric methods in chemistry* (Leningrad: Mir)
[17] Gershuni G and Zhukhovotskiy E 1972 *Convective stability of an incompressible fluid* (Moskow: Nauka)
[18] Hirschfelder J O, Curtiss Ch F and Bird R B 1954 *Molecular theory of gases and liquids* (New York: Wiley) p 1219
[19] Wilke C R and Lee C J 1955 *Ind. Eng. Chem.* **47** 1253
[20] Nevenkin S L, Nachev N I, Popov P V and Kuznetsov V M 1989 *Tables of standart reference data. Moist air. Thermophysical properties in the temperature range 5...95 C and at pressure 99325 Pa* (Bulgaria: Sofia higher machine-electrical engineering Institute)

IOP Conf. Series: Materials Science and Engineering **581** (2019) 012017 doi:10.1088/1757-899X/581/1/012017

Viscosity of Magnetic Fluid in Oscillation System in a Strong Magnetic Field

V M Polunin, P A Ryapolov[a], A I Zhakin, E V Shel'deshova

South-West State University, Kursk, Russia

E-mail: [a] r-piter@yandex.ru

Abstract. The article gives an assessment of viscosity and its increment ('magnetoviscous' effect) in a thin near-wall layer of the column of magnetic fluid that oscillates in the tube when applying a strong transverse magnetic field. Viscosity is calculated according to a formula derived from two different theoretical approaches. For calculation, previously published experimental results, which were discussed earlier on the assumption that there is no field dependence of viscosity, are used in the article. A comparative analysis of the "near-wall viscosity" estimates obtained using dynamic elasticity as well as the magnetization curve and the coefficient of static elasticity is carried out. The significance of the results obtained for 'magnetoviscous' effect for performing diagnosis of interparticle aggregation is indicated.

1. Introduction

Under certain conditions, in magnetic fluids (MFs), intense interaction of magnetic nanoparticles occurs with the formation and disintegration of aggregates, which leads to changes in their physical properties. Studying such processes in MFs is a difficult task, since conventional techniques for studying microscopic processes using electron or atomic force microscopes are applied under static conditions, while the aggregation processes are determined by both MF dynamics and magnetic fields structure. Therefore, studying aggregation processes in dynamic conditions is a topical task. Undoubtedly, when solving this task experimental and theoretical researches of the changes of MF viscosity in a magnetic field can be useful and productive.

[1, 2] present measurements and a theoretical analysis of the oscillations of MF-column confined by magnetic levitation in a tube in a strong magnetic field. Moreover, the calculations carried out using the ponderomotive elasticity model with the correction applied for the resistance of a moving viscous fluid are compared with the experimental magnetization curve. However, in the theoretical model, the MF viscosity invariance is assumed to be one of the approximations [2]. Consequently, the issue on the 'magnetoviscous' effect – the viscosity increment in the MF thin near-wall layer when applying a magnetic field – remains an open one.

In a shear flow, a moment of forces acts on a solid particle, which leads to its rotation. Magnetic field orients the magnetic moment of the particle and, if there is relation between the moment of the particle and the particle, hinders its free rotation. This leads to local gradients of the velocity of base-fluid near the particles and causes an increase of MF effective viscosity [3-7]. Saturation of the so-called 'rotational' viscosity occurs when a strong field rigidly orients the particles. The actual viscosity increment in strong magnetic fields for a magnetite sample with the volume concentration of 0.19–0.24 in Poiseuille flow through a capillary in a perpendicular magnetic field does not exceed 5–6% [8].

The lack of information on 'magnetoviscous' effect does not allow obtaining a comprehensive picture of the physical mechanisms of oscillatory motion of magnetic-fluid active elements in various

IOP Conf. Series: Materials Science and Engineering **581** (2019) 012017 doi:10.1088/1757-899X/581/1/012017

engineering devices and makes it difficult to draw on the acquired methodological experience to expand the means of controlling wear resistance and the magnetic colloid consumption.

In the present study, which is the continuation of work on low-frequency oscillations of MF in strong magnetic fields [1, 2], the 'magnetoviscous' effect in a thin near-wall layer when applying a strong transverse magnetic field is determined to expand physical representation of MF oscillatory flow in a magnetic field and to diagnose interparticle aggregation. The value of the 'magnetoviscous' effect was calculated by a formula derived from two different theoretical approaches. An array of published experimental results of comprehensive measurements of the oscillation frequency, oscillation damping coefficient, saturation magnetization and static displacement of MF-column for two samples of magnetic fluid discussed earlier on the assumption that there is no field dependence of viscosity is used in this work [2].

2. The samples parameters and the experiment brief description of the previously published srtidies

In relation to the task set, it is necessary to describe briefly the experimental setup and measurement procedure. The experimental setup designed to measure MF-column oscillations frequency was described in detail in [2] (Figure 1. Block diagram of the experimental setup No 1). In the work, we used a laboratory electromagnet FL-1; a tube with an internal diameter d = 12 mm made of plexiglass was placed between the pole tips of the electromagnet. The tube axis passes horizontally (or vertically) through the centre of the pole gap and is parallel to the surface of the pole tips. The axis OZ coincides with the axis of the tube, and its beginning is in the centre of the MF-column in the equilibrium.

A significant circumstance of the obtained dependences of the transverse value of the magnetic field $H_x(z)$ is the presence of a linear section on the curves at the level of z = 57.5 mm, which makes it possible to consider the magnetic field intensity gradient in this area $\Delta H_x / \Delta z$ = const. In a strong and inhomogeneous magnetic field, MF-column takes a form close to a cylindrical one. In the conducted experiment, the distance between the bases of the cylinder is b=115 mm. In the view of the assumptions concerning the tube 'thinness', magnetic field at the points of the MF free surface is tangential to it, i.e. it has only the tangential component H_x (perpendicularly to the surface of the pole tips) and the magnetic field intensity gradient is directed perpendicularly to the surface along the OZ axis to its beginning. The maximum magnetic field in the centre between the poles of the electromagnet is 900 kA/m.

Method of static displacement of the MF-column is also considered in [2]; the essence of this method is that static pressure is applied to the MF-column. Figure 2. shows a block diagram of the experimental setup No. 2. In this case, to create a hydrostatic pressure, a communicating vessel in the form of a U-shaped tube filled with distilled water is used and it is connected with a MF-column by a flexible pipe.

The samples of magnetic fluids MF-1 and MF-2 are synthesized on the basis of finely dispersed magnetite – Fe_3O_4 which is stabilized by a surfactant – oleic acid $C_8H_{17}CH{=}CH(CH_2)_7{-}COOH$. In the sample MF-1, aviation kerosene TS-1 was used as a dispersive medium – a carrier liquid; in the sample MF-2, undecane $C_{11}H_{24}$ (a hydrocarbon from the alkane class) was carrier liquid. The objects of the research were synthesized in the Fundamental Scientific Research Laboratory of Applied Ferrohydrodynamics of Ivanovo State Power Engineering University.

The densities of the samples MF-1 and MF-2 and their saturation magnetization are ρ = 1,245 kg/m^3 and ρ = 1,227 kg/m^3, M_s = 39.5 kA/m and M_s = 40.4 kA/m, respectively. The shear viscosity η is measured using Brookfield DV2T viscosimeter; its values for MF-1 and MF-2 is 34.8 mP·s and 30.4 mPa·s respectively at a shear rate of 79.200 1/s.

3. Theoretical justification

Within the framework of continuum mechanics, i.e. neglecting the dispersion of the system and the interaction of the dispersed particles with each other, the equation of damped harmonic oscillations can be written in the form:

$$M_\omega \frac{d^2\xi}{dt^2} + r'' \frac{d\xi}{dt} + k_p \xi = 0 \ , \tag{1}$$

where ξ is the displacement from the equilibrium position of the centre of gravity of the MF-column in the tube, r'' is the fluid active resistance on the inner surface of the tube, and k_p is the ponderomotive elasticity coefficient .

An expression for the ponderomotive elasticity coefficient k_p is obtained in [1]:

$$k_p = \mu_0 \frac{\pi d^2}{2}\left(M_x \frac{\partial H_x}{\partial z}\right)_{z=b/2}, \tag{2}$$

in which μ_0 is the magnetic constant, M_X and $\partial H_x / \partial z$ are the magnetization and the magnetic field intensity gradient at the MF-column base, respectively, d is the tube diameter, and b is the length of the MF-column.

An expression for fluid active resistance r'' is given in [9]:

$$r'' = \pi d b \sqrt{\rho \eta \omega / 2} \tag{3}$$

Expression (3) was first obtained by Helmholtz. The limitation is the ratio of the tube circumference πd to the length of the viscous wave $\lambda' = 2\pi\sqrt{2\eta / \rho\omega}$ at which it exceeds 10. In the case under consideration, at the maximum frequency of 15 Hz applied in the experiment, this ratio is ~7.5.

In equation (1), M_ω can be represented as follows [9]:

$$M_\omega = m\left(1 + \frac{2}{d}\sqrt{\frac{2\eta}{\rho\omega}}\right). \tag{4}$$

where m is the mass of fluid in the tube.

The second term in the brackets of formula (4) is relatively small (≈0.1); if it is multiplied by the mass of the MF, it represents the so-called 'added mass' determined by fluid viscosity.

Let us rewrite the equation of damped oscillations (1) in the standard form:

$$\frac{d^2\xi}{dt^2} + 2\beta\frac{d\xi}{dt} + \omega_0^2\xi = 0 \tag{5}$$

where $2\beta = r'/M_\omega = r' / m\left(1 + \frac{2}{d}\sqrt{\frac{2\eta}{\rho\omega}}\right) = 2\beta_{HE} / \left(1 + \frac{2}{d}\sqrt{\frac{2\eta}{\rho\omega}}\right)$, $\quad \beta_{HE} = \frac{1}{d}\sqrt{\frac{2\eta\omega}{\rho}}$.

In case of transition of $\omega \rightarrow \nu$:

$$\beta_{HE} = \frac{2}{d}\sqrt{\frac{\pi\eta\nu}{\rho}} \tag{6}$$

Expression (6) obtained on the basis of the Helmholtz formula (3) allows calculating oscillations damping coefficient.

In equation (5)

$$\omega_0^2 \equiv k_p / M_\omega . \tag{7}$$

The general solution of equation (5) has a known form $\xi = Ce^{-\beta t}\cos(\omega t + \psi)$, where C, ψ are arbitrary constants, β is the oscillations damping coefficient and the oscillation frequency ω is expressed as follows:

$$\omega = \sqrt{\omega_0^2 - \beta^2}\,, \tag{8}$$

After algebraic transformations, relation (8) takes the form:

$$\omega^2 = \frac{k_p}{m\left(1 + \frac{2}{d}\sqrt{2\eta/(\rho\omega)}\right)} - \frac{\beta_{HE}^2}{\left(1 + \frac{2}{d}\sqrt{2\eta/(\rho\omega)}\right)^2}. \tag{9}$$

or

$$4\pi^2\nu^2 = \frac{4k_p}{\pi\rho b d^2\left(1 + \frac{2}{d}\sqrt{\eta / \pi\nu\rho}\right)} - \frac{\beta_{HE}^2}{\left(1 + \frac{2}{d}\sqrt{\eta / \pi\nu\rho}\right)^2} \tag{10}$$

Neglecting the second term in the right-hand side of (10) and taking into account (2), we obtain:

$$\pi^3 v^2 d^2 b\rho + 2db\sqrt{\pi^5 \eta\rho v^3} = \mu_0 \frac{\pi d^2}{2}\left(M_x \frac{\partial H_x}{\partial z}\right)_{z=b/2} \tag{11}$$

There is another approach to solving this problem; it is based on the application of the law of energy conservation in the oscillatory system [2]. As oscillations occur, the oscillatory energy of MF-column with ponderomotive elasticity decreases. Let us take the value of kinetic energy as maximum E_{k0} at the initial time. Taking into account the formula for energy dissipation due to the reciprocating flow of a viscous fluid column in a tube [10], the potential energy of the oscillation motion after a quarter period can be represented as follows:

$$\frac{k_p \Delta z_{01}^2}{2} = \frac{k_p \Delta z_0^2}{2} - \frac{\pi^2 bd}{2}\sqrt{\frac{\omega^3 \eta\rho}{2}} \cdot \frac{\Delta z_0^2}{2}. \tag{12}$$

where Δz_0 and Δz_{01} are displacement amplitudes at the initial moment and after a quarter period.

If we make a transition in the left side of (12): $\Delta z_{01} \to \Delta z_0$ and $k_p \to (k_p + \delta_\eta)$, and require that the equality was not formally upset, we get:

$$\frac{(k_p + \delta_\eta)\Delta z_0^2}{2} = \frac{k_p \Delta z_0^2}{2} - \frac{\pi^2 bd}{2}\sqrt{\frac{\omega^3 \eta\rho}{2}} \cdot \frac{\Delta z_0^2}{2}, \tag{13}$$

In this case, δ_η is the correction for ponderomotive elasticity coefficient, resulting from the viscous fluid flow:

$$\delta_\eta = -\frac{\pi^2}{2\sqrt{2}} \cdot bd\sqrt{\omega^3 \eta\rho} \tag{14}$$

However, the formula obtained for δ_η gives an 'upper' estimate (the maximum value after a quarter period). The viscous elasticity coefficient k_η is the 'half-period average' value of the harmonic function of the maximum value, i.e. $k_\eta = \frac{2}{\pi}\delta_\eta$, therefore:

$$k_\eta = -\frac{2}{\pi} bd\sqrt{\pi^7 v^3 \eta\rho} \tag{15}$$

In addition, it was thought previously that MF viscosity does not depend on the magnetic field intensity. Assuming such a dependence, let us make a replacement of η by η_H:

$$k_\eta = -2bd\sqrt{\pi^5 v^3 \eta_H \rho} \tag{16}$$

Then equation (17) in [2] can be written as follows:

$$\mu_0 \frac{\pi d^2}{2}\left(M_x \frac{\partial H_x}{\partial z}\right)_{z=b/2} = \pi^3 v^2 \rho b d^2 + 2bd\sqrt{\pi^5 v^3 \eta_H \rho} \tag{17}$$

Equation (17) coincides with expression (11) obtained above. After elementary algebraic operations with equation (17), it is easy to obtain a formula for calculating viscosity:

$$\eta_H = \frac{1}{v^3}\left[\frac{\mu_0 d \cdot M_x}{4b\pi\sqrt{\pi\rho}} \cdot (\partial H_x / \partial z)_{z=b/2} - \frac{\sqrt{\pi\rho} \cdot d \cdot v^2}{2}\right]^2 \tag{18}$$

Under the action of hydrostatic pressure, due to the difference in water levels in the elbows of the U-shaped tube, the MF column shears to equilibrium:

$$\rho_w g\Delta h = 4k_s \Delta z / \pi d^2 \tag{19}$$

where k_s is the static elasticity coefficient, ρ_w is water density, Δh is the difference in water levels in the elbows of the U-shaped tube, Δz is the MF-column displacement under hydrostatic pressure, g is the gravity factor.

The coefficient of static elasticity k_s is a numerical equivalent of the coefficient of ponderomotive elasticity k_p and it can be used to determine the force impact of inhomogeneous magnetic field on a magnetized fluid.

Let us obtain from (19):

$$k_s = \pi d^2 \rho_w g(\Delta h / 4\Delta z) \tag{20}$$

Equation (17) can be rewritten as follows:

$$\pi d^2 \rho_w g(\Delta h / 4\Delta z) = \pi^3 v^2 \rho b d^2 + 2bd\sqrt{\pi^5 v^3 \eta_H \rho} \tag{21}$$

Then the formula for calculating viscosity in the fluid near-wall layer is as follows:

$$\eta_H = \left[\frac{d\rho_w g(\Delta h / \Delta z)}{8b\sqrt{\pi^3 v^3 \rho}} - \frac{d\sqrt{\pi v \rho}}{2} \right]^2 \tag{22}$$

4. The results of calculations

Figures 1 and 2 show the graphs of dependences of magnetization of fluids MF-1 and MF-2 on H^{-1} in the vicinity of magnetic saturation, taken from the article [2]. Experimental data of M are obtained for a magnetic field of ≤ 750 kA/m. The circles in the figures indicate magnetization values corresponding to the reciprocal magnitude of the magnetic field $1/H_0$: 1/800 m/kA and 1/900 m/kA. The crosses show the magnetization values corresponding to the reciprocal of the magnetic field intensity at the edges of MF-column $1/H_*$: 1/619 m/kA and 1/686 m/kA. The formulas in the figures, which analytically reflect the linear approximation of the dependence under consideration, allow us to calculate the numerical values of M_x and obtain these values M_x' and M_x'' for the values of $1/H_0$ and $1/H_*$. Taking into account the obtained values of M_x using formula (18) the viscosity values η_H' and η_H'' in the near-wall layer in a strong transverse magnetic field are calculated.

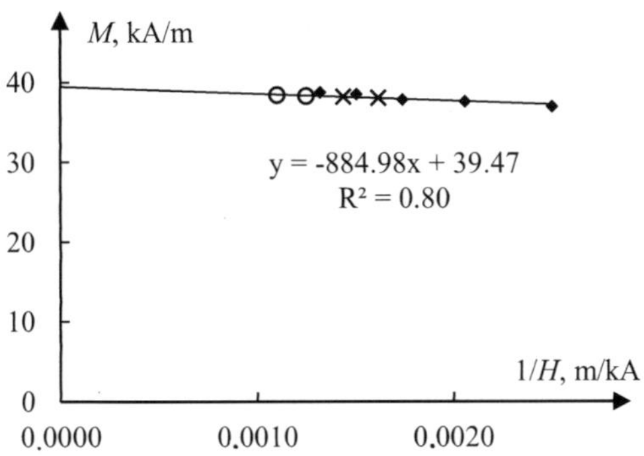

Figure 1. The dependence of M on $1/H$ for sample MF-1. Block diagram of the experimental setup.

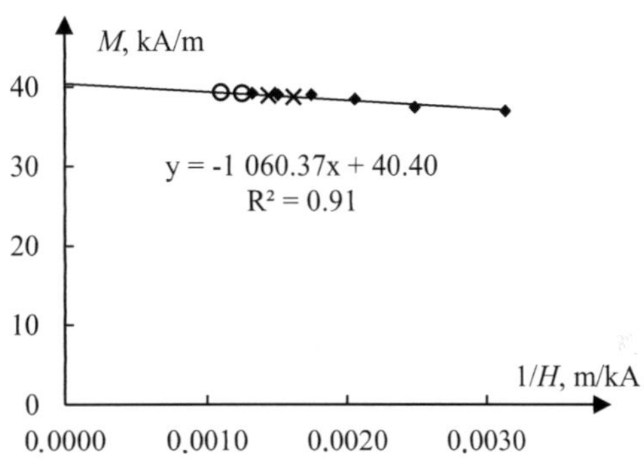

Figure 2. The dependence of M on $1/H$ for sample MF-2

The obtained results for η_H' give an 'upper' estimate, since in this case the values of M_x chosen for the maximum magnetic field in the interpolar gap H_0 are used. At the same time, the results η_H'' give a 'lower' estimate, since they are calculated for the minimum value of the magnetic field intensity related to the boundaries of the MF-column.

Table 1 gives the applied combination of values of magnetic field parameters: H_0 is the magnetic field intensity in the center between the electromagnet poles, H_* is the magnetic field intensity and $\Delta H_x/\Delta z$ is the intensity gradient at the base of the MF-column. It also shows the oscillation frequency of the MF-column v, M_x', M_x'', the calculated values η_H', η_H'', the increments of viscosity in the magnetic field ('magnetoviscous effect') $\Delta\eta_H'$, $\Delta\eta_H''$, and the relative values of these parameters $\Delta\eta_H'/\eta$, and $\Delta\eta_H''/\eta$.

Table 1. Physical parameters

Sample	H_0, kA/m	H_*, kA/m	$\Delta H_x/\Delta z$, MA/m^2	ν, Hz	M'_x, kA/m	M''_x, kA/m	η'_H, Pa·s	η''_H, Pa·s	$\Delta\eta'_H$, Pa·s	$\Delta\eta''_H$, Pa·s	$\Delta\eta'_H/\eta$	$\Delta\eta''_H/\eta$
MF-1	800	619	14.2	14.5	38.3	38.0	0.048	0.043	0.013	0.009	0.4	0.3
MF-2				14.6	39.1	38.7	0.069	0.059	0.039	0.029	1.3	1.0
MF-1	900	686	15.9	15.3	38.4	38.2	0.052	0.047	0.017	0.013	0.5	0.4
MF-2				15.5	39.3	38.9	0.066	0.057	0.036	0.027	1.2	0.9

Based on the 'upper' and 'lower' estimates of the 'near-wall viscosity', their contribution to the oscillation damping coefficient is calculated. Table 2 shows the experimental values of the damping coefficient β, taken from [2]. It also presents the values of β'_{HE} and β''_{HE}, calculated by formula (6) using η'_H and η''_H, respectively.

Table 2. System attenuation rating

Sample	H_0, kA/m	H_*, kA/m	$\Delta H_x/\Delta z$, MA/m^2	ν, Hz	β, c^{-1}	β'_{HE}, c^{-1}	β''_{HE}, c^{-1}
MF -1	900	686	15.9	15.3	9.5	7.5	7.1
MF -2				15.5	9.4	8.5	7.9

Table 3 shows H_0, which is the magnetic field intensity; v, which is the oscillation frequency of the MF-column; $\Delta h/\Delta z$, which is the ratio of the difference in water levels in the elbows of the U-shaped tube Δh to the displacement of the MF-column under the action of static pressure Δz; the values η_H calculated by formula (22); the increments of viscosity in the magnetic field ('magnetoviscous' effect) $\Delta\eta_H$; the relative value of this parameter $\Delta\eta_H/\eta$.

Table 3. Viscosity determination using statistical experiment data

Sample	H_0, kA/m	v, Hz	$\Delta h/\Delta z$	η_H, Pa·s	$\Delta\eta_H$, Pa·s	$\Delta\eta_H/\eta$	β, s^{-1}	β_{HE}, s^{-1}
MF-1	900	15.3	158	0.057	0.022	0.6	9.5	7.8
MF-2		15.5	160	0.063	0.033	1.1	9.4	8.3

Comparing the values of the parameters: η'_H, η''_H and η_H; $\Delta\eta'_H$, $\Delta\eta''_H$ and $\Delta\eta_H$; $\Delta\eta'_H/\eta$, $\Delta\eta''_H/\eta$ and $\Delta\eta_H/\eta$ given in Tables 1, 2, and 3, it is possible to see their quantitative close proximity, which confirms the validity of the theoretical approaches used when obtaining formulas (18) and (22) as well as the soundness of the assumption of quantitative close proximity of the coefficients of ponderomotive and static elasticity introduced in [1, 2].

Table 3 presents the values of the oscillation damping coefficients β and β_{HE} measured and calculated by formula (6) using η_H. The numerical values of these parameters are in the same correlation as the correlation of the values of β, β'_{HE}, and β''_{HE} given in Table 2. The calculation results indicate that, firstly, the obtained values of the 'near-wall viscosity' do not lead to an excess of the experimental result β, which, in its turn, can be regarded as physical justification of the proposed model theory; secondly, the presence of an excess of β over β_{HE}, β'_{HE}, and β''_{HE} by ~ 15 % actually reflects the presence of other energy dissipation mechanisms in the oscillatory system (in particular, the emission of elastic energy into the structural elements of the setup and the environment).

The limitation of the applied theories indicated above allows preferring the values of the calculated parameters belonging to the highest magnetic field intensity of 900 kA/m, at which the maximum value of the oscillation frequency and the minimum length of the viscous wave were obtained. At the same time, in order to interpret the differences in the values of the 'near-wall viscosity' of the samples

MF-1 and MF-2, data on the characteristic features of their structure, which are absent in the published work [2], are necessary.

There is one fundamental difference between the considered techniques for determining the 'near-wall viscosity' of a MF in a strong magnetic field, which is considered further. When applying a static (hydrostatic) technique for determining the elasticity coefficient of an oscillating system and combining it with a technique for determining the dynamic elasticity coefficient obtained using measurements of the oscillation frequency, there is no need to make complex measurements of magnetic field intensity and intensity gradient as well as to obtain a magnetization curve. Moreover, the measurements mentioned are characterized by a large total error. The only condition that must be fulfilled is that all magnetic, pressure and temperature parameters must be identical in the experiments for static and dynamic measurements.

Obtaining information on the 'near-wall viscosity' for MFs synthesized applying new technologies is essential in order to diversify carrier medium, concentration and sizes of magnetic nanoparticles as well as in connection with the problem of aggregation in magnetic colloids. Additional possibilities are emerging for the diagnosis of interparticle aggregation to the occurrence of which 'magnetoviscous' effect in the thin near-wall layer turns out to be very sensitive when applying a strong magnetic field to MF.

5. Conclusion

The results of the research are as follows:

– Estimation of the viscosity increment ('magnetoviscous' effect) in a thin near-wall layer of a MF-column oscillating in the tube when a strong transverse magnetic field is applied;

– Calculation of the value of 'near-wall viscosity' is obtained by a formula derived from two different theoretical approaches;

– An array of previously published experimental results was used; it includes complex measurements of oscillation frequency, oscillation damping coefficient, saturation magnetization, and static displacement of the MF-column for two fluid samples;

– Based on the 'upper' and 'lower' estimates of the 'near-wall viscosity' obtained using the magnetization curve as well as the results of static displacement of the MF-column, their contribution to the oscillation damping coefficient is calculated;

– Additional possibilities are emerging for the diagnosis of interparticle aggregation to the occurrence of which 'magnetoviscous' effect in the thin near-wall layer turns out to be very sensitive when applying a strong magnetic field to MF;

– The calculations do not take into account the transition thin layer of the viscous wave passage, since it is assumed that the entire MF-column is involved in the oscillatory motion;

– The applicability of the relations obtained for the evaluation of can be extended, for example, by increasing the magnetic field intensity gradient, which allows obtaining a higher oscillation frequency of a MF-column.

Acknowledgements

The investigation was carried out in the framework of the State Project of the Ministry of Education and Science of the Russian Federation. Project code is 3.2751.2017/PP.

References

[1] Polunin V M, Ryapolov P A, Platonov V V and Kuzko A E 2016 *Acoustic Journal* **62** 302-307
[2] Polunin V M, Ryapolov P A, Platonov V B, Sheldeshova E V, Karpova G V and Arefyev I M 2017 *ACOUST PHYS+* **63** 371–379
[3] Mc Taque J P 1969 *J. Chem. Phys.* **51** 133-136
[4] Rosensweig R E, Kaiser R and Miskolezy G. 1969 *J. Colloid Interface Sci.* **29** 680-686
[5] Shliomis M. I. 1971 *JETP.* **6 (12)** 2411-2418
[6] Shliomis M I 1974 *Adv. Phys.* **112** 427-459
[7] Cebers A O 1975 *MHD* **4** 37-44.
[8] Maiorov M M 1980 *MHD* **4** 11-18
[9] Rzhevkin S N 1960 *Lectures on the Theory of Sound* (Moscow: Publishing house of Moscow state University) 336
[10] Landau L D and Lifshits E M 1988 *Theoretical Physics. Hydrodynamics* (Moscow: Science) **6** 736

IOP Conf. Series: Materials Science and Engineering **581** (2019) 012018 doi:10.1088/1757-899X/581/1/012018

Grain Structure Refinement Description in the Two-Level Statistical Crystal Plasticity Model

T V Ostanina, A I Shveykin, P V Trusov and E S Makarevich

Perm National Research Polytechnic University, Perm, 614990, Russian Federation

E-mail: tv-ostanina@yandex.ru, alexey.shveykin@gmail.com, tpv@matmod.pstu.ac.ru, makareviches@inbox.ru

Abstract. As both self-consistent and direct multilevel models (suggesting solution of boundary value problems at the meso-level) are extremely resource intensive, nowadays statistical *crystal plasticity* models are supposed to be the most promising ones for modeling technological processes of thermo-mechanical treatment of materials. The statement of a boundary value problem at the current configuration in the rate form is preferable, as it is convenient for applying numerical methods. In this case, step-by-step solution with redefining the computational domain configuration (including contacting surfaces) is possible. The two-level (macro- and meso-level) statistical constitutive model for describing deformation of polycrystalline metals and alloys, being formulated in terms of the actual configuration in the rate form, is proposed. The flexible coordinate system at the meso-level is connected with the symmetry elements of the crystallites, which determines appropriate choice of the corotational derivative in the constitutive relation. The approximate model for describing grain structure refinement on the basis of considering the amount of accumulated inelastic deformation as an integral characteristic of the defect structure state is included into the model. The results of the test calculations for describing loading processes, being relevant for equal-channel angular pressing, are given.

1. Introduction

In recent decades, processing by intensive plastic deformation have been increasingly used in various branches of engineering. This is due to significant improvement in performance characteristics of materials and products during the materials' grain structure refinement in these processes [1], [2] and etc. For example, in the works [3], [4] it is noted, that decrease in the grain size below a certain threshold value can lead to a drastic change in the properties. Firstly, the microhardness of such materials can be 2-6 times higher than the microhardness of the coarse-grained analogues. Secondly, yield strength and tensile strength are also increased significantly. The possibility of optimal combination of mechanical properties in metals and alloys with a nanosubmicron structure opens up broad prospects for their usage as new structural and functional materials.

At low homologous temperatures, when recrystallization processes can be ignored, the key factor determining the course of grain structure refinement process is (localized) accumulation of defects. As a result of it, the polycrystalline grains are firstly divided into a multitude of related microareas (fragments), where misorientations is increased in the deformation process [5, 6, 7, 8, 9, 10]. When the misorientations between the fragments reach the values being characteristic for the high-angle boundaries of the intergranular type, division of the grain into the separate parts (that is, loss of perfect connectivity) may occur (in particular, a shift along the formed border becomes possible). In a sense, the process of grain fragmentation into separate parts resembles the process of viscous fracture, but with preservation of coherence between the resulting parts of the original crystallite. Thus, grain structure refinement is manifested at the mesoscale level (the grain level) and is due to implementation of the defect structure changing processes at the microscale level.

The most suitable apparatus to describe the changing internal structure of metals and alloys is multilevel models based on *crystal plasticity* [11, 12, 13, 14, 15, 16, 17]. In the models of that class, the internal variables [18, 19, 20, 21] are explicitly introduced to describe the structure elements and individual deformation mechanisms. To describe their changing, equations based on the known information from the solid-state physics, are formulated. There are three main groups of models of that class: statistical, self-consistent and direct ones. In statistical models [22, 23, 16] the meso-level elements (crystallites) are considered to be relatively independent on each other. Integration of meso-level elements into the macro-level element is carried out for a part of the characteristics based on the accepted kinematic or static hypotheses. The other characteristics are statistically averaged. In self-consistent models [24, 25, 26, 27] at the meso-level, the behavior of individual meso-level elements in the environment of the material matrix with effective characteristics is determined. These characteristics are defined iteratively by the properties of the meso-level elements using a particular averaging procedure for the last ones. In the direct models [28, 17, 29] the solution of boundary-value problems at the mesoscale level is implemented (in the vast majority of papers, the finite element method is used). Application of self-consistent and especially direct models is associated with extremely large computational costs, therefore, statistical models are the most promising ones for modeling real technological processes of the materials' thermo-mechanical treatment in the near future.

The two-level statistical model of polycrystalline metals and alloys, supplemented by correlations for describing grain structure refinement, is proposed.

2. Structure of the two-level statistical model

The two-level statistical model of polycrystalline metals and alloys [30] has been taken as a base constitutive model. The mechanisms of the intra-grain dislocation sliding and the crystallites' lattices rotation are taken into consideration in the model.

For the meso-level element (crystallite) the elastic law in the rate relaxation form in the actual configuration terms is as follows:

$$\mathbf{\kappa}^{cr} = \overline{\mathbf{\pi}} : \left(\mathbf{e}^{e}\right)^{cr} = \overline{\mathbf{\pi}} : (\mathbf{z} - \mathbf{z}^{in}) = \overline{\mathbf{\pi}} : \left(\mathbf{l} - \overline{\mathbf{\omega}} - \sum_{k=1}^{K} \dot{\gamma}^{(k)} \mathbf{b}^{(k)} \mathbf{n}^{(k)}\right), \quad (1)$$

where $\mathbf{\kappa} = \overset{o}{\rho} / \hat{\rho}\, \mathbf{\sigma}$ is the weighted Kirchhoff stress tensor, $\mathbf{\sigma}$ is the Cauchy stress tensor, $\overset{o}{\rho}$, $\hat{\rho}$ is density of the crystallite's material in the initial (unloaded) and the actual configurations, $\mathbf{\kappa}^{cr} = d\mathbf{\kappa}/dt + \mathbf{\kappa} \cdot \overline{\mathbf{\omega}} - \overline{\mathbf{\omega}} \cdot \mathbf{\kappa}$ is the corotational derivative of the Kirchhoff tensor, $\overline{\mathbf{\omega}}$ is the spin tensor of the movable coordinate system, $\overline{\mathbf{\pi}}$ is the tensor of elastic properties (which components are constant in the basis of the flexible coordinate system), $\left(\mathbf{e}^{e}\right)^{cr}$ is the elastic deformations' rate, being fixed by the observer in the flexible coordinate system, $\mathbf{l}$ is the rate gradient, $\mathbf{z} = (\mathbf{l} - \overline{\mathbf{\omega}})$ is the full strains velocity tensor, being defined by the observer in the flexible coordinate system, $\mathbf{z}^{in} = \sum_{k=1}^{K} \dot{\gamma}^{(k)} \mathbf{b}^{(k)} \mathbf{n}^{(k)}$ is the inelastic strains rate, $\dot{\gamma}^{(k)}, \mathbf{b}^{(k)}, \mathbf{n}^{(k)}$ are the shear rate, the shear direction and the normal to the shear plane in the actual configuration for the k-th intra-grain slip system (of the edge dislocations; the doubled quantity of the slip systems is used for convenience).

To define $\overline{\mathbf{\omega}}$, the rotation model, described in [30, 31], is used. According to this model, the flexible coordinate system is connected with symmetry elements of the crystallites, which determines the appropriate choice of the corotational derivative. In this case, the tensors, being a part of the constitutive model formulation, would change as indifferent ones for the observer in the fixed laboratory coordinate system, which allows satisfying the principle of the constitutive relation independence on the reference frame choice. In [32] the comparison results of using the specified spin and the alternative formulations, including classical models of constrained rotation by Taylor and rotation with $\mathbf{R}^{e}$ (orthogonal tensor from decomposition of the strain gradient elastic component $\mathbf{F}^{e}$), are described. The improvement in description of crystallites' lattice rotation by the including the force (moment) factors, appearing due to interaction of the neighboring grains' (subgrains') defects, into equations, can be noted as a development direction for the *crystal plasticity* models [33].

IOP Conf. Series: Materials Science and Engineering **581** (2019) 012018 doi:10.1088/1757-899X/581/1/012018

To determine the shear rates, the viscoplastic relation is used:

$$\dot{\gamma}^{(k)} = \dot{\gamma}_0 \left(\tau^{(k)} / \tau_c^{(k)}\right)^m H(\tau^{(k)} - \tau_c^{(k)}), \tag{2}$$

where $\tau^{(k)}$, $\tau_c^{(k)}$ are shear and critical shear stresses on the k-th slip system, $\tau^{(k)} = \mathbf{\kappa} : \mathbf{b}^{(k)}\mathbf{n}^{(k)}$, in common case $\tau_c^{(k)}$ is a function of the shifts in the slip system accumulated up to the current moment and history of their changing, $\dot{\gamma}_0$ is the shear rate in the slip system when the shear stress reaches the critical shear stress, m is the indicator of material rate sensitivity (in the calculations, the parameter values are assumed as follows $\dot{\gamma}_0 = 0.001\,s^{-1}$, $1/m = 0.012$), $H(\cdot)$ is the Heaviside function. The evolution relation for critical stresses $\tau_c^{(k)}$ and its parameters are supposed to be as in [34] and correspond to the technical copper.

To determine the impact at the meso-level the generalized Taylor hypothesis is used $\mathbf{l} = \mathbf{L}$, where $\mathbf{L}$ is the rate gradient at the macro-level.

The macro-scale variables are determined by averaging the meso-level variables using the procedure of the constitutive relations' consistency [33]. Note, that taking temperature into consideration should be added into the problem formulation for the common case. Here, consideration is limited by a test isothermal case.

Note, that for simulation of technological processes, the boundary value problem statement in terms of the actua configuration in the rate form is more preferable, being convenient for applying numerical methods. In this case, the step-by-step solution with redefining configuration for the calculated area (including the surfaces being in contact) is possible. The proposed formulation of the two-level constitutive model is organically incorporated into such a statement.

To describe the possible grain structure refinement, the model is supplemented by the following relationships.

3. The description of grain structure refinement within the multi-scale constitutive model

The process of grain structure refinement is determined by localized accumulation of defects. Grain refinement is supposed to be realized by analogy with viscous destruction as a result of high defect density accumulation [35, 36]. As a result, a large angle boundary appears between the grain parts, but with preservation of their connectivity. Within the two-level statistical model the accumulated shears by slip systems are integral characteristics of the defects' densities [37, 29]. Within the framework of the basic statistical model the grain deformation is supposed to be homogeneous. An explicit consideration of the defective substructures at the micro level requires large computational resources due to introduction of an additional level into the constitutive model and a significant number of internal variables. Therefore, in the developed model, planned to be used in simulating the real technological processes of mechanical treatment, a simplified description of refinement is proposed to be used in a first approximation. For this, the separation criterion described below and based on consideration of accumulated shifts in slip systems is proposed to be applied. A simplifying hypothesis has been adopted that two ellipsoidal grains of the same size and shape are formed as a result of the original grain fragmentation. The maximum size of a new grain corresponding to a part of the initial grain is equal to the half of the maximum linear size of the initial grain and corresponds to the perpendicular original direction. The dimensions in the transverse direction are proportional to the original transverse dimensions. The proportionality coefficient is determined from the condition of equality for the total volume of the new grains to the volume of the original one. The axes of the new grains are disoriented at sufficiently small angles according to a random law (the distribution can be established using experimental data).

In the fragmentation criterion formulation based on consideration of the accumulated inelastic deformation amount, a large dislocation density along the slip system is assumed to be increasing the probability of separating the grain along this plane. The fragmentation criterion is taken as $f = f_C$, where f_C is a model parameter, determined when identifying the model from the experimental data on the

IOP Conf. Series: Materials Science and Engineering **581** (2019) 012018 doi:10.1088/1757-899X/581/1/012018

average grain size changing in deformation processes (f_C depends on the average grain size, it is assumed to be equal $f_C = 0.7$ in the test calculations), where a significant grains fragmentation occurs. The function f is defined as follows

$$f = \max_{\mathbf{n},\boldsymbol{\tau}} (h_{\mathbf{n}} / h_{\boldsymbol{\tau}} \, \boldsymbol{\tau} \cdot \mathbf{E}^{in} \cdot \mathbf{n}) . \quad (3)$$

In the relation (3) $\mathbf{n}$, $\boldsymbol{\tau}$ are the normal to the shear plane and the direction of the shift in it (they are determined as a result of solving the optimization problem for ensuring the maximum value of the function), $h_{\mathbf{n}}$, $h_{\boldsymbol{\tau}}$ are the maximum grain-ellipsoid sizes in the directions $\boldsymbol{\tau}$ and $\mathbf{n}$, $\mathbf{E}^{in}$ is the plastic strain accumulated in the grain. Thus, (3) determines the maximum shear strain at the current time, but taking into consideration the grain elongation. The amount $\mathbf{E}^{in}$ is determined by integrating the equation $\dot{\mathbf{E}}^{in} - \bar{\boldsymbol{\omega}} \cdot \mathbf{E}^{in} + \mathbf{E}^{in} \cdot \bar{\boldsymbol{\omega}} = \mathbf{z}^{in}$, where $\bar{\boldsymbol{\omega}}$ is the spin of the flexible coordinate system, connected with the crystalline lattice [30], $\mathbf{z}^{\mathrm{in}} = \sum_{k=1}^{K} \dot{\gamma}^{(k)} \mathbf{b}^{(k)} \mathbf{n}^{(k)}$ is the grain inelastic deformations rate due to the intragranular dislocation sliding, the shear rates in the slip systems are determined by (2).

In a more general formulation, an explicit consideration of individual grain fragments misorientations [38] can be included into the refinement submodel by taking into consideration the moment factors and disclinations, but implementation of this model requires significantly more computational resources.

4. Results

To simulate equal-channel angular pressing, the analytical relation for the rate gradient given in [39] is used, according to which the $\mathbf{L}$ tensor components in the basis of the laboratory coordinate system $OX_1X_2X_3$ can be written as follows:

$$[\mathbf{L}] = [\nabla \mathbf{v}^{\mathrm{T}}] = L_0 \begin{bmatrix} -\sin\theta\cos\theta & \cos^2\theta & 0 \\ -\sin^2\theta & \sin\theta\cos\theta & 0 \\ 0 & 0 & 0 \end{bmatrix}, \quad (4)$$

where L_0 is a (negative) shear rate, θ is a half angle of articulation for the channels (the output channel is located along the axis OX_1, the entrance channel is angled at 2θ in the plane OX_1X_2). Thus, emulation of equal-channel angular pressing is a multiple repetition of the cycle: shear deformation (4) and unloading (the turn of the workpiece before the next pass is possible; in the calculations, which results are presented here, the turn was not considered).

In the initial state, all crystallites were assumed to be equal balls with the size 50 mkm. During the deformation, significant refinement of the grain structure is observed. The characteristics of its state are shown in figure 1 and figure 2.

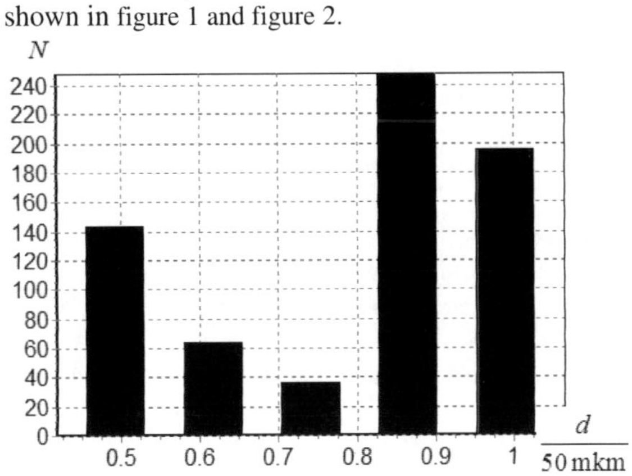

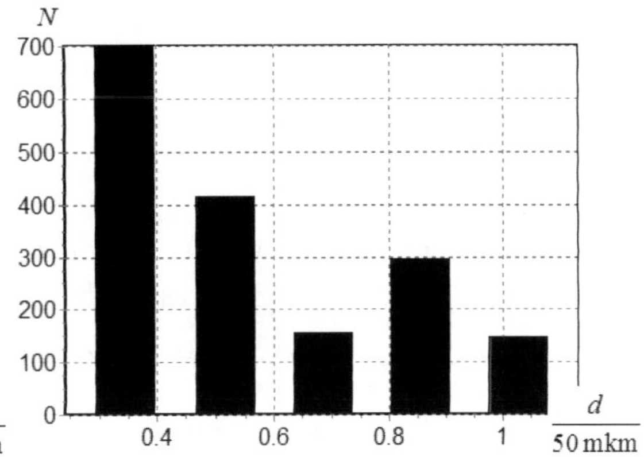

Figure 1. The histogram of distribution for the average grain semiaxis length, mkm (the left one is the result after the first pass, the right one is the result after the second pass)

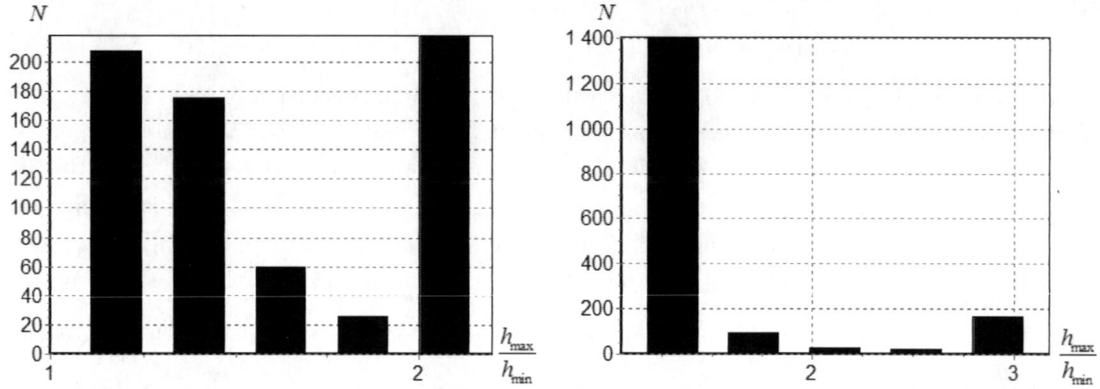

Figure 2. The histogram of distribution for the ratio of the greatest grain semiaxis to the smallest one, mkm (the left one is the result after the first pass, the right one is the result after the second pass)

It can be noted that after the second pass the structure is more fined, the shape of the grains becomes closer to the equiaxial one. It qualitatively corresponds to the data on the grain structure changing in the process of the equal-channel angular pressing [8, 9].

5. Conclusion

The two-level constitutive model of polycrystalline metals and alloys allowing taking into consideration the grain refinement process as a first approximation is developed. The obtained simulation results indicate the corresponding capabilities of the model. The proposed approach can be used for modeling the structure evolution of a polycrystalline material when describing the technological processes of mechanical treatment for metals and alloys being based on severe plastic deformations.

6. Acknowledgments

The work was supported by the Russian Science Foundation (grant No. 17-19-01292).

References

[1] Valiev R Z, Naimark O B 2007 *Innovations* **12** (110) pp 70–76 *(in Russian)*
[2] Gleiter H 1989 *Prog. Mater. Sci.* **33** pp 223–330
[3] Valiev R Z, Alexandrov I V 2000 Nanostructured materials obtained by severe plastic deformation (*Logos, Moscow*) *(in Russian)*
[4] Noskova N I, Mulukov R R 2003 Submicrocrystalline and nanocrystalline metals and alloys *(UrB RAS, Yekaterinburg) (in Russian)*
[5] Ribin V V, Vergazov A N and Lihachev V A 1974 *Physics of Metals and Metallography* **37** pp 620–624 *(in Russian)*
[6] Ribin V V 1986 Large plastic strains and destruction of metals (*Metallurgy, Moscow*) *(in Russian)*
[7] Horstemeyer M F McDowell D L 1998 *Mechanics of Materials* **27** pp 145–163
[8] Beyerlein I J, Lebensohn R A and Tome C N 2003 *Materials Science and Engineering A345* pp 122–138
[9] Torre F D, Lapovok R, Sandlin J, Thomson P F, Davies C H J and Pereloma E V 2004 *Acta Materialia* **52** pp 4819–4832
[10] Poortmans S, Duchene L, Habraken A M end Verlinden B 2009 *Acta Materialia* **57** pp 1821–1830

[11] Horstemeyer M F, Potirniche G P end Marin E B 2005 *Handbook of Materials Modeling* ed Yip S (Springer: Netherlands) pp 1133–1149
[12] Horstemeyer M F 2009 *Practical Aspects of Computational Chemistry* eds Leszczynski J and Shukla M K (Springer Science + Business Media B V) pp 87–135
[13] McDowell D L 2010 *Int. J. Plasticity* **26** pp 1280–1309
[14] Roters F, Eisenlohr P, Bieler T R and Raabe D 2010 Crystal Plasticity Finite Element Methods in Materials Science and Engineering *(WILEY-VCH Verlag GmbH & Co. KGaA)* p 209
[15] Roters F, Eisenlohr P, Hantcherli L, Tjahjanto D D, Bieler T R and Raabe D 2010 *Acta Materialia* **58** pp 1152–1211
[16] Trusov P V and Shveykin A I 2013 *Physical Mesomechanics* **16** (1) pp 23–33
[17] Trusov P V and Shveykin A I 2013 *Physical Mesomechanics* **16** (2) pp 99–124
[18] McDowell D L 2005 *Handbook of Materials Modeling* ed Yip S (Springer) pp 1151–1169
[19] Guo Y B, Wen Q and Horstemeyer M F 2005 *Int. J. Mech. Sci.* **47** pp 1423–1441
[20] Saï K 2011 *Int. J. Plasticity* **27** pp 250–281
[21] Maugin G A 2015 *Mechanics Research Communications* **69** pp 79–86
[22] Van Houtte P 2009 *Microstructure and Texture in Steels* eds Haldar A, Suwas S, Bhattacharjee D (Springer) pp 209–224
[23] Dancette S, Delannay L, Jodlowski T and Giovanola J 2010. *Int. J. Mater. Form.* **3** (1) pp 251–254
[24] Lebensohn R A and Tomé CN 1993 *Acta Metall.* **41**.pp 2611–2624
[25] Lebensohn R A, Tomé C N, and Ponte CastaÑeda P 2007 *Philosophical Magazine*.**87** (28) pp 4287–4322
[26] M'Guil S, Ahzi S, Youssef H, Baniassadi M and Gracio J J 2009 *Acta Mater.* **57** pp 2496–2508
[27] Beyerlein I J, McCabe R J and Tome C N 2011 *J. Mech. Phys. Solids.* **59** pp 988 –1003
[28] Roters F 2011 Advanced Material Models for the Crystal Plasticity Finite Element Method: Development of a general CPFEM framework *(RWTH Aachen, Aachen)*
[29] Ardeljan M, Beyerlein I J and Knezevic M 2014 *J. Mech. and Phys. Solids.* **66** pp 16–31
[30] Trusov P V, Shveykin A I and Kondratev N S 2017 *Nanoscience and Technology: An International Journal* **8** (2) pp 133–166
[31] Trusov P V, Shveykin A I and Yanz A Yu 2017 *Physical Mesomechanics* **20** (4) pp. 357–376
[32] Shveykin A I and Trusov P V 2019 *Nanoscience and Technology: An International Journal* **10**(1) pp.1–20
[33] Trusov P V, Shveykin A I Nechaeva E S and Volegov P S 2012 *Physical Mesomechanics* **15**(3–4) pp 155–175
[34] Anand L 2004 *Comp. methods in appl. mech. and eng.* **193** pp 5359–5383
[35] Nazarov A A, Enikeev N A, Romanov A E, Orlova T S, Alexandrov I V and Beyerlein I J 2006 *Nanostructured Materials by High-Pressure Severe Plastic Deformation* eds Zhu Y T and Varyukhin V pp 61-66
[36] Orlova T S, Romanova A E, Valiev R Z, Alexandrov I V and Nazarov A A 2005 *Solid state physics* **47** (5) pp 820-826 *(in Russian)*
[37] Kocks U F, Mecking H 2003 *Progress in Materials Science* **48** pp 171–273
[38] Ostanina T V, Telkanov M A, Makarevich E S, Trusov P V and Shveykin A I 2017 *AIP Conference Proceedings* **1909** 020160
[39] Li S 2008 *Acta Materialia* **56** pp 1018–1030

IOP Conf. Series: Materials Science and Engineering **581** (2019) 012019 doi:10.1088/1757-899X/581/1/012019

Plateau-Rayleigh Instability of Ferrofluid Drop-Like Aggregates in Zero Magnetic Field

A S Ivanov

Institute of Continuous Media Mechanics Ural Branch of Russian Academy of Sciences the affiliate of Perm Federal Scientific Research Centre, 1Korolev str., Perm, Russia

E-mail: lesnichiy@icmm.ru

Abstract. Experimental investigation was carried out in order to study the properties of drop-like aggregates – the needle-shaped drops of condensed phase, which appear in magnetic fluids undergoing field-induced phase transition of the gas-liquid type. When the applied magnetic field is removed, the needle-shaped aggregates demonstrate the Plateau-Rayleigh instability and disintegrate into series of separate spherical drops due to the surface tension σ at the interface between the gas and liquid phases. The surface tension entirely depends on interparticle interactions, thus experimental investigation of σ allows us to analyse the influence of temperature and magnetic field on the interparticle interactions. It is shown that drop-like aggregates condensed at high temperature demonstrate higher surface tension, than the aggregates condensed at low temperature. This anomalous behaviour can be attributed to the high magnetic field intensity needed to cause the phase transition at high temperature.

1. Introduction

Magnetic fluids (ferrofluids) are stable colloids of magnetic materials in nonmagnetic carrier liquids, which are used in technical applications due to the unique combination of contradictory physical properties: liquid state and high magnetic susceptibility (approx. 1-10 units). Ferrofluids are also widely used for fundamental studies, because they possess exaggerated dipole-dipole interparticle interactions, typical for dipolar media. It is well known, that any gas or liquid containing polar molecules is usually described by a complicated equation of state with numerous virial coefficients [1]. The simplest example of polar fluid is water, because there is no adequate equation of state for water and vapor in all ranges of temperature T and pressure p, which are interesting for technical applications, - and till nowadays the most accurate information about water and its vapor is given in the form of international experimental skeleton tables supplied with approximation fitting equations without any clear physical meaning [2]. Thus it is easier to make fine laboratory measurements and numerical simulations with magnetic fluids [3] than with other polar gases and liquids.

The typical experimental approach to study dipole-dipole interparticle interactions in magnetic fluids is to measure the magnetic susceptibility χ dependence on temperature T and volume concentration φ of particles, because χ is very sensitive to particle interaction and granulometric parameters of the ferrofluid [4, 5]. However there is an alternative way to study interparticle interactions using surface tension σ measurements. This idea comes from the analogy with classic one-component fluids, because in ordinary fluids the phenomenon of surface tension is explained by the energy of intermolecular attraction [1]. It is not excessive to remind the theory of corresponding states by Sugden [6] who suggested to compare molar volumes of liquids at equal surface tension (not at the critical point (T^*, p^*) as it was suggested by Van der Waals), because equal surface tensions guarantee equal intermolecular forces inside the compared samples.

To study the interparticle interactions with σ measurements in ferrofluids we should be able to create the liquid-vapor interface between the gas and liquid phases of colloidal particles: the surface tension at the ferrofluid-air surface is not suitable, of course, because it depends mainly on the interaction of carrier liquid molecules ($\sigma \sim 10^{-3}$ N/m) which exceeds the typical surface tension at the surface of gas-liquid phases of ferroparticles ($\sigma \sim 10^{-7}$-10^{-6} N/m) by 3-4 orders of magnitude. Fortunately, dilute magnetic fluids ($\varphi \sim 0.01$) demonstrate the field-induced first-order phase transition of the "gas-liquid" type under the action of applied magnetic field H. This effect was studied experimentally [7] and numerically [8]. The field-induced phase transition in magnetic fluids resembles phase transition in moist air as the temperature goes down, because the liquid phase condensates in the form of microdroplets (about 1 μm in diameter). The microdroplets grow and coalesce into macroscopic drops, the so-called drop-like aggregates (its length ~ 1 mm). Drop-like aggregates elongate along the magnetic field lines and become needle-shaped (see Fig. 1). The drop-like aggregate surface exhibits the measurable surface tension σ, which in general case is the function of temperature T and applied field H. When the applied field is nonzero, all particles interact with the field and with each other simultaneously, and the measurements of σ depend on both interactions. However, if we are interested in the interparticle interactions only, then we should exclude the applied field and study the behavior of drop-like aggregates in the zero field. When the external field is switched off, the needle-shaped drop-like aggregate disintegrates into series of separate spherical drops due to hydrodynamic Plateau-Rayleigh instability [9-11] (see Fig. 1).

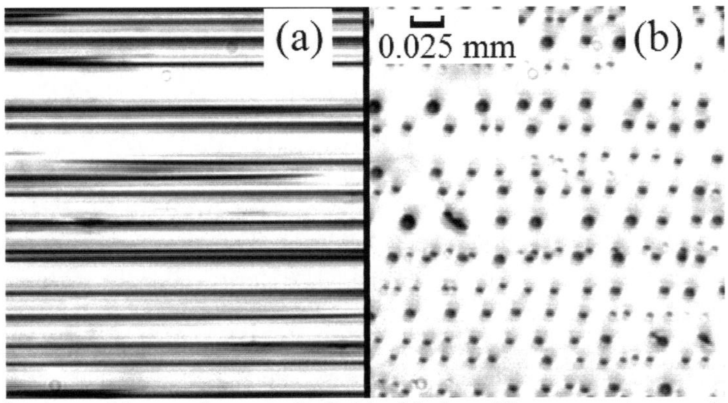

Figure 1. Thin horizontal layer of magnetic fluid with drop-like aggregates condensed during the field-induced phase transition stretch along the H field lines (a) disintegrate into series of separate spherical drops (b) due to the surface tension σ after the applied field is turned to zero.

To sum it up, the present paper delivers new experimental investigation of the surface tension σ at the ferroparticle liquid-gas interface, based on the analysis of Plateau-Rayleigh capillary instability of drop-like aggregates in the zero applied field, while the drop-like aggregates are preliminary condensed by the nonzero field. There is no contradiction between applying the nonzero field H to cause the field-induced phase transition, and removing the field in order to measure σ, which depends entirely on interparticle interactions.

2. Main and auxiliary experimental setups

New experimental setup was made in order to investigate Plateau-Rayleigh capillary instability of drop-like aggregates. The most convincing and simple are the direct visual measurements, that is why it was decided to study drop-like aggregates in a thin horizontal layer (Hele-Shaw cell) placed at the center of Helmholtz coils (Fig. 2). The Hele-Shaw cell was made of two parallel glass plates with a thin square copper wire frame placed between them. The base and the cover glasses were 10.0 mm and 1.3 mm thick, respectively. The wire diameter specified the thickness of the ferrofluid layer (0.2 mm), which was small compared to its length and width (30 mm) and large compared to the diameter of a drop-like aggregate (approx. 0.02 mm). The cell was sealed with hot (120 C) Canada balsam to prevent evaporation of the ferrofluid sample.

The ferrofluid temperature T was controlled by the liquid thermostat (not shown in the Fig. 2). The temperature was measured by two copper-constantan thermocouples. The hot junction of thermocouple TC_1 was embedded in a straight groove on the surface of the Hele-Shaw cell wall and its cold junction was placed in a Dewar flask with melting ice. Both junctions of the thermocouple TC_2 were located in the wall grooves as far from each other as possible. The signal from both thermocouples was simultaneously measured by multichannel 24-bit ADC. Thus, TC_1 measured the temperature of the fluid and the TC_2 measured the temperature inhomogeneity inside the ferrofluid layer. The inaccuracy of temperature measurements did not exceed 0.3 C.

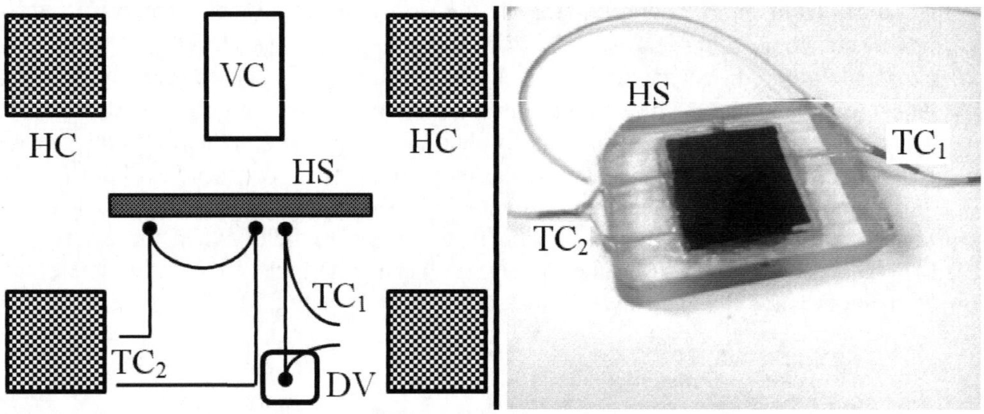

Figure 2. The experimental setup scheme (on the left) and the Hele-Shaw cell (HS) filled with magnetic fluid (on the right): TC – thermocouples, HC – Helmholtz coils, DV – Dewar flask, VC – video camera.

Video camera was integrated in the head of the instrumental microscope. Video camera recorded the disintegration process of drop-like aggregates at given T and use this raw experimental data for numerical analysis of the Plateau-Rayleigh instability. The numerical analysis allowed to calculate σ.

The procedure of experimental measurements was the following. First, the desired temperature T was set on the controls of the liquid thermostat, and after a while (approx. 1 hour) thermocouples registered the thermal equilibrium state of the system "thermostat – measuring cell" (TC_1 readings were steady and TC_2 readings registered temperature inhomogeneity along the cell less than 0.1 C). Second, the Helmholtz coils were powered by stabilized DC power supply in order to generate homogeneous magnetic field. The electric current flowing through the coils was slowly increased till it was enough to cause the first-order phase transition in the magnetic fluid sample. This process was observed on the computer screen attached to the video camera VC (Fig. 2). When the phase transition process was over (it usually took approx. 1 hour) the drop-like aggregates stopped to grow and coalesce. Third, the camera was switched on for recording and the applied field was abruptly switched off (the power source wire was detached). The disintegration process duration varied from several seconds to half a minute. Then the recording was stopped and everything repeated from beginning at the new desired T (however, sometimes it took several hours more to wait till all drops of condensed phase evaporated into the surrounding homogeneous colloid).

The main experiment (Fig. 2) required also the independent auxiliary experiment in order to measure the density and the viscosity temperature dependencies for the gas η_1(T) and liquid η_2(T) phases, because this experimental data is needed for numerical processing. For this purpose drop-like aggregates (the liquid phase) were separated from the surrounding gas phase in the auxiliary setup (Fig. 3(a)).

IOP Conf. Series: Materials Science and Engineering **581** (2019) 012019 doi:10.1088/1757-899X/581/1/012019

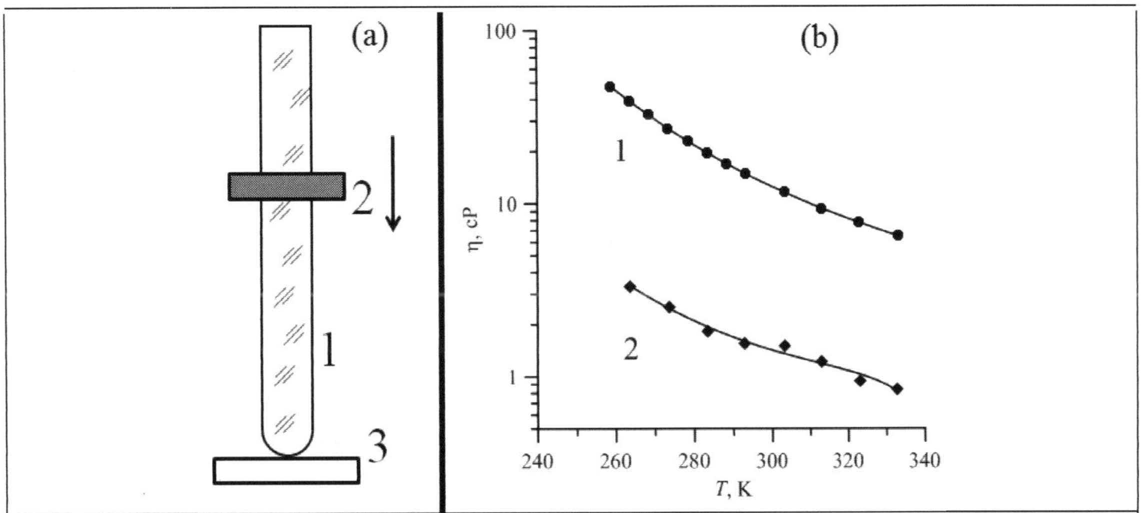

Figure 3. The auxiliary setup (a): 1 - test tube with ferrofluid; 2 - ring magnet collecting drop-like aggregates at the bottom of the tube; 3 - mount plate. The $\eta(T)$ dependence for the liquid and gas phases (curve 1 and 2 on (b) respectively): dots - experiment, solid lines - fitting curves.

The auxiliary setup was used to condensate and collect drop-like aggregates at the center of the ring magnet, which was dragged down the test tube with the original ferrofluid sample. The magnet moved slowly (6 cm per hour) to collect all amount of liquid phase at the bottom of the test tube and to avoid the hydrodynamic mixing of phases during sedimentation. The auxiliary setup allowed to collect the required amount of the liquid phase (5 ml) for Brookfield viscometer viscosity measurements (Fig 3(b)).

3. Calculation procedure

Video records (described in the previous section) supplied us with the raw experimental data at each given T for several (5-10) aggregates in the ensemble: the radius (half thickness) of the aggregate before disintegration R_0, the increment τ of perturbations of the aggregate surface $R(t) \sim \exp(t/\tau)$ and the average distance λ between separate drops after the disintegration process (Fig. 1(b)). The characteristic time τ and the characteristic distance λ are the temporal and spatial characteristics of the disintegration process, which are related by the functional dependence known as the dispersion equation [9]. However, the dispersion equation relates not the τ and λ values, but the temporal and spatial frequencies (characterized by the circular frequency ω and the wave number k) of the capillary waves (disturbances) on the surface of the drop-like aggregate. In the general case the disintegration process is oscillatory and the circular frequency in the dispersion equation $\omega = \omega(k)$ is complex $\omega = \mathrm{Re}(\omega) + \mathrm{i}\,\mathrm{Im}(\omega)$, where $\mathrm{i} = (-1)^{0.5}$. In our case disintegration is a relaxation (non-periodic) process, therefore the real part of ω equals zero $\mathrm{Re}(\omega) = 0$, and its imaginary part equals the characteristic relaxation time $\mathrm{Im}(\omega) = 1/\tau$.

It is interesting to recall that Chandrasekhar [10] also derived the dispersion equation, but his expression differs from the one obtained by Rayleigh [9], because Rayleigh used the "mechanics" approach that implies complex ω and real k, while Chandrasekhar used the "electrodynamics" approach with real ω and complex k. Probably this difference can be attributed, on the one hand, to the possibility of direct measurements of λ in mechanics, and, on the other hand, to the possibility of generating the electric oscillations of fixed frequency by a signal generator in the XXth century. It is also remarkable that the equivalence of equations published a century (Rayleigh) and half a century (Chandrasekhar) ago was demonstrated (and proved) only in 2018 [11]. However, the Rayleigh theory is too simple for the current study, because it neglects the influence of the surrounding media on the instability of the liquid thread. The appropriate dispersion equation $\omega = \omega(k)$ for the viscous liquid jet (drop-like aggregate) of the radius R_0 surrounded by another viscous immiscible fluid with the surface tension $\sigma(T)$ at the interface was derived in [12] from the zero determinant of the 4x4 matrix with functional elements. It is very complicated, thus it is not given in this paper, because it fully repeats the main result described in [12].

Let us now describe the calculation procedure, which was used to calculate the $\sigma(T)$ dependence. At each given T the central area of the Hele-Shaw cell was captured on video, which was then processed frame-by-frame in the image processing Comef OEG software in order to measure the disintegration of drop-like aggregates. The raw experimental data for each aggregate in the ensemble included the initial radius R_0, the wave-length $k = 2\pi\lambda$ calculated according to the averaged drop-to-drop distance, and $\mathrm{Im}(\omega) = 1/\tau$. The relaxation time τ for each aggregate was found according to the linear theory of the Plateau-Rayleigh instability, which defines the radius of the instable drop-like aggregate $R(t,z)$ as $R(t, z) = R_0(1 + \varepsilon \exp(\mathrm{i}\omega t - \mathrm{i}kz\})$, where z is the axis parallel to the major axis of the aggregate, ε is the arbitrary initial amplitude of the capillary wave at time $t = 0$, when H is switched off. The direct measurements of the anti-node amplitude (maximum radius R at time t) allows us to calculate τ.

After accomplishing the data preparation procedure, the experimental data for each aggregate was substituted into dispersion equation [12], which was then solved numerically using the Wolfram Matematica package similar to [11]. The final result values of $\sigma(T)$ averaged over an ensemble of aggregates at a given T are present in Fig. 4. Error bars in Fig. 4 are equal to the standard deviation, because all other measurement uncertainties are at least one order of magnitude less.

4. Results and discussions

The experimental study of $\sigma(T)$ described in this paper was carried out to investigate the Plateau-Rayleigh instability of the gas-liquid interface in a zero field. This instability was observed at the surface of drop-like aggregates elongated along the magnetic field lines, which disintegrated into series of separate drops after the applied magnetic field was switched off. The result measurements of $\sigma(T)$ show that the interparticle interactions in the condensed phase demonstrate the nonmonotonous temperature T dependence.

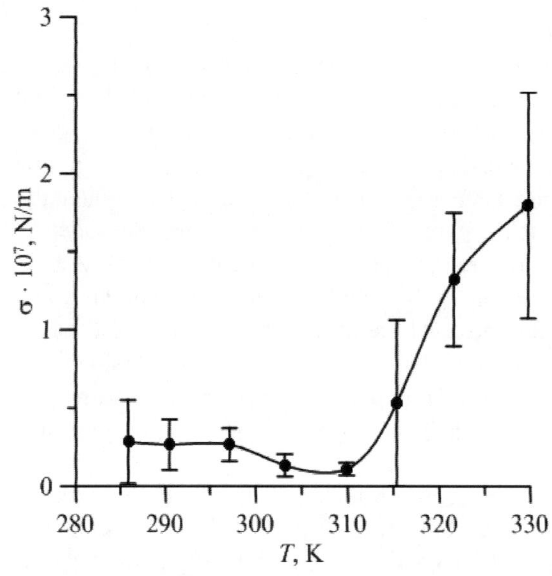

Figure 4. The experimental $\sigma(T)$ curve: dots - experiment, solid line - fitting curve.

The explanation of this anomalous phenomenon may be given in the framework of the Stockmayer fluid model [13], [14]. Thus, the fresh result [14] obtained by means of molecular-dynamics simulations is the following: the applied field H shifts the critical point and interfacial properties σ in different directions, e.g. an external field H parallel to the interface increases the critical temperature. This is actually what happens in our case: the most of drop-like aggregate's surface is parallel to the magnetic field lines, and we can observe the growth of interparticle interactions inside the aggregate, i.e. the increase of σ.

Another explanation of the observed phenomenon can be given by statistical thermodynamic methods developed in the framework of the molecular theories of liquids [1], [15]. This approach in case of ferrofluids gives the following power dependence $\sigma \sim (\rho_2 - \rho_1)^2$, where ρ_2 and ρ_1 are the mass densities of the liquid and gas phases, respectively. This power-law dependence of σ resembles classical experimental result for one-component fluids $\sigma \sim (\rho_2 - \rho_1)^4$ [16] and is in good qualitative agreement with the Zeldovich theory of surface tension at the interface between two mutually soluble fluids $\sigma \sim (\nabla\rho)^2$ [17]. The work [13] gives us the results of computer simulations of the Stockmayer fluid undergoing the field-induced phase transition, including the densities ρ_2 and ρ_1 at different T and H. As one can see, the density difference $(\rho_2 - \rho_1)$ [13] increases with increase of H and decreases with the increase of T. It means that the final result – the ordinary decrease or anomalous increase of $\sigma(H,T)$ depends on the order (magnetic condensation) -- disorder (thermal motion) competition in the colloid. Thus, the Stockmayer model gives a qualitative explanation of the experimental observations.

5. Conclusions

New experimental investigation was carried out in order to study the properties of the condensed phase drops, which appear in magnetic fluids undergoing field-induced phase transition of the gas-liquid type. When the applied magnetic field is removed, the needle-shaped drops demonstrate the Plateau-Rayleigh instability and disintegrate into series of separate spherical drops due to the surface tension σ at the interface between the gas and liquid phases. The surface tension entirely depends on interparticle interactions, thus experimental investigation of σ allows us to analyse the influence of temperature and magnetic field on the interparticle interactions. It is shown that drop-like aggregates condensed at high temperature demonstrate higher surface tension, than the aggregates condensed at low temperature. This anomalous behaviour can be attributed to the high magnetic field intensity needed to cause the phase transition at high temperature. This phenomenon indicates the increase of interparticle interactions inside the condensed phase with increase of temperature due to the simultaneous increase of the applied magnetic field needed to cause the phase transition.

References

[1] Hirschfelder J O, Curtiss Ch F and Bird R B 1954 *Molecular theory of gases and liquids* (New York: Wiley) p 1219
[2] Wagner W and Pruß A 2002 *J. Phys. Chem. Ref. Data* **31** 387
[3] Elfimova E A, Ivanov A O and Camp P J 2013 *Phys. Rev. E* **88** 042310
[4] Kuznetsov A A, Lebedev A V and Pshenichnikov A F 2018 *Magnetohydrodynamics* **54** 73
[5] Batrudinov T M, Nekhoroshkova Y E, Paramonov E I, Zverev V S, Elfimova E A, Ivanov A O and Camp P J 2018 *Phys. Rev. E* **98** 052602
[6] Bretschnider S 1966 *Properties of fluids and gases* (Leningrad: Himiya) p 536
[7] Peterson S A and Krueger A A 1977 *J. Coll. Int. Sci.* **62** 24
[8] Ivanov A O and Novak E V 2007 *Coll. J.* **69** 302
[9] Strutt J W (Lord Rayleigh) 1892 *Phil. mag.* **34** 177
[10] Chandrasekhar S 1961 *Hydrodynamic and Hydromagnetic Stability* (UK: Oxford)
[11] Pekker L 2018 *J. Imag. Sci. Tech.* **62**, 40405
[12] Tomotika S and Taylor G I 1935 *Proc. Roy. Soc. A* **150** 322
[13] Stevens M J and Grest G S 1995 *Phys. Rev. E* **51** 5976
[14] Moore S G, Stevens M J and Grest G S 2015 *Phys. Rev. E* **91** 022309
[15] Croxton C A 1974 *Liquid state physics: a statistical mechanical introduction* (Cambridge University Press) p 421
[16] Macleod D B, A M and Sc D 1923 *Trans. Faraday Soc.* **19** 38
[17] Zeldovich Y B 1949 *Zhur. Fiz. Khim.* **23** 931

IOP Conf. Series: Materials Science and Engineering **581** (2019) 012020 doi:10.1088/1757-899X/581/1/012020

Field-Induced Deformation and Structure Changes in a Magnetic Polymersome: Many-Particle Simulation

A V Ryzhkov, Yu L Raikher
Institute of Continuous Media Mechanics, Russian Academy of Sciences, Ural Branch, 614013 Perm, Russia

E-mail: ryzhkov.a@icmm.ru

Abstract. Coarse-grained molecular dynamics simulation is performed to analyse structure and deformation response of a magnetic polymersome – a submicron hollow capsule whose membrane is made of amphiphilic block-copolymer, and the intramembrane space is filled with magnetic nanoparticles. The major prospects of these objects are related to controlled drug release. The developed many-particle model is used to calculate equilibrium magnetized configurations of polymersomes with different intensity of magnetic interparticle coupling. The analysis of those data reveals that under external magnetic field, an initially spherical capsule stretches along the field direction assuming a spheroid-like shape. This behaviour is caused by the field-driven tendency of the nanoparticles to aggregate inside the polymersome membrane. At weak interparticle magnetic interaction no extended chains are observed. Accordingly, the deformation and density redistribution are relatively small. In a polymersome with strong interparticle coupling the nanoparticles self-organize in long chains aligned with the field, and this induces considerable elongation of the polymersome along with accumulation of the majority of the nanoparticles in the "equatorial" zone of the membrane of the capsule.

1. Introduction

The promising prospects of actively developing nanomedicine and theranostics are connected with synthesis of remotely-controlled polymeric microcarriers for drugs. Embedding of magnetic nanoparticles (MNPs) in submicron-size entities open vast opportunities of induced transport and drug release. A particular type of such microcontainers are magnetic microgels (microferrogels) [1, 2] comprising several hundred MNPs grafted to or trapped in a hydrogel molecular mesh. External magnetic fields enhance the interparticle magnetic coupling and make the MNPs to move both translationally and orientationally. The particles, experiencing resistance on the mesh, transfer some part of their motion to the network, thus, making a microferrogel sample to deform (shrink or swell) and to change its structure.

A fascinating example of remotely-driven containers is *magnetic polymersomes* (MPSs) [3]. These objects are hollow submicron capsules whose membranes consist of two layers of amphiphilic block-copolymer molecules with hydrophobic groups of each layer looking inward. The intramembrane space of an MPS has enough room to accommodate a monolayer of MNPs [4]. Figuratively speaking, the MPS membrane contains a small sample of a 2D ferrofluid. In the absence of external magnetic field, the intramembrane MNPs are either in a "gaseous" state or form chain-like structures due to their magnetic dipole-dipole interaction. However, independently of the particle aggregation, the pressure distribution in the membrane is isotropic on the average. Such an MPS, when suspended in a fluid, minimizes its surface energy and assumes spherical shape maybe to within some fluctuations.

An applied field aligns the MNP magnetic moments, this straightens the particle chains and makes the intramembrane pressure distribution non-uniform. Responding to those changes, the membrane deforms, and the MPS assumes anisometric shape. Elongation in the direction of the field was detected in experiments for synthesized MPSs [5] as well as in observations of model micron-sized colloidosomes [6].

The field-induced deformation is accompanied by the change of the volume of the internal cavity of the MPS, so that, depending on the particle and membrane properties, the polymersome might either release its content (e.g., a drug solution) or to suck in the fluid it is suspended in.

Along with this direct field-driven release/uptake mechanism, the field might have an implicit effect on the mass transfer of the carried chemical through the MPS wall. Namely, the filed-induced redistribution of MNPs inside the membrane modulates local permeability of the latter and, due to that, might quite strongly impede the diffusion transport through the wall. This issue, formerly, never addressed in theoretical literature, is the main goal of the present paper. In below we show that in an MPS elongated in the field direction, the "polar" zones are virtually free of the MPSs whereas the "equatorial belt" is enriched in them.

The description that we use here is based on the coarse-grained approach of many-particle dynamics. It had been successfully applied for simulation of magnetic gels [7–9]. One of the main its advantages is that it binds explicitly the degrees of freedom of all the elements of the system: both the polymer molecules and MNPs. In below, before presenting our particular results, we briefly outline the coarse-grained molecular model that we use; a more detailed account on that technique could be found in [10].

The geometry of the model is shown in figure 1. Two concentric shells form the membrane walls. The monolayer of magnetic beads with built-in magnetic moments is placed inside the membrane. As the interactions between the particles belonging to either shell are purely elastic, the initial equilibrium shape of the MPS is spherical. Note that the constructed MPS possess also some amount of intershell elastic bonds that helps to keep the membrane thickness approximately constant. The polymer shells are impenetrable for MNPs. As to the MNPs themselves, they are taken to be spheres of identical size bearing identical magnetic moments of constant magnitude. The particles are magnetically isotropic and interact with each other by way of dipole-dipole potential.

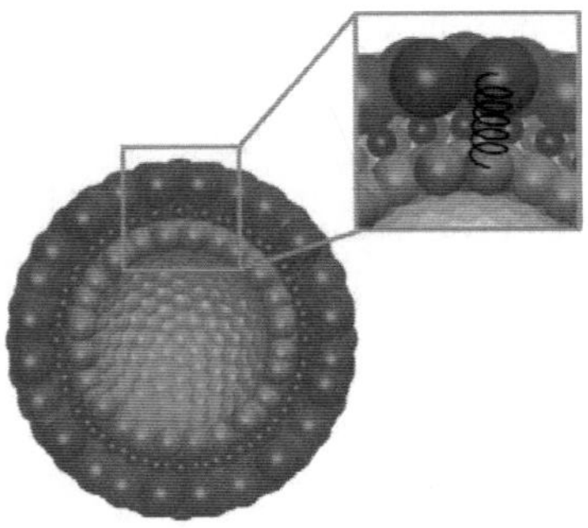

Figure 1. Geometry scheme (cross-section) of the model MPS. Blue and red beads are coarse-grained shells (layers) of the amphiphilic membrane. Black beads denote the monolayer of MNPs enclosed in the membrane. In the inset the elastic bonding between outer and inner shells is depicted.

The equilibrium configuration, i.e., positions of all the particles (beads) under any set of given conditions (the level of MNPs interaction, magnitude of the external field, etc.), is calculated from Newton equations of motion for each i-th bead:

$$m_i \ddot{\boldsymbol{r}}_i = \boldsymbol{F}_i - \zeta \dot{\boldsymbol{r}}_i + \boldsymbol{f}_i(t), \tag{1}$$

where m_i is the bead mass; the two last terms in the right-hand side are: the translational friction force with coefficient ζ and the random force produced by Langevin thermostat [11], respectively. The force $\boldsymbol{F}_i$ is derived from interaction of i-th particle with other elements and is denoted as $\boldsymbol{F}_i^{\text{shell}}$ for the shell beads and as $\boldsymbol{F}_i^{\text{MNP}}$ for the MNPs, see [10] for details.

Computer simulations of the polymersome deformation in external field were performed in software ESPResSo [12]. The major parameters which define the magnetodeformation effect are:

- ratio of the pairwise dipolar magnetic interaction energy of particles of diameter d_{MNP} and magnetic moments μ to thermal energy: $\lambda = \mu^2 \left(d_{\text{MNP}}^3 k_{\text{B}} T\right)^{-1}$;

IOP Conf. Series: Materials Science and Engineering **581** (2019) 012020 doi:10.1088/1757-899X/581/1/012020

- ratio of the energy of interaction of magnetic moment μ with external field of strength H_0 to thermal one: $\xi = \mu H_0 \left(k_B T\right)^{-1}$.

2. Results and Discussion

The above-described coarse-grained model was applied to investigate structural changes which cause the deformation of an MPS in a uniform external magnetic field. The membrane of the polymersome contains a fixed number of MNPs. In the presented here simulation template it equals 393 that is comparable to that in actually synthesized samples [13]. In the geometry scale of the model, the nanoparticle volume fraction in the intramembrane space is about $\varphi_{\text{vol}} = 11\%$.

Two types of model polymersomes were studied. The MPS of the first type are filled with MNPs characterized by weak interparticle magnetic coupling: their dipole-dipole interaction parameter is $\lambda = 1$. The MPS of the second type contains the MNPs which interact much stronger, for them $\lambda = 5$. In the performed computer simulations, the objects of both types were exposed to gradually increasing magnetic field: parameter ξ grew from 0 to 10.

Under these conditions, the randomly oriented magnetic moments of MNPs are stimulated to align with the direction of the applied field. Notably, the weakly interacting MNPs do not display a pronounced tendency for aggregation. To the contrary, the strongly interacting MNPs readily form elongated columns (chains) parallel to the field direction. This difference in structuring clearly manifests itself if to compare the extent of elongation of the MPS of both types.

To assess the MPS deformation, we define the aspect ratio of an anisometric object as

$$R = 2b/\left(a + c\right), \tag{2}$$

where the dimension parameters are presented in figure 2 as follows from equation (2), $R > 1$ for a prolate spheroid and $R < 1$ for an oblate one. In simulations, the aspect ratio is calculated by evaluating the diametrically opposed extreme points in three perpendicular directions.

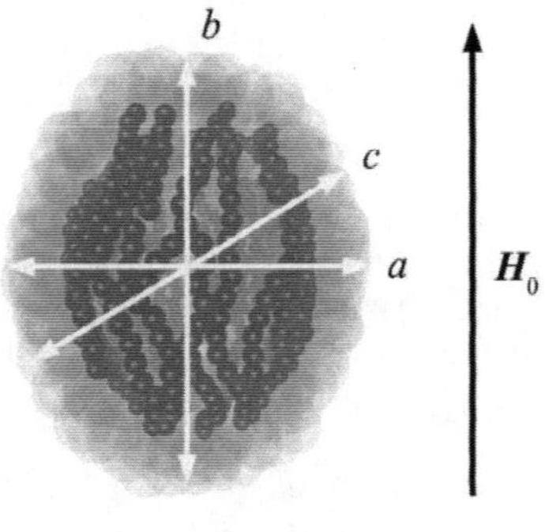

Figure 2. Scheme of finding of deformed polymersome axes to define aspect ratio as it is specified in equation (2).

The obtained dependences of the MPS aspect ratio on the nondimensional strength ξ of applied field are plotted in figure 3(a) for $\lambda = 1$ and in figure 3(b) for $\lambda = 5$. The numerical data (dots) were approximated by sigmoid curves (red lines). For weak dipolar interactions the energy of thermal fluctuations is comparable with the energy of MNPs magnetic attraction, and, therefore, the extent of aggregation is rather low. As figure 3(a) shows, up to $\xi \approx 3$ the data most probably just fluctuates. Out of this region, i.e., at higher fields, the fluctuations fade out, however, so that one may infer that the attained aspect ratio is indeed different from unity. Nevertheless, it always remains quite small and saturates at the level of about 5% above $R = 1$.

For the particles with pronounced tendency for field-induced aggregation ($\lambda = 5$), the effect is several times greater. The fluctuation region occupies a more narrow interval (up to $\xi \approx 2$); whereas beyond this range the aspect ratio grows as high as 17% at $\xi \approx 10$ and does not seem completely saturated.

IOP Conf. Series: Materials Science and Engineering **581** (2019) 012020 doi:10.1088/1757-899X/581/1/012020

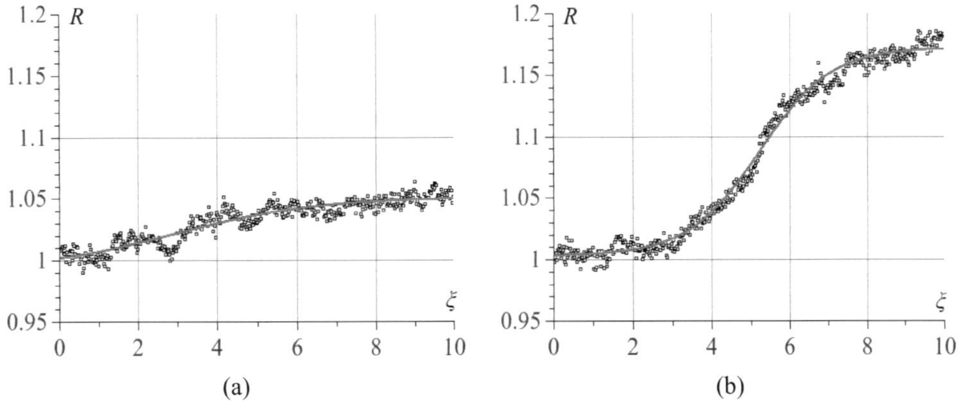

Figure 3. Aspect ratio dependences on the strength of the applied field ξ for model polymersomes with intramembrane content of MNP filler $\varphi_{vol} = 11\%$; pane (a) – weakly interacting MNPs ($\lambda = 1$); pane (b) – strongly interacted MNPs ($\lambda = 5$). Dots resemble the data obtained in numeric experiments, red lines are sigmoid curves used as eye guides.

The shape changing process in the MPS with $\lambda = 5$ is accompanied by rearrangement of the MNPs chains. Under the action of the field, previously existing randomly oriented chains straighten and strive to settle along the field direction. Being confined within a quasi-spherical membrane, they have to bend assuming the curved contours resembling the Earth meridians. Those arrangements are never perfect, however, due to the thermal motion of the MNPs constituting the chains.

As the chains are magnetized in the same direction, they repel each other and strive to find the positions which are most remote from each other. Therefore, the azimuthal distribution of the particles (we take the direction of the field as the polar axis) becomes angular-modulated. This effect is easy to foresee, and the spatial scale of this type of the MNP density non-uniformities caused by it is small in comparison with the overall size of the MPS. Indeed, those perturbations is of the same order of magnitude as they are in a non-magnetized MPS where the chains are oriented at random.

The distribution of MNPs in a magnetized polymersome with respect to polar angle is a more complicated problem. If to consider it from the magnetostatic viewpoint, one encounters two opposing tendencies. On the one hand, any chain reduces its energy the more the longer it is. Thus, the most favorable pattern for a single chain is to stretch along "meridian" from pole to pole. On the other hand, the mutual repulsion enhances the closer are the neighboring chains, and this means that the chains would avoid the places where their parts come too close to one another. In the given confined geometry those places are located in the poles of the MPS. Therefore, to within the inaccuracy caused by Brownian motion, one would, in general, expect that the "polar" zones are depleted of the MNPs.

To get quantitative notion of the MNPs distribution between the polar and equatorial zones of the considered MPS, we do the following. Two boundary meshed surfaces of a simulated magnetized MPS are replaced by the corresponding spheroids whose semi-axes are taken from evaluation of the aspect ratio.

We denote the major and minor axes of the inner spheroid as L_{in} and l_{in}, respectively, and for the outer one as L_{out} and l_{out}, see figure 4. With allowance for the symmetry of the MPS body, it suffices to consider only half of the object. The residual hemisphere is crossed by a stack of planes perpendicular to the direction of external field. The lines where these planes intersect the inner and outer spheroid surfaces make respective "parallels" if to stick to the above-adopted geographical terminology.

Any pair of neighboring planes cut out of the space between the nested half-spheroids a circular ring of specified thickness and height (measured along the polar direction). Then the number of MNPs in a given ring is counted and divided by the volume of the ring. Taking the volumes of the inner and outer half-spheroids as

IOP Conf. Series: Materials Science and Engineering **581** (2019) 012020 doi:10.1088/1757-899X/581/1/012020

$$V_{\text{out}} = \pi\left(l_{\text{out}}^2 \left(x_2 - x_1\right) - \frac{l_{\text{out}}^2}{3L_{\text{out}}^2}\left(x_2^3 - x_1^3\right) \right), \quad V_{\text{in}} = \pi\left(l_{\text{in}}^2 \left(x_2 - x_1\right) - \frac{l_{\text{in}}^2}{3L_{\text{in}}^2}\left(x_2^3 - x_1^3\right) \right), \tag{3}$$

the resulting number density is presented in the form

$$n = \frac{N_{x_1,x_2}}{V_{\text{out}} - V_{\text{in}}}, \tag{4}$$

where N_{x_1,x_2} is the number of the MNPs located in the ring bounded by the planes crossing the longer semi-axis in the points with coordinates x_1 and x_2, see figure 4.

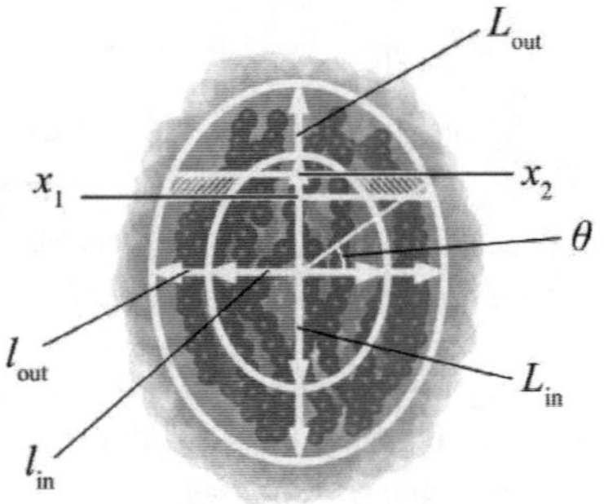

Figure 4. To determining the concentration distribution. Cross-section view of inner and outer surfaces of the polymersome membrane described by spheroids with semi-axes L_{in}, l_{in}, L_{out} and l_{out}. The analysed layer of thickness $x_2 - x_1$ has polar angle coordinate θ. The shaded regions denote the ring where concentration is analyzed.

Concentration distributions n of MNPs as functions of the polar angle θ for the polymersomes with weak and strong magnetic dipolar interaction are presented in figures 5 and 6 for several values of nondimensional external field ξ. As well in those figures are shown the snapshots of top view for each configuration.

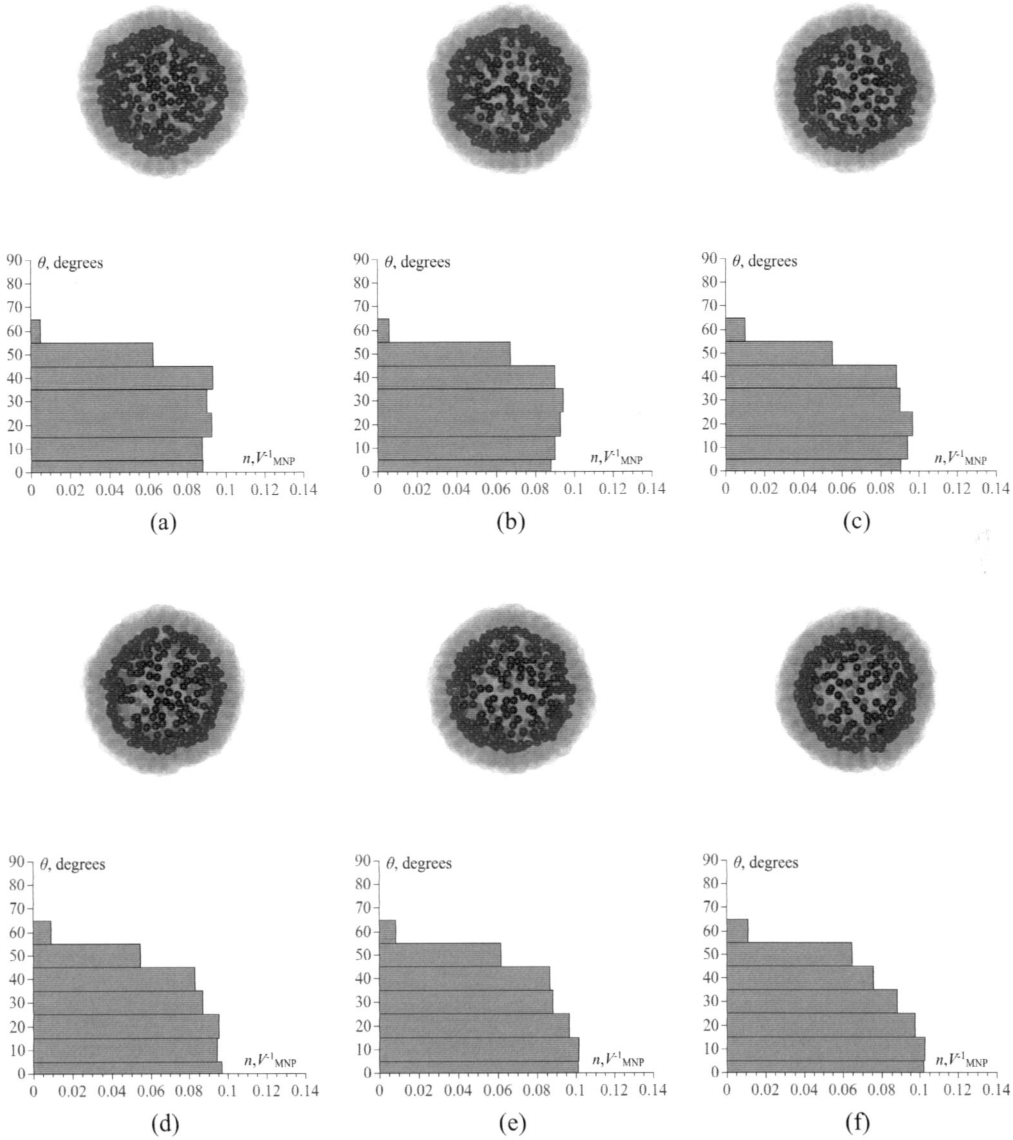

Figure 5. An MPS with volume content of the nanoparticles $\varphi_{\text{vol}} = 11\%$ and dipolar parameter $\lambda = 1$; snapshots of top views in the field directed perpendicularly to the plane of view) and polar angle dependences of the MNP concentration distribution (scaled in inverse MNP volume V_{MNP}^{-1}) for the field strength $\xi = 0$ (a), 2 (b), 4 (c), 6 (d), 8 (e), 10 (f).

IOP Conf. Series: Materials Science and Engineering **581** (2019) 012020 doi:10.1088/1757-899X/581/1/012020

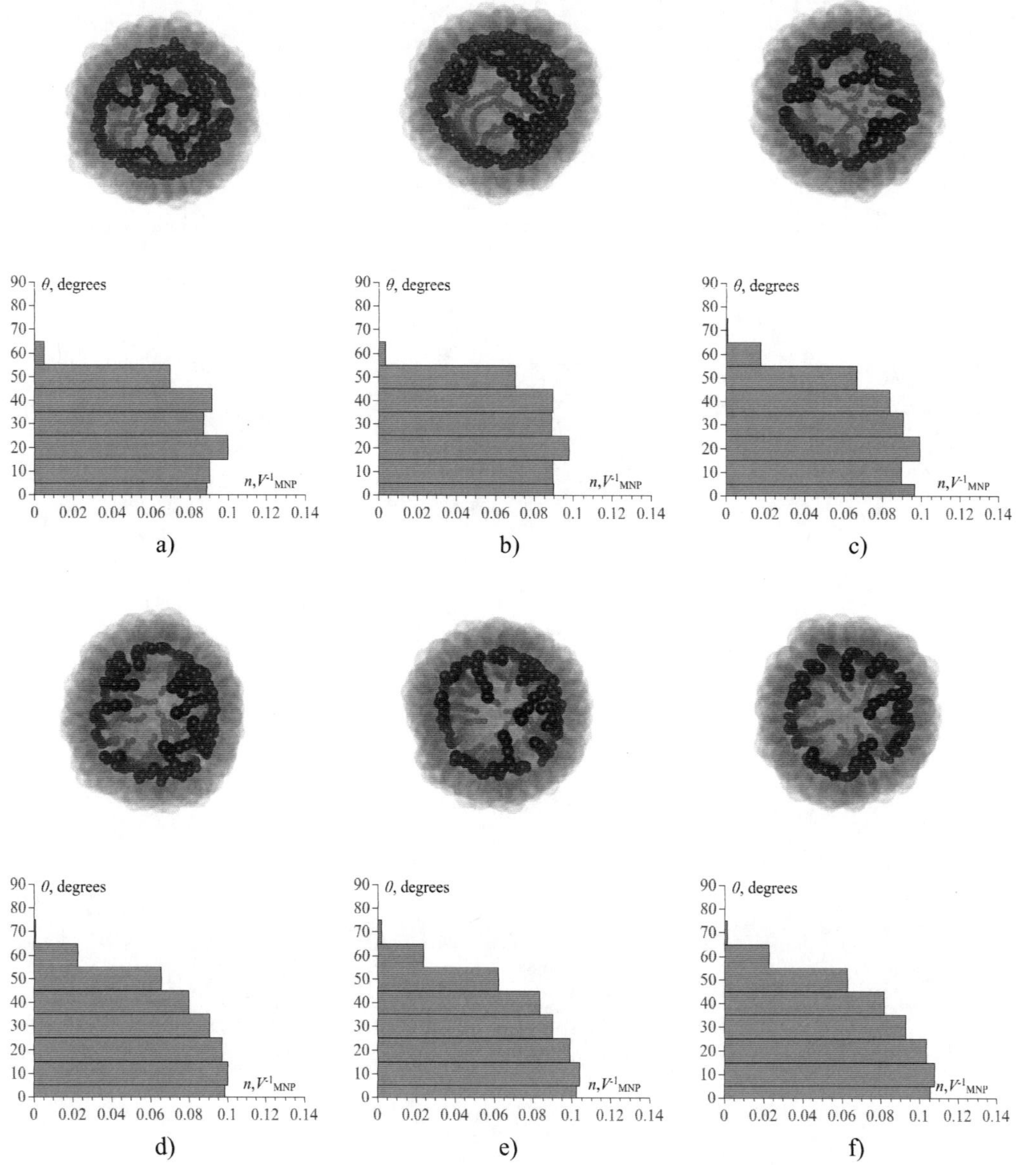

Figure 6. An MPS with volume content of the nanoparticles $\varphi_{\text{vol}} = 11\%$ and dipolar parameter $\lambda = 5$; snapshots of top views in the field directed perpendicularly to the plane of view) and polar angle dependences of the MNP concentration distribution (scaled in inverse MNP volume V_{MNP}^{-1}) for the field strength $\xi = 0$ (a), 2 (b), 4 (c), 6 (d), 8 (e), 10 (f).

Before commenting figures 5 and 6 we would like to make an important remark. One should consider the data presented in the plots as reliable only approximately up to $\theta \approx 60°$. Above that angle (in the near-polar zone), the slicing procedure that we use does not work well: the volume of the sliced rings becomes comparable with that of an MNP, and calculation errors (numeric fluctuations) become huge. To get notion of the particle distribution in the near-polar zone it is better to consult the snapshots: although qualitatively, but they are more correct for that case.

The snapshots of figures 5 and 6 illustrate how, under enhancement of the field, the MNP concentration in polar zones of an MPS gradually changes from notably dense pattern, as in figure 5(a), to more rarefied distribution, as in figure 5(f). This tendency is rather weak at $\lambda = 1$ and very well pronounced at $\lambda = 5$. As the angle plots show, the concentration distribution at zero field is uniform for any λ.

In real MPSs uneven concentration distribution may strongly affect to permeability of membrane for low molecular weight content. Namely, acting on the capsule by external field it is possible to change structural organization of magnetoactive particles and thereby cause deformation and control the release of substance. For the model MPSs with intensively interacted MNPs (figure 6) decreasing character of concentration expressed even stronger. At zero field, see figure 6(a), the MNPs are organized in randomly oriented chains encircling MPS membrane, that is why the large-scale initial distribution is uniform. With enhancement of the field strength, the aggregates rebuilt in the chains stretched along meridional direction (simultaneously, the aspect ratio changes significantly). In this case the poles are indeed strongly depleted of the MNPs, see the corresponding snapshots.

3. Conclusions

The developed coarse-grained model of a magnetic polymersome was applied to study the field-induced deformation and internal restructuring. The advantage of the employed many-particle approach is that it allows one to estimate (and partially quantify) the mechanical and structural changes in a magnetopolymer capsule. Visualization of the equilibrium nanoparticles configurations makes it possible to observe and confirm the redistribution of nanoparticles in the intramembrane space. The main conclusion is that the effect of nanoparticle accumulation in the equatorial zone is found, and the dependence of the effect on the intensity of intraparticle magnetic interaction is revealed.

Acknowledgments
This study was funded by Russian Foundation for Basic Research in the frameworks of research projects #18-31-00326 and #17-52-590504.

References

[1] Backes S, Witt M U, Roeben E, Kuhrts L, Aleed S, Schmidt A M and von Klitzing R 2015 Loading of PNIPAM Based Microgels with CoFe 2 O 4 Nanoparticles and Their Magnetic Response in Bulk and at Surfaces *Journal of Physical Chemistry B* **119** 12129–12137.

[2] Turcu R, Socoliuc V, Craciunescu I, Petran A, Paulus A, Franzreb M, Vasile E and Vekas L 2015 Magnetic microgels, a promising candidate for enhanced magnetic adsorbent particles in bioseparation: synthesis, physicochemical characterization, and separation performance *Soft Matter* **11** 1008–1018.

[3] Hickey R J, Haynes A S, Kikkawa J M and Park S-J 2011 Controlling the Self-Assembly Structure of Magnetic Nanoparticles and Amphiphilic Block-Copolymers: From Micelles to Vesicles *Journal of American Chemical Society* **133** 1517–1525

[4] Lecommandoux S, Sandre O, Chécot F, Rodriguez-Hernandez J and Perzynski R 2005 Magnetic Nanocomposite Micelles and Vesicles *Advanced Materials* **17** 712–718

[5] Sanson C, Diou O, Thévenot J, Ibarboure E, Soum A, Brûlet A, Miraux S, Thiaudière E, Tan S, Brisson A, Dupuis V, Sandre O and Lecommandoux S 2011 Doxorubicin Loaded Magnetic Polymersomes: Theranostic Nanocarriers for MR Imaging and Magneto-Chemotherapy *ACS Nano* **5** 1122–1140

[6] Long Y, Liu C, Zhao B, Song K, Yang G and Tung C-H 2015 Bio-inspired controlled release through compression–relaxation cycles of microcapsules *NPG Asia Materials* **7** e148–e148

[7] Weeber R, Kreissl P and Holm C 2019 Studying the field-controlled change of shape and elasticity of magnetic gels using particle-based simulations *Archive Applied Mechanics* **89** 3–16

[8] Minina E S, Sánchez P A, Likos C N and Kantorovich S S 2018 The influence of the magnetic filler concentration on the properties of a microgel particle: Zero-field case *Journal of Magnetism and Magnetic Materials* **459** 226–230

[9] Ryzhkov A V, Melenev P V, Balasoiu M and Raikher Y L 2018 The effect of magnetic nanoparticle concentration on the structure organisation of a microferrogel *Journal of Physics:*

Conference Series **994** 012004
[10] Ryzhkov A and Raikher Y 2018 Coarse-Grained Molecular Dynamics Modelling of a Magnetic Polymersome *Nanomaterials* **8** 763
[11] Berendsen H J C, Postma J P M, van Gunsteren W F, DiNola A and Haak J R 1984 Molecular dynamics with coupling to an external bath *Journal of Chemical Physics* **81** 3684–3690
[12] Arnold A, Lenz O, Kesselheim S, Weeber R, Fahrenberger F, Roehm D, Košovan P and Holm C 2013 ESPResSo 3.1: Molecular Dynamics Software for Coarse-Grained Models *Lecture Notes in Computational Science and Engineering* vol 89, ed M Griebel and M A Schweitzer (Berlin, Heidelberg: Springer Berlin Heidelberg) pp 1–23
[13] Hickey R J, Koski J, Meng X, Riggleman R A, Zhang P and Park S J 2014 Size-controlled self-assembly of superparamagnetic polymersomes *ACS Nano* **8** 495–502

IOP Conf. Series: Materials Science and Engineering **581** (2019) 012021 doi:10.1088/1757-899X/581/1/012021

Numerical Calculation of the Process Removal of Localized Convective Structures in a Layer of a Porous Medium

T N Zagvozkin[1,a], **T P Lyubimova**[1,2]

[1] Institute of Continuous Media Mechanics UB RAS, Perm, Russian Federation
[2] Perm State University, Perm, Russian Federation

E-mail: [a] timofey@zagvozkin.ru

Abstract. Thermal convection in a flat horizontal layer of a porous medium with solid impermeable boundaries on which the heat flow is given is considered. The porous medium is saturated with a viscous incompressible fluid pumped along the layer. In the system under discussion, with a vertical heat flow inhomogeneous along the layer, localized convective structures may occur in the region where the heat flow exceeds the critical value corresponding to homogeneous heating from below and corresponding to the beginning of convection in the layer. With an increase in the rate of longitudinal pumping of the liquid through the layer, a transition from a state in which the localized convective structures are stable to a state in which the localized convective flow is completely washed out of the region of its excitation occurs. Calculations were performed in the framework of Darcy-Boussinesq. Results of the numerical calculation of the process removal of localized convective structures from the zone of its excitation with an increase in the rate of longitudinal pumping of liquid through the layer are presented. The map of the system state modes is obtained.

1. Introduction

Thermal convection in porous media is of interest both for applied problems related to technical and natural processes, and from the point of view of mathematical physics. In connection with many practical manifestations, special attention is drawn to the development of convective flows from localized heat sources. Thus, for a homogeneous liquid in different versions of the problem statement, a convective torch from a point source in an infinite medium and from a horizontal linear source was taken into account [1]. The stability of the convective flow of a homogeneous liquid caused by heating inhomogeneity was studied in [2]. For the first time, the problem of thermal convection in a horizontal layer of a porous medium bounded by solid impermeable isothermal boundaries was theoretically considered in [3], and experimentally in [4]. The study of a convective torch from a point source of heat in an unlimited array of porous media was first carried out in [5], from a horizontal linear source — in [6]. Free and forced convection in the vicinity of a linear heat source or a heated cylinder was discussed in [7]. Large-scale convection in a liquid-saturated horizontal layer of a porous medium bounded by solid boundaries, with a given heat flow and fluid pumping through the layer was the object of study in [8, 9].

In this paper, we study the process of leaching localized convective structures in a horizontal layer of a porous medium from the region of their excitation.

2. Problem statement

The paper considers a horizontal layer of porous medium saturated with viscous incompressible liquid. The layer is limited to solid surfaces, which is specified inhomogeneous horizontal vertical heat flux. The coordinate system is chosen so that the plane (x,y) is horizontal, and the lower and upper boundaries of the layer correspond to the coordinates $z = 0$ and $z = h$. Temperature differences are considered small, so we can assume that the density of the liquid depends on them linearly. Figure 1 shows the scheme of the problem.

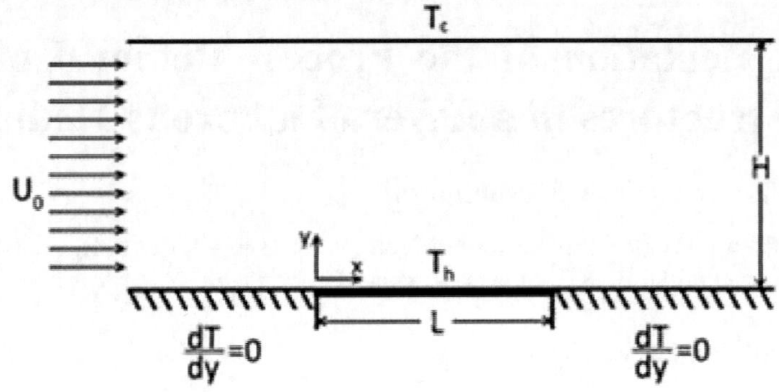

Figure 1. Schematic representation of the problem.

Convective filtration in the layer was described in terms of the current function ψ and the temperature T by the equations

$$\left[\frac{\partial^2\psi}{\partial^2 X}+\frac{\partial^2\psi}{\partial^2 Y}\right]=\frac{Ra}{Pe}\frac{\partial T}{\partial X}. \tag{1}$$

$$\frac{\partial T}{\partial t}=\left[\frac{\partial^2 T}{\partial^2 X}+\frac{\partial^2 T}{\partial^2 Y}\right]+Pe\left[\frac{\partial\psi}{\partial Y}\frac{\partial T}{\partial X}-\frac{\partial\psi}{\partial X}\frac{\partial T}{\partial Y}\right]. \tag{2}$$

The equations are written in dimensionless form and contain the following dimensionless parameters: the filtration Rayleigh number and the Péclet number

$$Ra=Kg\beta(T_h-T_c)H/\alpha\nu. \tag{3}$$

$$Pe=U_0H/\alpha. \tag{4}$$

where K - the permeability of the medium, g - acceleration due to gravity, β - coefficient of thermal expansion, T_h - the temperature of the heater, T_c - the temperature of the upper boundary, H - the layer thickness, α - the effective thermal conductivity of the saturated medium, ν - the kinematic viscosity of the fluid, U_0 - the speed of longitudinal pumping of the liquid through the layer.

The initial conditions for the fields of the current function and the temperature were set in the form:

$$T(X,Y,0)=0. \tag{5}$$

$$\psi(X,Y,0)=Y. \tag{6}$$

The boundary conditions on the upper and lower boundaries of the layer:

$$\psi(X,1,t)=1. \tag{7}$$

$$T(X,1,t)=0. \tag{8}$$

$$\psi(X,0,t)=0. \tag{9}$$

$$T(X,0,t)=1,\ \text{for}\ -L/2\le X\le L/2. \tag{10}$$

$$\frac{\partial T}{\partial Y}(X,0,t)=0,\ \text{for}\ X<-L/2\ \text{и}\ X>L/2. \tag{11}$$

The problem was solved numerically using the finite difference method. The calculations were carried out in a horizontal layer cell with a length of 10 (sufficient to exclude the influence of lateral boundaries at the considered lengths of the excitation region L). On the vertical boundaries periodicity condition were set.

3. Results

Figure 2 shows the dependence of the maximum value of the current function on time, obtained in the calculations, illustrating the temporal evolution of the initial perturbations of the ground state corresponding to the regime of homogeneous pumping, for the Rayleigh number equal to 100, and different values of the Péclet number (for clarity, the maximum values of the function $\bar{\psi} = \psi - \psi_0$, are presented, where $\psi_0 = 1$ corresponds to the regime of homogeneous pumping). The inserts show the structure of the steady flow for each of these values of the Péclet number. As can be seen from the figure, with a small Péclet number (Pe=1), the intensity of the convective flow increases with time and after the transition process reaches a constant value, i.e. a stationary convection regime in the layer is observed. At a higher Péclet number (Pe=8) at the initial stage, the perturbations also increase, but after the transition process, a mode of stationary oscillations is established. With an even greater Péclet number (Pe=15), the initial perturbations of the ground state fade with time and after the transition process, the maximum value of the current function goes to a value equal to 1, which corresponds to the regime of homogeneous pumping (the value $\bar{\psi} = \psi - \psi_0$ decreases to zero), i.e. it can be said that the convective structures are completely washed out of the layer. Note that in this situation, for any number of Rayleigh, there is a weak convective flow near the boundaries of the heater caused by the heterogeneity of heating (the presence of a horizontal temperature gradient).

Figure 3 shows a map of modes on the Pe(L) plane obtained by processing numerical data for different values of the Rayleigh number (for the construction of curves, data on the time evolution of initial disturbances similar to those shown in figure 2, in different ranges of parameters). Crosses in the figure marked the points corresponding to the parameters for which the figure 2 shows the dependence of the maximum value of the current function on time. For Ra=100 above the curve with points marked by circles, a uniform pumping mode is implemented - the excited convective flow is completely washed out of the region of its excitation by longitudinal pumping. Below the curve with the points shown by the squares, after the transition process a stationary regime is established. At values of parameters in an interval between the specified curves the stationary oscillation mode is observed. For Ra=80 and Ra=60, the lower curves are absent, at these values, the pumping of the liquid through the layer at any speed leads to the appearance of an oscillating convection mode. Above the curves with the points shown by the circles, the behavior is similar to the case Ra=100.

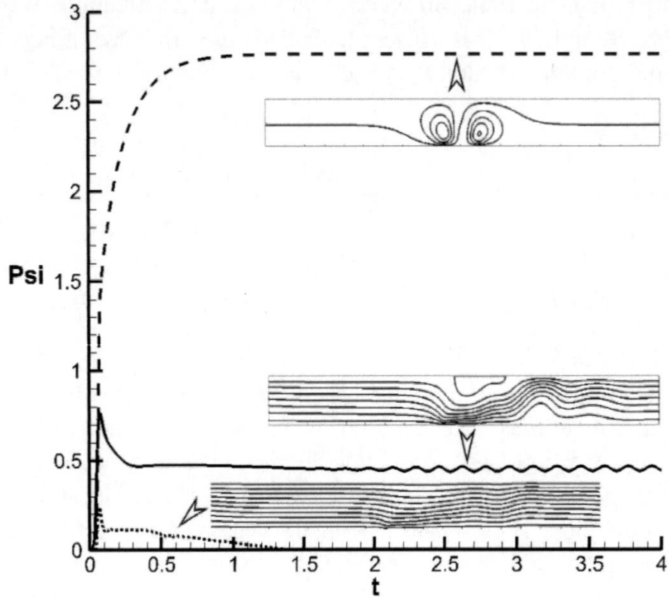

Figure 2. The dependence of the maximum value of the current function on time at L=1, Ra=100, Pe=1 (dashed), Pe=8 (solid), Pe=15 (dotted).

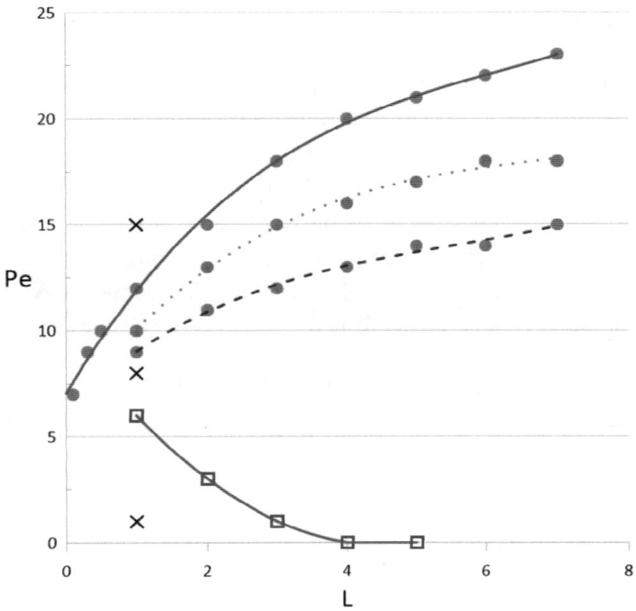

Figure 3. Convection mode map at Ra=60 (dashed), Ra=80 (dotted), Ra=100 (solid).

Figure 4 shows the dependence of the wave number on the Péclet number for the oscillatory modes of convection. The inserts show the contours of the current function, corresponding to the extreme points of the range of values of the Péclet number, shown in the figure. As can be seen, the wave number of structures monotonically increases with the increase in the number of Péclet by a law close to linear.

Figure 5 presents the dependence of the oscillation frequency on the Péclet number. As can be seen, with the increase in the Péclet number, a monotonic increase in the frequency of oscillations is observed by a law close to linear.

Figure 6 shows the dependence of the maximum of stream function on the Péclet number. The asymptotic approximation of the curve to the value $\psi = 1$, corresponding to the regime of homogeneous pumping, indicates the attenuation of convection with an increase in the pumping rate, associated with the leaching of convective structures from the region of their excitation.

Figure 7 shows the typical structure of the convective flow occurring in the layer. The vortices washed out by the main current from the field of excitation (moving to the right) in the direction of fluid pumping are visible.

Figure 8 shows the temperature field corresponding to the same parameter values.

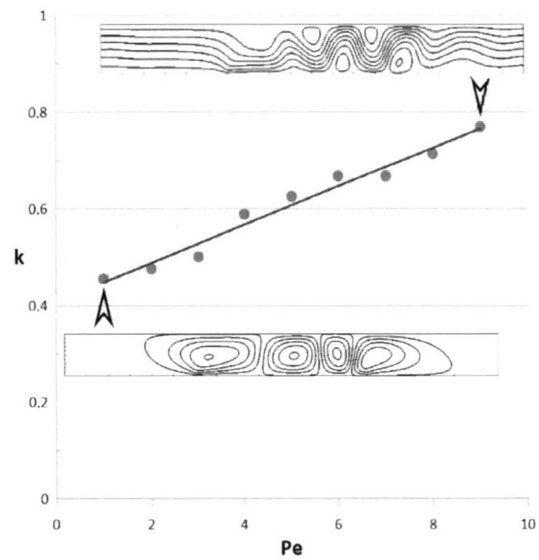

Figure 4. The dependence of the wave number from the Péclet number at L=4, Ra=100.

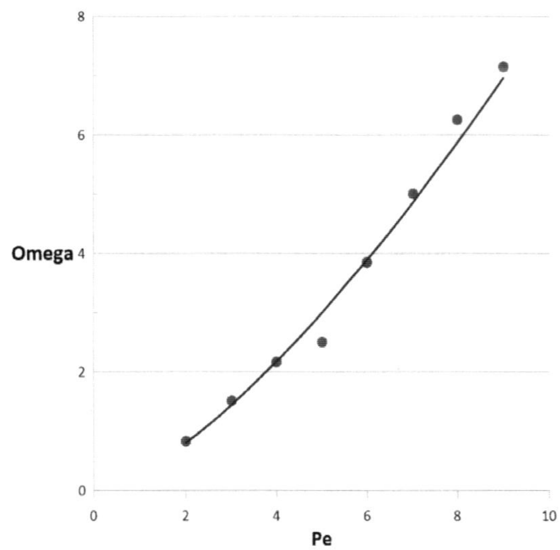

Figure 5. The dependence of the frequency from the Péclet number at L=4, Ra=100.

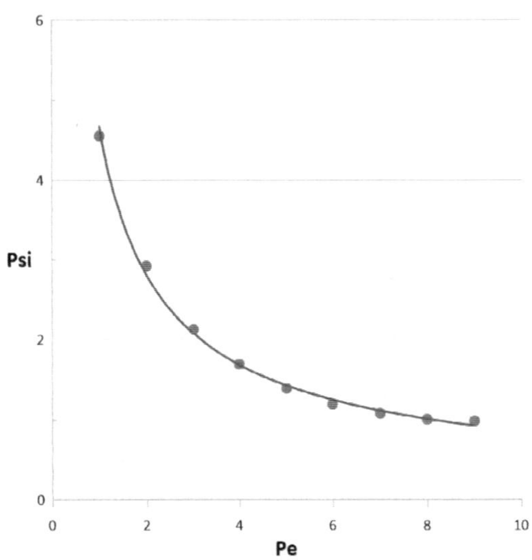

Figure 6. The dependence of the maximum value of the current function from the Péclet number at: L=4, Ra=100.

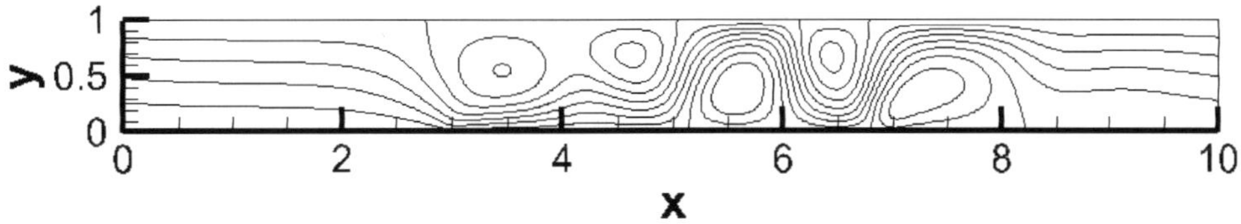

Figure 7. Contours of stream function for a set of parameters: L=4, Ra=100, Pe=5, t=2.5.

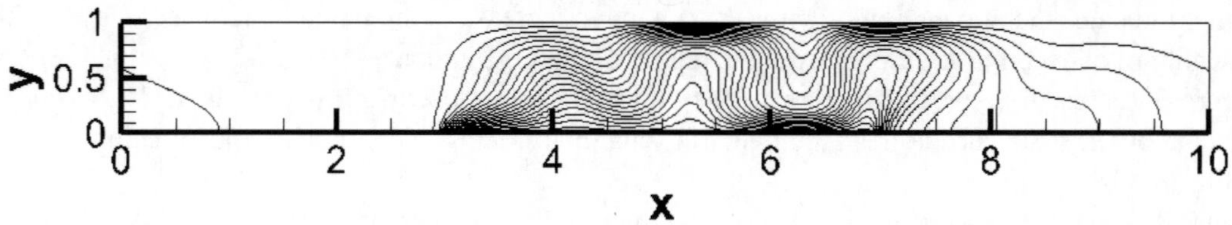

Figure 8. Temperature contours for a set of parameters: L=4, Ra=100, Pe=5, t=2.5.

4. Conclusion

Convection in a horizontal layer of a porous medium saturated with a liquid in the presence of an inhomogeneous heat flow at the boundaries and liquid pumping along the layer is studied. The map of convection regimes in the parameter plane of the length of the region of excitation - the Péclet number, for several Rayleigh numbers, is obtained. It is found that at low Péclet numbers a stable localized stationary convective flow is observed. At high Péclet numbers a localized oscillating current is implemented. At even higher Péclet numbers localized convective structures are completely washed out of the region of their excitation and there is a regime of homogeneous fluid pumping.

Thus, with the increase in the rate of liquid pumping through the layer (at a given length of the excitation region), the system from the mode of stationary convection passes to the vibrational mode, and then to the mode of homogeneous pumping, similar to what was observed in the scenarios of "leaching" of localized structures arising from random spatial inhomogeneity of the parameters in [10].

References

[1] Brand R.S., Lahey F.J. The heated laminar vertical jet // *J. Fluid Mech.* – 1967. – Vol. **29**, no. 2. – P. 305-15.

[2] Lyubimov D.V., Cherepanov A.A. Stability of convective flow induced by inhomogeneous heating. *Konvektivnye techeniya – Convective flows*, Perm State Pedagogical University, 1991. Pp. 17-26.

[3] Horton C.W. Rogers Jr. F.T. Convection currents in a porous medium // *J. Appl. Phys.* – 945. – Vol. **16**, no. 6. – P. 367-70.

[4] Morrison H.L., Rogers Jr. F.T., Horton C.W. Convection currents in porous media: II. Observation of conditions at onset of convection // *J. Appl. Phys.* – 1949. – Vol. **20**, no. 11. – P. 1027-9.

[5] Wooding R.A. Convection in a saturated porous medium at large Rayleigh number or Peclet number // *J. Fluid Mech.* – 1963. – Vol. **15**, no. 4. – P. 527-44.

[6] Nakayama A. Free Convection from a horizontal line heat source in a power-law fluid-saturated porous medium // *Int. J. Heat Fluid Flow.* – 1993. – Vol. **14**, no. 3. – P. 279-83.

[7] Kurdyumov V.N., Liñán A. Free and forced convection around line sources of heat and heated cylinders in porous media // *J. Fluid Mech.* – 2001. – Vol. **427**. – P. 389-409.

[8] Goldobin D.S., Shklyaeva E.V. Large-scale thermal convection in a horizontal porous layer // *Phys. Rev.* E. – 2008. – Vol. **78**, no. 2. – 027301.

[9] Goldobin D.S., Lyubimov D.V. Soret-driven convection of binary mixture in a horizontal porous layer in the presence of a heat or concentration source. *J. Exp. Teor. Phys.*, 2007, Vol. **104**, no. 5, pp. 830-6.

[10] Goldobin D.S., Shklyaeva E.V. Localization and advectional spreading of convective flows under parametric disorder // *Journal of Statistical Mechanics: Theory and Experiment*. – 2013. – Vol. **2013**.

IOP Conf. Series: Materials Science and Engineering **581** (2019) 012022 doi:10.1088/1757-899X/581/1/012022

Patient-Specific Bile Flow Simulation to Evaluate Cholecystectomy Outcome

A Kuchumov

Department of Computational Mathematics, Mechanics, and Biomechanics, Perm National Research Polytechnic University, 29 Komsomolskiy Prospect, Perm 614990, Russia

E-mail: kychymov@inbox.ru

Abstract. Gallbladder disease has one of the leading positions by patients' number in the World. About 16% of the adult population is suffering now from this disease. Cholecystectomy is believed to be a general surgical method of the gallbladder disease treatment, but the success rate is quite low because the surgeons do not take into account the patient-specific features during the treatment and cannot predict operation results.The main purpose of the paper is to create a computational tool for numerical evaluation of cholecystectomy outcome compared with healthy state and current pathological state based on the patient-specific patients' data. The patient-specific features of the biliary tree were studied by 1-way FSI bile flow simulation. The extra-hepatic biliary tree geometries were extracted from MRI and after that imported to ANSYS CFX for the subsequent fluid dynamics analysis.It was revealed that in the pathology state, velocities were found to have lower magnitude while the pressures were higher. The patient-specific features have a dramatic influence on the bile flow patterns. Cholecystectomy leads to the decrease of bile flow rate in the extra-hepatic biliary tree. The proposed computational approach can be applied to medical practice to evaluate the circumstances of surgical interventions.

1. Introduction

Personalized medicine, taking into account the individual approach to the treatment of a particular patient, has considerable potential for improving the quality and effectiveness of medical care. The development of non-invasive diagnostic methods, modern mathematical and computer models allows us to describe the biomechanical processes occurring in the body with an increasing degree of accuracy. This circumstance increases the possibility of their use in improving existing and developing new personalized methods for diagnosing and predicting treatment.

In Russia, about 25 million people suffer from cholelithiasis and diseases of the biliary tract [1]. Gallstone disease is a multifactorial and multi-stage disease associated with the formation of stones in the gallbladder and / or biliary tract. The pathology of the biliary system is in third place in the World in the number of patients after cardiovascular diseases and cancer [2]. The presence of stones in the biliary tract and gallbladder can lead to various complications, ranging from inflammation of the ducts, ending with a fatal outcome due to malignant tumors of the biliary system [3].

For the treatment of patients with cholelithiasis, surgical removal of the gallbladder is usually used (cholecystectomy). There are about 2 million cholecystectomies (operations to remove the gallbladder) per year, including 320 thousand in Russia [4]. However, in 15% of cases, the results of this operation lead to postoperative complications [5,6]. One of the reasons is the use of subjective experience and the insufficient number of individualized biomechanical models for the analysis of surgical interventions [7]. It should be noted that there are few works on biomechanical modeling of the functioning of the biliary system in normal and pathological conditions [8].

Compliance of soft tissues walls plays a crucial role in the biofluids transport. In particular, the motor function alterations of the extra-hepatic ducts have an influence on the bile flow. Fluid flow in the elastic and compliant vessels can be described by various models (for example, Windkessel model,peristaltic flow model, fluid–structure interaction, Navier–Stokes equations, e.t.c.).

Study of the bile flow in the cystic duct

The cystic duct geometry effect on the flow resistance with using of 2D- and 3D-models of bile flow was studied inpaper [9]. The cystic duct was represented in the form of a straight pipe with baffles. Bile was assumed to be Newtonian, a fluid with a viscosity in the range of 1–4 mPa·s. Further, the results were compared with a more realistic two-dimensional model obtained as a result of processing the patient's cholangiogram. It was found that both the height and the number of baffleshave a dramatic effect on the flowresistance. According to calculations, the channel geometry has a greater effect on the distribution of velocities and pressures in the duct than the change in viscosity.

Ooi et al. [9] and Al-Altabi, et al. [10, 11] conducted a series of experimental studies with a uniform arrangement of partitions. In experiments, the Reynolds number is Re> 50, which is significantly higher than in the human biliary system (Re~10). Thus, only qualitative comparisons between experiment and numerical simulation are currently available. However, for large Reynolds numbers, a good quantitative agreement between numerical results and experimental measurements was obtained by Al-Altabi et al. [12–14]. In these studies, the cystic duct was considered as a separate channel with rigid walls. Li et al. [15] constructed two one-dimensional T-shaped models of the cystic duct, common hepatic and common bile duct with rigid and elastic walls. The phases of emptying and filling were considered.

Using the developed one-dimensional model, the effect of geometry, the elasticity of the wall of the cystic duct, the flow rate and viscosity of bile on the pressure gradient was studied in detail.

It was found that many factors, including the elastic modulus and bile viscosity, can influence the pressure gradient. However, the influence of changes in the geometry of the cystic duct was still dominant. It was found that the number of baffles has a significant effect, because it causes large changes of the channel equivalent diameter. The pressure gradient also increases with a decrease in the Young's modulus of the cystic duct during emptying.

A series of works [7, 16–18] was devoted to aspects of the flow of bile in the cystic duct during pathology (presence of a stone, narrowing of the ductal lumen). Bile was considered as a non-Newtonian fluid (Herschel–Bulkley model) by analogy with [19].

Al-Atabi et al. [20] presented an individualized model of the flow of bile in the cystic duct. Bile was considered a Newtonian fluid. The geometry was formed from plastic castings of real cystic ducts, surgically removed from patients and scanned using Model Maker W (3D Scanners, UK) to create three-dimensional geometric profiles. Simulations were performed to obtain quantitative indications for comparison with clinical measurements. The results showed that the dimensionless pressure gradient in the cystic duct is 4 times higher than in a straight round tube of equivalent length and average diameter.

Study of the flow of bile in the common bile duct

Maiti and Misra [21] presented a model of peristaltic flow in a porous channel to study the influence of various factors (critical pressure, porosity parameter and bile velocity) on reflux. It was found that with the presence of gallstones, the rate of bile increases as the porosity parameter increases, and the critical pressure for reflux decreases as the porosity increases. A mathematical model of the pathological flow of bile as a Casson fluid in a channel with a stone was presentedin paper [22]. A bile velocity profile versus time was obtained depending on the size of the lumen. In addition, the dependence of the pressure in the duct on the size of the stone was obtained. The created mathematical model allowed us to estimate the dynamics of the postoperative period and to predict the development of specific complications based on the values of bile pressures. Information about the size of the stone, obtained using cholangiography or ultrasound, makes it possible to calculate the daily consumption of bile entering the duodenum.

Construction of three-dimensional individualized geometry of the bile ducts

It should be noted that in the majority of works on modeling the flow of bile, only the movement of biofluid in individual segments was considered, and the complete flow model was not taken into account. Some attempts to examine the flow in the complete biliary system were made in [23, 24], but they did not take into account the individualized geometry of the ducts, their ductility, the influence of the gallbladder, and the peristalsis of the sphincter of Oddi. The first papers [25, 26] on the construction of individualized geometry were published in 2013.

IOP Conf. Series: Materials Science and Engineering **581** (2019) 012022 doi:10.1088/1757-899X/581/1/012022

Summary

Based on the above, when analyzing the numerical and mathematical modeling of the flow of bile in thebiliary system, it was noted that only the individual elements of the biliary system were considered, and the entire system was not considered completely. The main segments in the simulation of the flow of bile were the cystic and common bile ducts. In this paper we present patient-specific modeling of bile flow to analyze bile flow in the healthy state, pathology, and cholecystectomy.

2. Materials and Methods

2.1. Image acquisition

MRI data of 14 patients were obtained using 1.5T Siemens scanner. The in-plane spatial resolution was 0.5 m. ITK-Snap open source software was used to create 3D models of extra-hepatic ducts for subsequent simulations.

2.2. Problem formulation

Mass and momentum conservation equations for an incompressible fluid can be expressed as

$$\nabla \cdot u = 0, \tag{1}$$

$$\rho_f \left(\frac{\partial u}{\partial t} + ((u - u_g) \cdot \nabla) u \right) = -\nabla p + \nabla \cdot \tau \tag{2}$$

where is ρ_f is the fluid density, p is the pressure, u is the fluid velocity vector and u_g is the moving coordinate velocity. In the ALE formulation, $(u-u_g)$ is the relative velocity of the fluid with respect to the moving coordinate velocity. Here τ is the deviatoric stress tensor. This tensor is related to the strain rate tensor; if the fluid is incompressible and viscosity is constant across the fluid, this equation can be written in terms of an arbitrary coordinate system as

$$\tau_{ij} = \mu \left(\frac{\partial v_i}{\partial x_j} + \frac{\partial v_j}{\partial x_i} \right), \tag{3}$$

where x_j is the th spatial coordinate, v_i is the fluid's velocity in the direction of axis i, τ_{ij} is the jth component of the stress acting on the faces of the fluid element perpendicular to axis .

In the case of pure shear this relation reduces to

$$\tau = \eta \dot{} \tag{4}$$

For a healthy bile, which is considered as Newtonian fluid (η=0.001 Pa·s) [9]

$$\eta = \text{const.} \tag{5}$$

The Carreau's equation, which is used for simulation of lithogenic bile, is written as follows

$$\eta = \frac{\eta_0 - \eta_\infty}{(1 + (a \cdot \dot{\gamma})^2)^k} + \eta_\infty . \tag{6}$$

The parameters of the Carreau's equation were taken from [24]. The momentum conservation equation for solid body can be written as follows

$$\nabla \cdot \sigma_s = \rho_s \dot{}_s \tag{7}$$

where ρ_s, σ_s, and $\dot{}_s$ are solid density, solid tensor, and local acceleration of the solid, respectively.

It is known that blood vessels can be described as hyperelastic material [27]. Due to similar anatomical composition, the bile ducts can be also considered as hyperelastic materials. For hyperelastic materials, the stress-strain relation is written as follows

$$\sigma_s = \frac{\partial W}{\partial \varepsilon}. \tag{8}$$

The Mooney-Rivlin hyperelastic potential is taken in the form

$$W = c_{10}\left[\bar{I}_1 - 3\right] + c_{01}\left[\bar{I}_2 - 3\right] + \frac{1}{d}[J-1]^2, \tag{9}$$

The FSI interface conditions are written as follows. At first, it is necessary that displacements of the fluid and solid domain must be compatible:

$$\delta_s = \delta_f,$$

secondly, the tractions at this boundary must be at equilibrium:

$$\sigma_s \cdot \hat{n}_s = \sigma_f \cdot \hat{n}_f,$$

and at the last, the no-slip condition for the fluid is

$$u = u_g,$$

where δ, σ and $\hat{n}$ are, respectively, displacement, stress tensor and boundary normal with the subscripts 'f' and 's' indicating a property of the fluid and solid, respectively.
Boundary conditions are shown in figure 1. Boundary conditions correspond to emptying phase. During the gallbladder emptying, the bile comes out from the gallbladder and the liver at the same time. Bile pressures and initial velocity were applied to inlets and outlet. Following Howard et al. [28], who measured the mean flow rate of the bile from the gallbladder (2.0–3.0 ml/min) after meal, we recomputed it as the boundary condition as 3 mm/s. Pressure in the outlet was taken equal to the pressure in the duodenum (0.96 kPa [29])

2.3. Mesh and numerical procedure
Fluid mesh consisted of 120869 nodes and 609969 elements. Solid mesh consisted of 78830 nodes and 78660 elements.

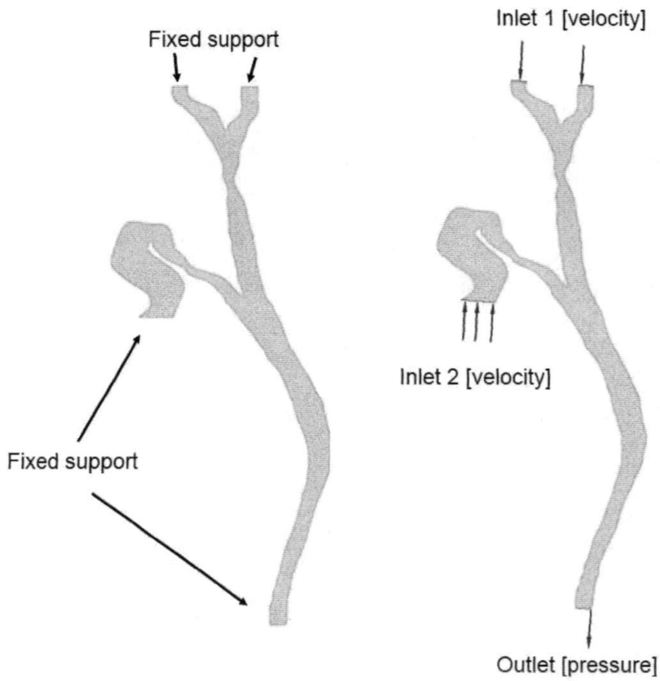

Figure 1. Boundary conditions.

One-way FSI (fluid–structure interaction) algorithm was adopted. The FSI simulation was performed using ANSYS Workbench 13.0 (ANSYS, Inc., Canonsburg, PA, USA). The two solvers (ANSYS Mechanical and ANSYS CFX) were coupled and solved iteratively within each time step until the specified maximum residual is reached. The convergence criterion for the fluid solver and the solid

IOP Conf. Series: Materials Science and Engineering **581** (2019) 012022 doi:10.1088/1757-899X/581/1/012022

solver were 10 and 10 , respectively. Within one-time step of the FSI simulation, the forces acting on the body were interpolated from the converged solution provided by the fluid solver, and transferred to the solid domain to attain the effect of the acting forces, namely the wall displacement, from the converged solution of the solid solver.

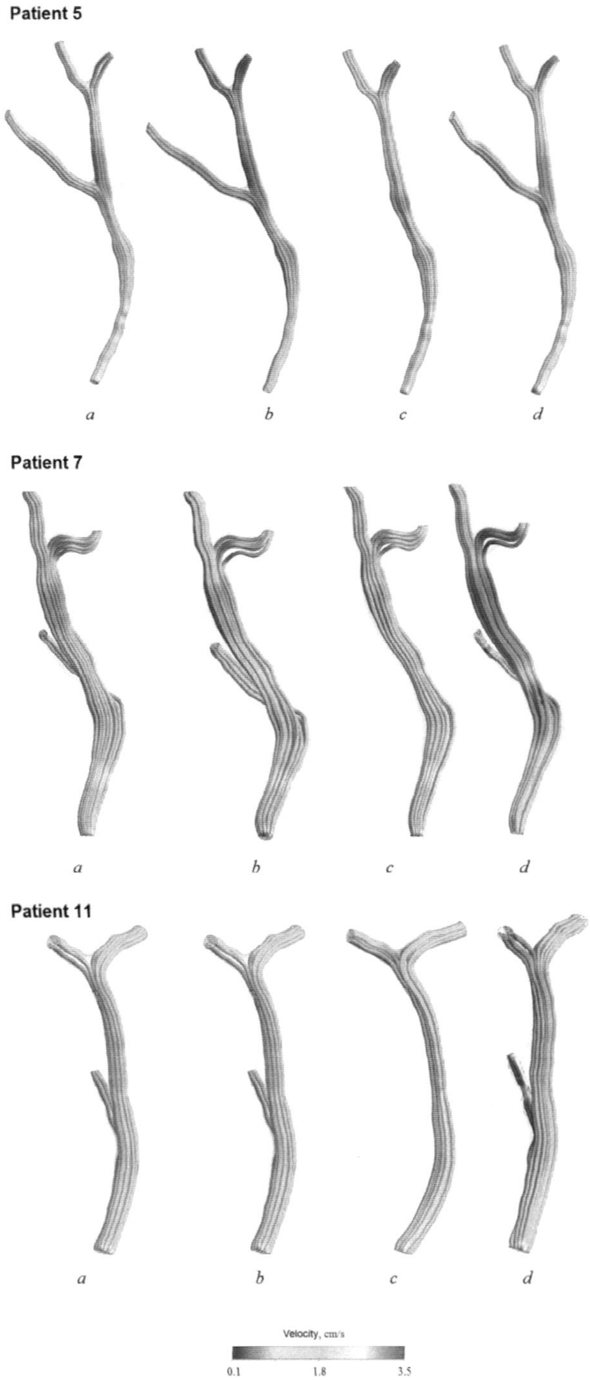

Figure 2. Velocity distribution during the gallbladder emptying. (a) healthy bile, (b) lithogenic bile, (c) cholecystectomy, (d) in the case of the stone in the cystic duct.

The displacements at the wall boundary between solid and fluid domain were then interpolated to the fluid domain to generate a deformed fluid mesh. This step closed one inner loop of the FSI simulation and, these steps were repeated until the changes in the flow forces and the structural displacements fell below the convergence criterion.

3. Results and Discussion

The difference between velocity and pressure distributions was examined for four cases in different patient-specific biliary tree geometries. The considered cases were: healthy bile flow (which is considered as Newtonian fluid) and lithogenic bile (which is considered as non-Newtonian Carreau fluid) at the same boundary conditions. Moreover, we tested cases of lithogenic bile flow in the biliary tree with a stone in the cystic duct and flow in the biliary tree after cholecystectomy.
Velocity distribution for lithogenic and healthy bile is shown in figure 2. When gallbladder empties, the bile flows out the gallbladder via cystic duct and common bile duct. At the same time, bile flows out the liver via hepatic ducts and common bile duct. We simulated this physiological phenomena in the model. From figure2 one can see that the peak of velocity is observed in the distal part of the common bile duct, where bile accelerates to come to the duodenum. It can be noticed also that the geometry of the biliary tree has a dramatic influence on the bile flow character.

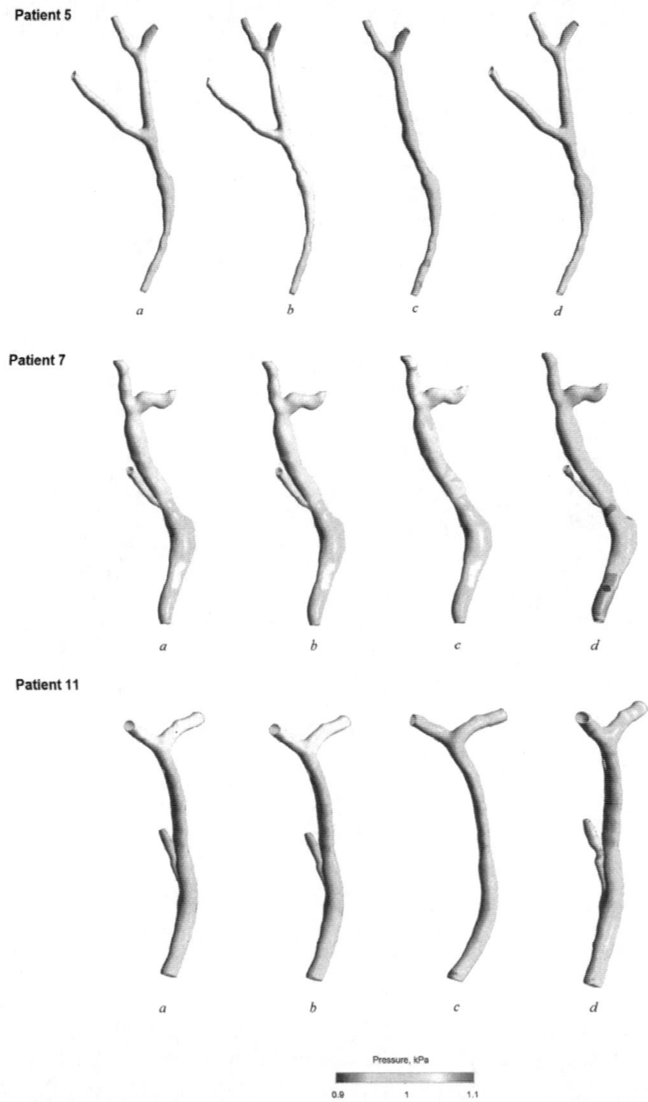

Figure 3. Pressure distribution during the gallbladder emptying. (a) healthy bile, (b) lithogenic bile, (c) cholecystectomy, (d) in the case of the stone in the cystic duct.

The pressure distribution for lithogenic and healthy bile flow is shown in figure 3. The pressure values in the case of lithogenic bile flow are higher. The maximum pressure occurs in the gallbladder neck, and it decreases along the cystic duct. In the case of healthy bile flow, the pressure in the common bile duct is equal to approximately 1.1 kPa, which corresponds to the known medical data [30]. The changes in bile viscosity and gallbladder presence play a great role in pressure distribution. The lithogenic bile case shows that pressure values are higher in the case of healthy bile flow because of bile viscosity increase in the case of pathology. The stone presence leads to an increase in pressure values in the extrahepatic biliary tree, which is approved by medical evidence in the paper [31].

4. Conclusion

The effect of bile pathology on the velocity and pressure distributions in the patient-specific extra-hepatic biliary tree was investigated by FSI approach. Comparisons were made for healthy and lithogenic bile in terms of fluid flow. There was no significant difference for velocities between the healthy bile and lithogenic bile cases. Thus, bile rheology changes play important role in the changes of the pressures, which can lead to the gallbladder disease progress.Cholecystectomy leads to the decrease of bile flow rate in the extra-hepatic biliary tree. The proposed computational approach can be applied to medical practice to evaluate the circumstances of surgical interventions.The proposed model may help to surgeon to evaluate velocity and pressure distributions to asess the cholecystectomy results for the current patientand compare them to normal values.

5. Acknowledgments

The research was performed under financial support of the Government of Perm Region.

References

[1] Reshetnyak V I 2012 *World J Hepatol.***4** 18–34
[2] Stinton L M and Shaffer E A 2012 *Gut Liver***6** 172–87
[3] DumonceauJ-M, HeresbachD,DevièreJ, CostamagnaG, BeilenhoffU, Riphaus A2011*Endosc.* **43** 617–26
[4] Pucher PH, Brunt LM, Davies N 2018*Surg Endosc.* **32**2175–83
[5] Taki-Eldin A, Badawy A E2018 *Arq. Bras. Cir. Dig.* **31**e1347
[6] Radunovic M, Lazovic R, Popovic N 2017 *Maced. J. Med. Sci.* **4** 641–46
[7] Agarwal S, SinhaAK, SinghSP 2012 *Int. J. Appl. Math Mech.* **8** 92–103
[8] Luo X Y, Li W,Bird N, Chin S B, Hill N A, Johnson A G2007*World J. Gastroenterol.* **13** 1384–92
[9] OoiRC, Luo X Y, Chin S B, Johnson A G, Bird N C 2004 *J. Biomechanics***37**1913–22
[10] Al-Atabi MT, Chin S B, Luo X Y 2010 *ASME J. of Biomech. Eng.* **132** 4
[11] Al-Atabi M T, Chin S B, Beck S, Luo X Y 2007*New Trends in Fluid Mechanics Research***1** 610–15
[12] Al-Atabi MT, Chin S B, Luo X Y 2006 *J. of Eng. Sci. Tech.* **1** 1–9
[13] Al-Atabi, MT, Chin S B, Luo X Y2006*JMMB***6** 249–60
[14] Al-Atabi, MT, Chin S B, Luo X Y, Beck S.B.M.2008*J. Biomech. Sci. Eng.***3** 411–18
[15] LiWG, LuoXY, JohnsonAG, HillNA, Bird N, Chin S B 2007*ASME. J. Biomech. Eng.* **129** 164–73.
[16] AgarwalS, Singh S P 2016 *International Journal of Engineering Trends and Technology***32** 309– 14.
[17] Agarwal S, Sinha A K, Singh S P 2012*Adv. Appl. Sci. Res.* **3** 1098–1106.
[18] Agarwal S, Singh S P2014 *Int. J. of Sci. and Inn. Math. Res.* **2** 180–85.
[19] Cho S W, Kim S W, Sung M H, Ro K C, Ryou H C 2011*Korea-Aust. Rheol.* **23** 7–16.
[20] Al-Atabi M, Ooi R C, Luo X Y, Chin S B, Bird N2012*Med. Eng. Phys.* **34** 1177–83.
[21] Maiti S, Misra J C2011 *J. Eng. Sci.* **49** 950–66.
[22] Kuchumov A G, Nyashin Y I, Samarcev V A, Gavrilov V A 2013 *Acta Bioeng. Biomech.* **15**9–17
[23] Lo RC, Huang W L, Fan Y M2015*Comput. Biol. Med.* **60** 51–65
[24] Kuchumov A G, Gilev V, Popov V, Samartsev V, Gavrilov V 2014*Korea Aust. Rheol. J.***26** 81–90.

[25] Bhuvana R,Anburajan M 2013 *Conf. Commun. Signal Process. ICCSP***1** 658–62.
Kuchumov, A.G.CFD approach to bile flow problems solution /
[26] Kuchumov A G2013 *18th International Symposium on Computational Biomechanics in Ulm* **1**67– 68.
[27] Holzapfel GA, Gasser TC, Ogden RW 2000*J. of Elast.***61**, 1-48
[28] Howard PJ, Murphy GM, Dowling RH 1991Gut **32** 1406–11.
[29] Kuchumov AG 2016*Russ J Biomech.* **20**77–96.
[30] Geenen J E, Hogan W J, Dodds W J, Stewart E T, Arndorfer R C 1980 *Gastroenterology***78** 317–24
[31] Csendes A, Kruse A, Funch-Jensen P, Oster M J, Ornsholt J, Amdrup E 1979 *Gastroenterology***77** 1203–10

IOP Conf. Series: Materials Science and Engineering **581** (2019) 012023 doi:10.1088/1757-899X/581/1/012023

Evaluation of the Residual Strength of Structures Made of Composite Materials Based on a Conservative Distribution of Damage Parameters

B N Fedulov[1, a], A N Fedorenko[1] and E V Lomakin[2]

[1] CDM2, Skolkovo Institute of Science and Technology, Bolshoy Boulevard 30, bld. 1 Moscow, Russia 121205, www.skoltech.ru
[2] Lomonosov Moscow State University, GSP-1, Leninskie Gory, Moscow, Russia, 119991, www.msu.ru

E-mail: [a] fedulov.b@mail.ru

Abstract. The purpose of this study is to develop the method that conservatively estimates the residual strength of a composite structure with barely visible impact damage (BVID), without direct modelling of impact process. The basis of the research is in the conservative transformation of kinetic energy of impactor into damage of the material, which is associated with the system of microcracks in the matrix, fiber and interface. A special algorithm that distributes damage parameters in worst way for the composite structure provides conservative estimation of residual strength. Thus, bvid allowables can be set using fewer experiments.

1. Introduction

Barely visible impact damage (BVID) is the key factor for thick laminate composites. Commonly it is associated with the damage due to a low velocity impact as possible incidents during structure lifecycle: falling of tools, hailstones, runway debris etc. This type of damage may significantly reduce the mechanical characteristics without any visible marks on impacted surface. General certification procedure of aerospace composite structures based on demonstrating of "no growth" concept for such defects after impact under operational loads [1 - 15]. This leads to increase of safety margins and as a result to an essential reduction of weight efficiency. Compression after impact (CAI) is a common test for evaluation of damage tolerance of composite laminates. Practice indicates that prediction methods show unsatisfactory results in many cases and physical testing still prevails for evaluation of composite residual strength and damage growth in industry for today [1]. To reduce the amount of expensive physical tests, a reliable predictive method is required for engineering practice.

This work proposes the approach for estimation of the minimum possible residual strength, based on "the worst" distribution of matrix, fiber and interface damages and its accumulation after impact. The minimum possible residual strength condition is fulfilled by decreasing the stiffness of the laminate via energy absorption from the impact. It is assumed, that the portion of the kinetic energy of an impactor is transformed into microcracks of fractured matrix materials (by several damage modes) but the other portion is transformed back to the impactor or to elastic and kinetic energies of the specimen.

2. Transformation of impact energy into damage

The ratio of kinetic energy, which is transformed into the damage, can be estimated by tests or can be taken conservatively. The transformation of unit of energy into unit of damage can be derived from material damage model. For example, using composite material failure model described in [16, 17], the energy spent on degradation can be approximated as an area taken by close loop of loading curve (fig. 1).

IOP Conf. Series: Materials Science and Engineering **581** (2019) 012023 doi:10.1088/1757-899X/581/1/012023

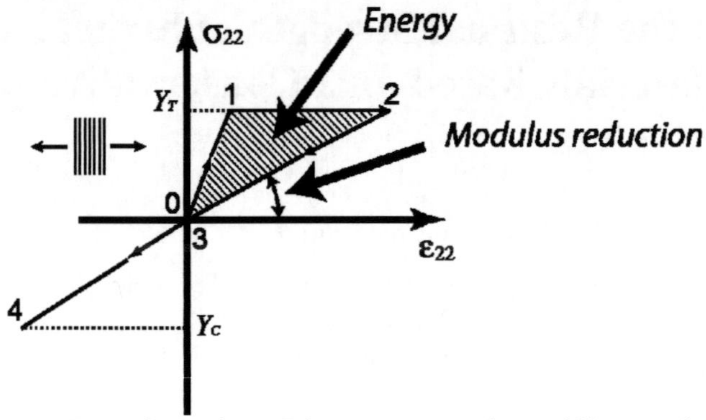

Figure 1. Transformation of damage energy into stiffness reduction.

This idea gives a formal rule for transformation of energy (En) into damage, for example, in case of transversal tension loading:

$$En = \frac{(\varepsilon_{22}^1 - \varepsilon_{22}^2)Yt}{2} = \frac{\left(\frac{Yt}{E_2^1} - \frac{Yt}{E_2^2}\right)Yt}{2} = \frac{Yt^2}{2E_2^0}\left(1 - \frac{1}{\psi}\right) \tag{1}$$

where:
ε_{22}^1 – deformation at point 1 (fig. 1)
ε_{22}^2– deformation at point 2 (fig. 1)
Yt – failure stress in case of transversal tension
E_2^1 – transversal modulus at point 1 (fig. 1)
E_2^2 – transversal modulus at point 2 (fig. 1)
E_2^0 transversal modulus of not damaged material
ψ – damage parameter, associated with stiffness reduction ($0 \le \psi \le 1$)

The equation (1) gives the relation between damage parameter ψ and required for this damage energy En.

For simplicity we can assume, that the damage of the material was obtained by a compression loading and use parameters corresponding to a compressive failure - stress Yc and transversal modulus E_2.

Another source to spend impact energy is the delamination. The energy spent on delamination can be estimated as

$$En_{delam} = SG_I \tag{2}$$

where:
S – area of delamination
G_I- fracture toughness for open type crack growth
First mode of fracture toughness G_I is chosen in (2) due to conservative fact that $G_I < G_{II}$ and $G_I < G_{III}$.

3. Damage distribution algorithm

Avoiding the direct impact modelling, we can analyze variants with one, two, three and so on delaminated layers with different damage distribution for corresponding energy taken from impactor (fig. 2) and choose the worst variant of them.

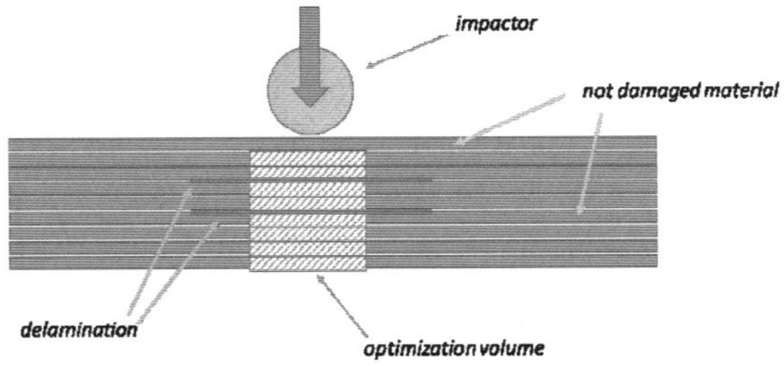

Figure 2. Delamination placement and volume with possible damage

Having estimated value of energy spent on stiffness degradation, we need to distribute damage parameters in the volume Ω, which is assumed to be affected by an impactor. Conservative distribution is the case when the structure has the lower residual strength. It is questionable point, but for the first step we can assume that the worst case is the one with maximum deformation of analyzing volume Ω. Formally, we can assume that it is deformation energy in the possible affected volume Ω. Now it is possible to formulate the problem for damage distribution:

$$Max_{\psi(x,y,z)}\left(1/2\int_{\Omega} E_{ijkl}\,\varepsilon_{ij}\varepsilon_{kl}d\Omega\right), \tag{3}$$

where following [16]

$$E_{ijkl} = \begin{bmatrix} \frac{1}{E_{11}} & -\frac{\psi v_{21}}{E_{22}} & -\frac{\psi v_{31}}{E_{33}} & 0 & 0 & 0 \\ -\frac{\psi v_{12}}{E_{11}} & \frac{1}{\psi E_{22}} & -\frac{\psi v_{32}}{E_{33}} & 0 & 0 & 0 \\ -\frac{\psi v_{13}}{E_{11}} & -\frac{\psi v_{23}}{E_{22}} & \frac{1}{\psi E_{33}} & 0 & 0 & 0 \\ 0 & 0 & 0 & \frac{1}{\psi G_{12}} & 0 & 0 \\ 0 & 0 & 0 & 0 & \frac{1}{\psi G_{13}} & 0 \\ 0 & 0 & 0 & 0 & 0 & \frac{1}{\psi G_{23}} \end{bmatrix}^{-1} \tag{4}$$

subjected to constrains:

$$\begin{gathered} 0 < \psi \le 1 \\ En = \int_{\Omega} e_n d\Omega = \int_{\Omega} \frac{Yc^2}{2E_2^0}(1 - 1/\psi)d\Omega = Const. \end{gathered} \tag{5}$$

Consider gradient of the objective function for finite volume used in topology optimization method:

$$\frac{\partial \int_{\Omega} E_{ijkl}\,\varepsilon_{ij}\varepsilon_{kl}d\Omega}{\partial \psi_e} = \int_{\Omega_e} \frac{\partial E_{ijkl}}{\partial \psi_e}\varepsilon_{ij}\varepsilon_{kl}d\Omega \tag{6}$$

IOP Conf. Series: Materials Science and Engineering **581** (2019) 012023 doi:10.1088/1757-899X/581/1/012023

where ψ_e- damage of the material at volume Ω_e (undamaged material corresponds to $\psi = 1$), index e is associated with volume and damage of one element in FEM analysis. Each tensor component $\frac{\partial E_{ijkl}}{\partial \psi_e}$ can be calculated directly using matrix of (4).

The SIMP optimization algorithm described in [18] has been implemented into Abaqus software using special user subroutines. The iterative procedure for finding of element damage value ψ_e^K at step K has the form

$$\psi_e^{K+1} = \begin{cases} max\{(1-\zeta)\psi_e^K, \psi_{min}\}\; if\; \psi_e^K B_K^{-\eta} \le max\{(1-\zeta)\psi_e^K, \psi_{min}\} \\ min\{(1+\zeta)\psi_e^K, 1\}\; if\; min\{(1+\zeta)\psi_e^K, 1\} \le \psi_e^K B_K^{-\eta} \\ \psi_e^K B_K^{-\eta}\; otherwise, \end{cases} \tag{7}$$

where $B_K = \Lambda_K^{-1} \int_{\Omega_e} \frac{\partial E_{ijkl}}{\partial \psi_e} \varepsilon_{ij} \varepsilon_{kl} d\Omega$, and Λ_K is a multiplier that satisfies the total damage energy constraint and is found by a bi-sectioning algorithm. The variable η is a tuning parameter with typical value of 0.5 and ζ is a move limit with reasonable value of 0.2.

4. Example problem

To demonstrate the algorithm of damage parameters distribution to get the worst case, let us consider the compression problem of composite plate with dimensions 200mm×200mm and total thickness of 8mm formed by 40 layers of quasi-isotropic layup [0°/45°/-45°/90°]$_{5S}$. Fig. 3 shows the placement of optimizing volume, which is a cylinder of diameter 50mm with one non-damaged, and consequently, not optimized layer at the side of impact. The loading was applied through displacement to get better convergence of numerical analysis. Displacements were chosen to get deformation of the layup to be 2%, which is near the failure one. The direction of compression loading is 0° degree. Solid elements with incompatible modes (perform close to second-order elements for regular element shape) were used with grid density of one element per layer.

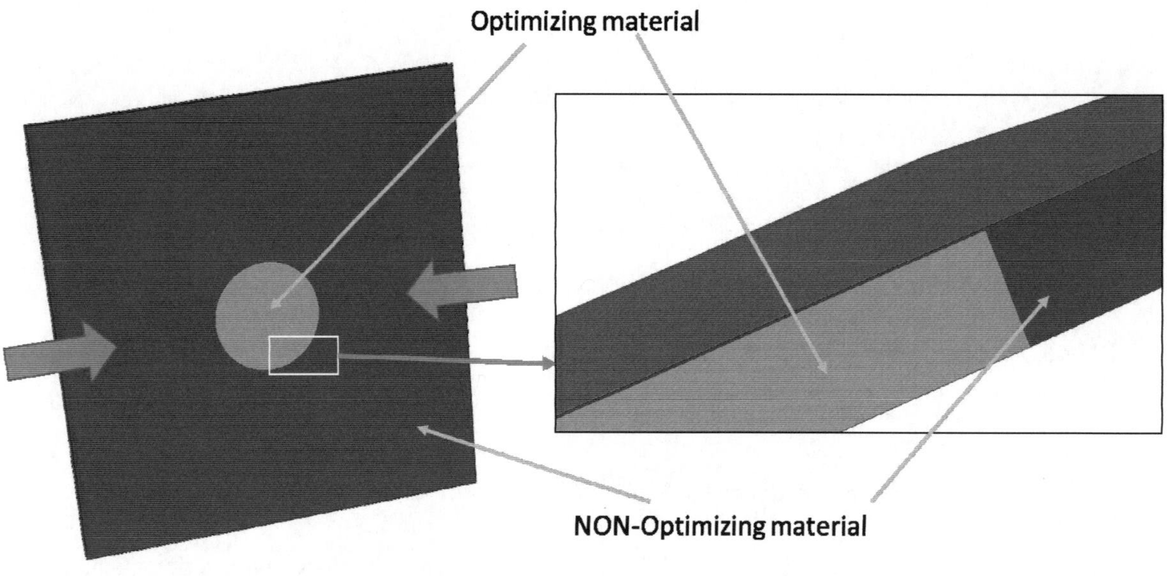

Figure 3. Problem statement scheme

Model has two materials. First one represents optimizing volume, and second one has no influence from damage variation. Nonlinear geometry changes during deformations were taken into account. Material properties used in the analysis are shown in Table 1 and 2.

Table 1 Mechanical properties [16]

E_{11} (GPa)	E_{22}(GPa)	G_{12}(GPa)	ν_{12}	ν_{23}
126	11	6.6	0.28	0.4

Table 2 Strength properties [16]

X_t(MPa)	X_c(MPa)	Y_t(MPa)	Y_c(MPa)	S_{12}(MPa)
1950	1480	48	200	79

Having done few program runs it was found, that the algorithm tries to spend all energy to the most imperfect elements. This led to idea to restrict minimum damage value to 0.5 what looks reasonable for physical meaning. Thus, in this study further models have a constraint for damage parameter:

$$0.5 < \psi \leq 1$$

Using new restriction, after 20 iterations damage has been distributed by program only into ±45 layers and completely removed from layers with 0° and 90° degree (Fig.4).

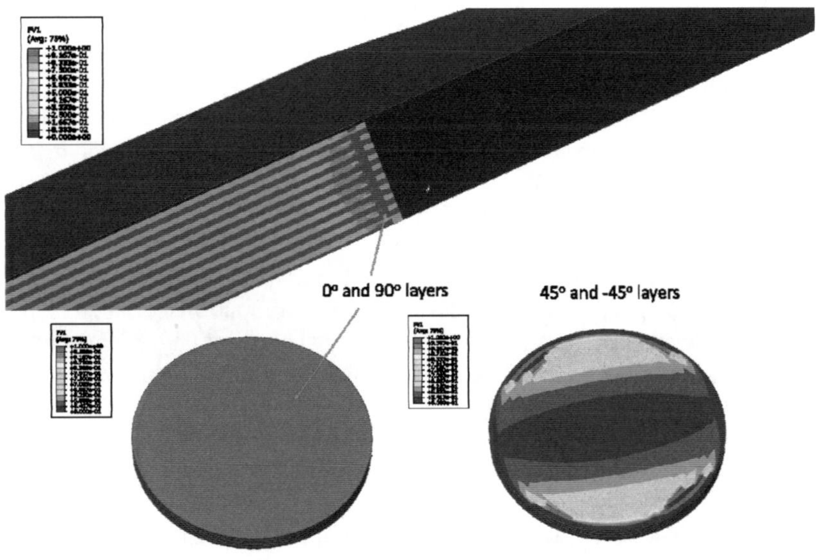

Figure 4. Damage parameter distribution

The same result was obtained for values 10, 20, 40 and 60 Joules spent on damage energy. Thus, we can state that for pure damage defect the worst case for uniaxial compression load is the reduction of matrix stiffness in ±45° layers.

5. Conclusions

The algorithm to estimate residual strength of the composite part subjected to the low velocity impact was performed. The advantage is that the approach avoids direct modelling of the impact process. The core idea of the method is the search of the worst damage distribution inside the area of the material affected by impact stresses. It was shown that the problem is identical to the topology optimization one but with opposite purpose to minimize stiffness of the structure. Example with pure damage was analyzed and result showed that for compression load degradation of ±45° layers is the most critical case. The corresponding damage parameter distribution was shown. The proposed modular formulation of the performed approach lets us to take into account more knowledge about defects from experimental statistics and reduce conservatism systematically, and eventually build a perfect BVID defect for design and development purposes in structural engineering.

6. Acknowledgement

This research was supported by the Russian Foundation for Basic Research (Grant no. 18-31-20026).

References

[1] Composite Materials Handbook – 17 CMH-17-3G Volume 3 2012
[2] AC 20-107B - Composite Aircraft Structure 2009
[3] Abrate S. Impact on composite structures. Aeronaut J 1998;108:541–63.
[4] Richardson M.O.W., Wisheart M.J. Review of low-velocity impact properties of composite materials. Compos Part A Appl Sci Manuf 1996; 27:1123-1131
[5] Seltzer R, González C, Muñoz R, Llorca J, Blanco-Varela T. X-ray microtomography analysis of the damage micromechanisms in 3D woven composites under low-velocity impact. Compos Part A Appl Sci Manuf 2013;45:49–60.
[6] Rivallant S, Bouvet C, Abi Abdallah E, Broll B, Barrau JJ. Experimental analysis of CFRP laminates subjected to compression after impact: The role of impact-induced cracks in failure. Compos Struct 2014;111:147–57.
[7] González E V., Maimí P, Camanho PP, Turon A, Mayugo JA. Simulation of drop-weight impact and compression after impact tests on composite laminates. Compos Struct 2012;94:3364–78.
[8] Menna C, Asprone D, Caprino G, Lopresto V, Prota A. Numerical simulation of impact tests on GFRP composite laminates. Int J Impact Eng 2011;38:677–85.
[9] Elias A, Laurin F, Kaminski M, Gornet L. Experimental and numerical investigations of low energy/velocity impact damage generated in 3D woven composite with polymer matrix. Compos Struct 2017;159:228–39.
[10] Burlayenko VN, Sadowski T. A numerical study of the dynamic response of sandwich plates initially damaged by low-velocity impact. Comput Mater Sci 2012;52:212–6.
[11] Ullah H, Harland AR, Lucas T, Price D, Silberschmidt V V. Finite-element modelling of bending of CFRP laminates : Multiple delaminations. Comput Mater Sci 2012;52:147–56.
[12] Rivallant S, Bouvet C, Hongkarnjanakul N. Failure analysis of CFRP laminates subjected to compression after impact: FE simulation using discrete interface elements. Compos Part A Appl Sci Manuf 2013;55:83–93.
[13] Hongkarnjanakul N, Bouvet C, Rivallant S. Validation of low velocity impact modelling on different stacking sequences of CFRP laminates and influence of fibre failure. Compos Struct 2013;106:549–59.
[14] Bouvet C, Rivallant S, Barrau JJ. Low velocity impact modeling in composite laminates capturing permanent indentation. Compos Sci Technol 2012;72:1977–88.
[15] Yang L, Wu Z, Gao D, Liu X. Microscopic damage mechanisms of fibre reinforced composite laminates subjected to low velocity impact. Comput Mater Sci 2016;111:148–56.
[16] Fedulov B. N., Fedorenko A. N., Kantor M. M. and Lomakin E. V.. Failure analysis of laminated composites based on degradation parameters. Meccanica, 2017;53:359–372.
[17] Fedulov B. N., Fedorenko A.N., Safonov A.A. and Lomakin E.V., Nonlinear shear behavior and failure of composite materials under plane stress conditions. Acta Mechanica. 2017 Jun 1;228(6):2033-40.
[18] Bendsoe, Martin Philip, and Ole Sigmund. Topology optimization: theory, methods, and applications. Springer Science & Business Media, 2013.

IOP Conf. Series: Materials Science and Engineering **581** (2019) 012024 doi:10.1088/1757-899X/581/1/012024

Amplitude Dependence of Dynamic Susceptibility of a Magnetic Fluid at Acoustic Frequencies

M A Koskov[1,2,a], **A F Pshenichnikov**[1,2]
[1] Institute of Continuous Media Mechanics Ural Branch of Russian Academy of Sciences the affiliate of Perm Federal Scientific Research Center, Korolev str. 1, Perm, Russia, 614013
[2] Perm State University, Bukireva str. 15, Perm, Russia, 614990

E-mail: [a] koskov.m@icmm.ru

Abstract. The energy dissipation in a magnetic fluid in the alternating magnetic field at the acoustic frequency was investigated experimentally. The purpose of the work is to gain information concerning the amplitude dependence of the dynamic susceptibility and the applicability range of the simplest linear relaxation equation for magnetization. Two independent methods were used: direct measurement of the dynamic susceptibility by a mutual inductance bridge in a weak sounding field and measurement of heat release in a magnetic fluid at high magnetic field amplitudes. In the latter case, the measuring technique is based on the well-known formula relating the heat release rate to the imaginary part of the dynamic susceptibility, which directly follows from the linear relaxation equation. A comparison of the results obtained by different methods revealed a systematic discrepancy between the results of two series of experiments, which is explained by the existence of multi-particle clusters (aggregates) in the magnetic fluid with uncompensated magnetic moments. At frequencies of about 10^4 Hz, the brownian mechanism for the relaxation of magnetic moments is blocked due to the large size of the aggregates ($\omega\tau_B \gg 1$). In addition, at small amplitudes of the sounding field, the Neel relaxation mechanism is also blocked due to the high energy of inter-particle interactions inside the aggregate and the appearance of high potential barriers. The situation changes with an increase in the field amplitude, when the energy of interaction of the aggregated particles with the external field becomes comparable with the height of the potential barrier. The aggregates are involved in the process of magnetization reversal, increasing the dynamic susceptibility and energy dissipation in a magnetic fluid.

1. Introduction

Magnetic fluids — colloidal solutions of ferromagnetic nanoparticles in a carrier fluid — are typical representatives of superparamagnetic systems [1–4]. When the fluid is in the state close to a thermodynamic equilibrium, the magnetization dynamics of M is well described by a linear equation [1–3]

$$\frac{d\mathbf{M}}{dt} = -\frac{1}{\tau}(\mathbf{M} - \mathbf{M}_0) \quad (1)$$

where $\mathbf{M}_0$ is the equilibrium magnetization, τ is the characteristic relaxation time. Since in weak fields the equilibrium magnetization is proportional to the magnetic field strength ($\mathbf{M}_0 = \chi_0\mathbf{H}$, χ_0 is the initial susceptibility), and equation (1) is linear, the dynamic magnetization will be proportional to the magnetic field strength. In the most general case of alternating, linearly polarized magnetic field $H = H_0\exp(i\omega t)$, the solution of equation (1) can be written as

$$\widetilde{M} = (\chi_1 - i\chi_2)H_0\exp(i\omega t)$$

where the real and imaginary parts of the dynamic susceptibility are defined by the Debye formulas [1, 5].

$$\chi_1 = \frac{\chi_0}{1+\omega^2\tau^2}, \qquad \chi_2 = \frac{\chi_0\omega\tau}{1+\omega^2\tau^2} \tag{2}$$

The real part of the susceptibility χ_1 determines the component of magnetization, which is in phase with the field strength, whereas the imaginary part χ_2 is responsible for the dissipation of magnetic energy due to the alternating processes of the liquid magnetization reversal. The corresponding component of magnetization has a 90 degree lag from the phase of the field strength. According to [3, 5], the power dissipated in an arbitrary volume V of the magnetic fluid in a weak alternating magnetic field is determined by the formula

$$Q = \mu_0\omega\chi_2 V\frac{H_0^2}{2} \tag{3}$$

where $\mu_0 = 4\pi\cdot10^{-7}$ H/m is the magnetic constant, ω and H_0 are the cyclic frequency and amplitude of the alternating magnetic field, respectively.

A characteristic feature of real magnetic fluids is a wide spectrum of relaxation times associated with the polydispersity of single-domain particles, the existence of clusters and two independent mechanisms for the relaxation of the magnetic moment of a particle - the Brownian and Neel relaxation mechanisms [1, 6]. For this reason, formulas (2), corresponding to a system with the same relaxation time, are represented as the expansion in terms of the Debye function. The low-frequency part of the dynamic susceptibility spectrum we are interested in is well described by series with a relatively small ($N = 4\ldots7$) number of terms, each of which corresponds to a separate fraction of particles (or clusters) with a fixed time τ_i of the magnetic moment relaxation [6].

$$\chi_1 = A_0 + \sum_{i=1}^{N}\frac{A_i}{1+\omega^2\tau_i^2}, \qquad \chi_2 = \sum_{i=1}^{N}\frac{A_i\omega\tau_i}{1+\omega^2\tau_i^2} \tag{4}$$

The spectral amplitudes A_i denote the contribution of the i-th "Brownian" fraction to the equilibrium susceptibility, and the first term in (4) describes the frequency-independent (at frequencies up to 10^5 Hz) contribution of the smallest particles, in which relaxation is governed by the Néel mechanism. Obviously, the sum of all spectral amplitudes is nothing else but an initial equilibrium susceptibility. If the fraction is represented by individual particles, the spectral amplitudes can be described by the well-known expression for the initial susceptibility of a magnetic fluid in the framework of the modified effective field model [7]:

$$A_i \approx \frac{\mu_0\langle m_i^2\rangle n_i}{3kT}\left(1+\frac{\chi_L}{3}\right), \qquad \chi_L = \frac{\mu_0\langle m^2\rangle n}{3kT}, \tag{5}$$

where n and n_i are the numerical densities of the particles for the system as a whole and for the specified fraction, respectively, m is the magnetic moment of the particle. Angular brackets denote averaging over the ensemble. According to (5), the spectral amplitude is proportional to the mean square of the magnetic moment, i.e. the sixth power of the diameter of the magnetic core of particles contained in the fraction. The appearance of clusters in a magnetic fluid does not lead to a qualitative change in the situation: the main contribution to the equilibrium susceptibility is by coarse fractions.

In the low-frequency limit, the quasi-equilibrium susceptibility does not depend on the field amplitude, if the Langevin parameter, corresponding to coarse particles, is sufficiently small. If we assume that the error of one percent is acceptable, then the corresponding condition is written as

$$\xi = \frac{\mu_0 m H_0}{kT} \leq 0.3\,.$$

For typical magnetic colloids this condition is fulfilled at the field amplitudes as high as 400–500 A/m [8] and is actually the condition for the realization of formula (5) for spectral susceptibilities A_i. In the dynamic mode ($\omega\tau_i \sim 1$), the relaxation times and their dependence on the field strength and inter-particle interactions, including the formation of aggregates, play a key role. Although the effect of inter-particle interactions on the dynamics of magnetization has been the focus of several recent papers, the problem is still far from being solved [9]. The purpose of this work is to obtain an experimental estimate of the range of sounding fields, in which the dynamic susceptibility can be considered constant quantity, which is independent of the field amplitude, and formulas (4) can be used to calculate the energy dissipation in a magnetic fluid. From the viewpoint of applied problems, the work may be of special interest in connection with the ongoing attempts of using magnetic nano particles for hyperthermia of tumors [10, 11]. In particular, in [10], the emphasis is placed on the problems associated with the duration of the therapeutic procedure and control of tumor temperature. The development of solutions to both problems requires an appropriate selection of the frequency and amplitude of the alternating magnetic field and correct estimation of the maximum heat release rate.

2. Experiment details

To create a uniform magnetic field, a long solenoid with a length of $L = 374$ mm and a diameter of $D = 33$ мм ($L \gg D$) was used. The solenoid was formed by winding an enameled wire (EWW) of a diameter of 0.31 mm on a tube made of polyvinylchloride. The number of turns is N = 1080. A test tube with a magnetic fluid was located axially in the middle part of the solenoid. The length of the test tube was 107 mm, and the internal diameter was 8 mm. The ends of the test tubes were tightened with plastic washers, the diameters of which coincide with the internal diameter of the tube. This prevents the shift of the tube inside the solenoid during measurements. To measure the strength of the solenoid current, the solenoid was connected in series to a non-inductive resistor $R_1 = 1.05$ Ohms. The half-period average voltage of the resistor was measured with a F5053 voltmeter. The solenoid was connected to an a. f. amplifier intended for receiving a sinusoidal signal from a low-frequency generator.

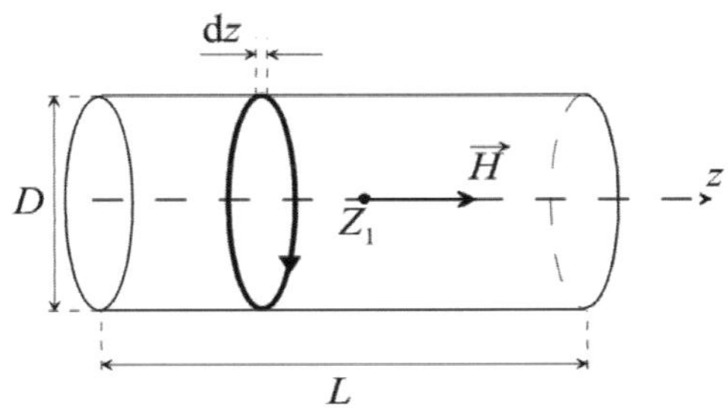

Figure 1. Schematic representation of the solenoid and coordinate axis

The magnetic field along the axis of the solenoid can be easily calculated by breaking the solenoid into thin rings and using the superposition principle (for example, [12]).

$$H = \frac{IN}{L}\left(\frac{Z_1}{\sqrt{4Z_1^2 + D^2}} + \frac{L - Z_1}{\sqrt{4(L - Z_1)^2 + D^2}}\right), \tag{6}$$

where N is the number of turns of the solenoid, I is the strength of the winding current, z is the distance from the ring plane to an arbitrary point Z_1 on the solenoid axis, at which the field strength is evaluated (Fig. 1). In the middle part of a long solenoid, formula (6) predicts the existence of a region with a uniform field $H = IN{\cdot}L^{-1}$. In particular, for the above mentioned geometry of the solenoid, the field strength in the tube was quite correctly described by the formula $H = \alpha I$, $\alpha = (2.88{\pm}0.02){\cdot}10^3$ m^{-1}, and the field heterogeneity did not exceed 1 %.

Temperature measurements were taken using a copper-constantan thermocouple, the hot junction of which was glued to the wall of the tube, and the cold one – to the inner wall of the solenoid. The EMF of the thermocouple was measured with a voltmeter SHCH–300.

In the experiments, we studied two samples of a typical magnetic fluid, which is a composition of magnetite particles suspended in kerosene and stabilized by the oleic acid. The sample densities were measured with a densitometer at room temperature. The volume fraction of the solid phase was estimated by the formula $\varphi = (\rho - \rho_b)/(\rho_m - \rho_b)$, where $\rho_b = 0.78$ g/cm^3 is the kerosene density at room temperature, and $\rho_m = 5.2$ g/cm^3 is the magnetite density. The fluid viscosity was evaluated by the Chow formula [13]

$$\frac{\eta}{\eta_0} = \exp\left(\frac{2.5\varphi}{1-\varphi}\right) + \frac{A\varphi^2}{1-AP\varphi^2}.$$

Here, $A = 4.67$, $P = 0.605$ is the coefficient of particle packing, at which the fluid loses its fluidity, $\eta_0 = 1\cdot10^{-3}$ Pa·s is the viscosity of kerosene at room temperature. The operating frequency range was determined from the condition $\omega\tau_B \sim 1$, where

$$\tau_B = \frac{\pi\eta d^3}{2kT}$$

is the Brownian relaxation time of the magnetic moments [4] corresponding to the fraction of colloidal particles with a hydrodynamic diameter $d = 15$ nm prevailing in the solution [6]. An estimate of the Brownian relaxation time gives $\tau_B \sim 10^{-5}$ s, which corresponds by the order of magnitude to the frequency of the sounding field $\nu = 10$–16 kHz. For this frequency range, the dynamic effects should be most pronounced.

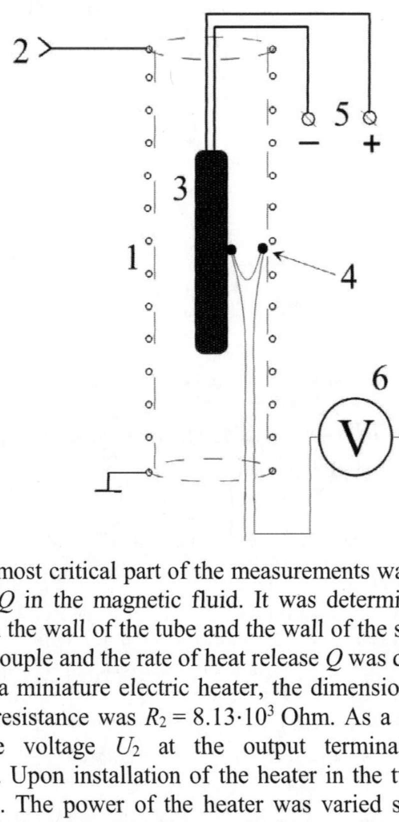

Figure 2. Schematic representation of the experimental setup used in the calibration test. 1' – solenoid, 2' – output terminal of the amplifier, 3' – a test tube with a heater, 4' – a thermocouple, 5'–output terminals of the GPR–7550D source, 6' – a voltmeter SHCH–300.

The most critical part of the measurements was finding the magnitude of the maximum rate of heat release Q in the magnetic fluid. It was determined by the radial temperature difference in the gap between the wall of the tube and the wall of the solenoid. The relationship $\varepsilon(Q)$ between the EMF of a thermocouple and the rate of heat release Q was derived from the calibration tests. For this purpose, we created a miniature electric heater, the dimensions of which allowed its location inside the test tube. Heater resistance was $R_2 = 8.13\cdot10^3$ Ohm. As a power source we used a GPR–7550D model, whose effective voltage U_2 at the output terminals was measured with a GDM–8246 voltmeter (Fig. 2). Upon installation of the heater in the tube, the remaining space of the tube was filled with glycerin. The power of the heater was varied stepwise, and the stationary temperature regime was established within several tens of minutes after changing. To ensure maximum approximation of the calibration conditions to the conditions of magnetic fluid experiment, the alternating current passing through the coil of the solenoid was of the same strength as the current in real tests. Examples of calibration curves for two values of the solenoid current strength are shown in figure 3. As expected, the calibration curves are well approximated by a linear relationship.

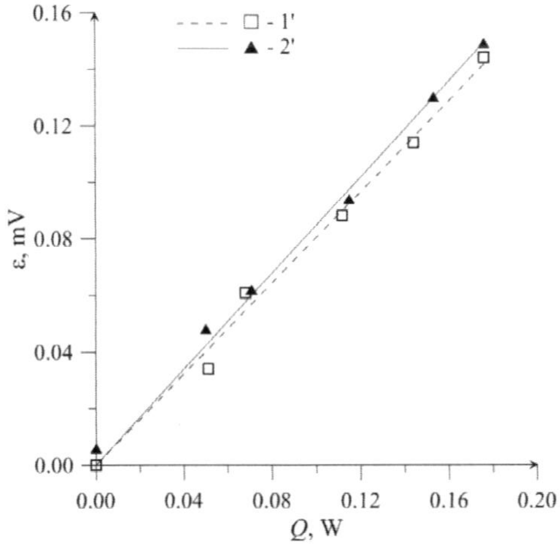

Figure 3. EMF of the thermocouple versus heater power at two values of the solenoid current strength (1' – 0.19 A, 2'– 0.29 A). The dots correspond to the experimental data, the straight lines correspond to the linear approximation.

Another systematic error in the calculation of the imaginary part χ_2 of susceptibility was associated with its temperature dependence and the variation of the average temperature with a change in the frequency and amplitude of the field. To control the temperature of the magnetic fluid, an additional thermocouple was used. Its hot junction was in the liquid, and the cold one in a vessel with melting ice. The temperature coefficient χ_2 was determined based on the results of measurements by the mutual inductance bridge in the limit of small amplitudes [8]. The dependence $\chi_2(t)$ was approximated by a linear function $\chi_2(t) = mt+c$. The results of the approximation are given in Table 1.

Table 1. Coefficients of linear approximation $\chi_2(t)$ of the ferrofluid sample $\varphi = 0.13$ at different frequencies of the sounding field.

v, kHz	$m \cdot 10^{-2}$, °C^{-1}	c
10.0	-1.39	0.920
12.0	-1.36	0.917
14.0	-1.30	0.911
16.0	-1.40	0.950

3. The results of measurements

The results of calculating the imaginary part of the dynamic susceptibility by formula (3) at higher magnetic field amplitudes (up to 1.3 kA / m) were compared with the results of measurements by a mutual inductance bridge at small amplitudes (up to 200 A / m) taking into account the corrections for temperature according to Table. 1. The results of this comparison are shown in figure 4 for a sounding field frequency of 10 kHz. These results look very unusual, since an increase in the field amplitude would have led to a decrease in susceptibility due to a decrease in the spectral amplitudes. That was not the case. Moreover, there is even a slight increase of about 15% in the imaginary part of the susceptibility. In our opinion, such a behavior of the dynamic susceptibility is associated with two factors: the presence of nanoscale aggregates in the magnetic fluid with the Brownian mechanism for the relaxation of magnetic moments and the selected frequency range. At frequencies of the order of 10^4 Hz, the Brownian relaxation mechanism of aggregates is blocked because of their large sizes ($\omega\tau_B \gg 1$). Moreover, at small amplitudes of the sounding field, the Neel mechanism of magnetization relaxation is also blocked due to the high energy of inter-particle interactions inside the aggregate. The magnetic moments of individual particles in an aggregate are not capable of overcoming the existing potential barriers, and the aggregate "is not involved" in the processes of magnetization reversal. The situation changes with an increase in the field amplitude, in which case the energy of interaction of the aggregated particles with the external field is comparable with the height of the potential barrier. The aggregates are included in the process of magnetization reversal, increasing the dynamic susceptibility and energy dissipation in the magnetic fluid. In the quasi–static mode ($\omega\tau_B \ll 1$) the magnetization reversal occurs due to the Brownian rotation of the aggregates and the existence of potential barriers inside the aggregate does not affect the dynamics of the process.

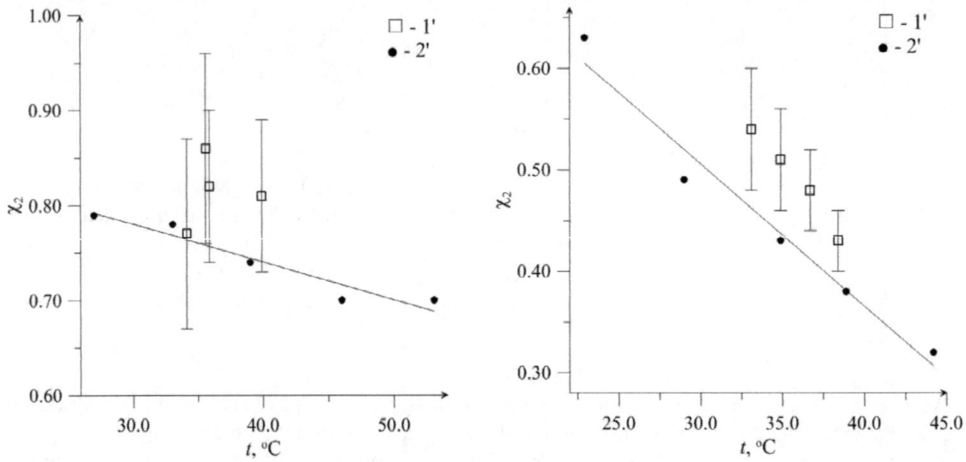

Figure 4. The imaginary part of the dynamic susceptibility of magnetic fluid samples with $\varphi = 0.13$ and $\varphi = 0.17$, respectively, at a frequency of 10 kHz. 1' – corresponds to the calculation by formula (3), 2' – corresponds to the mutual inductance bridge measurements. The solid line shows the linear approximation of the results obtained from the mutual inductance bridge measurements.

4. Conclusion

In general, our results suggest that the dynamic effects associated with the unblocking of the Néel mechanism for the relaxation of magnetic moments inside the aggregates almost completely compensate for a decrease in the spectral amplitudes with decreasing strength of the sounding field, in compliance with equations (4). The contributions of separate terms in the right-hand side of (4) substantially depend on the field amplitude (at a fixed frequency), but their sum varies only slightly. As a result, the amplitude dependence of the dynamic susceptibility at the acoustic frequencies is very weak, and the range of magnetic fields, at which equation (3) is valid at the acoustic frequencies, can be extended at least to 1.3 kA / m.

References

[1] Shliomis M I Magnetic fluids 1974 *J. Sov. Phys. Usp.* **17** 153–169.
[2] Blums E Ya, Maiorov M M , Cebers A O 1997 *Magnetic fluids* (Berlin: Walter de Gruyter)
[3] Starodubtsev Yu. N. *Magneto soft materials. Encyclopedic dictionary-reference* 2011 (Moscow: Technosphere)
[4] Rosensweig R E *Ferrohydrodynamics* 1985 (Cambridge: Cambridge University Press)
[5] Rosensweig R E Heating magnetic fluid with alternating magnetic field 2002 *Journal of magnetism and magnetic materials* **252** 370–374
[6] Lakhtina E V, Pshenichnikov A F Dispersion of magnetic susceptibility and the microstructure of magnetic fluid 2006 *Colloid Journal* **68** N3 294–303
[7] Ivanov A O, Kuznetsova O B Magnetic properties of dense ferrofluids: an influence of interparticle correlations 2001 *J. Physical Review E.* **64** N4 401–405
[8] Pshenichnikov A F A mutual-inductance bridge for analysis of magnetic fluids 2007 *J. Instruments and Experimental Techniques* **50** N4 509–514
[9] Ivanov A O, Kantorovich S S, Zverev V S, et al Temperature-dependent dynamic correlations in suspensions of magnetic nanoparticles in a broad range of concentrations: combined experimental and theoretical study 2016 *J. Physical Chemistry Chemical Physics* **18** 18342–18352

[10] Brusnetsov N A, Shevelev A A, Brusnetsova T N, et al Magneto-liquid regional inductionhyperthermia of sarcoma 2002 *J. Chemical-pharmaceutical magazine* **36** Issue 3 P. 10 (In Russian)

[11] Hergt R, Dutz S Magnetic particle hypertermia – biophysical limitations of a visionary tumour therapy 2007 *Journal ofmagnetism and magnetic materials* **311** 187–192

[12] Yavorsky B M, Detlaf A A, Lebedev A K *Handbook on physics for engineers and university students* 2006 (Moscow: Mir and Education Publishers) (In Russian).

[13] Chow T S Viscoelasticity of concentrated dispersions 1994 *J. Physical Review E.* **50** 1274

IOP Conf. Series: Materials Science and Engineering **581** (2019) 012025 doi:10.1088/1757-899X/581/1/012025

Testing Mechanical Features of Rubber Composites under Biaxial Loading

Oleg K Garishin[1,a], **Vladimir V Shadrin**[1,2]

[1] Institute of Continuous Media Mechanics UB RAS, Russia
[2] Perm State University, Russia

E-mail: [a] gar@icmm.ru

Abstract. The main object of studies were composites on the base of synthetic rubber filled with various nano- and micro particles, which are already used or intended to be used in the tire industry. Mechanical behavior of such materials under biaxial cyclic loading was studied experimentally. The tests were carried out using the four-vector test rig Zwick/Roell. It allows to define complex deformation trajectories in two mutually perpendicular directions independently. The test program consisted of two or three cycles of load-unload (with stops for relaxation) in one direction and then two or three more such cycles in the perpendicular one. Experiments made it possible to study viscous-elastic properties as well as the effects of softening and the occurrence of induced anisotropy in filled elastomers under the action of a biaxial load. In the case of pure rubber, they are practically absent, so unfilled system can be considered as elastic and isotropic. A different picture is observed for filled systems. All the tested composites become viscoelastic after filler input: a hysteresis loop appeared and its area (dissipative losses) enlarged with increasing amplitude of cyclic deform. Also the dependences of induced anisotropy of mechanical properties and softening on the type of filler in the composite were investigated.

1. Introduction

Disperse-filled elastomeric composites are systems consisting of a highly elastic low modulus rubber matrix (continuous phase) into which solid particulate filler particles (dispersed phase) are embedded. Their industrial analogs can be considered rubbers for various purposes (from automobile tires to current-conducting high-elastic gaskets), solid rocket propellants, etc. At present they are the object of intensive research both theoretical and experimental [1, 2].

Such materials are characterized by complicated mechanical behavior (finite deformations, nonlinear elasticity and viscoelasticity), which is due to various reversible and irreversible structural changes occurring during deformation [3-11]. In particular, they are characterized by a phenomenon such as "softening" during repeated deformation (the Patrikeev-Mullins effect) [12-15], which causes certain problems in their operation.

It is well known from practice that structurally inhomogeneous materials, when used in real devices, often behave far from what was observed during standard testing of samples. Biaxial tests can significantly approach the solution of this problem by setting complex loading trajectories along different axes. This means that it is possible to test samples in a wider range of loads under conditions closer to reality.

2. The object of study

The main object of biaxial studies was the synthetic butadiene-styrene rubber SBR-1500, in which various fillers different in their mechanical and physicochemical properties were added such as: 1) microshugite (MS); 2) nanoshugite (NS); 4) white soot (WS); 4) carbon black (CB); 5) carbon nanofibers (CNF) with carbon black. Also the similar tests were carried out for the pure elastomer without filler. For the convenience of the results comparing, the mass concentration for all fillers was taken 65 phr (filler weight parts per hundred rubber weight parts).

A brief description of the fillers used is given below.

Shungite is a clay-like mineral consisting mainly of fullerene-like carbon (30%) and silicon dioxide SiO_2 (60%) [16-18]. It is fairly widely distributed in nature, inexpensive and characterized by high ecological safety. As for their structure, shungites are natural composites with a uniform distribution of highly disperse crystalline silicate particles in a carbon matrix. There is a strong bond between the carbon and silicate components. The particles of the shungite powder contain different phases with respect to polarity. Due to the bipolarity, powders of shungite rocks mix well practically with all known substances (aqueous suspensions and fluoroplastics, rubbers, resins and cements, etc.). Therefore, they are one of the most promising fillers in terms of universality. Nanoparticles of the globular type are formed when shungite is dispersed. Rubbers filled with shungite nanoparticles are characterized by increased wear resistance. Currently, they are used in the tire industry [19, 20]. In our case, the composite samples contained two types of shungite filler: 1) microshungite particles with an average characteristic size about 500 nm; 2) nanoshungite particles size of 60-80 nm.

White soot ($mSiO_2 \cdot nH_2O$) is a particulate hydrated silicon dioxide. Depending on the method of obtaining the hydrate bond can vary from a strong chemical to a weak adsorption. This filler is usually added into rubber products operating in difficult conditions. White soot grade BS-120 particles with an average size of 20-30 nm [21] was used in our samples.

Carbon black (or technical carbon) is one of the most popular fillers in tire industry. Carbon black grade N220 (ASTM standard) was taken: the average particles size is about 30 nm.

Carbon nanofibers are also one of the very promising nanostructured materials, which are currently becoming widely used in various industries (including tire manufacturing). According to their structure, these ones consist of distorted cones with graphene grids (similar to layers in graphite) nested into each other. We used carbon nanofibers grade VGCF [22, 23]. They have an average length from 10 to 20 μm and a diameter about 150-200 nm. Their weight concentration was taken equal 5 phr. This concentration is rather significant for fibers, because, these ones (by specific shape and size) "extend their influence" to much larger distances than the granular inclusions. Carbon black (phr = 60) was also added to these samples with nanofibers.

3. Experiment and results discussion

The biaxial tests were carried out using the four-vector test rig Zwick/Roell (the only one in Russia). This test bench is capable of performing experiments with a complex trajectory of loading along two mutually perpendicular directions independently (Figure 1). The original cross-shaped samples were used (Figure 2). Their shape and dimensions were set on the basis of special theoretical studies carried out in ICMM UB RAS. A corresponding patent was obtained [24].

The fan-folding links connecting cruciform specimen with grippers (Figure 2) were used to provide the uniform stress-strain field in the working region of the sample. Such fastening provides the good uniformity of stresses not less than for 87% of the sample area.

When stretching the cruciform specimens only along one axis, care must be taken to avoid distorting influence from the grippers that are responsible for the loading in perpendicular direction. Therefore, when the specimen was uniaxially stretched, a simultaneous displacement of the perpendicular grips was carried out, so that their "lateral" mechanical effect on the sample was minimized.

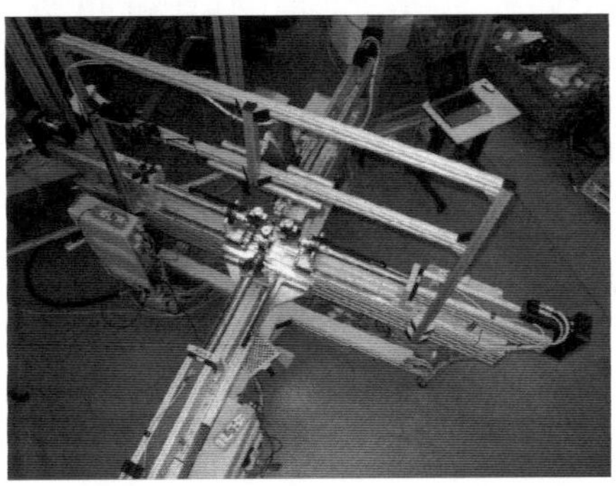

a

b

Figure 1. The four-vector test rig Zwick/Roell.

IOP Conf. Series: Materials Science and Engineering **581** (2019) 012025 doi:10.1088/1757-899X/581/1/012025

The cruciform samples used in the experiments had a working zone in the form of a square with a side of 35 mm. The links connecting this zone with the grippers of the tensile machine (10 pieces per side) had the form of long thin rods of rectangular cross-section (length 45 mm, thickness 3×2 mm). They were manufactured of the same material as the whole specimen. Actually, all the measurements of displacements in the sample were made only in the central region of the working area with a size of 30×30 mm (it is highlighted by white lines in Figure 2). This additional restriction allowed to provide practically 100% guarantee of the stress-strain state uniformity [25].

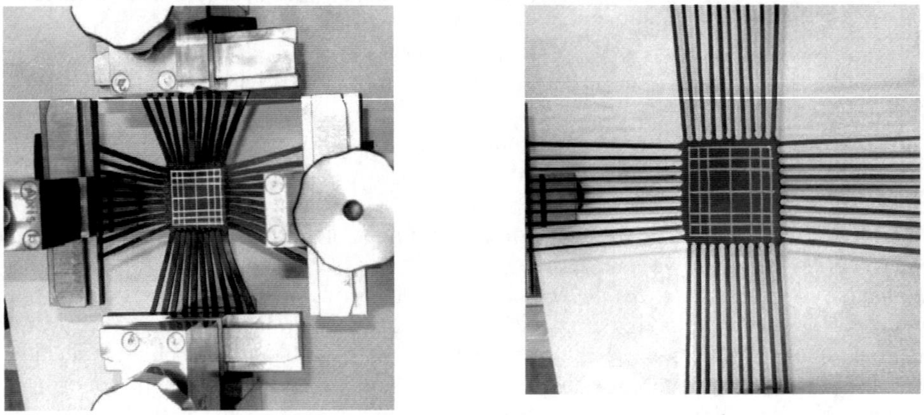

a *b*

Figure 2. General view of cross-shaped specimens used in biaxial tests;
(*a*) – initial unloaded state, (*b*) – the sample is deformed by 80% along each of the axes.

Two programs were used for biaxial testing.

Tests for pure elastomer and composites filled with microshungite, nanoshungite and white soot were performed according to Program I (four cycled). Each test cycle consisted of the following steps:

1) Stretching along one axis to a given deformation.
2) Stopping for stress relaxation.
3) Return to the initial state and again stopping for relaxation.

The rate of deformation was 25%/min, the stop for relaxation lasted 30 min. In the first and second cycles, the sample was stretched along the *X* axis to a deformation of 25% and 50%, respectively. In the third and fourth cycles, the same procedure was repeated along the perpendicular *Y* axis.

Program II (four cycled) was applied to the specimens with two fillers: 1) carbon black particles; 2) carbon nanofibers with carbon black particles:

1) Stretching along the *X* axis to a deformation of 150% and stopping for relaxation.
2) Return to the initial state along the *X* axis to the initial state and stopping for relaxation.

Procedures 1) and 2) were repeated 3 times.

3) Stretching along the *Y* axis to a deformation of 150% and stopping for relaxation;
4) Return to the initial state along the *Y* axis and stopping for relaxation.

Procedures 3) and 4) were also repeated 3 times.

Deformation rate in Program II was 25 %/min, stop time for relaxation was 20 min.

Testing of pure rubber SBR-1500 according to Program I established (Figure 3) that uniaxial cyclic deformation practically does not affect its mechanical properties in other directions. The hysteresis loops in the "load-unload" mode are also very weakly expressed. That is, such a material can be considered practically elastic and isotropic.

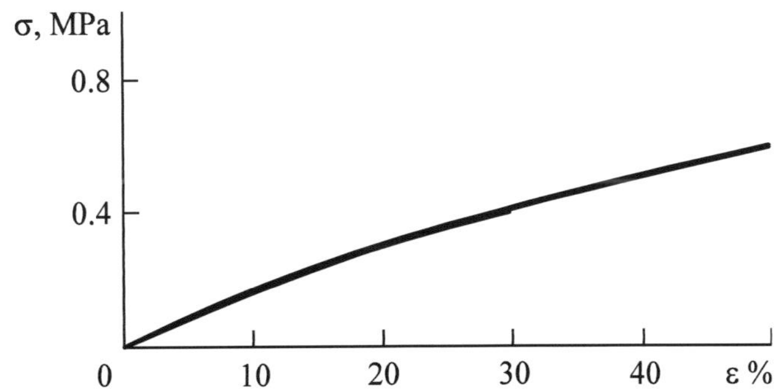

Figure 3. Biaxial tests of pure elastomer; black lines – tensile curves along the *X* axis, gray lines – along the *Y* axis. σ – nominal stress, ε – stretching deformation (0 phr).

A different pictures were observed for composite samples with micro- and nanoshungite fillers (Figure 4 and Figure 5). The graphs show that the addition of these micro or nanoparticles enhances the viscosity properties of the composite: a hysteresis loop appeared and its area (dissipative losses) enlarged with increasing amplitude of cyclic deformation.

For nanoshungite dependences of stresses on deformation, constructed under loading along the *X* and *Y* axes, practically coincide, i.e. the appearance of anisotropy induced by the deformation is not observed. But in for the case of microshungite filler cycle curves along *Y* axis lie below than those along *X*. So, microparticles contribute to the material softening in the perpendicular direction (unlike nanoparticles) and to appearance of induced anisotropy.

It should also be noted that at the same filler concentration, nanoshungite causes greater material enhancement than microshungite.

Figure 6 shows the results of tests for composites with white soot as filler. In this case, the hysteresis curves turned out to be larger than in the previous experiment which indicates a significant increase in the viscosity properties of the composite. The stiffness decreases with repeated loading (the Patrikeev-Mullins effect) for curves corresponding to deformation along the same axis. Also, the appearance of an induced anisotropy of properties is observed for these samples. The curves σ(ε) obtained under loading along the *Y* axis lie much lower than those constructed for the *X* axis.

Experiments for composites filled by carbon black and nanofibers with carbon black (Program II) are represented in Figure 7 and 8. Figure 7 depicts the dependences of nominal stresses on deformation for the composite with the carbon black filler only. These plots show that the material under cyclic loading along the first axis (*X*) undergoes considerable softening after the first cycle, in the second and third cycles the situation stabilizes (Figure 7*a*). The hysteresis loop on the first cycle is also much larger than in the second and third ones. Most interestingly, almost exactly the same curves were obtained for subsequent loading in the perpendicular *Y* direction (Figure 7*b*). That is, the load history for *X* practically has no effect on the *Y* loading history for a given composite.

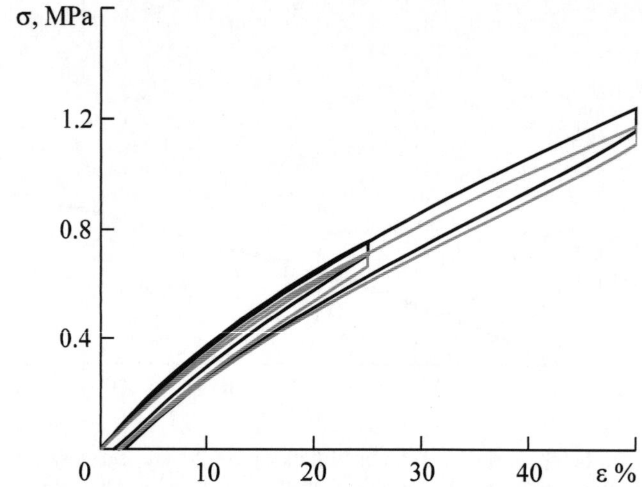

Figure 4. Biaxial testing of an elastomer filled by shungite microparticles (65 phr); black lines – tensile curves along the *X* axis, gray lines – along the *Y* axis.

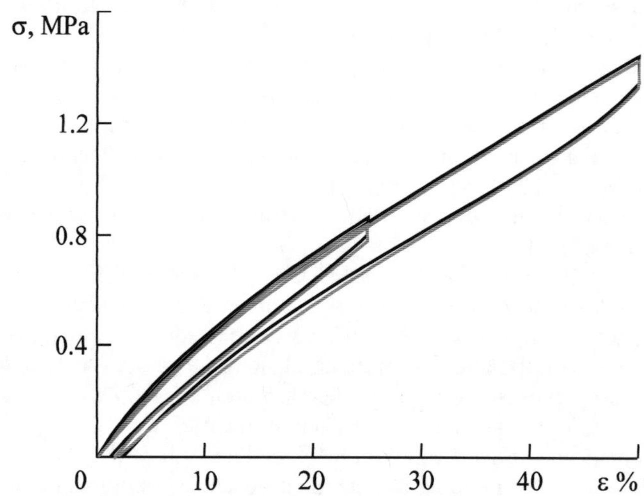

Figure 5. Biaxial testing of an elastomer filled by shungite nanoparticles (65 phr); black lines – tensile curves along the *X* axis, gray lines – along the *Y* axis.

The addition of just 5 phr of carbon nanofibers to this composite significantly changes the behavior of the material. Figure 8*a* shows the results of the first three cyclic loads along the *X* axis, and in Figure 8*b* – the three subsequent cycles along the *Y* axis. When loaded along *X*, the plot is qualitatively the same as in the case of only carbon black filler: the first cycle is characterized by the largest area of hysteresis and the greatest softening of the material. The curves corresponding to the second and third cycles lie nearby. In the subsequent deforming along *Y*, the material behaves quite differently. The previous deformation in the perpendicular direction led to the fact that all three cycles in *Y* coincide with the second and third cycles in *X*, that is, the uniaxial "training" of this composite causes its isotropic softening.

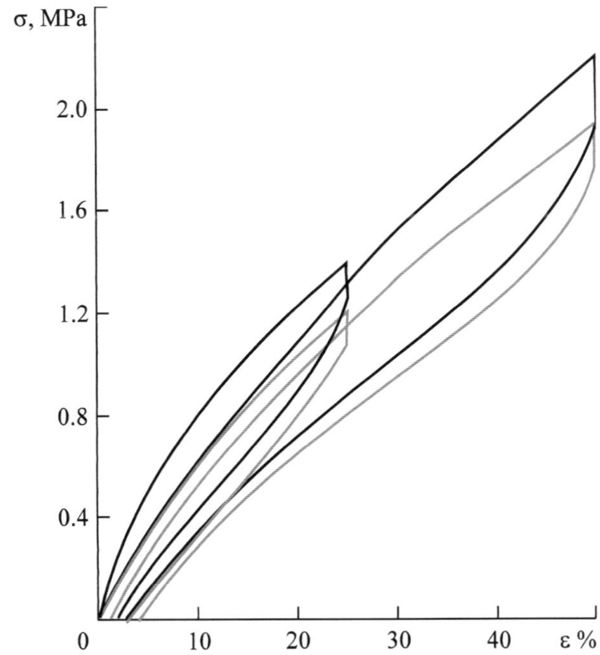

Figure 6. Biaxial testing of an elastomer filled by white soot nanoparticles (phr = 65); black lines – tensile curves along the X axis, gray lines – along the Y axis.

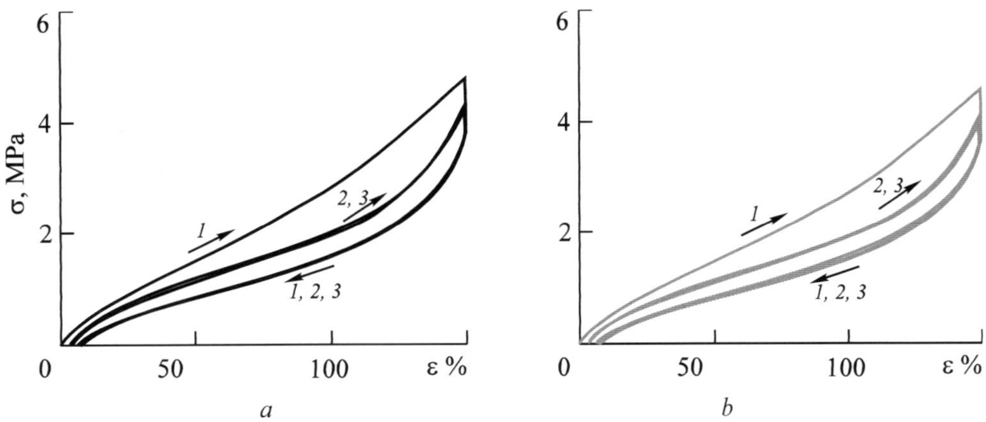

Figure 7. Biaxial testing of an elastomer filled by carbon black nanoparticles: a – tensile curves along the X axis, b –along the Y axis.

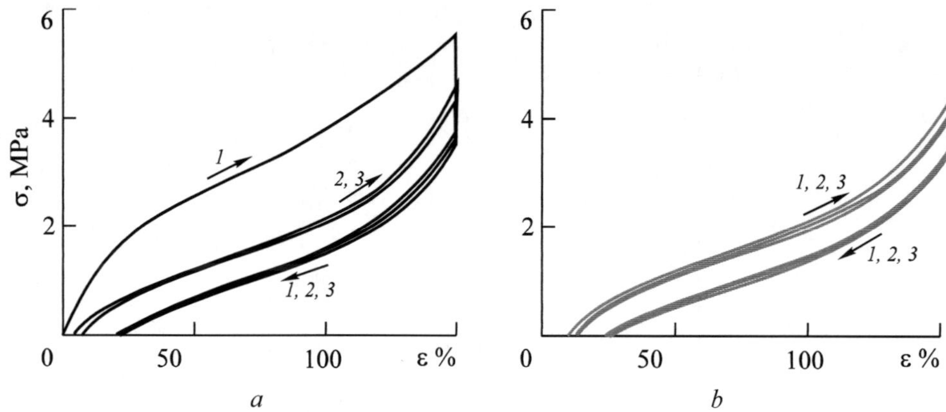

Figure 8. Biaxial testing of an elastomer filled by nanofibers and carbon black; (*a*) – deformation along the *X* axis, (*b*) – deformation along the *Y* axis. 1, 2 и 3 – cycle numbers.

4. Conclusions

Experimental studies showed that the mechanical behavior of dispersed filled elastomers (softening, induced stiffness anisotropy and viscoelastic properties) at biaxial tests depends on what materials are used as filler. Pure rubber remains practically elastic and isotropic material, regardless of the type of applied load. But the addition of the filler to the elastomer leads to a significant change in its mechanical behavior. The composite begins to exhibit viscoelastic properties (a hysteresis loop is), also the induced anisotropy of mechanical features and softening of the material during cyclic loading may occur.

Thus, it makes us possible to change the mechanical properties of the composite in the desired direction by varying the composition and type of the filler, thereby making the system more or less anisotropic or viscoelastic.

Acknowledgements
This work is executed at the financial support of RFBR Grant 17-08-01118.

References

[1] Moshev V V and Garishin O K 2005 Structural mechanics of dispersed filled elastomeric composites *J. Achievements in Mechanics.* **4, No 2** 3–36 (in Russian)
[2] Garishin O K 2012 Mechanical properties and destruction of dispersely filled elastomers. Structural modeling. *Germany. Saarbrucken: Palmarium Academic Publishing (LAP).* 286 (in Russian)
[3] Rooj S, Das A, Stuckelhuber K W, Reuter U and Heinrich G 2012 Highly exfoliated natural rubber/clay composites by "propping-open procedure": the influence of fatty acid chain length on exfoliation *J. Macromol. Mater. Eng.* **297**. 369–83
[4] Rooj S, Das A, Stuckelhuber K W, Mukhopadhyay N, Bhattacharyya A R and Jehnichen D 2012 Pre-intercalation of long chain fatty acid in the interlayer space of layered silicates and preparation of montmorillonite/natural rubber nanocomposites *J. Appl. Clay. Sci.* **67–68** 50–56
[5] Rooj S, Das A, Morozov I A, Stuckelhuber K W, Stocek R and Heinrich G 2013 Influence of "expanded clay" on the microstructure and fatigue crack growth behavior of carbon black filled NR composites *J. Composites Science and Technology* **76** 61–68
[6] Lvov Y, Fakhrullin R, Wang W and Zhang L 2016 Hallousite clay nanotubes for loading and sustained release of functional compounds *J. Advanced Materials* **28, No 6** 1227–50
[7] Garishin O K, Shadrin V V, Svistkov A L, Sokolov A K and Stöckelhuber W K 2017 Visco-elastic-plastic properties of natural rubber filled with carbon black and layered clay nano-particles. Experiment and simulation *J. Polymer Testing* 63 133–40
[8] Garishin O K, Shadrin V V, Svistkov A L and Sokolov A K 2017 Experimental studies of rubbers filled by layered clay nanoparticles *XLV Summer School-Conference "Advanced problems in mechanics": Proceedings, St.-Petersburg* 168–74

[9] Mokhireva K A, Svistkov A L, Solod'ko V N, Komar L A and Stöckelhuber K W 2017 Experimental analysis of the effect of carbon nanoparticles with different geometry on the appearance of anisotropy of mechanical properties in elastomeric composites *J. Polymer Testing* **59** 46–54
[10] Kornev Yu V, Chirkunova C V, Boiko O V, Simonova M A, Shugaeva V O and Yanovsky Yu G 2012 Investigation of the effect of dispersion of mineral shungite on the complex properties of elastomeric compositions based on rubber SKTN-A *J. Caoutchouc and rubber* **6** 10–14 (in Russian)
[11] Kornev Yu V, Yanovskiy G, Boiko V, Semenov N A and Chirkunova S. 2013 Investigating the influence of the degree of dispersion of mineral shungite on the properties of elastomeric materials based on butadiene-styrene rubber *J. Int. Polymer Sci. and Technol.* **40, No 3** 17–22
[12] Mullins L 1947 Effect of stretching in the properties of rubber *J. of Rubber Research* **16, No 12** 245–89
[13] Shadrin V V 2005 Recovery of the mechanical properties of rubber under thermal treatment *J. Polymer Science* B **47, No 7–8** 220–22
[14] Diani J, Fayolle B and Gilormini P 2009 A review on the Mullins effect *European Polymer Journal* **45** 601–12
[15] Sokolov A K, Svistkov A L, Komar L A, Shadrin V V and Terpugov V N 2016 The manifestation of the softening effect of the material in the change in the stress-strain state of the tire *J. Computational Continuum Mechanics* **9, No 3** 358–65 (in Russian)
[16] Buseck P R 2002 Geological fullerenes: review and analysis *Earth and Planetary Science Letters* **203** 781–92
[17] Kwiecinska B, Pusz S. Krzesinska M and Pilawa B 2007 Physical properties of shungite *Int. J. of Coal Geology* **71, No 4** 455–61
[18] Ignatov I, Mosin O V 2012 Composition and structural properties of the natural fullerene-containing shungite mineral *J. Nanoengineering* **6** 17–23
[19] Kornev Yu V, Yanovskiy Yu G, Boiko O V, Chirkunova S V and Guseva M A 2013 The effect of carbon nanotubes on the properties of elastomeric materials filled with the mineral shungite *Int. J. Polymer Sci. and Technology* **40, No 2** 29–32
[20] Kornev Yu V, Yanovskiy Yu G, Boiko O V, Semenov N A and Chirkunova S V 2013 Investigating the influence of the degree of dispersion of mineral shungite on the properties of elastomeric materials based on butadiene-styrene rubber *Int. J. Polymer Sci. and Technology* **40, No 3** 17–22
[21] 1990 *Chemical Encyclopedia* (Moscow: Soviet encyclopedia) p. 671 (in Russian)
[22] Koyama T and Endo M 1973 *J. Structure and Growth Processes of Vapor-Grown Carbon Fibers* **42** 690–96 (in Japanese)
[23] Tibbetts G 1985 Lengths of carbon fibers grown from iron catalyst particles in natural gas *J. Crystal Growth* **73, No 3** 431–38
[24] Mokhireva K A, Svistkov A L and Shadrin V V *RU Patent 2552121*
[25] Mokhireva K A, Svistkov A L and Shadrin V V 2014 Determination of the optimal shape of the samples for experiments on biaxial stretching *J. Computational Continuum Mechanics* **7, No 4** 353–62 (in Russian)

IOP Conf. Series: Materials Science and Engineering **581** (2019) 012026 doi:10.1088/1757-899X/581/1/012026

Non-isothermal Steady Flow of Non-Newtonian Fluid in an Axisymmetric Channel

E I Borzenko, K E Ryltseva and G R Shrager

National Research Tomsk State University, 36 Lenin Avenue, Tomsk 634050, Russian Federation

E-mail: borzenko@ftf.tsu.ru

Abstract. Non-isothermal steady-state flow of a viscoplastic fluid in an axisymmetric channel is studied with account for viscous dissipation at a specified constant flow rate. The rheology of the medium is described by the Herschel-Bulkley law with a temperature dependence of yield stress and consistency defined by exponential law. On the solid wall, the no-slip boundary condition and the assigned temperature are used. The mathematical statement of the problem includes the dimensionless motion and energy equations and boundary conditions. The problem is solved numerically using a finite-difference approach. The difference equations are solved by sweep method. When applying a shock-capturing method for calculating the flow, the rheological model is regularized in order to eliminate stress singularity in the regions of zero shear rates. The steady-state distributions of velocity, temperature, viscosity, and dissipative function are obtained. A limiting value of pressure drop defining the existence domain of a steady solution is proved to exist. Two problem solutions are obtained at a specified pressure drop, which are referred to as high- and low-temperature flow regimes. As a result of computations, different flow structures with unyielded regions occurring near symmetry line and in the dead zone next to a solid wall are revealed.

1. Introduction

When processing the polymeric materials and metals in a liquid state, the steady-state flows in the plane and axisymmetric channels are frequently encountered in the technologies for various purposes. Moreover, the fluid medium behavior is characterized by non-Newtonian properties and temperature-dependent rheological parameters. In most cases, a successful solving of the problems of non-isothermal steady-state flows of non-Newtonian media in the channels is possible only by using numerical methods. Since the middle of the last century, the above mentioned flows have been attracting the attention of many researchers. One of the first efforts to consider one-dimensional non-isothermal flows of a viscoplastic fluid has been made in [1,2]. Stability of the steady-state solutions to one-dimensional problems and the associated phenomenon of hydrodynamic thermal explosion for the case of pseudoplastic fluid are discussed in papers [3-5]. In experimental studies [6,7], investigation of the viscoplastic fluid flows in a circular pipe at a specified flow rate is carried out for both laminar and turbulent regimes. The results of numerical studies on non-isothermal viscoplastic fluid flows in a flat channel, circular pipe, and concentric annular duct are presented in [8-12]. In these works, the heat flux on the solid wall is studied in details in terms of the governing parameters, but there is very scant data on the structure of the flow.

The purpose of this work is to study the flow structure occurred in the case of steady-state one-dimensional flow of a viscoplastic fluid with account for viscous dissipation and dependence of the rheological parameters on temperature.

2. Formulation of the Problem

A steady-state non-Newtonian incompressible fluid flow in an axisymmetric channel at a given constant flow rate is considered taking into account the effect of viscous dissipation. The fluid rheology is described by the Herschel-Bulkley law with a temperature dependence of the rheological parameters defined by exponential law. On the solid wall, the no-slip boundary condition is used, and the temperature is assigned as T_1. The flow is assumed to be one-dimensional and laminar. The solution domain and the coordinate system are presented in figure 1.

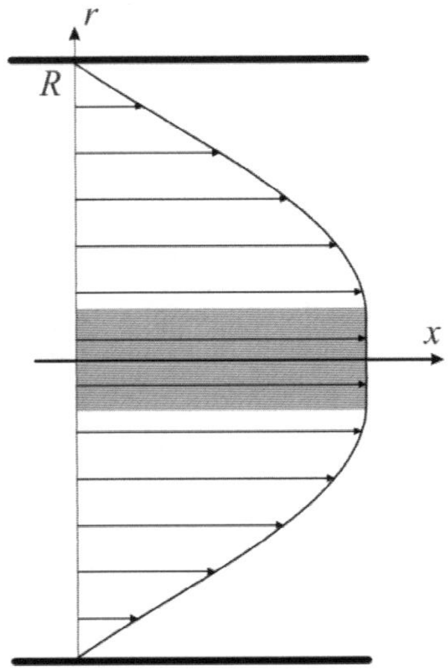

Figure 1. Solution domain.

The flow is described by the motion and energy equations which are written with due regard to the above mentioned assumptions

$$\frac{1}{r}\frac{\partial}{\partial r}\left(r\eta\frac{\partial v}{\partial r}\right)=\frac{\partial p}{\partial x},\quad \frac{\lambda}{r}\frac{\partial}{\partial r}\left(r\frac{\partial T}{\partial r}\right)+\eta\left(\frac{\partial v}{\partial r}\right)^2=0. \tag{1}$$

Here, v is the axial velocity, T is the temperature, r is the radial coordinate, $\partial p/\partial x$ is the axial component of the pressure gradient, and λ is the thermal conductivity.

System (1) is enclosed by the Herschel-Bulkley rheological law according to which the apparent viscosity η is defined as

$$\eta(T,\dot{\gamma})=\left(\tau_0\,\mathrm{e}^{-\beta_1(T-T_0)}+k_0\,\mathrm{e}^{-\beta_2(T-T_0)}\,\dot{\gamma}\right)/\,\dot{\gamma}$$

where $\dot{\gamma}$ is the shear rate, τ_0 is the yield stress at temperature T_0, k_0 is the consistency at temperature T_0, β_1 and β_2 are the constants in the rheological law, and n is the power-law index. The boundary conditions are assigned as follows:

$$r=0:\ \frac{\partial v}{\partial r}=0,\ \frac{\partial T}{\partial r}=0;\ r=R:\ v=0,\ T=T_1.$$

The system of equations (1) is written in terms of the dimensionless parameters as

$$\frac{1}{\zeta}\frac{\partial}{\partial\zeta}\left(\zeta\eta\frac{\partial u}{\partial\zeta}\right)=\mathrm{Bn}\,\delta p,\ \frac{1}{\zeta}\frac{\partial}{\partial\zeta}\left(\zeta\frac{\partial\theta}{\partial\zeta}\right)+\mathrm{Br}\eta\left(\frac{\partial u}{\partial\zeta}\right)^2=0, \tag{2}$$

where ζ is the dimensionless coordinate ($\zeta = r/R$), u and θ are the dimensionless velocity and the temperature, respectively, $\mathrm{Br}=\dfrac{\beta_1 k_1 U^{n+1}}{\lambda L^{n-1}}$ is the Brinkman number, $\mathrm{Bn}=\dfrac{\tau_1 L^n}{k_1 U^n}$ is the Bingham number,

δp is the dimensionless pressure drop per unit-length, k_1 is the consistency at wall temperature and τ_1 is the yield stress at wall temperature. The typical scales for velocity and pressure are the average velocity U and τ_1, respectively. The dimensionless temperature is referred to as $\theta = \beta_2 (T - T_1)$. The dimensionless apparent viscosity is defined by formula

$$\eta = \left(\mathrm{Bn}\, \mathrm{e}^{-\beta\theta} + \mathrm{e}^{-\theta} \left| \frac{\mathrm{d}u}{\mathrm{d}\zeta} \right|^n \right) \left| \frac{\mathrm{d}u}{\mathrm{d}\zeta} \right|^{-1}. \tag{3}$$

The boundary conditions are written as

$$\begin{gathered} \zeta = 0: \ \frac{\mathrm{d}u}{\mathrm{d}\zeta} = 0, \ \frac{\mathrm{d}\theta}{\mathrm{d}\zeta} = 0; \\ \zeta = 1: \ u = 0, \ \theta = 0. \end{gathered} \tag{4}$$

Here, $\beta = \beta_1 / \beta_2$ is the dimensionless parameter. The value δp is taken in accordance with the equality of volume flow rate per unit-area to unity

$$2\int_0^1 u\zeta \mathrm{d}\zeta = 1. \tag{5}$$

Consequently, the problem solving is reduced to the obtaining of velocity and temperature profiles which satisfy the system of equations (2),(3) and conditions (4),(5).

3. Method of Solving

The formulated problem is solved numerically. The solution domain is covered by a computational grid presented in figure 2. The velocity is calculated in the grid nodes denoted by integer indices, the viscosity and temperature, in the grid nodes denoted by fractional indices.

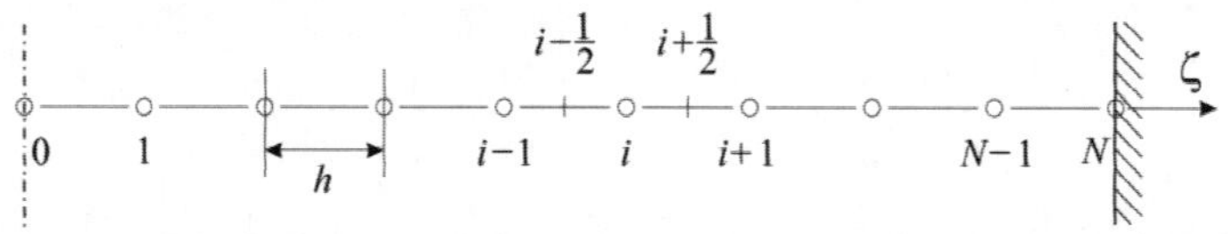

Figure 2. Computational grid.

The finite-difference analogs for system (2) are written in compliance with employment of a sweep method [13]

$$\begin{gathered} \zeta_{i+0.5}\eta_{i+0.5}u_{i+1} - \left(\zeta_{i+0.5}\eta_{i+0.5} + \zeta_{i-0.5}\eta_{i-0.5}\right)u_i + \zeta_{i-0.5}\eta_{i-0.5}u_{i-1} = \mathrm{Bn}\,\delta p\, h^2 \zeta_i, \\ \zeta_{i+1}\theta_{i+1.5} - \left(\zeta_{i+1} + \zeta_i\right)\theta_{i+0.5} + \zeta_i\theta_{i-0.5} + æ\,\eta_{i+0.5}\left(u_{i+1} - u_i\right)^2 = 0. \end{gathered} \tag{6}$$

The boundary conditions (4) in terms of the finite-differences are presented as follows:

$$\begin{gathered} u_0 = u_1 - \frac{\mathrm{Bn}\,\delta p\, h^2}{4\eta_{0.5}}, \ \theta_{0.5} = \theta_{-0.5}; \\ u_N = 0, \ \theta_{N+0.5} = -\theta_{N-0.5}. \end{gathered} \tag{7}$$

The Herschel-Bulkley model has a feature of “infinite” apparent viscosity in the regions where the stress level is less than yield stress. When using a shock-capturing method for calculating viscoplastic medium flows without explicit separating of the unyielded regions, a regularization of the rheological model is implemented to ensure the stability and accuracy of the calculations in the regions of low strain rate. In this work, a modified rheological model [14] is utilized which is characterized by apparent viscosity defined by formula

$$\eta = \mathrm{Bn}\,\mathrm{e}^{-\beta\theta}\left[1-\exp\left(\left|\frac{\mathrm{d}u}{\mathrm{d}\zeta}\right|\varepsilon^{-1}\right)\right]\left|\frac{\mathrm{d}u}{\mathrm{d}\zeta}\right|^{-1} + \mathrm{e}^{-\theta}\left|\frac{\mathrm{d}u}{\mathrm{d}\zeta}\right|^{n-1},$$

where ε is the small parameter. In the region of shear flow, the viscosity values calculated using the initial and regularized models are almost coincide with each other; in the unyielded region, regularization provides a large but final viscosity value. The separation of zones with stress level less than yield stress is performed by means of the following condition:

$$\eta\left|\frac{\mathrm{d}u}{\mathrm{d}\zeta}\right| < \mathrm{Bn}\,\mathrm{e}^{-\beta\theta}.$$

Difference analogues (6), (7) approximate differential equations (2) and conditions (4) with the second order of accuracy. The resulting systems of algebraic equations (6) are solved by sweep method. A simultaneous solving of the systems of difference equations (6) intended to obtain the steady-state fields of velocity and temperature requires an iterative process to be organized. A zero temperature is used as initial approximation for iterative process.

The value of parameter δp, which ensures the fulfillment of condition (5) with a desired accuracy, is selected by bisection method [15]. When calculating the steady-state fields of velocity and temperature, the following conditions are used as convergence criteria

$$\max_i\left|\frac{u_i^{k+1}-u_i^k}{u_i^{k+1}}\right| < \varepsilon_u,\ \max_i\left|\frac{\theta_i^{k+1}-\theta_i^k}{\theta_i^{k+1}}\right| < \varepsilon_\theta,\ 0 \le i \le N-1,$$

where the superscript number corresponds to the iteration index, and the subscript number, to the computational node index.

The convergence of the algorithm for calculating δp is verified in accordance with condition

$$\left|1-2\sum_{i=1}^{N}\frac{u_{i-1}\zeta_{i-1}+u_i\zeta_i}{2}h\right| < \varepsilon_Q.$$

Testing the computational method, a benchmark problem of isothermal steady-state flow of a viscoplastic Herschel-Bulkley fluid in a circular pipe is solved. The corresponding analytical solution is written as [6]

$$u = \begin{cases} u_0,\ 0 \le \zeta \le \zeta_0; \\ u_0\left[1-\left(\dfrac{\zeta-\zeta_0}{1-\zeta_0}\right)^{1+\frac{1}{n}}\right],\ \zeta_0 < \zeta \le 1. \end{cases}$$

Here, $\zeta_0 = -2/\delta p$ is the radial coordinate of unyielded region boundary, $u_0 = \left(-\dfrac{\delta p\,\mathrm{Bn}}{2}\right)^{\frac{1}{n}}\dfrac{n}{n+1}\left(1-\zeta_0\right)^{\frac{1+n}{n}}$ is the velocity value in the unyielded region. The pressure drop value ensuring the flow rate equal to unity is determined by equation

$$\left(1-\zeta_0\right)^{\frac{3n+1}{n}} - \frac{3n+1}{n}\left(1-\zeta_0\right)^{\frac{2n+1}{n}} + \frac{6n^2+5n+1}{2n^2}\left(1-\zeta_0\right)^{\frac{n+1}{n}} = \frac{6n^3+11n^2+6n+1}{2n^3}\left(-\frac{2}{\delta p\,\mathrm{Bn}}\right)^{\frac{1}{n}}.$$

Table 1. The values of u_0, δp, and ζ_0 versus grid step h
(Br = 0, Bn = 1, n = 0.5, $\varepsilon_u = \varepsilon_\theta = \varepsilon_Q = 10^{-6}$, and $\varepsilon = 10^{-3}$).

h	1/10	1/20	1/50	1/100	1/200	Analytical solution
u_0	1.4562	1.4434	1.4399	1.4394	1.4393	1.4389
$-\delta p$	6.9629	6.9360	6.9285	6.9274	6.9272	6.9271

ζ_0	0.2872	0.2882	0.2886	0.2887	0.2887	0.2887

Table 2. The values of u_0, δp, and ζ_0 versus regularization parameter (Br = 0, Bn = 2, n = 0.5, $\varepsilon_u = \varepsilon_\theta = \varepsilon_Q = 10^{-6}$, and h = 1/200).

ε	10^{-1}	10^{-2}	10^{-3}	10^{-4}	10^{-5}	Analytical solution
u_0	1.3713	1.3415	1.3384	1.3381	1.3380	1.3379
$-\delta p$	4.6453	4.6526	4.6532	4.6532	4.6532	4.6532
ζ_0	0.4305	0.4299	0.4298	0.4218	0.4123	0.4298

The approximation convergence of the calculated results to the analytical values in the isothermal case is shown in table 1, which presents the calculated values of unyielded region size, velocity, and pressure drop versus grid step. The effect of regularization parameter on the calculated results is reflected in table 2. It is evident that, as ε decreases, the values of u_0 and δp tend to the results of analytical solution. However, starting with $\varepsilon = 10^{-4}$, the error in the calculated coordinate of unyielded region boundary increases which may be attributable to an increase in the approximation errors for apparent viscosity in the region of unyielded motion. Further calculations are carried out using the grid step of h = 1/200 with the following computational parameters: $\varepsilon_u = \varepsilon_\theta = \varepsilon_Q = 10^{-6}$ and $\varepsilon = 10^{-3}$.

Table 3. The values of u_0, θ_0, δp, and ζ_0 versus grid step h (Br = 2, Bn = 1, n = 0.5, and β = 1).

h	1/10	1/20	1/50	1/100	1/200
u_0	2.5484	2.5407	2.5384	2.5381	2.5380
θ_0	1.3863	1.3892	1.3900	1.3901	1.3901
$-\delta p$	3.4395	3.4500	3.4528	3.4533	3.4534
ζ_0	0.1454	0.1445	0.1443	0.1442	0.1442

The approximation convergence of the computational algorithm developed for non-isothermal case is shown in table 3 where θ_0 is the temperature on the symmetry axis.

4. Results and Discussion

A non-isothermal case with account for dependence of the rheological parameters on temperature is considered. A dependency diagram of the dimensionless pressure drop versus Brinkman number is shown in figure 3, while the Bingham number is obtained by formula

$$\mathrm{Br}\,\mathrm{Bn}^2 = C.$$

In accordance with this equality, an increase in Br may be interpreted as an increase in the dimensional flow rate under otherwise equal conditions. Variation in the constant C on the right side of the formula leads to a change in the quantitative characteristics of the curve in figure 3. The plot presented in figure 3 demonstrates the relation between pressure drop and average flow rate at C = 25.

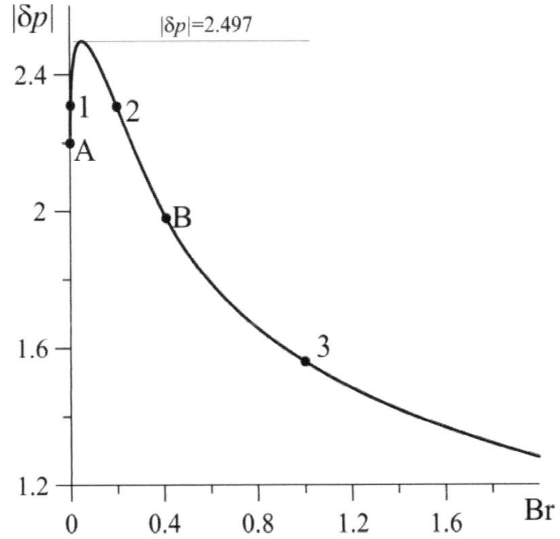

Figure 3. Dependence of $|\delta p|$ on Br (n = 1 and β = 1).

It is notable that the value of δp is limited over the whole range of flow rate variation. If the flow with a specified pressure drop is considered, there is no steady-state solution at $|\delta p| > 2.497$ which means that the heat generated due to viscous dissipation is not able to be removed from the pipe through the walls in time, and a phenomenon, which is referred to as a hydrodynamic thermal explosion (an analogue of the thermal explosion), occurs [16,17].

In the region of $|\delta p| < 2.497$, two regimes are possible for a specified pressure drop: low- and high-temperature regimes at a flow rate equal to unity. For example, the case when the governing parameters take the values corresponding to points 1 and 2 in figure 3. For these two regimes, the velocity and temperature profiles are presented in figures 4(a) and 4(b), respectively. The flow region can be separated into unyielded regions near the axis (painted in gray) and a shear flow in the rest part. The temperature and velocity both have constant values in the unyielded regions.

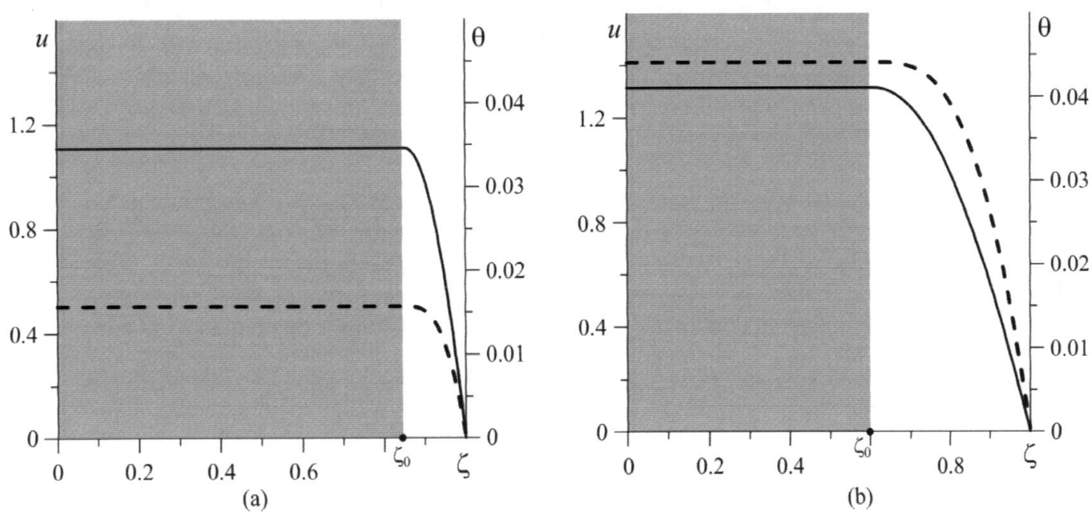

Figure 4. Velocity (solid line) and temperature (dashed line) profiles at n = 1, β = 1: (a) Br = 0.003, Bn = 91.3 and (b) Br = 0.2, Bn = 11.2.

With an increase in Br (starting with point B, figure 3), a high-temperature flow regime with an unyielded region near the axis and a dead zone near the wall is observed. The profiles of unknown variables corresponding to Br and Bn numbers at point 3 (figure 3) are shown in figure 5. The fluid is at rest, and the temperature decreases linearly in the wall-adjacent zone. The dissipative function (figure 5(b), solid line) is equal to zero both in the unyielded region and in the wall-adjacent zone. The viscosity behavior in the shear flow region is demonstrated in figure 5(b) (dashed line). The pressure drop at point A (figure 3) corresponds to a minimum value at which the flow is possible, i.e., the stress level is higher than the yield stress.

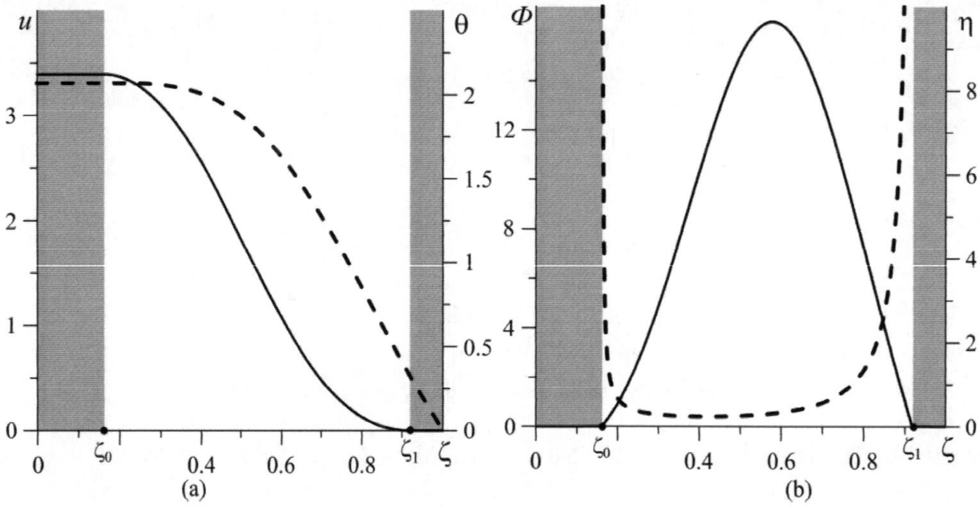

Figure 5. Profiles of the (a) velocity and temperature and (b) dissipative function and viscosity at $n = 1$, $\beta = 1$, Br = 1, and Bn = 5.

Figure 6 illustrates the plots of the coordinates of unyielded region boundaries and stagnant zone versus Bn number for three values of Br parameter. Dissipative effects are weak at Br = 0.1, and the temperature has little impact on the viscosity. As a result, the influence of the Bn number in the considered range is similar to that in the isothermal case: with an increase in Bn number, the central unyielded region expands, and the stagnant zone does not occur. Increase in Br up to 0.5 leads to the qualitative and quantitative changes in the flow pattern. On the one hand, large Bn numbers entail high viscous stresses in the shear flow region and, as a consequence, high dissipative function values are achieved. On the other hand, dissipation leads to an increase in temperature and to a decrease in viscosity. As a result, the flow regime with central unyielded region and dead zone occurs in the pipe at Bn > 6.65. At the same time, with an increase in Bn, the size of unyielded region decreases, while the dead zone increases. The flow regime with a dead zone is observed for Br = 2.5 at Bn > 2.

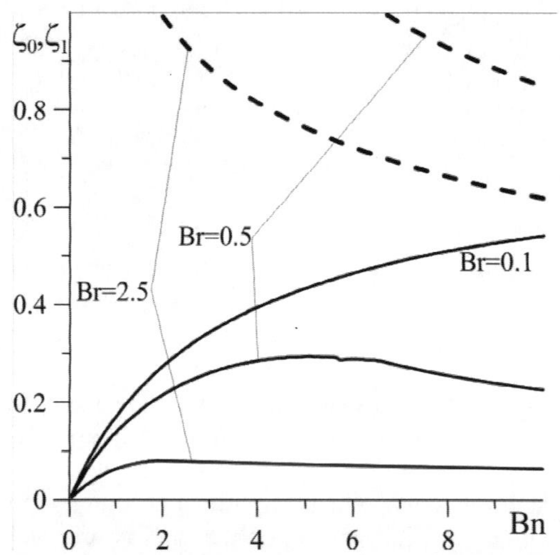

Figure 6. Coordinates of unyielded flow regions ($n = 1$ and $\beta = 1$).

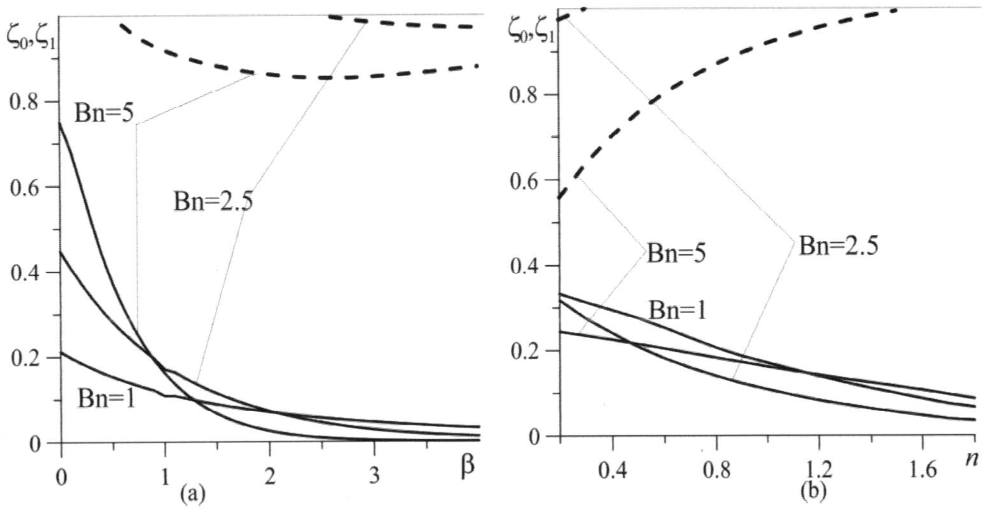

Figure 7. Coordinates of unyielded flow regions (Br = 1: (a) $n = 1$ and (b) $\beta = 1$).

The effect of parameter β on the flow structure is shown in figure 7(a). The flow is characterized by the presence of unyielded region in the central part at $\beta = 0$. Increasing this parameter, the size of the unyielded region decreases, and the regime with both unyielded region and dead zone is observed. The central unyielded region does not occur at Bn = 5 and $\beta > 2.7$, and only a dead zone is distinguished in the flow. Investigation of the flow structure as a function of the power-law index n is presented in figure 7(b). In the considered range of variation in n, which covers dilatant and pseudoplastic regions, it is possible to obtain the above mentioned regimes depending on the governing parameters.

5. Conclusion

The mathematical statement of the problem of a steady-state non-isothermal flow of a viscoplastic Herschel-Bulkley fluid in a circular pipe is formulated in a one-dimensional approximation with account for viscous dissipation and dependence of the yield stress and consistency of the medium on temperature. The numerical algorithm based on the finite-difference method is developed to solve a stated problem. A set of test calculations is carried out. The kinematic and thermophysical characteristics of the flow are presented. The flow structure is studied in terms of the governing parameters. The flow regimes characterized by the presence of unyielded region in the vicinity of symmetry axis and a dead zone are revealed.

Acknowledgment

The research is implemented at the expanses of the Russian Science Foundation (project No. 18-19-00021).

References

[1] Pervushin V E 1974 *J. Appl. Mech.Tech. Phys.* **15** 350–4
[2] Shul'man Z P, Khusid B M and Benderskaya S L 1979 *J. Eng. Phys.* **37** 1410–4
[3] Bostandzhiyan S A, Merzhanov A G and Khudyaev S I 1965 *Dokl. Akad. Nauk SSSR* **163** 133–6 (in Russian)
[4] Bostandzhiyan S A and Chernyaeva S M 1966 *Dokl. Akad. Nauk SSSR* **170** 301–304 (in Russian)
[5] Borzenko E I and Shrager G R 2018 *Tomsk State University Journal of Mathematics and Mechanics* **52** 41–52 (in Russian)
[6] Peixinho J, Nouar C, Desaubry C and Theron B 2005 *J. Non-Newton. Fluid Mech.* **128** 172–84
[7] Peixinho J, Desaubry C and Lebouche M 2008 *Int. J. Heat Mass Transf.* **51** 198–209
[8] Labsi N, Benkahla Y K and Boutra A 2017 *J. Braz. Soc. Mech. Sci. & Eng.* **39** 267–77
[9] Coelho P M and Poole R J 2018 *Heat Transfer Eng.* **39** 1753–69
[10] Berabou W, Benkahla Y K, Labsi N, Ouyahia S E and Boudiaf A 2018 *J. Braz. Soc. Mech. Sci. & Eng.* **40** 404
[11] Boualit A, Zeraibi N, Boualit S and Amoura M 2011 *Int. J. Therm. Sci.* **50** 36–43

[12] Fusi L 2017 *Meccanica* **52** 3577 92
[13] Godunov S K and Ryabenkii V S 1987 *Difference Schemes: An Introduction to the Underlying Theory* (Amsterdam: Elsevier Science Publishers) p 489
[14] Papanastasiou T C 1987 *J. Rheol.* **31** 385–404
[15] Demidovich B P and Maron I A 1981 *Computational Mathematics* (Moscow: Mir Publishers) p 688
[16] Frank-Kamenetskii D A 2015 *Diffusion and Heat Exchange in Chemical Kinetics* (Princeton: Princeton University Press) p 384
[17] Hudyaev S I 2003 *Threshold Phenomena in Nonlinear Equations* (Moscow: Fizmatlit) p 271 (in Russian)

IOP Conf. Series: Materials Science and Engineering **581** (2019) 012027 doi:10.1088/1757-899X/581/1/012027

Experimental study of electrical impedance in airways filled with electrolyte solution aerosol

A L Zuev[1, 2, a], Ia V Mishlanov[3] and V B Polyakov[3]

[1] Institute of Continuous Media Mechanics, Perm Federal Research Center of Ural Branch of Russian Academy of Sciences, Perm, Russia
[2] Perm National Research Polytechnic University, Perm, Russia
[3] Perm State National Research University, Perm, Russia

E-mail: [a] zal@icmm.ru

Abstract. This article presents the results of an experimental study of electrical impedance in a model cell through which the air flow containing aerosol particles of the 10% sodium chloride solution generated by an ultrasonic nebulizer passes. The active and capacitive components of admittance were determined at different values of alternating current frequency, flow rate and aerosol atomization intensity. Measurement results were compared with the most probable electrical conductivity models. An equivalent electrical circuit in the form of a parallel *RC*-circuit, which corresponds well to the obtained frequency dependences of impedance magnitude and phase shift angle, is proposed. In addition, a linear regression model describing the dependence of the impedance module logarithm on two parameters of the flow of aerosol particles at different electrical current frequencies is developed.

1. Introduction

Rheographic methods based on registration of bioelectrical impedance (resistance of biological media and tissues to an alternating current) found apllication in the last decades in modern medicine for noninvasive diagnosis of human diseases due to their exceptional avalaibility, simplicity and cheapness. These methods can be used to evaluate the status of the cardiovascular system (rheocardiography and the other analogous techniques), to examine the composition of human body tissues (the assessment of the ratio of fat, musculoskeletal and water masses), to analyze the molecular composition of biological fluids, to apply a conductometric method for counting blood cells, etc.

Researchers from the Department of Propaedeutics of Internal Medicine at Perm Medical University were the first who suggested using impedancemetry in clinical diagnostics of respiratory diseases [1-3]. They have developed a novel method of electrical impedance spirometry to record variations in the electrical impedance of the entire set of biological objects that constitute an electrical circuit when an electrical current propagates from the respiratory tract to the skin of the human chest: lung tissues, aerosol exhaled by the patient during the respiratory cycle, bronchial secretion electrolyte, etc.

The underlying concept of the method consists in preliminary saturation of the respiratory tract with small droplets of a conductive fluid (0.9% sodium chloride solution) generated by an ultrasonic nebulizer. Electrical impedance measurements were performed using a bipolar method, in which one electrode was installed in the nebulizer mouthpiece (without direct contact with the patient body), and another electrode was located on the chest skin in the region of 4th intercostal space. The electrical circuit between the electrode in the nebulizer mouthpiece and the mucous membrane of the respiratory tract was closed due to the flow of aerosol particles of the conducting fluid exhaled by the patient.

Analysis of the results of clinical studies has revealed a significant difference in the average modular value of electical impedance between the groups of patients with various disorders of the external respiration function and the group of practically healthy subjects. A signficant growth in the impedance magnitude was found for patients with bronchial asthma (BA) in comparison with controls. The method is also able to detect early pneumonia signs; it shows approximately 90% sensitivity when compared with conventional spirometry. In some cases (e.g. in the case of allergic inflammation of the respiratory tract), the sensitivity of the method exceeds that of the common spirometry by 30%, which makes it possible to diagnose bronchial asthma in the patients who had no yet clinically significant symptoms of this disease.

However, further development of electrical impedance spirometry and its introduction into clinical practice is impossible, without revealing physical mechanisms responsible for changes in the electrical conductivity of the respiratory tract and lung parenchyma. Real physical factors causing the differences in bioimpedance values in patients with disorders in large and small respiratory tracts have not been sufficiently studied and fully understood. The increase of electrical resistance can be attributed to the changes in blood vessels, small airways, lung parenchyma, bronchial tree diameter, bronchial secretions, which may arise in damaged lung tissues, during the development of fibrosis and pneumocirrhosis, pulmonary edema, and the formation of hydrothorax. A reduction in the respiratory tract diameter can lead to a decrease in ventilation indicators, which is typical of patients with bronchial asthma. Another important indicator of the medical assessment of respiratory organs is the strength of the respiratory musculature, determined by the rate of air flow expired by the patient during the respiratory cycle. A typical symptom of respiratory disorders in BA and diseases of lung parenchyma is a decrease in the rate of this flow. These findings suggest that the sensitivity of the electrical impedance spirometry method is probably caused by the dependence of the electrical impedance of the aerosol moving in the tube on the rate of its flow. Meanwhile, the mechanism governing such dependence still remains unclear, and hence this assumption requires further experimental verification.

In this paper, we present the results of the model experiment carried out to test an assumption that the value of electrical impedance depends on the diameter of the respiratory tract and the velocity of aerosol particles in it. Distribution of aerosol particles was generated by a medical ultrasonic nebulizer. Important parameters that affect the properties of the obtained aerosol are the acoustic power of the ultrasonic radiator and the airflow velocity above the liquid surface. The concentration of particles in the aerosol thus generated depends on both parameters [4]. The conductivity of the aerosol flow is determined by several mechanisms. The aerosol particles obtained in the nebulizer can have an electrical charge when particle ionization occurs due to the accidental collision (friction) between particles or between particles and surrounding surfaces during the aeration of the solution of a conducting liquid by ultrasonic waves [5-6]. Therefore, the magnitude of the resulting electric current must be proportional to the velocity of the flow of charged particles, i.e. the expiratory flow rate. Due to temperature difference, the aerosol can condense on the walls of the tube in which it moves. In the case of good wettability of the walls, a thin liquid film is formed on them; its thickness also depends on the flow rate and the nebulization intensity. When the electrodes touch the walls, the conductivity of such a condensation film must also be taken into account.

2. Experimental procedure

The electrical conductivity of the aerosol flows of saline solutions was investigated using an experimental setup consisting of an aerosol generator, a measuring cell and a gauge for measuring the impedance magnitude and phase shift angle.

The measuring cell was a polystyrene tube of diameter 1 mm and length 70 mm, through which the aerosol flow of the conducting fluid was blown. Silver chloride (AgCl) plate electrodes of length 25 mm and width 8 mm (spaced 3mm apart in one plane) were placed inside the measuring cell in one plane near the internal wall of the tube. It is exactly this design of the measuring cell and electrodes that has been used earlier in clinical studies of the electrical impedance spirometry method [3] and has demonstrated its effectiveness in analyzing the respiration of patients.

Aerosols were generated by the Omron Ultrasonic Nebulizer Ne-u17. To this end, the nebulizer reservoir was filled with a 10% sodium chloride (NaCl) or calcium chloride ($CaCl_2$) solution. The volume of the liquid was 50 ml. The aerosol in the nebulizer is obtained by exposing the liquid to ultrasonic vibrations at different nebulization (spraying) intensity, corresponding to ten levels from 1 to 10. The average particle size is ~4-6 μm. The nebulizer has a fan to evacuate the aerosol from the reservoir chamber. The nebulizer air flow rate can vary from 0 to 17 l/min (11 levels – from 0 to 10) [7].

In order to simultaneously determine the impedance magnitude and phase shift angle, we used the RCL-meter GW-Instek LCR-8105G with the following technical characteristics: resolution – 6 decimal places, frequency range – 20 Hz to 5 MHz, impedance measuring range – 0.1 Ohm to 100 MOhm, voltage amplitude – 10 mV to 2.0 V, and basic accuracy of measurements – 0.1% [8]. The tests were carried out according to a bipolar scheme: the unipolar current and potential electrodes of the RCL-meter were jointly connected to the electrodes of the measuring cell. The voltage amplitude was assumed to be equal to 2.0 V. The LCR-meter operation was controlled by a personal computer. The original computer program enables the computer to execute a series of tests consisting of 10 measurements (doubles) of electrical impedance sequentially for each electric current frequency with saving the results in a data file. All measurements were done at 1 min interval after the aerosol begins to travel through the measuring cell.

Mathematical treatment of the experimental results is frequently performed subject to the normal law of random measurement distribution, i.e. the result of each measurement is influenced by a great number of factors, and each factor makes a small contribution to the value of error. At the same time, in a number of works, the necessity of mandatory verification of the hypothesis concerning the normal distribution of a random error is pointed out. This problem is solved by applying fit criteria, including Pearson's chi-square criterion and Kolmogorov's criterion. In our investigation, we use the Shapiro-Wilk criterion, which is well suited for assessing a small number of data [9]. A statistical distortion was excluded from the obtained series of electrical impedance values. In each statistical sampling, the average value of every factor and its standard deviation were calculated. The measurement result was excluded if its value did not fall within the interval of three standard deviations from the average value of the remaining sample [10].

3. Results and discussion

In the experiments with the 10% sodium chloride solution, the frequency dependencies of impedance magnitude and phase shift angle were obtained at frequencies 20, 100, 500 Hz, 1, 5, 10, 20, 30, 40 and 50 kHz for six values of the air flow rate V (levels 1, 3, 5, 7, 8, 10) and five values of the nebulization intensity W (levels 2, 4, 6, 8, 10), respectively. Thus, the total number of the analyzed combinations of the parameters of the electrical current and the aerosol particle flow was 300 different variants. The measured values of the impedance magnitude reached rather high values ~ 1-100 MOhm. Typical curves of the frequency dependencies for the impedance magnitude and phase shift angle with the other fixed parameters are shown in figure 1 and figure 2. It was found that at frequencies below 5 kHz no significant change in the impedance magnitude is observed in the case when the frequency increases (the active component of the resistance remains constant), and the phase angle values are close to zero (the capacitive component of the resistance is practically absent). At higher frequencies, a significant decrease in the impedance magnitude and an increase in the negative value of the phase shift angle from 0 to –78 degrees are observed. The obtained results indicate that at lower frequencies (20–5000 Hz) the aerosol impedance magnitude is mainly determined by the active resistance, and at higher frequencies (10–50 kHz) the conductivity increases because of the capacitive component. In this case, the curves of the corresponding frequency components in a logarithmic scale have a constant slope.

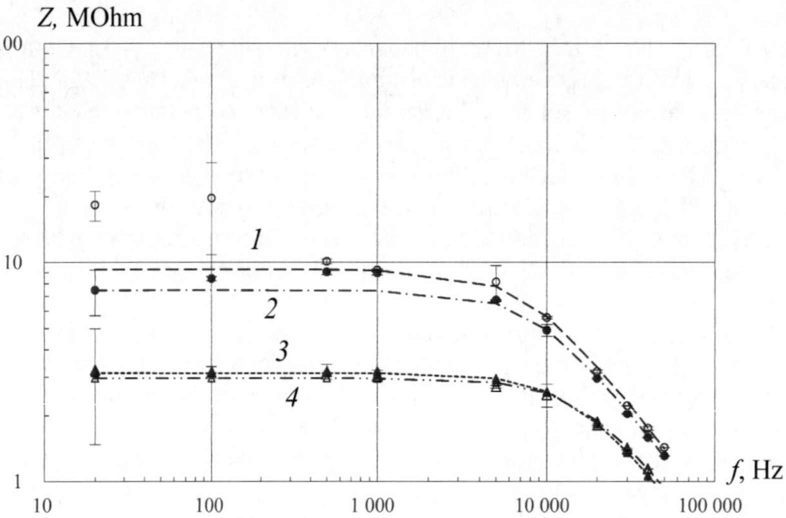

Figure 1. Impedance magnitude Z versus frequency f at $W = 4$, $V = 5$ (*1*); $W = 4$, $V = 8$ (*2*); $W = 8$, $V = 5$ (*3*); $W = 8$, $V = 8$ (*4*). Lines – calculated dispersion dependencies for the corresponding parallel *RC*-circuits.

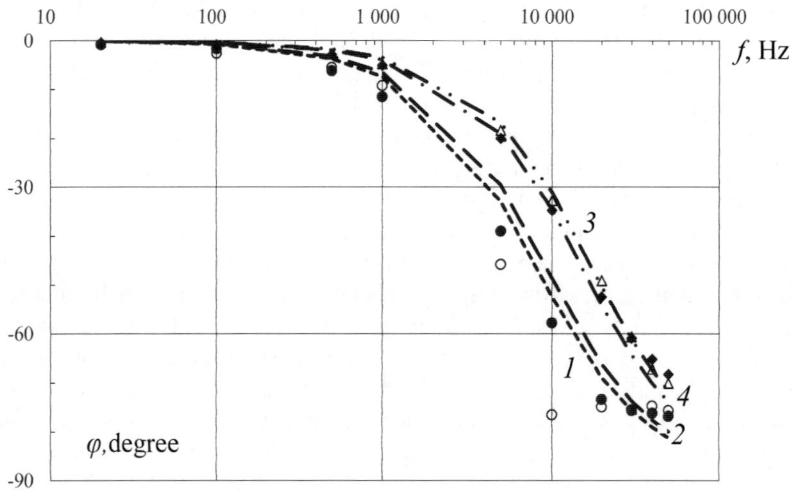

Figure 2. Impedance phase angle φ versus frequency f at $W = 4$, $V = 5$ (*1*); $W = 4$, $V = 8$ (*2*); $W = 8$, $V = 5$ (*3*); $W = 8$, $V = 8$ (*4*). Lines – calculated dispersion dependencies for the corresponding parallel *RC*-circuits.

Dispersion dependencies of such type are typical of the parallel *RC*-circuit [11-13], which allows us to offer this type of an equivalent circuit and to calculate the corresponding values of the active resistance R and capacitance C at the given flow rate and nebulization intensity.

The parameters R and C are calculated as follows. The impedance magnitude Z for the parallel *RC*-circuit is described by the expression:

$$Z = \frac{R}{\sqrt{1 + (\omega RC)^2}}$$

After transformation, this equation becomes

IOP Conf. Series: Materials Science and Engineering **581** (2019) 012027 doi:10.1088/1757-899X/581/1/012027

$$\frac{1}{Z^2} = \frac{1}{R^2} + C^2\omega^2$$

Using the method of linear regression [14] (the dependent variable Z^{-2} , and the independent variable ω^2), from the above expression we can obtain the coefficients R^{-2} and C^2 .

The values R lie within the range from 3.1 to 90 MOhm, and the values C within the range from 0.56 to 4.5 pF. The solids lines in figures 1 and 2 are plotted using the calculated data. It is seen that the theoretical results coincide well with the experimental data. Within the entire measurement range, the mean error was about 30% for R and 5% for C.

The typical dependences of the calculated value of the capacity C on the airflow rate and the nebulization intensity are represented by the curves given in figures 3 and 4. It is seen from the curve $C(V)$ (figure 3) that at the different prescribed values of nebulization intensity the capacitance increases in a monotonic fashion as the rate of the aerosol flow generated by the nebulizer fan increases, and the dependence is close in this case to the linear one. Besides, the capacity also increases with increasing nebulization intensity (see $C(W)$ in figure 4).

However, both the values of the capacitance itself and its changes are very insignificant being of only a few pF, which cannot make a significant contribution to the impedance due to its capacitive component at relatively low AC frequencies used in the experiments.

In turn, the experimental data on the dependence of the active resistance on the aerosol velocity $R(V)$ and on nebulization intensity $R(W)$ indicate that the resistance decreases monotonically with both aerosol velocity and nebulization intensity. In this case, the active resistance changes in much more significant limits, decreasing tens of times, in comparison with the capacitance. In particular, at $W = 4$, $V = 8$ are as follows: $R = 8.5 \pm 0.6$ MOhm, $C = 2.50 \pm 0.04$ pF; at $W = 8$, $V = 5$: $R = 2.91 \pm 0.05$ MOм, $C = 3.26 \pm 0.03$ pF. Figure 5 presents the graph showing the dependence of the impedance magnitude obtained at low frequencies of 20 Hz and 50 kHz (when the total resistance is caused by the active component only) on the aerosol flow velocity at different nebulization intensities. It is shown that the straight lines with a negative slope coefficient best represent the obtained experimental data. The graphs showing the dependence of the impedance magnitude logarithm on the intensity of spraying aerosol particles at different values of the airflow velocity are also the straight lines with a negative slope coefficient (figure 6).

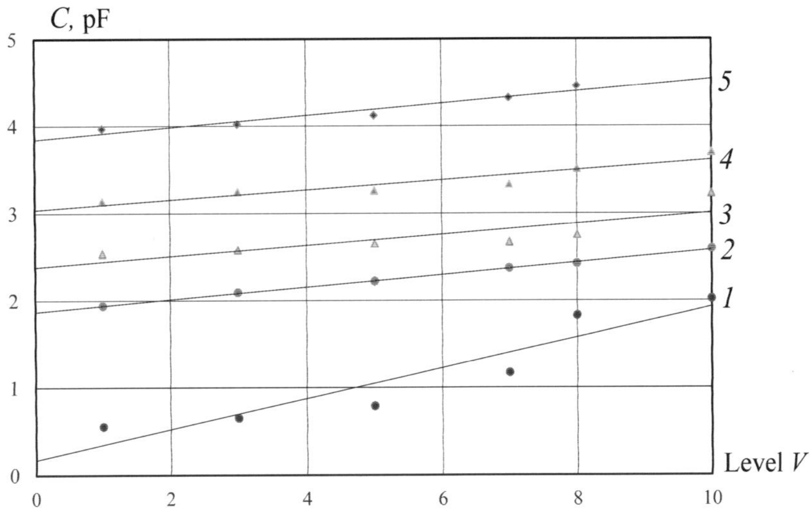

Figure 3. Capacitance C of the equivalent RC-circuit versus aerosol particle flow rate V at nebulization intensity $W = 2$ (*1*); $W = 4$ (*2*); $W = 6$ (*3*); $W = 8$ (*4*); $W = 10$ (*5*).

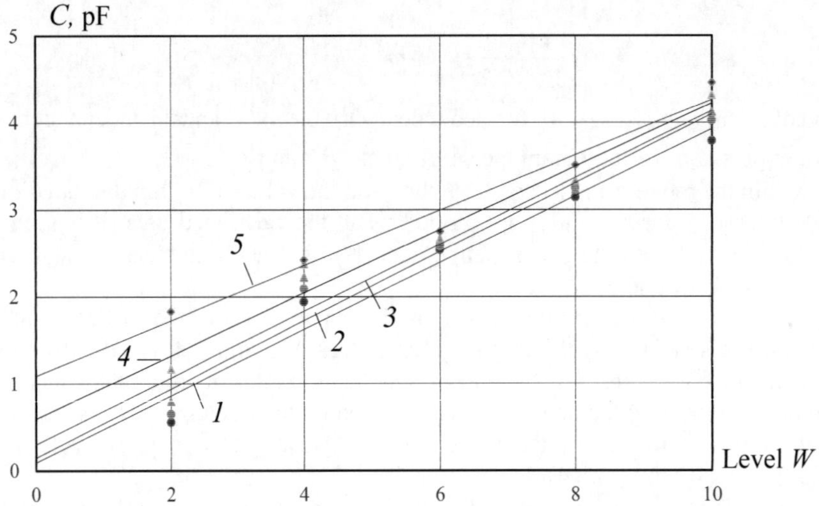

Figure 4. Capacitance C of the equivalent RC-circuit versus nebulization intensity W at flow rate $V = 1$ (*1*); $V = 3$ (*2*); $V = 5$ (*3*); $V = 7$ (*4*); $V = 9$ (*5*).

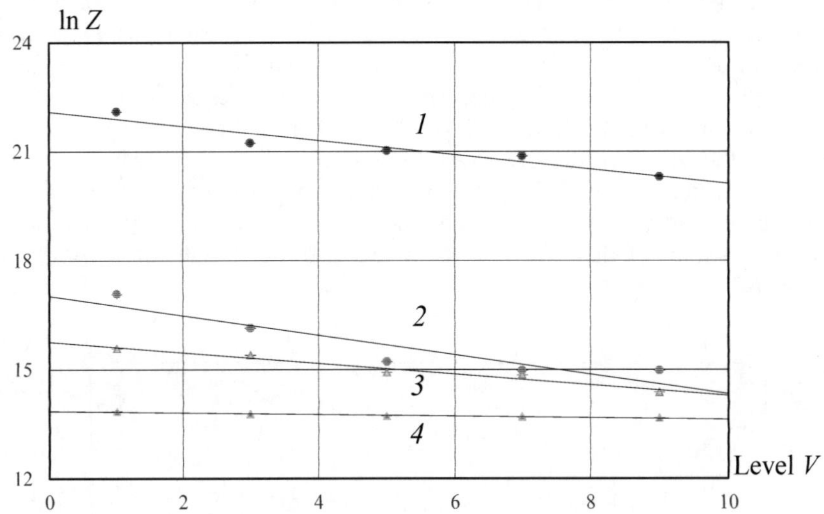

Figure 5. Logarithm of impedance magnitude Z versus aerosol velocity V for frequency $f = 20$ Hz at $W = 2$ (1); $W = 8$ (*2*); and for frequency $f = 50$ kHz at $W = 2$ (*3*), $W = 8$ (*4*).

Based on the experimental data obtained in this study, we have suggested the regularities in the form of expressions (1) и (2) [15]:

$$Z \sim e^{-aV}, \tag{1}$$

$$Z \sim e^{-bW} \tag{2}$$

where a, b are coefficients.

In order to determine the dependences of the aerosol impedance magnitude Z on V and W for each electric current frequency, we derive, on the basis of relations (1) and (2), the linear regression [14] (a_1, a_2 and a_3 are some constants):

$$\ln Z = a_1 V + a_2 W + a_3 \tag{3}$$

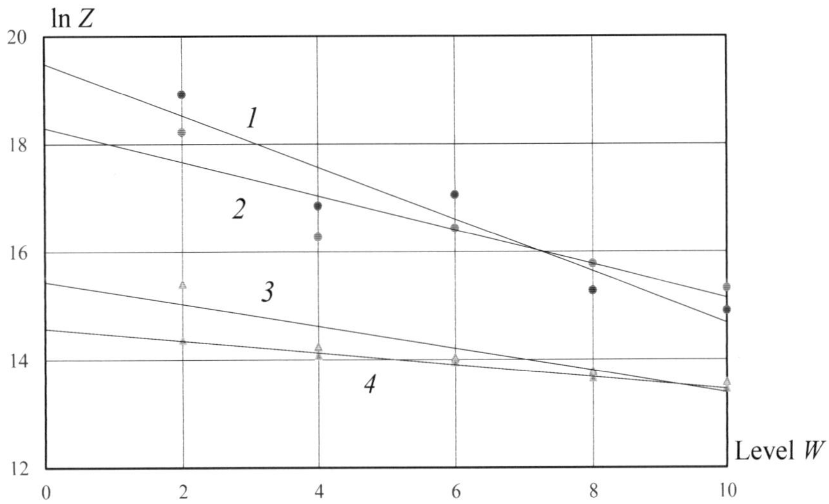

Figure 6. Logarithm of impedance magnitude logarithm Z versus intensity of spraying of aerosol particles W for frequency $f = 20$ Hz at $V = 2$ (*1*); $V = 8$ (*2*); and for frequency $f = 50$ kHz at $V = 2$ (*3*), $V = 8$ (*4*).

By applying a linear regressive analysis for the model describing dependencies of the impedance magnitude of the aerosol (equation 3), we obtained the results given in Table 1:

Table 1. Values of parameters of regression equation (3), square of the adjusted coefficient of determination and the relative mean square error at frequencies 20 Hz, 1, 10 and 50 kHz

Frequency, Hz	Parameter	Value	R^2_{adj}	Mean square error S_n, %
20	a_1	–0.153 ± 0.009	0.709	8.84
	a_2	–0.618 ± 0.011		
	a_3	17.67 ± 0.09		
1 000	a_1	–0.131 ± 0.006	0.739	5.49
	a_2	–0.386 ± 0.006		
	a_3	15.53 ± 0.05		
10 000	a_1	–0.097 ± 0.003	0.756	3.39
	a_2	–0.219 ± 0.004		
	a_3	13.28 ± 0.03		
50 000	a_1	–0.077 ± 0.003	0.797	2.92
	a_2	–0.189 ± 0.003		
	a_3	11.66 ± 0.02		

The calculated values of the adjusted coefficient of determination and the mean square error indicate that equation (3) is in good agreement with the experimental data, and with increasing frequency the mean square error decreases and the proportion of explained variance increases.

IOP Conf. Series: Materials Science and Engineering **581** (2019) 012027 doi:10.1088/1757-899X/581/1/012027

4. Conclusion

Thus, the assumption regarding the interaction between electric and kinetic parameters of the flow of aerosol particles of conducting liquid was confirmed in our model experiments. The impedance measurement data on the air flow composed of aerosol particles of the 10% sodium chloride solution traveling through the measurement cell and the calculated results obtained at an alternating flow of different frequency have indicated that an equivalent electric circuit (a parallel *RC*-circuit in this study) provides an adequate description of the frequency dependence of impedance parameters. The coefficients of the calculated capacitance were found to increase linearly with increasing velocity of the airflow generated by the ultrasonic nebulizer and with the intensity of spraying aerosol particles. The inverse dependencies of the impedance magnitude on the aerosol velocity V and the nebulization intensity W were determined, and the model describing the dependence of impedance magnitude logarithm on two variables was developed. The coefficients of a linear regression model were calculated for each frequency. It is shown that the proposed model describes satisfactorily the experimental data. The obtained results suggest that the velocity of the exhaled air flow can be determined by applying the electrical impedance method, which paves the way for designing a new electrical impedance sensor required to measure the velocity of an expired airflow, as well as methods for medical diagnosis of bronchial obstructive diseases.

References

[1] Mishlanov V Yu 2011 Lung and airways electrical impedance measurement on different probing current frequences or lung functional testing study *The Bulletin of Contemporary Clinical Medicine* **4(4)** 24-8.

[2] Mishlanov V Yu *et al* 2013 A method for diagnosing respiratory function using impedance spirography and the BIA-Lab Spiro hardware and software system for its implementation *Patent for invention* RUS 2487662 20.07.2013. In Russian.

[3] Mishlanov V Yu, Zuev A L, Ustyanzeva T L, Mishlanov Ja V and Savkin V V 2013 Lung functional testing by electroimpedance spirometry: experimental and clinical study *Neuroscience and Behavioral Physiology - Sechenov Physiology Journal* **99(12)** 1425-34.

[4] Barreras F, Amaveda H and Lozano A 2002 Transient high-frequency ultrasonic water atomization *Experiments in Fluids* **33**. 405-13.

[5] Flament M P, Leterme P and Gayot A 2001 Study of technological parameters of ultrasonic nebulization *Drug Development and Industrial Pharmacy* **27(7)** 643-49.

[6] Simones M P, Loyalka S K and Duffy C 2014 Measurement of size and charge distribution of sodium chloride particles generated by Aeroneb Pro pharmaceutical nebulizer *Eur. J. Nanomed* **6(1)** 29-36.

[7] Ultrasonic Nebulizer Omron NE-U17: Manual 2012 (Kyoto^ Omron Healthcare Co Ltd) 24 pp.

[8] Precision LCR-Meter LCR-8000G Series: User Manual 2013 (Taiwan: GW Instek) 144 pp.

[9] Rebrova O Yu 2002 *Statistical analysis of medical data. Application software package Statistica* (Moscow: Mediasphere) 312 pp. In Russian.

[10] Yavorsky V A 2011 *Planning a scientific experiment and processing experimental data* (Moscow: Nauka) 44 pp. In Russian.

[11] Zuev A L, Mishlanov V Yu, Sudakov A I, Shakirov N V and Frolov A V 2012 Equivalent electrical models of biological objects *Russian Journal of Biomechanics* **16(1)** 98-107.

[12] Horowitz P and Hill U 1993 *The art of circuit engineering* (Moscow: Mir) 413 pp. In Russian.

[13] Zuev A L, Sudakov A I and Shakirov N V 2014 Identical electrical models of biological objects *Russian Journal of Biomechanics* **18(4)** 427-32.

[14] Radchenko S G 2011 *Methodology of regression analysis: Monograph* (Kiev: Korniychuk) 376 pp. In Russian.

[15] Lvovsky E N 1988 *Statistical methods for constructing empirical formulas: Textbook for high schools* 2nd ed (Moscow: Vysshaya shkola) 239 pp. In Russian.

IOP Conf. Series: Materials Science and Engineering **581** (2019) 012028 doi:10.1088/1757-899X/581/1/012028

Mechanical Behavior of Polycrystalline Rhenium under 3-Points Bending at a Low Homological Temperature

Peter Panfilov[1,a], **Yuri N Gornostyrev** [1,2], **Dmitry Zaytsev** [1,3], **Gleb P Panfilov** [1], **Vitalii P Pilyugin** [1,2]

[1] Ural Federal University, Lenin Ave. 51, 620000 Yekaterinburg, Russia;
[2] Institute of Metalphysics of Ural Branch of Russian Academy of Science, Kovalevskaya Str. 18, 620219 Yekaterinburg, Russia;
[3] Institute of High-Temperature Electrochemistry of Ural Branch of Russian Academy of Science, Akademicheskaya Str. 20, 620990 Yekaterinburg, Russia.

E-mail: [a] peter.panfilov@urfu.ru

Abstract. Mechanical behaviour of polycrystalline rhenium under 3-points bending at such low homological temperature as room is discussed. Two metallurgical technologies (electron beam melting and powder metallurgy) were used for the samples preparation. Fine-grained samples (PM metal) exhibit some plasticity prior the failure, while coarse-grained ones (EBM metal) behave like a brittle solid. The intergranular fracture is the fracture mode of rhenium in both cases. Basal slip and prismatic slip of dislocations are not active in rhenium at low homological temperature, but the grain boundary sliding occurs under these conditions. Therefore, polycrystalline rhenium cannot be machined at room temperature despite the growth of grain boundary cracks are braked in the samples due to grain boundary sliding.

1. Introduction

Mechanical behavior of rhenium, which is the refractory metal with HCP lattice, is still unclear [3], in spite of deformation mechanisms in the majority of HCP-metals are well-known [1,2]. The main feature of rhenium is its strongest work hardening under loading that makes it unworkable materials at least at room and elevated temperatures [3,4]. It should be noted that room temperature is a low temperature at the homological scale for rhenium in comparison with other HCP-metals. Mechanical twinning is considered as a possible cause of this anomaly in rhenium [5]. However, rhenium single crystals demonstrate considerable elongation (more than 100%) at low homological temperatures. Both basal slip and prismatic slip are active in the single crystals under these conditions, while mechanical twinning is active, as well [6]. The interaction between twin lamella and grain boundary (GB) may be the reason why rhenium in the polycrystalline state exhibits poor plasticity at low homological temperature [7]. On the other hand, at 1200°C polycrystalline rhenium behaves like an HCP-metal demonstrating considerable plasticity with usual work hardening [8]. Another source of poor workability of rhenium may be its complicated metallurgy namely refining from non-metallic impurities. Indeed, the powder metallurgy technology (PM) is applied for producing of the majority of rhenium goods, such as wire, foil, and sheets [4], despite mechanical properties of a PM metal are lower than properties of a metal from the re-melted ingot. In addition, it is unclear what means a work hardening of the metal, which cannot be plastically deformed, as it takes place in polycrystalline rhenium at room temperature? The rhenium problem continuous to be open till the present time [9].

Such deformation scheme as 3-points bending allows reaching a considerable level of plastic deformation in a metallic sample without its fixing in the grips of a testing machine. It is a very important feature for polycrystalline rhenium, which possesses poor deformability and, hence, it can fail in the grips under loading. In addition, it gives an opportunity to build a stress-strain curve and, hence, obtains information on a work hardening of a material. The aim of this work is the study of the deformation behavior of polycrystalline rhenium under 3-points bending at room temperature that to looking for the possible cause of its poor workability at low homological temperatures.

IOP Conf. Series: Materials Science and Engineering **581** (2019) 012028 doi:10.1088/1757-899X/581/1/012028

2. Experimental procedure

Two technologies, which are used by manufacturers around the world, were chosen for the producing of the high pure polycrystalline rhenium in this work. Compacted rhenium sponge or PM rhenium (grain size is about 10 μm) and rhenium sponge subjected by the electron beam melting (EBM) or EBM rhenium (grain size is 5÷10 mm) were chosen as the model materials for mechanical testing. Samples with a shape of a parallelepiped having the size of 30x2x1 mm^3 were cut by means of spark-erosion technique from both PM rhenium and EBM metal. Their deformation behavior was compared with coarse grain Ti4Al alloy (grain size is about 10 mm) because c/a ratio for titanium alloys is close to this parameter for rhenium. Working surfaces of the samples were abraded with a help of abrasive papers and pastes. 3-points bending at room temperature was carried out on Shimadzu™ AG-X 50kN testing machine with Trapezium™ software on Shimadzu™ bending device (the distance between the prisms was 10 mm; the traverse rate was 0,1 mm/min). The parcel consisted of ten samples were tested for each model material. "Stress-strain" curves were built for samples from each parcel. Working surfaces and fracture surfaces of each sample were examined with the help of the light microscope (LM) and the scanning electron microscope (SEM) in the initial / undeformed state and after bending.

3. Results

"Stress-strain" curves of the model materials are shown in figure 1. The total deformation means the deformation prior to the failure of the sample under bending, while the point of maximum on the curve is accepted as the maximal strength of the material. The elastic moduli and the hardening coefficients are automatically calculated by Trapezium™ software. These parameters are given in table 1. No plastic flow stage is observed on the "stress-strain" curve of EBM rhenium, which takes apart after insignificant elastic deformation. Therefore, its mechanical behavior is estimated as brittle. On the contrary, PM rhenium exhibits ductile deformation behavior like Ti4Al alloy, because the plastic flow stage is on the "stress-strain" curve, while its total plasticity is about 4% (curve 2).

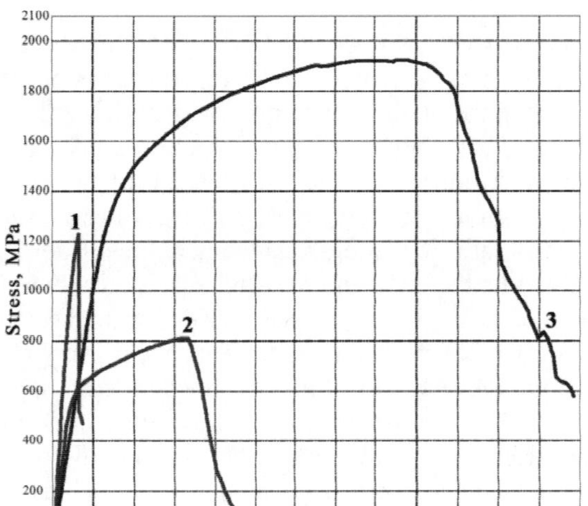

Figure 1. *"Stress-Strain" curves for 3-points bending at room temperature: curve 1a – EBM rhenium; curve b – PM rhenium; curve c –Ti4Al alloy.*

The elastic modulus of EBM rhenium is in two times more than this parameter for PM rhenium, which are similar to the modulus of Ti4Al alloy. This is in agreement with the difference in the mechanical behavior of EBM and PM metals. Ti4Al alloy is considerably more plastic material than PMT rhenium, despite its work hardening is higher. Hence, at low homological temperatures, PM rhenium behaves like a plastic material, while EBM rhenium demonstrates the brittle mechanical behavior. It should be especially noted that PM rhenium does not exhibit considerable work hardening in comparison with the hcp-metal possesses similar c/a ratio. No work hardening occurs in EBM rhenium because it behaves like a brittle solid.

The elastic modulus of EBM rhenium is in two times more than this parameter for PM rhenium, which are similar to the modulus of Ti4Al alloy. This is in agreement with the difference in the mechanical behavior of EBM and PM metals. Ti4Al alloy is considerably more plastic material than PMT rhenium, despite its work hardening is higher. Hence, at low homological temperatures, PM rhenium behaves like a plastic material, while EBM rhenium demonstrates the brittle mechanical behavior. It should be especially noted that PM rhenium does not exhibit considerable work hardening in comparison with the hcp-metal possesses similar c/a ratio. No work hardening occurs in EBM rhenium because it behaves like a brittle solid. These findings confirm the conclusion that PM rhenium is able to thermomechanical treatment, whereas EBM rhenium is undeformed material at low homological temperatures [3,4,8].

Table 1. Mechanical properties of polycrystalline rhenium and Ti4Al alloy under 3-point bending at room temperature.

Metal	Total deformation, %	Maximal stress, GPa	Elastic modulus, GPa	Hardening coef., GPa
EBM Rh	0.5	1.2	250	-
PM Rh	4	0.8	120	7.5
Ti4Al	12.5	1.9	115	10.5

The elastic modulus of EBM rhenium is in two times more than this parameter for PM rhenium, which are similar to the modulus of Ti4Al alloy. This is in agreement with the difference in the mechanical behavior of EBM and PM metals. Ti4Al alloy is considerably more plastic material than PMT rhenium, despite its work hardening is higher. Hence, at low homological temperatures, PM rhenium behaves like a plastic material, while EBM rhenium demonstrates the brittle mechanical behavior. It should be especially noted that PM rhenium does not exhibit considerable work hardening in comparison with the hcp-metal possesses similar c/a ratio. No work hardening occurs in EBM rhenium because it behaves like a brittle solid.

Metallographic (LM) examination of the back surfaces has shown that all cracks are localized in the middle part of the area of bending in the sample, where tensile stress is maximal. There are long cracks consisted of merging short cracks in the area of bending (figure 2).The length of the long cracks is compared with the width of the sample. Few long cracks are observed in the samples of PM rhenium, while only one of them transforms into the dangerous crack (figure 2a). The short cracks in PM rhenium are oriented almost normally to the tensile axis of the sample. SEM study of these cracks has shown that they appear on the grain boundaries (GBs), which are oriented normally to the tensile axis of the sample. Morphology of the fracture surface of PM rhenium samples agrees with this conclusion. Fracture mode of PM rhenium under bending is GB fracture, where the surface of grains is rough, that points to some plastic deformation of GBs. However, no any deformation tracks (slip bands and twin lamellas) are revealed inside GBs in the bending area of the sample. The elongation of the PM samples under bending is comparable with the sum of the widths of cracks in the area of bending.

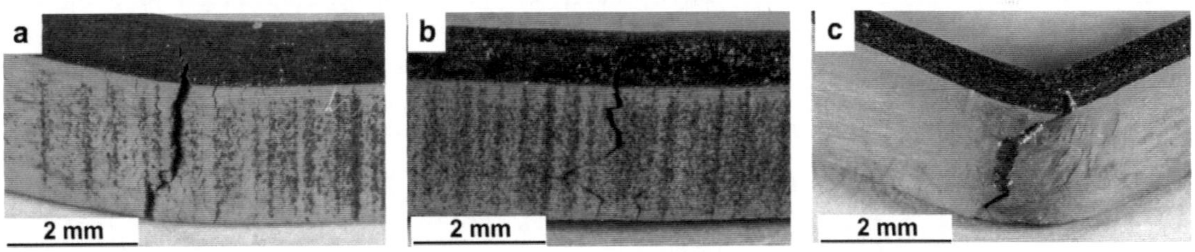

Figure 2. *Cracks in the area of bending under 3-point bending at room temperature: a - PM rhenium; b - EBM rhenium; c - Ti4Al alloy.*

The similar picture takes place in the samples of EBM rhenium under bending. There are 1÷3 long cracks in the vicinity of the middle part of the bending area (figure 2b). Such crack has a broken profile because it consists of GB cracks. The width of GB cracks depends on their orientation to the tensile axis of the sample. The maximal width has the crack oriented normally. Deformation tracks are not observed in the grains near the cracks. It agrees with the brittle mechanical behavior of EBM rhenium under bending. However, SEM examination of its fracture surface has shown that both slip bands and twin lamellas are observed on the almost smooth GBs. The same fracture behavior occurs in the coarse-grained samples of Ti4Al alloy under bending. Indeed, the dangerous crack also has a broken profile insomuch as it consists of cracks advancing along GBs (figure 2c). The difference between EBM rhenium and Ti4Al alloy is in their ability to plastic deformation. The first HCP-material is unable to macroscopic plastic deformation, while the second one is ductile material.

IOP Conf. Series: Materials Science and Engineering **581** (2019) 012028 doi:10.1088/1757-899X/581/1/012028

4. Discussion

According to the morphology of the fracture surface, the fracture mode of EBM rhenium under bending is the brittle intergranular fracture (BIF) with some features of GB plasticity. As a rule, BIF is caused by some non-metallic impurities (oxygen, carbon etc), which influence dramatically on the cohesive strength of GBs in a metallic material [10]. In the case of polycrystalline rhenium, this is not the impurity induced GB brittle fracture, because GB cracks are stable at least in PM metal, which demonstrates macroscopic plasticity prior the failure under considerable ultimate strength. Therefore, there should be an additional mechanism (the main one is the cracking) of the stress accommodation in polycrystalline rhenium under bending at room temperature. This is GB sliding that sometimes contributes to the mechanical behavior of coarse-grained HCP-metals [1,2]. Although its contribution to the total plasticity of metallic material is minor, the competition between GB sliding and the cracking can lead to the arrest of GB cracks. Perhaps the same picture takes place in polycrystalline rhenium.

It was shown that the mechanical behavior of polycrystalline rhenium determines by metallurgical technology: coarse-grained EBM rhenium is a brittle solid, while fine-grained PM rhenium is a plastic material. However, in both cases, polycrystalline rhenium cannot be machined at room temperature. No manifestations of plasticity such as slip bands are observed in both EBM and PM rhenium on the microscopic scale. It seems to be that the total elongation of the rhenium samples provides by the sum of the cracks in the area of bending. This parameter for PM metal is considerably high then for EBM metal and, as a result, PM rhenium looks like a ductile substance on the macroscopic scale. The cause of such behavior is that both basal and prismatic slip of dislocations, which provide considerable plasticity in the majority of HCP-metals, do not activate in polycrystalline rhenium at room temperature, whereas GB sliding occurs here. This effect cannot be considered as the contradiction with the empirical knowledge on HCP-metals, which based on the deformation behavior of HCP-metals having melting points lower than 2500°C [1,2], insomuch as room temperature is very low temperature on the homological scale for refractory rhenium. The study of electronic structure of chemical bonding in rhenium needs for the better understanding of mechanism of its mechanical behavior.

5. Conclusion

Mechanical behavior of polycrystalline rhenium under bending at room temperature is discussed. This refractory HCP-metal cannot be machined at room temperature because dislocation slip is not active in it under these experimental conditions. However, grain boundary sliding occurs in polycrystalline rhenium even at low homological temperature that causes to the stoppage of crack growth in the grain boundaries.

Acknowledgment
This work is supported by the Russian Science Foundation (#18-19-00217).

References

[1] R.W.K. Honeycombe, The Plastic Deformation of Metals, Edward Arnold, London, 1972.
[2] A.S. Argon, Strengthening Mechanisms in Crystal Plasticity, Oxford University Press, Oxford, 2007, - 404 p.
[3] R.I. Jaffee, D.J. Maykuth, R.W. Douglass, Rhenium and Refractory Platinum-Group Metals, in M. Semchyshen and J.J. Harwood, eds., Refractory Metals and Alloys, AIME Metallurgical Society Conferences, vol. 11 (Interscience, New York, 1961) pp. 383-463.
[4] C.T. Sims, C.M. Craighead, R.I. Jaffee, JOM, 1955, 1, pp. 168-179.
[5] R. A. Jeffery, E. Smith, Phil. Mag. 1966, 13, 126, pp. 1163-1168.
[6] T.V. Kopetskii, M.M. Myshlyaev, N.I. Novochatskai, V.A. Yukhanov, Physics of Metals and Metallography, 39 (1975) 164.
[7] J. Kacher, A.M. Minor, Acta Materialia 81 (2014) 1-8.
[8] I.S. Servi, J. Less-Common Metals 17 (1969) 340-343.
[9] I.L. Shabalin, Rhenium, in Ultra-High Temperature Materials I, Springer, London, 2014, pp. 317-358.
[10] I.M. Robertson, P. Sofronis, A. Nagao, M.L. Martin, S. Wang, D.W. Gross, K.E. Nygren, Metallurgical and Materials Transactions A, 2015, 46A,6, pp.2323-2341.

IOP Conf. Series: Materials Science and Engineering **581** (2019) 012029 doi:10.1088/1757-899X/581/1/012029

Diffraction Resonances of Acoustic Stresses in the Crystal Layer

Y N Belyayev

Department of Mathematical Modelling and Cybernetics, Syktyvkar State University, 55 Oktyabrskii pr., Syktyvkar, 167001, Russia

E-mail: ybelyayev@mail.ru, ynbelyayev@gmail.com

Abstract. The six-beam scattering of elastic waves by a crystal film is investigated. Cases of symmetric diffraction of waves inside the crystal layer are analyzed. In this scattering geometry, the projections of the wave vectors on the normal to the film surface are pairwise equal to each other up to the sign. Under certain conditions, quasi-shear waves modulate the structure in the path of quasi-longitudinal waves. If the period of the modulated structure is close to the length of a quasi-longitudinal wave, the amplitude of oscillations of the latter is amplified. The explanation of this effect is similar to the description of parametric resonance.This effect is illustrated by examples of calculations of the scattering spectra of elastic waves by cubic and tetragonal crystals in a three-layer model. Calculations are based on the transfer matrix method.

1. Introduction

Acoustic stresses in a solid medium depend on the type of elastic waves. As is known [1], in any direction of isotropic solid medium waves of compression and shear can propagate. Compression waves are also called longitudinal, since oscillations of particles of the medium occur along the direction of wave propagation. In shear wave, the amplitude vector oscillates perpendicular to the wave vector. The speed of longitudinal wave propagation is greater than the shear wave speed; therefore, the former is called P-type (*p*rima) wave, and the latter is S-type wave (secunda). If oscillations in a shear wave are directed perpendicularly or parallel to the plane of incidence, the wave is called a horizontally polarized (SH-type) and a vertically polarized (SV-type) shear wave, respectively. In any direction of an anisotropic medium, the propagation speeds of elastic waves may differ from each other and their polarizations are neither purely longitudinal nor purely transverse. A wave whose direction of oscillation is closest to the direction of the wave vector is called quasi-longitudinal. The other two waves are called quasi-transverse [2,3].

The transformations of various types of elastic waves underlie many methods of geophysical research, acoustic microscopy and various acousto-optic transducers, electromagnetic-acoustic instruments. This stimulates the development of new methods of elastic waves conversions, in particular the study of nonlinear interactions and the propagation of waves in inhomogeneous media [4-8]. Mutual conversions of the waves of all the above types occur in anisotropic media. In the anisotropic layer of finite thickness d, as a result of a plane elastic wave action, in the general case, six waves are formed (three in the direction of reflection and three in the direction of transmission). This paper investigates the possibility of resonant transformation of these waves and the stresses generated by them as a result of six-beam diffraction.

2. Basic equations and their solution method

Let the deformations in an anisotropic layered structure (the density ρ and elastic parameters of the medium depend only on one coordinate x_3) are generated by incident wave $\boldsymbol{u}_0 = \boldsymbol{A}_0\exp[i(\boldsymbol{k}_0 \cdot \boldsymbol{r} - \omega t)]$, where $\boldsymbol{u}_0$, $\boldsymbol{A}_0$, $\boldsymbol{k}_0$ and ω are displacement vector, amplitude vector, wave vector and cyclic frequency of incident wave respectively; $\boldsymbol{r}$ is a radius vector and i is the imaginary unit. Then the wave vectors $\boldsymbol{k}_j$ of all scattered waves lie in the plane of incidence (x_3, ξ) (see figure 1a).The dependence of the stress-strain state of the medium on the coordinate x_3 is determined by the matrix equation

$$\frac{d}{dx_3}\boldsymbol{\Psi} = \boldsymbol{W\Psi}, \quad \boldsymbol{\Psi} = \begin{Vmatrix} u_2 \\ \sigma_{23} \\ u_1 \\ u_3 \\ \sigma_{13} \\ \sigma_{33} \end{Vmatrix}, \quad \boldsymbol{W} = \begin{Vmatrix} w_{11} & w_{12} & w_{13} & w_{14} & w_{15} & w_{16} \\ w_{21} & w_{11} & w_{23} & 0 & w_{25} & w_{26} \\ w_{25} & w_{15} & w_{33} & w_{34} & w_{35} & w_{36} \\ w_{26} & w_{16} & w_{43} & 0 & w_{36} & w_{46} \\ w_{23} & w_{13} & w_{53} & 0 & w_{33} & w_{43} \\ 0 & w_{14} & 0 & w_{64} & w_{34} & 0 \end{Vmatrix}, \tag{1}$$

where u_1, u_2, u_3 are projections of displacement vector $\boldsymbol{u}$ on the Cartesian axes x_1, x_2, x_3 and σ_{13}, σ_{23}, σ_{33} are components of stress tensor. The elements w_{gh} of the matrix $\boldsymbol{W}$ depend on angles of incidence θ_0, α (figure 1b), cyclic frequency ω, the density ρ and components of the elastic compliance tensor S_{gh} of the medium [9].

The value $\boldsymbol{\Psi}(x_3)$ of the vector function $\boldsymbol{\Psi}$ at some depth x_3 is expressed in terms of the value $\boldsymbol{\Psi}(0)$ of this function on the surface $x_3 = 0$ by the following equality

$$\boldsymbol{\Psi}(x_3) = \boldsymbol{T}(x_3)\boldsymbol{\Psi}(0), \tag{2}$$

The transfer matrix $\boldsymbol{T}(x_3) = \exp(\boldsymbol{W}x_3)$ is calculated in this paper by the formula:

$$\boldsymbol{T}(x_3) = \left[\sum_{h=0}^{5}\left(\frac{Wx_3}{m}\right)^h\left(\frac{1}{h!} + \sum_{j=6}^{6+N}\frac{1}{j!}\sum_{g=0}^{h} p_{6-h+g}B_{j-1-g}(6)\right)\right]^m. \tag{3}$$

The scaling parameter m in the formula (3) is selected according to the condition $m = 2^\ell$ and ℓ is a minimal integer that provides the inequality

$$\eta \equiv 8.5\frac{\omega}{\min v}\frac{x_3}{m} < 1\,. \tag{4}$$

The minimum possible speed min v of elastic wave propagation in a layer can be estimated by the formula: $\min v \approx 1/(\rho \max|S_{gh}|)^{1/2}$. Coefficients p_h, $h = 1,\dots,6$ of the characteristic equation of the matrix $\boldsymbol{W}d/m$ are computed by the method of Le Verrier [10, p. 87]. The polynomials $B_j(6)$ are calculated using recurrence formulas

$$B_0(6) = B_1(6) = \cdots = B_4(6) = 0; \quad B_5(6) = 1; \quad B_j(6) = \sum_{h=1}^{6} p_h B_{j-h}(6),\ j \geq 6. \tag{5}$$

It was proved [10], that if condition (4) is satisfied then the relative truncation error in the calculation of the transfer matrix $\boldsymbol{T}$ by the formula (3) does not exceed

$$\varepsilon_m = \frac{m\eta^{N+1}}{(N+6)\prod_{j=1}^{N}(6+j)}\,. \tag{6}$$

For example, for a crystal layer of calcium molybdate 0.000636 m thick and frequency $\omega = 2\pi 10^7$ Hz scaling parameter should be $m > 115$. Therefore, if calculations with quadruple arithmetic accuracy are required ($\varepsilon_m \approx 10^{-32}$), it is advisable to take in the formula (3) $m = 128$ ($\ell = 7$) with $N = 25$, or $m = 256$ ($\ell = 8$) with $N = 21$, or $m = 512$ ($\ell = 9$) with $N = 18$.

As a result of the incident wave interaction with the crystal in the general case six scattered waves are generated outside the layer: $\boldsymbol{u}_j = \boldsymbol{A}_j\exp[i(\boldsymbol{k}_j\cdot\boldsymbol{r} - \omega t)]$, $j = 1,\dots,6$. The elements of the vector functions $\boldsymbol{\Psi}(0)$ and $\boldsymbol{\Psi}(d)$ are expressed in terms of the amplitudes $A_j = |\boldsymbol{A}_j|$ of the waves scattered in the region $x_3 < 0$ and $x_3 > d$, respectively. After that, from equation $\boldsymbol{\Psi}(d) = \boldsymbol{T}(d)\ \boldsymbol{\Psi}(0)$ it is easy to express the amplitudes A_j of the waves as a functions of the elements of the transfer matrix $\boldsymbol{T}(d)$.

Obviously, the amplitudes of the scattered waves are determined by the stresses of the crystal layer. Inverse dependences can be easily obtained from equation (2) with $x = d$. If regions $x_3 < 0$ and $x_3 > d$ are isotropic and the incident wave is a horizontal shear wave (figure 1a), then

$$\begin{aligned} \sigma_{13}(0,d) &= \mu_{0,d}\sum_{h=0,4}^{3,6}(c_{h1}k_{h3} + c_{h3}k_{h1})A_h, \\ \sigma_{23}(0,d) &= \mu_{0,d}\sum_{h=0,4}^{3,6}(c_{h2}k_{h3} + c_{h3}k_{h2})A_h, \\ \sigma_{33}(0,d) &= \sum_{h=0,4}^{3,6}\left[2\mu_{0,d}c_{h3}k_{h3} + \lambda_{0,d}\sum_{g=1}^{3}c_{hg}k_{hg}\right]A_h. \end{aligned} \tag{8}$$

Here c_{h1}, c_{h2}, c_{h3} and k_{h1}, k_{h2}, k_{h3} denote, respectively, the direction cosines of the amplitude vectors $\boldsymbol{A}_h$ and the wave vectors $\boldsymbol{k}_h$, $h = 0,\ldots,6$.They are determined by the geometry of scattering (figure 1*a*,*b*): $c_{01} = c_{11} = c_{41} = \sin\alpha$, $c_{02} = c_{12} = c_{42} = -\cos\alpha$, $c_{03} = c_{13} = c_{43} = 0$, $c_{21} = -\cos\theta_1\cos\alpha$, $c_{22} = -\cos\theta_1\sin\alpha$, $c_{23} = -\sin\theta_1$, $c_{31} = \sin\theta_3\cos\alpha$, $c_{32} = \sin\theta_3\sin\alpha$, $c_{33} = \cos\theta_3$, $c_{51} = \cos\theta_4\cos\alpha$, $c_{51} = \cos\theta_4\sin\alpha$, $c_{53} = -\sin\theta_4$, $c_{61} = \sin\theta_6\cos\alpha$, $c_{62} = \sin\theta_6\sin\alpha$, $c_{63} = \cos\theta_6$;$k_{j1} = k_0\sin\theta_0\cos\alpha$, $k_{j2} = k0\sin\theta0\sin\alpha$, $j = 0,\ldots,6$; $k_{03} = k_0\cos\theta_0$,$k_{13} = k_{23} = -k_1\cos\theta_0$, $k_{33} = -k_3\cos\theta_3$, $k_{43} = k_{53} = k_4\cos\theta_0$, $k_{63} = k_6\cos\theta_6$.

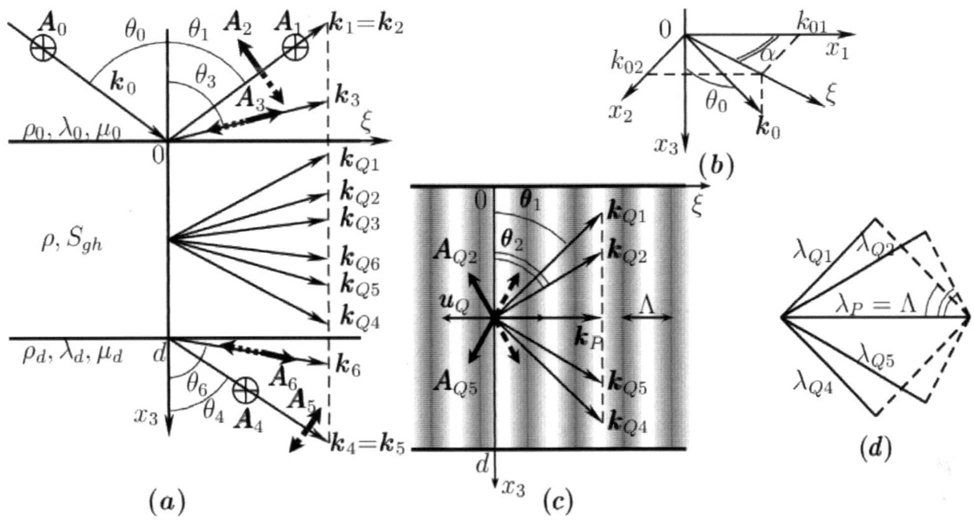

Figure 1. The geometry of scattering.

3. Symmetric scattering

Eigenvalues λ_j of the matrix $\boldsymbol{W}$ have a simple physical meaning, namely $\lambda_j = ik_{Qj3}$, $j = 1,\ldots,6$, where k_{Qj3} is the third component of wave vector $\boldsymbol{k}_{Qj}$ of quasi-transverse or quasi-longitudinal wave in the crystal layer. Wave scattering is called symmetric if $k_{Q13} = -k_{Q43}$, $k_{Q23} = -k_{Q53}$, $k_{Q33} = -k_{Q63}$ (see figure 1*a*). In this case characteristic equation of the matrix $\boldsymbol{W}$ has coefficients $p_1 = 0$, $p_3 = 0$, $p_5 = 0$. This situation is realized, in particular, if the surface (1 0 0) of a cubic or tetragonal crystal, coincides with the plane $x_3 = 0$.

Table 1. Crystal parameters.

Crystal	ρ	S_{11}	S_{12}	S_{13}	S_{16}	S_{33}	S_{44}	S_{66}
GaAs	5320[a]	11.72[b]	-3.65[b]	-3.65[b]	0[b]	11.72[b]	16.83[b]	16.83[b]
$CaMoO_4$	4255[a]	9.597[b]	-4.391[b]	-1.424[b]	-4.041[b]	8.684[b]	27.1[b]	24.56[b]

[a]The units of density ρ are kg/m^3.
[b]The units of the compliance tensor components S_{gh} are $10^{-12}Pa^{-1}$.

The scattering spectra of SH waves and the stresses were simulated using the examples of a GaAs and $CaMoO_4$ crystal layers for $\omega = 2\pi\cdot 10^7$ Hz. The calculations were performed for a three-layer model in which the regions $x_3<0$ and$x_3>d$ were assumed to be solid with Lame parameters $\lambda_0 = \lambda_d = 1.67\cdot 10^{10}$ Pa,$\mu_0 = \mu_d = 3.27\cdot 10^{10}$ Pa and densities $\rho_0 = \rho_d = 2650$ kg/m^3.The elements t_{gh} of the transfer matrix T depend on the thickness d of the anisotropic layer and the oscillation frequency ω as a product $\kappa = \omega d$. This parameter κ ranged from 10^2 m/s to 10^4 m/s.

One of the results of such calculations is shown in the figure 2. Here, the stresses amplitudes $|\sigma_{j3}(0,d)|$ ($j = 1,2,3$) at the boundaries of the GaAs layer are presented as functions of the angle of incidence θ_0 for values $\alpha = 75^\circ$, $d = 0.000582$ m. All amplitudes of the stress tensor components are in units of $A_0\cdot 10^{15}$Pa. The figure 2 also shows the dependences of values $C_j = |A_j/A_0|^2$, $j = 1,\ldots,6$, on the angle of incidence θ_0. The coefficients C_j characterize the conversions of the incident wave into scattered waves of SH- (j = 1,4), SV- (j = 2,5) and P-types (j = 3, 6).

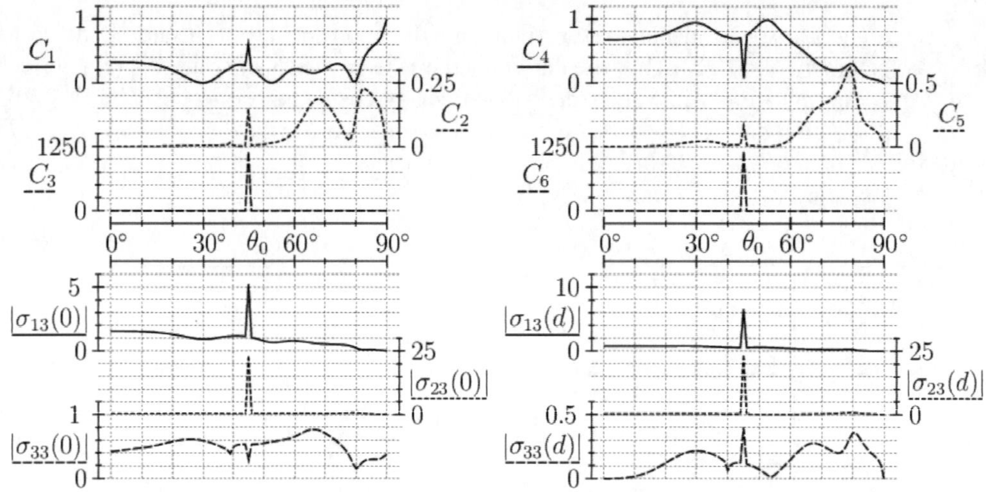

Figure 2. Scattering spectra and stress amplitudes in the gallium arsenide crystal layer.

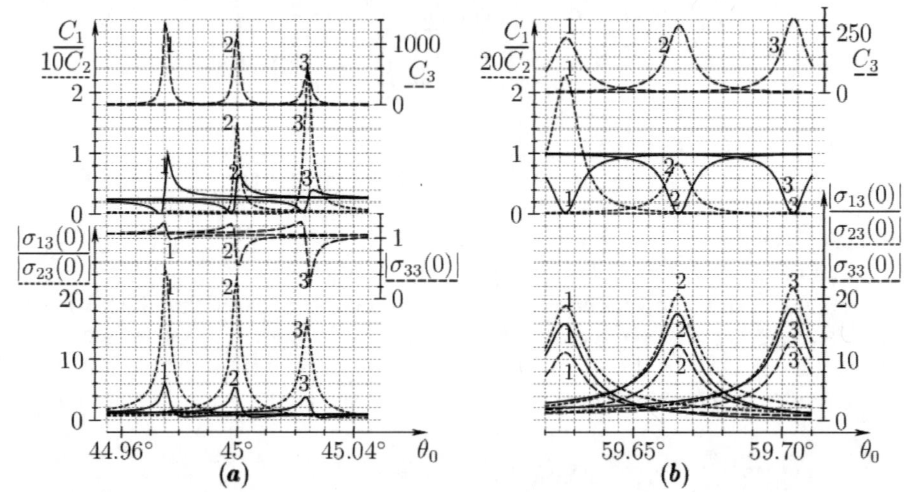

Figure 3. Acoustic resonances: ***(a)*** crystal layer GaAs; ***(b)*** crystal layer $CaMoO_4$.

If the angle θ_0 greater than the critical angle θ_{0C}, then longitudinal waves generated by a crystal in the isotropic half-spaces are inhomogeneous. The critical angle θ_{0C} is determined from the condition: $\sin\theta_{0C}=k_j/k_0, j=3,6$. In the considered three-layer model $\theta_{0C}=39.13^\circ$. These inhomogeneous longitudinal waves don't transfer energy from the crystalline layer. The amplitudes of these waves, as calculations show, can have a resonant character and be much larger than unity. In this case, the quasi-P waves in the crystal become inhomogeneous and propagate along the axis ξ. If the amplitude vectors A_{Q2} and A_{Q5} (or A_{Q1} and A_{Q4}) of two quasi-SV waves oscillate as shown in figure 1*c*, then as a result of their interaction a modulated structure appears inside the crystal. The period of this modulated structure is $\Lambda = \lambda_P$ (see figure 1*d*), where λ_P is the length of the quasi-P wave. This is the cause of the diffraction amplification of the quasi-P waves.

A sharp increase of functions $C_3(\theta_0.)$, $C_6(\theta_0.)$ and, accordingly, $|\sigma_{13}(0,d)|$ and $|\sigma_{23}(0,d)|$ in the layer of GaAs take place in figure 2 at the angle θ_0= 45. This example of diffraction resonance is shown in more detail in the figure 3*a.* The position of the diffraction resonance can be varied due to small changes in the values of the second angle of incidence α, the crystal thickness d and the oscillation frequency of the incident wave. The numbers 1, 2, 3 on the figure 3*a* indicate curves corresponding to values d = 0.000581 m, d = 0.000582 m, d = 0.000583 m respectively. Another one example of the influence of the thickness of the crystal layer on the position of the diffraction maximum is shown in figure 3*b*. Here resonances for $\alpha = 50^\circ$ in the calcium molybdate crystal layer with thickness d = 0.0006355 m (curves marked with

IOP Conf. Series: Materials Science and Engineering **581** (2019) 012029 doi:10.1088/1757-899X/581/1/012029

number 1), d = 0.0006360 m (curves marked with number 2), d = 0.0006365 m (curves marked with number 3) are demonstrated.

4. Conclusion

The resonances of acoustic stresses for the case of an incident SV wave are similar to those considered. The performed modeling of the spectra revealed that the largest increase in the coefficients C_3, C_6 and, accordingly, in the stresses amplitudes $|\sigma_{j3}|, j = 1,2,3$, occurs when during scattering the polarization of the shear wave is completely changed from one type to another.

Since the transfer matrix $\boldsymbol{T}$ is a function of $\kappa = \omega d$, therefore, displacements $\Delta\theta_0$ of the maximum stress as a result of a change in thickness d by an amount Δd and frequency ω by an amount $\Delta\omega = -\Delta d\,\omega/d$ are equivalent.

The maximum amplitude of a quasi-longitudinal wave is proportional to the number of the modulated structure periods Λ and, in a physical experiment, should depend on the width of the beam of the incident wave.

References

[1] Ewing W M, Jardetzky W S, Press F 1957 *Elastic waves in layered media* (New York: McGraw-Hill Book Company)

[2] Dieulesaint E, Royer D 1974 *Ondes élastiques dans les solides. Application au traitment du signal* (Paris: Masson)

[3] Brekhovskikh L M, Godin O A 1990 *Acoustics of layered media* (Berlin: Springer-Verlag)

[4] Zarembo L K, Krasil'nikov V A 1971 Nonlinear phenomena in the propagation of elastic waves in solids*Physics-Uspekhi* **13** 778–797

[5] Zarembo L K, Koshkina E N, Chunchuzov I P 1984 Mutual conversion of longitudinal, shear waves in an inhomogeneous isotropic medium *Akusticheskij Zhurnal* **30** 781-784

[6] Korneev V A, Demcenko A 2014 Possible second-order nonlinear interactions of plane waves in an elastic solid *J Acoust Soc Am* **135**(2) 591-598

[7] Gallot T, Malcolm A, Szabo T L, Brown S, Burns D 2015 Characterizing the nonlinear interaction of S- and P-waves in rock sample *Journal of applied physics* **117** 034902-1 –034902-13

[8] Wallen S P, Boechler N 2017 Shear to longitudinal mode conversion via second harmonic generation in two-dimenshional microscale granular crystal *Wave motion* **68** 22-30

[9] Belyayev Yu N 2018 Calculation of the six-beam diffraction in layered media using polynomials of principal minors *Journal of theoretical and computational acoustics* **26** 1850017-1–1850017-21

[10] Gantmacher F R 1959 *The Theory of Matrices* vol 1 (New York: Chelsea)

The Study of Impact Loading on GFRP Plates Using a Network of Piezoceramic Sensors

I N Shardakov [1,2,a], A P Shestakov [1,2,b], G S Serovaev [1,2,c], N A Kosheleva [1,d], V V Epin [2,e]

[1] Perm National Research Polytechnic University, Perm, Russia
[2] Institute of Continuous Media Mechanics UB RAS, Perm, Russia

E-mail: [a] shardakov@icmm.ru, [b] shap@icmm.ru, [c] serovaev@icmm.ru, [d] nataly.kosheleva@gmail.com, [e] epin@icmm.ru

Abstract. Structures made of multilayered composite materials are designed to work under different types of external loads, both static and dynamic. Often, under dynamic impact, material failure occurs in the inner layers of the material and is not noticeable during external visual inspection. Untimely detection of such damage may lead to a complete failure of the structure. Therefore, the substantial problem in the control of objects made of laminated composite materials is the determination of various characteristics of external dynamic impact, which will allow to estimate the danger of an impact and localize the damage in the structure. The object of this study is a multilayered plate made of fiberglass composite, subjected to impact loads of different energy. This paper investigates material failure after an impact load and the sensitivity of a network of piezoelectric sensors to external dynamic loading.

1. Introduction

The use of polymer composite materials can significantly reduce the weight of the designed structures while maintaining the necessary performance characteristics. Nowadays safety requirements intensify at all stages of the production and operation of the designed objects, which, in turn, requires monitoring of the mechanical state of the structure during its operation to evaluate its working life. The question of correct selection of monitoring methods [1] of the mechanical state of an object in real time is extremely popular these days. The use of effective monitoring systems allows to identify defects at an early stage of their occurrence and to assess their severity, thereby preventing breakages and dangerous accidents.

Various methods are used to identify defects in products, among which methods of non-destructive evaluation have gained particular popularity. Methods of non-destructive evaluation intended for the detection, localization and classification of defects include visual inspection [2]; optical methods [3]; ultrasound testing [4]; acoustic emission [5]; vibration analysis [6,7]; radiography and thermography [8,9].

However, none of the existing approaches is universal. The need for a deeper study of both the methods themselves and the possibility of combining them with each other to achieve greater efficiency opens up new prospects for improving monitoring systems, thereby increasing the level of evaluation of the mechanical state of a structure without taking it out of service.

One of the most promising is the vibrational approach. Such methods use vibrational processes excited and measured in the structure to detect damage. When developing such monitoring systems, piezoelectric elements are widely used, thanks to the direct and reverse piezoelectric effects, which allow both, to excite oscillations, that is, to act as an actuator, and to record the response of the system to external influences.

There are many scientific studies on the problems related to the development of monitoring system among which are: finding the best location for the piezoelectric sensor on the object under study for reliable detection of defects [10], issues of numerical simulation of wave propagation and it's interaction with damage area [11], signal processing methods [12] and detection of location and size of the defects [13]. That is why the study of various aspects of developing a monitoring system, which allows evaluating the mechanical state of products made of different types of materials, is a crucial task.

It is well known that among the most dangerous types of external loadings are impacts which often cause severe damage of the structure. The distinctive feature of such external influence is that under impact loading (usually under low velocity impacts) an internal damage can occur which cannot be detected on the surface of an object. Constructions made from composite materials are especially vulnerable to this type of damage due to the layered structure.

In this paper, the effect of dynamic loading on multilayered composite plates made of fiberglass material in a wide range of impactor speeds (6-109 m/s) was investigated. The designed and manufactured experimental facilities, which allow reaching such speed range of external loading, are described. The analysis of damage types which appear in the composite material at the considered impact speeds was carried out. A system of dynamic response measurement using a network of piezoceramic sensors, which allows recording the response of a structure in a synchronized mode of operation, was experimentally implemented. Such system can be used as a foundation for building monitoring system which allows to detect and localize impact and identify possible occurrence of defects.

2. Description of impact loading experiments

2.1. Object of study

The object of this study is a multilayered plate made of fiberglass material, subjected to impact loads of different energy. In the experiments, square plates made from glass fiber reinforced polymer (GFRP) with a size of 500×500×4 mm and consisting of 20 layers in thickness (Figure 1) were used. According to the diagram in Figure 2, 6 piezoelectric sensors are glued to the surface of the plate. The sensors are located around a circle with a radius of 160 mm and measure the response of the system to external dynamic loading.

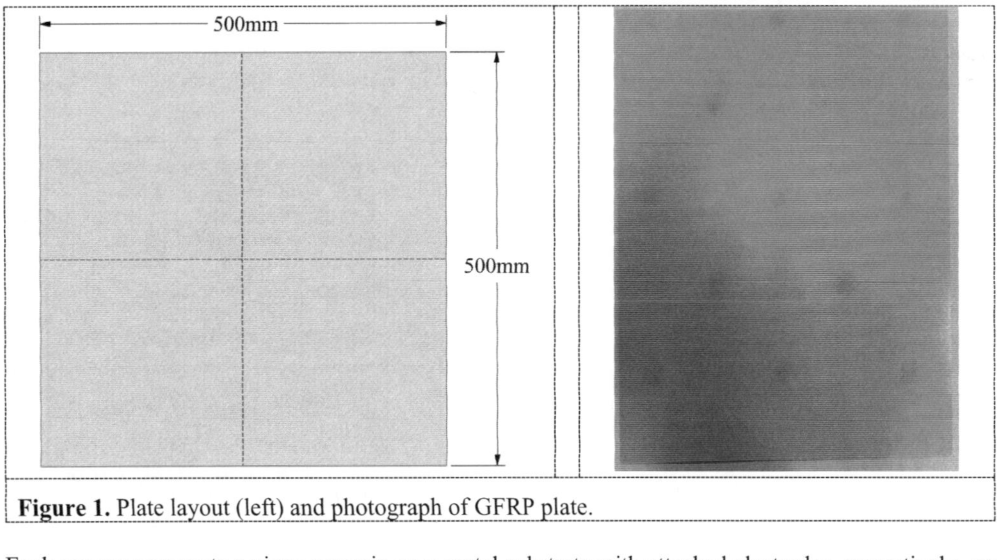

Figure 1. Plate layout (left) and photograph of GFRP plate.

Each sensor represents a piezo-ceramic on a metal substrate with attached electrodes, respectively, on the substrate and on the electro-plated surface of the piezo-ceramic (Figure 3). The diameter of the substrate is 12 mm, the diameter of piezoelectric ceramics is 9 mm, the total thickness of the sensor with the substrate is 0.21 mm, and the thickness of a single piezoceramics is 0.1 mm. A steel ball with a diameter of 25.4 mm and a weight of 66.5 g was used as an impactor

IOP Conf. Series: Materials Science and Engineering **581** (2019) 012030 doi:10.1088/1757-899X/581/1/012030

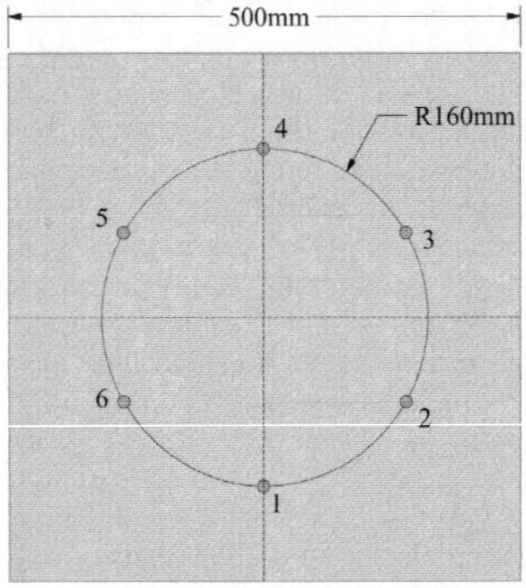

Figure 2. The layout of the piezoelectric sensors on the plate.

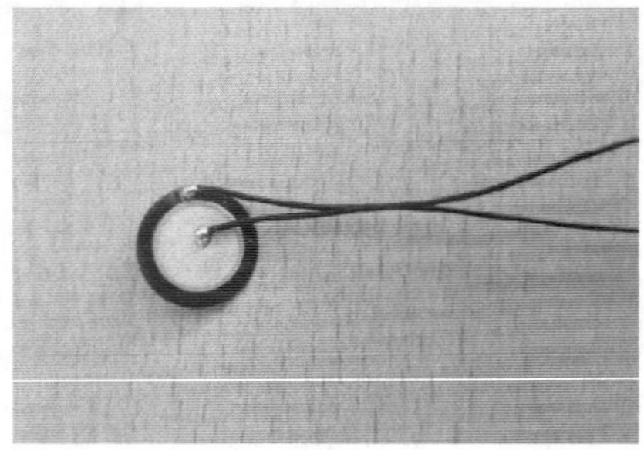

Figure 3. View of piezoelectric sensor used in experiments

2.2. Description of impact loading experiments

In the course of the study, experiments on applying a localized impact load to the sample were carried out. The tests were performed in two ranges of impactor speeds: from 6–13 m/s and speeds in the range of 29–109 m/s. These ranges correspond to 2 different test schemes. In the first velocity range, the impact load was carried out by a metal ball with given values of radius r and mass m falling in the gravity field. In accordance with the experimental scheme of Figure 4, the ball (4) is suspended to the point O by means of an inextensible thread (10). The ball is positioned in relation to the plate by means of a hole in the bracket (9). The distance H is determined by the relative position of the bracket and the plate (5). The plate is suspended on elastic, flexible filaments (7,8) at the four corner points of the plate. The plate is located in the horizontal plane (perpendicular to the direction of the field of gravity). The ball begins to move by cutting the thread (10). The positioning of the bracket with respect to the plate in the horizontal plane was carried out using a laser level GSL Professional BOSCH. Piezoelectric sensors (1, 2) are fixed to the bottom of the plate. After cutting the thread (10), the ball flies at a speed v_0 under the action of the field of gravity and hits the plate. The experimental scheme and the actual view of the tests are presented in Figures 4 and 5.

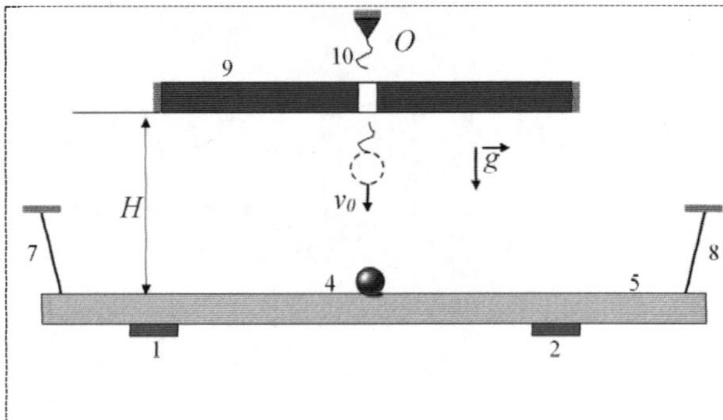

Figure 4. The experimental scheme.

Figure 5. Actual view of the tests

To achieve higher impactor speeds, a pneumatic gun was used. The block diagram of this type of test is shown in Figure 6, where 1 is a compressor, 2 is an accumulation chamber, 3 is a crane, 4 is a ball, 5 is a barrel, 6, 7 are lasers, 8, 9 are photodiodes, 10 is a catcher, 11 - test specimen, 12 - stopping unit, 13 - sample fastening.

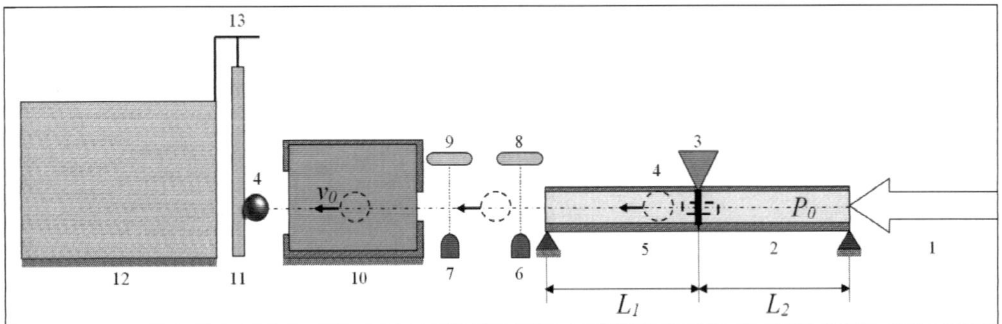

Figure 6. The experimental scheme of pneumatic gun.

In accordance with the scheme on Figure 6, the plate (11) was fixed in the bracket (13). Next the positioning of the plate (11) was carried out in order to ensure that the impactor hit the given point. The crane (3) was closed. The ball (4) was set in the barrel (5) before touching the crane (3). The required level of pressure in the chamber (2) was pumped which provides a set value of the velocity at the outlet of the barrel (5). After that the recording equipment should be turned on and the valve (3) opened to release the pressure and push the ball.

2.3. Impact loading in the speed range of 6-13 m/s

Impact loading in the speed range of 6-13 m/s, with the help of a metal ball falling in the field of gravity, corresponds to a height interval of 1.9 - 8.6 m. The tests were carried out according to the scheme in Figure 4. An individual point on the composite plate corresponded to each impactor speed. The marking of impact points on the plate is presented in Figure 7. The distance between the points is 10 mm. Data on the experiments is presented in Table 1

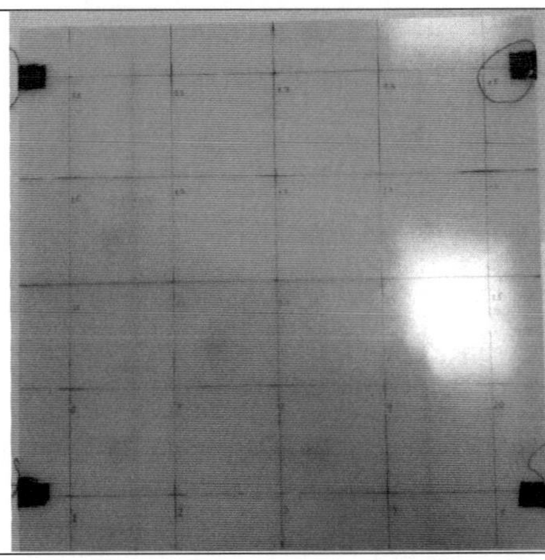

Figure 7. Layout of the plate during impact tests using a metal ball falling in the gravity field

Table 1. Experimental data on impact loading in the speed range of 6-13 m/s, using a metal ball falling in the field of gravity.

height (m)	velocity (m/s)	Point number
1.925	6.1	11
3.395	8.2	12
4.267	9.2	7
5.073	10.0	17
5.857	10.7	9
6.651	11.4	14
7.443	12.1	19
8.520	12.9	13

After the impact tests, a plate was visually inspected for damage. Photographs of impact areas where the damage is visually recognizable are presented in Figure 8.

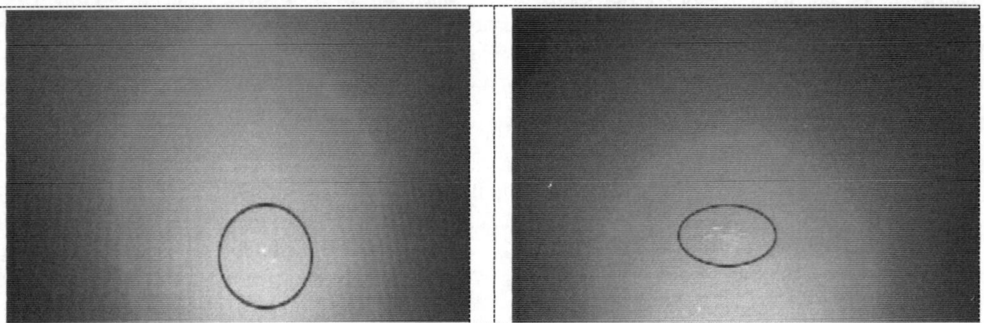

Point 17 (10.0 m/s) Point 13 (12.9 m/s)

Figure 8. Photos of impact areas with impactor speeds in the range of 6-13 m / s

The first visually noticeable damage is detected upon impact at a speed of 10 m/s. The damage area increases with the speed of the impactor. The size of the maximum visually detected defect is 8.3×4.8 mm. All visually detected damages are characterized by slight cracking in the local area from the side opposite to the impact.

2.4. Impact loading in the speed range of 29-109 m/s

A series of impact tests was carried out with the impactor speed range of 29-109 m/s using a pneumatic gun. Data on the speeds of individual tests and the size of the visually discernible damages are presented in Table 2. Photographs on both sides of the plate for some tests are shown in Figure 9.

Table 2. Experimental data on impact loading in the speed range of 29-109 m/s.

velocity (m/s)	Diameter of damage area, mm
29	8
43	11
54	35
69	36
85	68
109	78

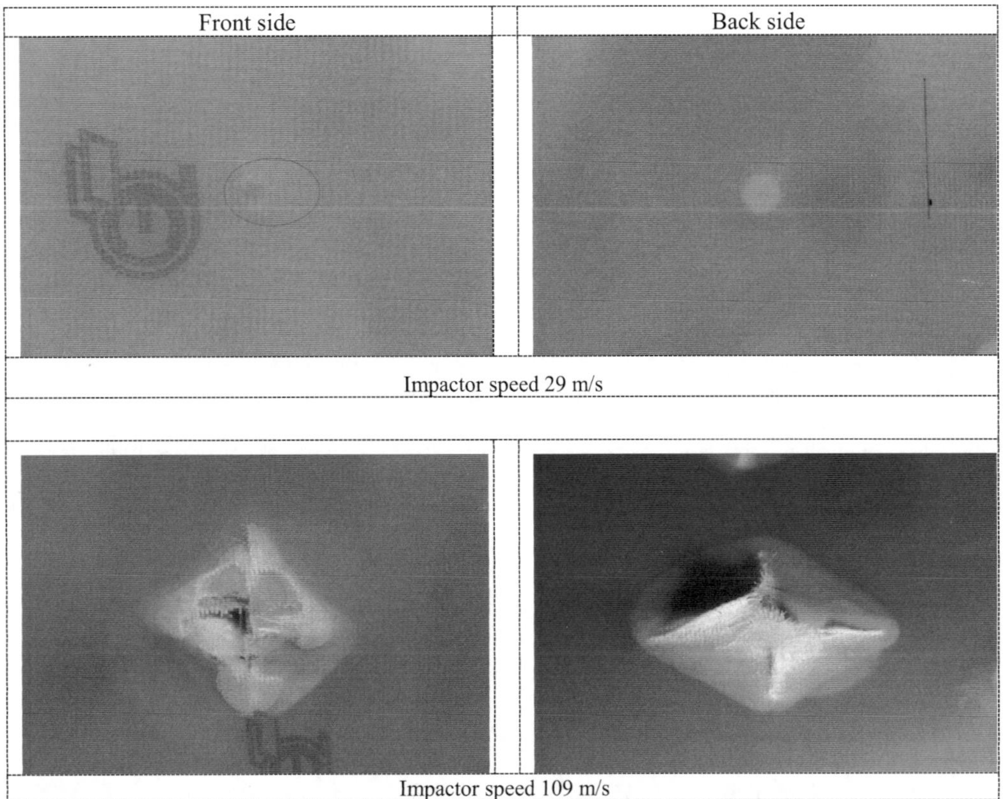

Figure 9. Images of impact areas with impactor speeds in the range of 29-109 m/s

In the considered speed range of the impactor, the damage on the plate is easily identified visually. Starting at 54 m/s, the threads of reinforcement of the composite material break in the form of a cross. In addition to breaking of the reinforcement filaments, zones of internal delamination and the area of cracking of the matrix are observed. At the last test at a speed of 109 m/s, a plate was pierced by the impactor.

3. Registration of impact loading by a network of piezoelectric sensors

As part of this work, a system for recording elastic waves arising from the interaction of an impactor and a plate was developed and constructed. It consists of 6 synchronized channels, which provide registration of vibrations at 6 points with the help of piezoceramic sensors. The registration of electrical signals from sensors during impact at various points of the plate and at different speeds of the impactor was performed.
A series of experiments of throwing a spherical weight from a height of 767 mm onto a plate, at the points of location of the piezoelectric sensors, but from the opposite side of the plate was carried out. The resulting signals demonstrate the greatest response of those sensors to the coordinates of which the impact was made. Figure 10 shows the signals of sensors 1 and 2 under impacts at corresponding points on the plate.

IOP Conf. Series: Materials Science and Engineering **581** (2019) 012030 doi:10.1088/1757-899X/581/1/012030

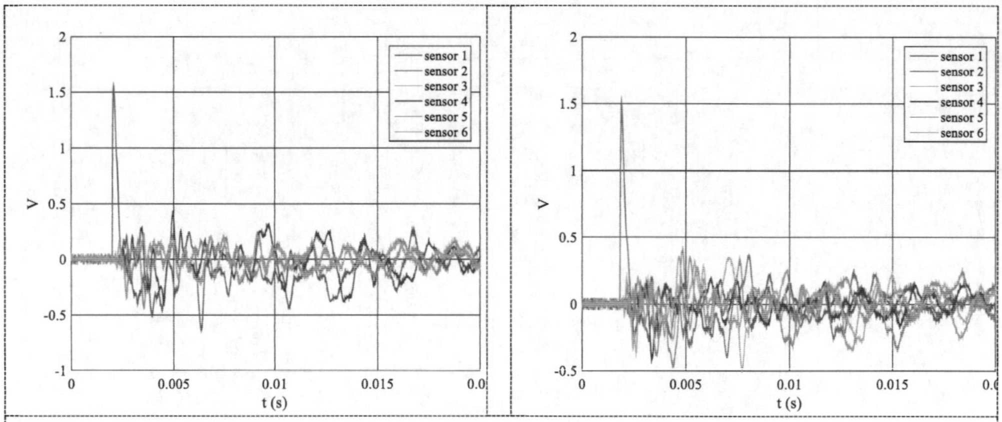

Figure 10. Sensor response upon impact at the location of piezoelectric elements.

Similarly, an impact was made to the center of the plate with the signals recorded by all the sensors (Figure 11). The received responses demonstrate the synchronous behavior of sensors 2, 3, 5, 6 and some shift of signals from sensors 1 and 2, which is clearly visible in the initial area of the graph in Figure 12.

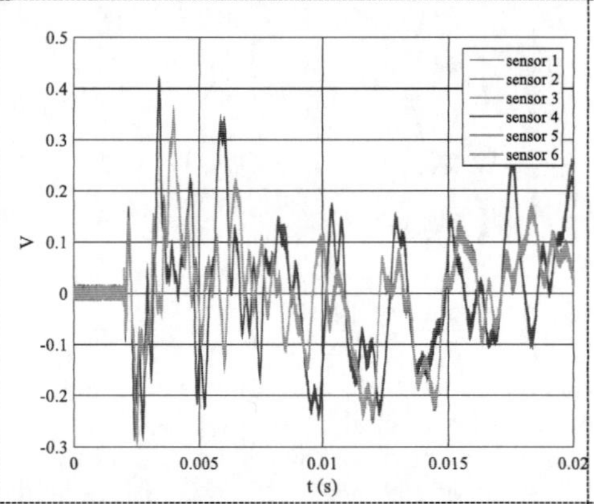

Figure 11. The response of the sensors upon impact to the center of the plate.

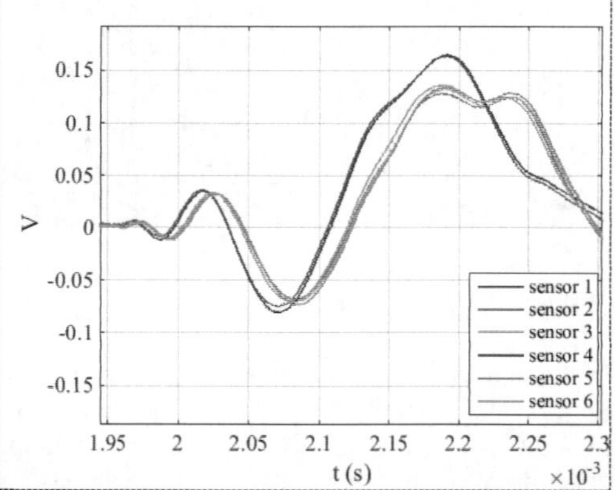

Figure 12. The initial area of the signals from piezoelectric sensors upon impact to the center of the plate.

4. Conclusions

In this work, the impact velocity leading to the formation of primary defects in the form of cracking of the matrix on the back surface of a fiberglass laminate was determined. Which is 10 m/s. With further increase in speed, the cracking area increases. Experimental setup for testing composite plates on the impact of a spherical body at high impactor speeds (29-109 m/s) was designed and tested. The following patterns were found with increasing of impactor speed. At speeds from 29 to 43 m/s, a significant increase in the area of matrix destruction throughout the entire thickness of the plate is observed. Starting at 54 m/s, the threads of reinforcement of the composite material break and form the shape of a cross. At speeds of the impactor over 109 m/s, the impactor goes through the plates. A system of elastic waves measurement resulting from the interaction of an impactor and a plate was developed. It consists of 6 synchronized channels, which provide registration of vibrations at 6 points of object under study.

The registration of electrical signals from piezoceramic sensors under impact loadings at various points of the plate and at different speeds of the impactor was performed which confirmed the sensitivity of developed system.

Acknowledgments
The results were obtained within the framework of the state task of Russian Ministry of Education and Science (project no. 11.2391.2017/4.6).

References

[1] Duchene P, Chaki S, Ayadi A and Krawczak P 2018 A review of non-destructive techniques used for mechanical damage assessment in polymer composites *J. Mater. Sci.* **53** 7915–38
[2] Komorowski J P, Simpson D L and Gould R W 1990 A technique for rapid impact damage detection with implication for composite aircraft structures *Composites* **21** 169–73
[3] Diamanti K and Soutis C 2010 Structural health monitoring techniques for aircraft composite structures *Prog. Aerosp. Sci.* **46** 342–52
[4] Grimberg R, Savin A, Steigmann R, Brauma A and Barsanescu P 2009 Ultrasound and eddy current data fusion for evaluation of carbon-epoxy composite delaminations *Insight Non-Destructive Test. Cond. Monit.* **51** 25–31
[5] Brunner A J 2018 Identification of damage mechanisms in fiber-reinforced polymer-matrix composites with Acoustic Emission and the challenge of assessing structural integrity and service-life *Constr. Build. Mater.* **173** 629–37
[6] Serovaev G S, Shestakov A P and Oshmarin D A 2018 Numerical Study of Vibrational Processes in Composite Material for the Development of a Delamination Control System *J. Appl. Mech. Tech. Phys.* **59** 1261–70
[7] Bykov A A, Matveenko B P, Serovaev G S, Shardakov I N and Shestakov A P 2015 Mathematical modeling of vibration processes in reinforced concrete structures for setting up crack initiation monitoring *Mech. Solids* **50** 160–70
[8] Ciampa F, Mahmoodi P, Pinto F and Meo M 2018 Recent Advances in Active Infrared Thermography for Non-Destructive Testing of Aerospace Components *Sensors* **18** 609
[9] Katunin A, Wronkowicz-Katunin A and Wachla D 2019 Impact damage assessment in polymer matrix composites using self-heating based vibrothermography *Compos. Struct.* **214** 214–26
[10] Boller C, Mahapatra D R, Sridaran Venkat R, Ravi N B, Chakraborty N, Shivamurthy R and Simon K M 2017 Integration of Non-Destructive Evaluation-based Ultrasonic Simulation: A means for simulation in structural health monitoring *Struct. Heal. Monit.* **16** 611–29
[11] De Luca A, Caputo F, Sharif Khodaei Z and Aliabadi M H 2018 Damage characterization of composite plates under low velocity impact using ultrasonic guided waves *Compos. Part B Eng.* **138** 168–80
[12] Kudela P, Wandowski T, Malinowski P and Ostachowicz W 2017 Application of scanning laser Doppler vibrometry for delamination detection in composite structures *Opt. Lasers Eng.* **99** 46–57
[13] Soleimanpour R, Ng C-T and Wang C H 2017 Higher harmonic generation of guided waves at delaminations in laminated composite beams *Struct. Heal. Monit.* **16** 400–17

IOP Conf. Series: Materials Science and Engineering **581** (2019) 012031 doi:10.1088/1757-899X/581/1/012031

Comparative Analysis of the Contact Deformation of the Spherical Sliding Layer of the Bearing with and without Taking into Account the Grooves with Lubricant

A A Adamov[1, a], A A Kamenskikh[2, b]

[1] Laboratory of Nonlinear Mechanics of a Deformable Solid, Institute of Continuous Media Mechanics of the UB of RAS, Perm, Russia

[2]Department of Computational Mathematics, Mechanics and Biomechanics, Perm National Research Polytechnic University, Perm, Russia

E-mail: [a] adamov.aa@ya.ru, [b] anna_kamenskih@mail.ru

Abstract. In this paper, the analysis of the contact deformation of elements of spherical bearing through anti-friction polymeric layer made of modified fluoroplastic with and without taking into account the grooves with lubricant is carried out. The effect of grooves with lubricant on contact zone parameters is considered. Comparison of the results of the spherical bearing draft obtained within the framework of numerical and full-scale experiments with a constant coefficient of friction of 0.04 is made: at a nominal vertical load of 1000 kN, the difference between the results of numerical and full-scale experiments is minimal and is ~ 7.13% for the model taking into account the grooves with lubricant. The analysis of frictional properties of modified fluoroplastic under the conditions of contact interaction with and without taking into account the lubrication of the mating surfaces is performed. Approximating functions have been selected to describe the dependence of the friction coefficient on the pressure acting on the contact section. The effect of the friction coefficient obtained experimentally on the contact deformation of the sliding layer is considered. Several qualitative and quantitative regularities have been established.

1. Introduction

Researchers note the main actual problems of transport and logistics systems related to bridge construction: geometric configuration and technology of deformation joints [1, 2], bearings [3, 4], bridge spans [5], hoisting structures of draw bridges [6], as well as other elements of bridge structures. We can note a special interest in the deformation behavior of the bearing of bridge spans [7-11, etc.]. The most of the works are aimed at the analysis of bearing capacity, strength, wear resistance of bridge structure elements, as well as the assessment of the stress-strain state of bearing structures in general [7,10] and the identification of patterns of change in contact zones in particular [10, 12]. Several works are devoted to the studies of the possibility of using modern anti-friction polymeric materials in the structures of bearing parts [9, 10, etc.]. Methods of mathematical modeling as well as effective numerical methods are used in the study of the deformation behavior of the bearings of bridge spans. For qualitative creation of numerical models of bearings, it is necessary to have experimentally confirmed models of behavior of contact sections' materials, in connection with this concomitant direction of research are experimental researches of materials with the purpose of reception of their thermomechanical, frictional and rheological properties. Thus, the actual line of research is the numerical analysis of deformation behavior of elements comprising the bearings of bridge structures taking into account properties of materials from which they are made.

In this paper, we consider contact deformation of an anti-friction layer made of modern polymeric material (modified fluoroplastic) taking into account its physical-mechanical and friction properties. The direction of the research is the analysis of how the design features of a spherical bearing effect on parameters of a contact zone of steel plates with a spherically thin sliding layer made of an anti-friction polymeric material.

2. Statement of a problem

Friction contact interaction of upper (1) and lower steel plate (2) of the spherical bearing through an elastoplastic polymeric layer (3) is considered in the paper with (Fig. 1, b) and without (Fig. 1, a) grooves with lubricant (4).

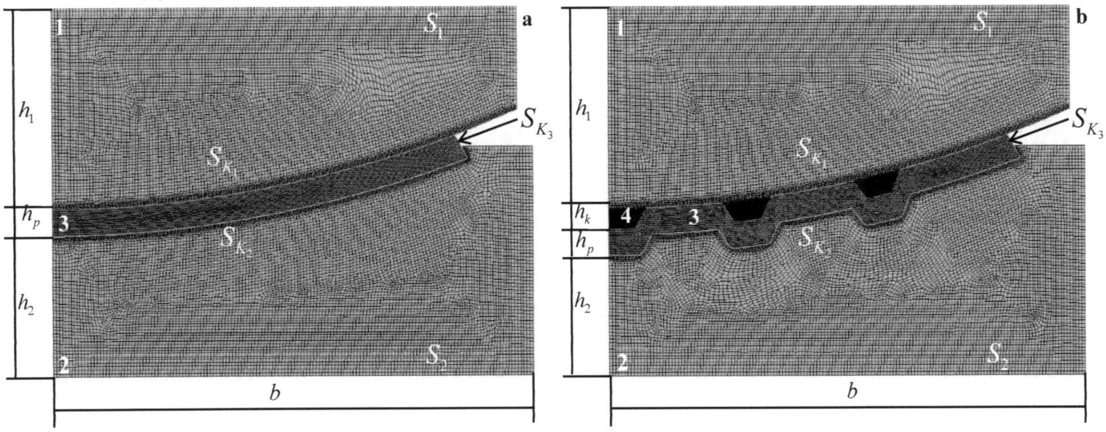

Fig. 1. Finite-element models of the spherical bearing: with (b) and without (a) taking into account the grooves with lubricant.

Contact deformation of the spherical bearing with the L-100 (Figure 1) anti-friction layer manufactured by AlphaTech LLC (Perm) is considered. The spherical bearing is designed for a nominal vertical load of 1000 kN. Two models of the bearing are considered, we introduce the following designations: model A is model of the spherical bearing without taking into account the grooves with lubricant; model B is model of the spherical bearing without taking into account the grooves with lubricant. Typical dimensions of the spherical bearing: maximum height of the top plate $h_1 = 0.03$ m, minimum height of the bottom plate with a spherical neckline $h_2 = 0.02$ m ($h_2 = 0.0179$ m for the model with taking into account the grooves), maximum width of the structure $b_k = 2b = 0.155$ m, thickness of the sliding polymeric layer $h_p = 0.004$ m, thickness of the grooves with lubricant $h_k = 0.003$ m. The deformation behavior of bridge bearings under the action of indentation force of 500-1250 kN, which corresponds to the indentation force of full-scale experiment on deformation of the spherical bearing, is considered in the paper. Constant indentation force and the bending is prohibited on S_1, on S_2 the y-movements are prohibited.

Previously, the Institute of Continuous Media Mechanics of the Ural Branch of the Russian Academy of Sciences carried out a cycle of experimental studies of the physical and mechanical characteristics of anti-friction materials at complex multistage deformation stories with offloading using a Zwick Z100SN5A testing machine. A series of full-scale experiments included: tests to determine the Brinell hardness of materials by penetrating a ball with a diameter of 5 mm; studies under free compression conditions, as well as studies of constrained compression by pressing cylindrical specimens with a diameter and length of 20 mm in a special fixture with a rigid steel cage. As a result, the physical and mechanical characteristics of materials were obtained: modulus of elasticity E , Poisson's coefficient ν , deformation diagrams $\sigma_z - \varepsilon_z$. In this paper, we consider a spherical bearing with an anti-friction layer of modified fluoroplastic with modulus of elasticity $E = 863.8$ MPa and Poisson's coefficient $\nu = 0.461$ of elastic region. Further, the scientific team of Alfa-Tech LLC and IMSS of the Ural Branch of the Russian Academy of Sciences prepared special equipment and carried out a series of full-scale experiments aimed at determining the friction properties of polymeric materials. Within the limits of a series of tests, the friction coefficient μ was determined as a function of pressure. Fig. 2 shows the results of full-scale experiments with and without taking into account lubricant on the mating surfaces at contact deformation of cylindrical samples with a diameter $0.097 \div 0.103$ m and height 0.01 m by steel plates of the press.

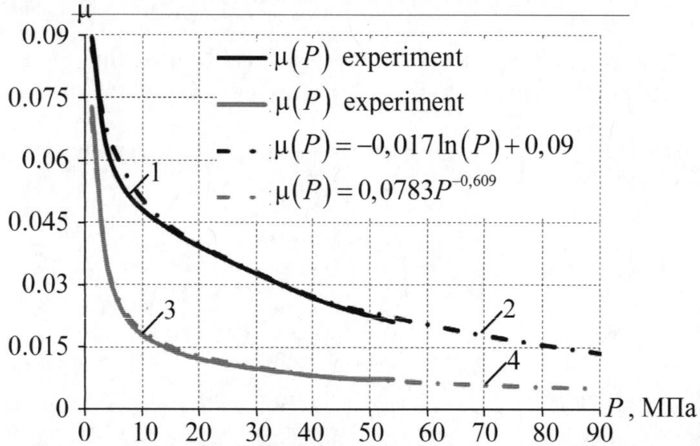

Fig. 2. Dependence of μ on P: 1 is experimental data without taking into account lubricant; 2 is approximation of experimental data (1); 3 is experimental data with taking account into lubricant; 4 -is approximation of experimental data (3).

The full-scale experiment was carried out in the pressure range of 1.2-54 MPa, and the working pressure range of the spherical bearings can reach 90 MPa. Within the limits of the analysis of experimental data, the approximation of results of full-scale experiments was executed, the approximating functions describing results of the experiment with the maximum error of 5 and 8 % for contact with and without taking into account lubrication on mating surfaces, respectively, were selected. The selected functions are required to calculate the friction coefficient for pressure > 54 MPa. In this paper, we consider a spherical bearing of L-100 with a nominal operating vertical load of 1000 kN (~55.5 MPa). The general mathematical statement of the problem of contact interaction of two elastic bodies through the anti-friction polymeric layer taking into account all types of friction contact is described in [12]. The problem is considered in the axisymmetric statement; the deformation theory of elasticity was chosen to describe the model of the behavior of the anti-friction layer material. Lubrication is modeled as a low compressible material with $E = 200$ MPa, $\nu = 0.4999$ and $\mu = 0.007$. The contact between the lubricant and anti-friction layer is not taken into account.

3. Findings

The problem is implemented in the ANSYS software package using axisymmetric finite elements with the Lagrangian approximation. Previously, it was established that numerical solutions for finite element mesh with 8 and 16 elements in the thickness of the layer for the model without taking into account the grooves with lubricant have minor differences [10]. When introducing the grooves with lubricant into the model, it was decided to choose a mesh with 16 elements in the thickness of the layer for a better solution to the problem. On the selected finite element model, the deformation behavior of the anti-friction layer of modified fluoroplastic was analyzed. At the first stage of research, the contact of elements of a spherical bearing with a constant friction coefficient of 0.04, which corresponds to the reference data of the manufacturer, was modeled. At the second stage of research for nominal load 1000 kN, the contact deformation with friction coefficients of 0.006787 and 0.02713, which corresponds to the friction coefficients obtained experimentally with and without taking into account lubrication on the mating surfaces, respectively, was modeled.

Findings of the numerical modeling of contact interaction of elements of a bearing through an anti-friction layer at a constant friction coefficient of 0.04 are given in Fig. 3-4. The introduction of grooves with lubricant into the model of the bearing has a significant influence on the qualitative and quantitative distribution of contact parameters: contact parameters change irregularly, there is a decrease in the level of contact pressure and contact tangential stress in the area of grooves with lubricant. The contact tangential stress increases near the interface between the lubricant and the sliding layer.

Fig. 3 shows the relative contact pressure distribution for the two spherical bearing models under consideration.

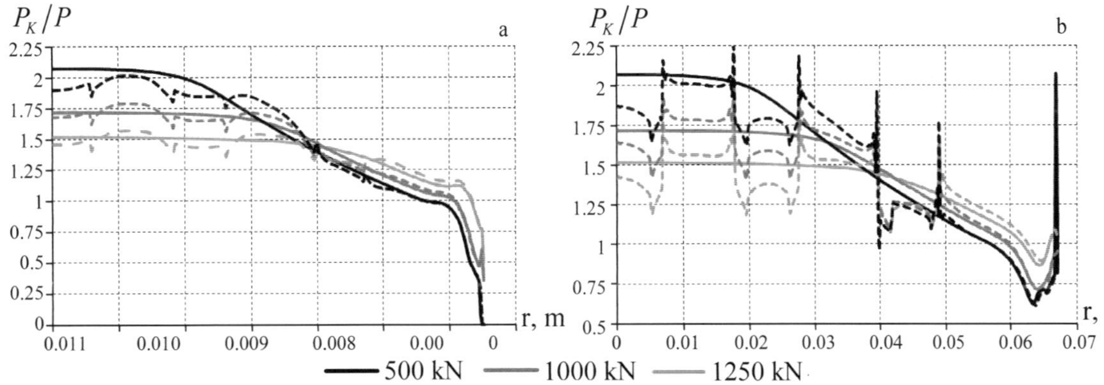

Fig. 3. Relative contact pressure on S_{K_1} (a) and S_{K_2} (b): solid line is Model A; dotted line is Model B.

The average maximum contact pressure does not change significantly at the contact boundary S_{K_1}, where the spherical segment of the bearing is rotated: there is a slight decrease in the level of P_K/P in the contact area with the lubricant, with a further increase in the contact area of the anti-friction material, whereas the value of P_K/P in the contact area of the anti-friction material is close to P_K/P of the model, without taking into account the grooves with the lubricant. The maximum level of P_K/P for model B differs on average by approximately 3% from model A.

Fig. 4 shows the distribution of the relative contact tangential stress.

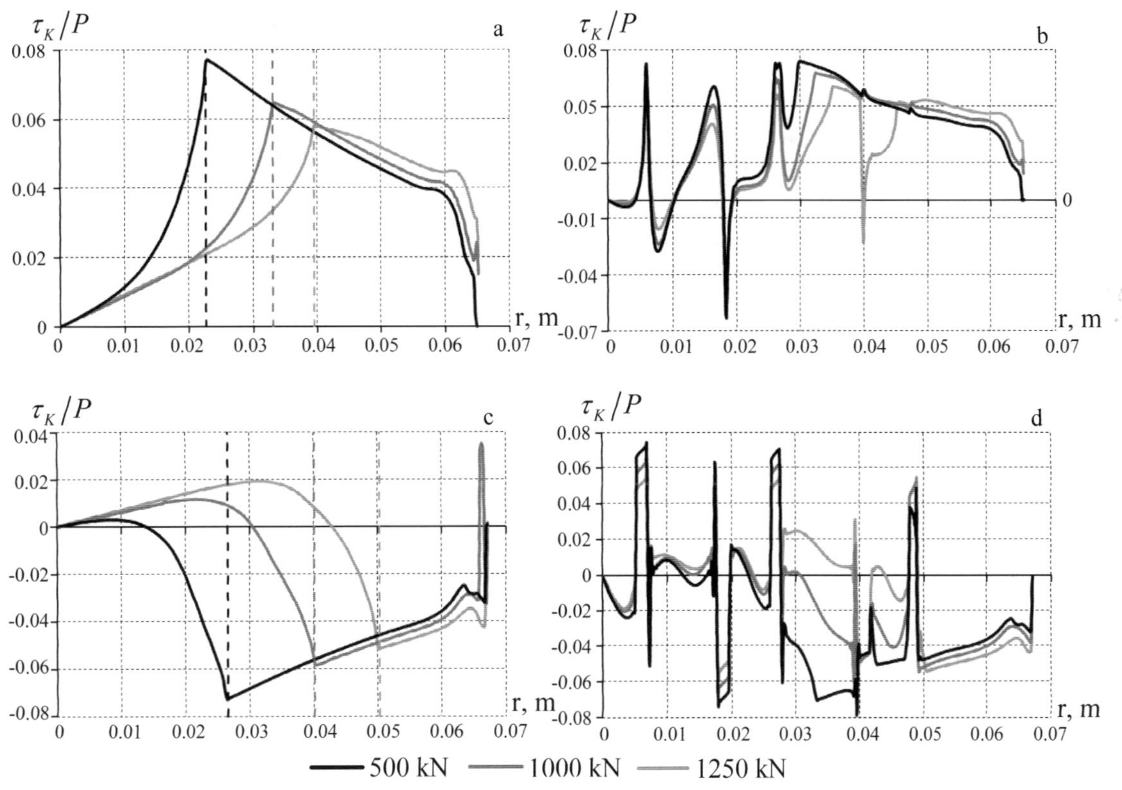

Fig. 4. Relative contact tangential stress at S_{K_1} (a,b) and S_{K_2} (c,d): a, c are model A; b, d are model B; dotted lines is "stick-slip" contact state zone.

The greatest effect of the grooves with the lubricant on the parameters of the contact area is observed in the interface area of the layer with the lower steel plate: the maximum contact pressure on the main contact surface is on average by 8.7% higher, the contact pressure in the contact area with the lubricant is less than 20% lower than that of model A.

On the contact surface S_{K_1}, there is a non-uniform distribution of contact tangential stress: there is a strong decrease τ_K/P in the areas of contact with lubricant, and negative values τ_K/P are observed in the region of grooves with lubricant. The non-uniform distribution τ_K/P with zones of positive and negative parameters has a strong influence on the wear of the anti-friction layer. The maximum contact tangential stress at S_{K_1} does not differ significantly for the two models of the bearing under consideration. Similarly to the contact pressure, the greatest influence of the grooves with lubricant is observed in the contact zone with the lower steel plate: the maximum level of contact tangential stress on the main surface of the anti-friction material is greater by 2.5; 6.5; 24.2 times for loads of 500, 1000, 1250 kN, respectively. The influence of the grooves with lubricant on the relatively free edge of the anti-friction layer is not significant.

Within the framework of the paper, the comparison of numerically obtained values of the spherical bearing draft with the results of a series of full-scale experiments on deformation of the real structure of the spherical bearing L-100 by the vertical load 500-1250 kN, was carried out (Fig. 5).

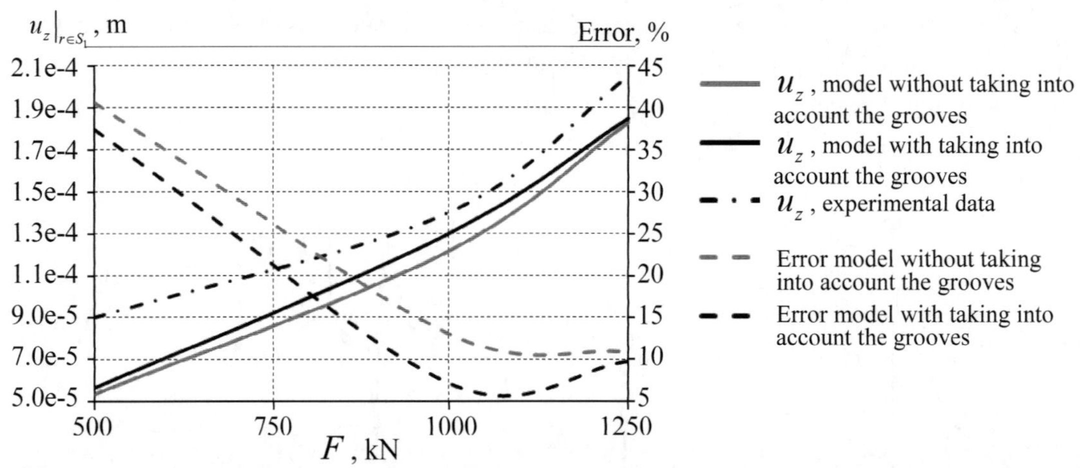

Fig. 5. Draft of the spherical bearing and the error of the numerically obtained draft values from the full-scale experiment.

It is found out that at the nominal vertical load of 1000 kN, the difference between the results of the numerical solution of the problem for the model taking into account the grooves with lubricant and the results of the full-scale experiment is minimal and makes ~ 7.13%, for the model without taking into account the grooves ~ 13.05%.

Further, within the limits of research, the comparative analysis of parameters for the zones of contact of a spherical bearing at its deformation with the friction coefficients received as a result of the full-scale experiments and the reference friction coefficient of 0.04, was executed.

Fig 6. and 7. show the relative contact pressure and the relative contact tangential stress, respectively.

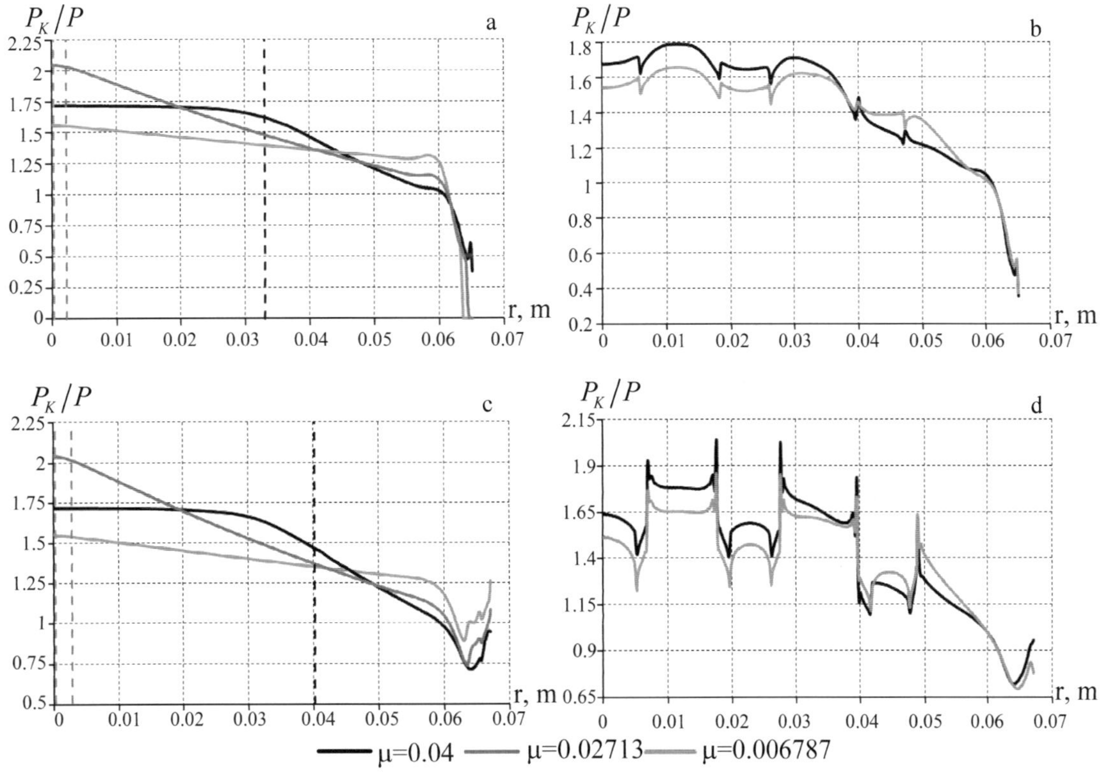

Fig. 6. Relative contact pressure at S_{K_1} (a,b) and S_{K_2} (c,d): a, c are model A; b, d are model B; dotted lines is "stick-slip" contact state zone.

For the contact deformation behavior of model A, the contact pressure at the friction coefficient corresponding to the contact with the lubricant is set at more than 9 % lower, and without taking into account the lubricant – at more than 19 % higher than the corresponding parameter at the contact deformation of the model with the friction coefficient of 0.04. In the case of model B, the maximum level P_K/P with the friction coefficient obtained experimentally is lower by more than 7 % and 9 % at S_{K_1} and S_{K_2}, respectively, than with the contact interaction with the friction coefficient of 0.04.

In model A, the contact tangential stress is more than 83% lower than the friction coefficient declared by the manufacturers of thin sliding bearing materials. When considering the influence of frictional properties of modified fluoroplastic on the contact parameters, we can conclude that the nature of the contact pressure and tangential stress distribution in general does not differ much from the calculations with a coefficient of friction of 0.04.

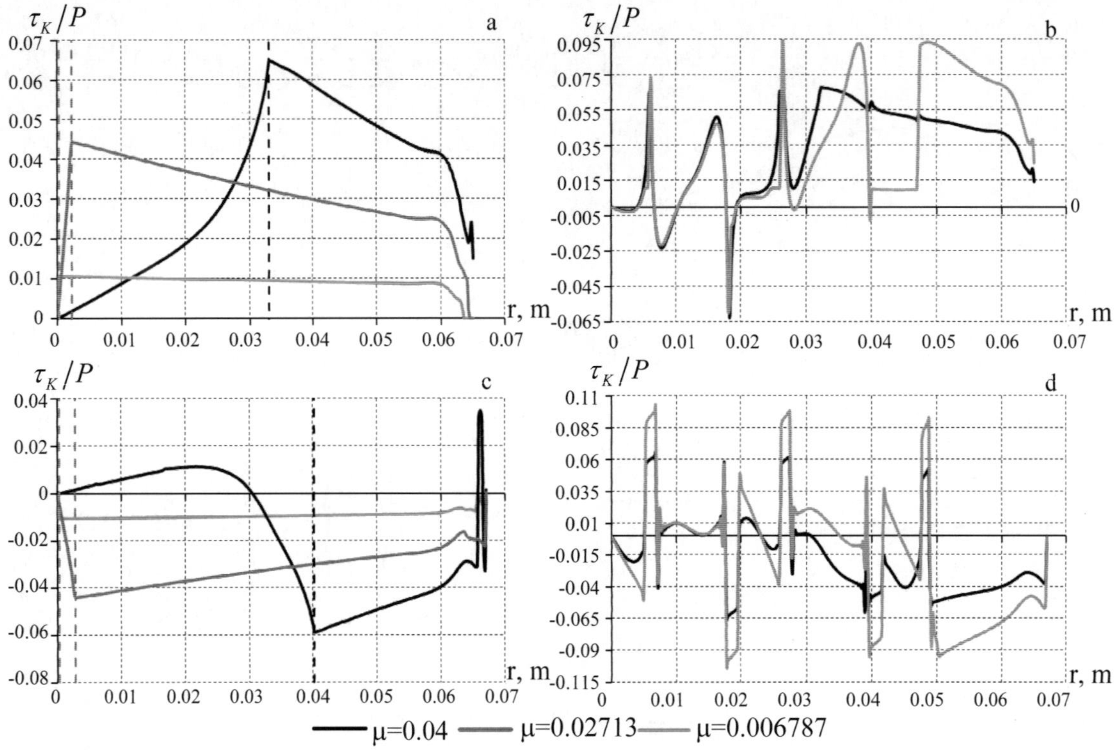

Fig. 7. Relative contact tangential stress at S_{K_1} (a,b) and S_{K_2} (c,d): a, c are model A; b, d are model B; dotted lines is "stick-slip" contact state zone.

It should be noted that the coefficient of friction has a significant impact on the contact state zones: for model A, full contact surface adhesion is observed at 0.4 and 3.5% of the contact surface for μ with and without taking into account the lubricant, respectively, with 1 to 2.5% of the contact surface near the edge of the layer is not involved in the work of the structure due to the loss of contact.

In the case of the model with taking into account the grooves with lubricant, there is no loss of contact between the top plate of the spherical bearing and the anti-friction layer. At the same time, the level of contact tangential stress in model B is 49% and 23% higher at S_{K_1} and S_{K_2}, respectively, than in case of contact interaction with the coefficient of friction of 0.04. The distribution of the relative contact pressure differs from the value previously obtained by a large drop of τ_K/P in the area of the groove near the edge of the anti-friction layer.

4. Conclusions

Within the framework of the study, the deformation behavior of modified fluoroplastic as a material of anti-friction layer of spherical bearing was modeled. The comparative analysis of deformation behavior of two models of the spherical bearing with and without taking into account the grooves with lubricant was carried out. Influence of lubricated grooves on the parameters of the contact zones of the spherical bearing elements with the anti-friction layer was analyzed.

Upon analysis of the results of a series of numerical experiments, it was found out:

- At contact interaction with a coefficient of friction of 0.04, the greatest influence of grooves with lubricant on contact parameters is observed in a contact zone with the bottom steel plate: the maximum contact pressure on the average is more by 8.7 %; the maximum contact tangential stress on the main surface of an anti-friction material is by 2.5; 6.5; 24.2 times higher for loadings 500, 1000, 1250 kN, respectively
- At nominal vertical load of 1000 kN, the difference of results of numerical and full-scale experiment on deformation of the spherical bearing with constant coefficient of friction of 0.04 for the model with

taking into account the grooves lubricant is minimal and makes ~ 7.13%, for the model without taking into account the grooves ~ 13.05%.

- For the model without taking into account the grooves with lubricant, the contact pressure at the coefficient of friction obtained experimentally with the lubrication of the mating surfaces is by more than 9 % lower and the contact tangential stress is by more than 83 % lower than at the coefficient of friction declared by the manufacturers of thin sliding materials of the bearings.
- For the model with taking into account the grooves with lubricant, the contact pressure at the coefficient of friction obtained experimentally with taking into account the lubricant on the mating surfaces, is by more than 7-9% lower, and the contact tangential stress is by more than 23-49% higher than at the coefficient of friction declared by the manufacturers of materials of thin sliding layer of bearings.
- When modeling the contact interaction of spherical bearing elements without taking into account the grooves with lubricant at the experimentally obtained coefficients of friction, there is a loss of contact near the edge of the layer by a maximum of 2.5% of the contact surface. The contact with taking into account the grooves with lubricant results in no contact near the edge of the layer.

The study supported by a grant of Russian Science Foundation (project No. 18-79-00147).

References

[1] Yankovsky L V, Kochetkov A V, Ovsyannikov S V and Trofimenko Yu A 2014 *Technical regulation in transport construction* **3(7)** 6-12.
[2] Yefanov A V and Ovchinnikov I G 2006 *Vestnik SSTU* **1(16)** 82-87.
[3] Ovchinnikov I G, Rasporov O N and Ovchinnikov I I 2015 *Modernization and scientific research in the transport complex* **1** 437-442.
[4] Choi E, Lee J S, Jeon H-Kw, Park T and Kim H-T 2010 *Nonlinear Dyn.* **62** 73-93.
[5] Ivanov B G 2006 *Diagnostics of damage to the span of metal bridges: a monograph* (Moscow: Marshrut) 208.
[6] Thrall A P, Adriaenssens S, Paya-Zaforteza I and Zoli T P 2012 *Engineering Structures* **37** 214–223
[7] Freire L and De Brito J 2006 *Proceedings of the 3rd International Conference on Bridge Maintenance, Safety and Management, Life-Cycle Performance and Cost* 205-206
[8] Wancheng Yu, Binbin W, Pakchiu Ch, Xinjian C and Zhaojun R 2012 *Earthq. Eng. & Eng. Vib.* **11** 173-183
[9] Fomin D V 2014 *Scientific Journal. Engineering systems and facilities* **2(15)** 80-90.
[10] Kamenskih A A and Trufanov N A 2013 *Comp. Continuum Mech.* **1** 54-61.
[11] Peel H, Luo S, Cohn A G, Fuentes R Automation In Construction **94** 244-256
[12] Kamenskih A A and Trufanov N A 2015 *Friction and Wear* 2 170-176.

IOP Conf. Series: Materials Science and Engineering **581** (2019) 012032 doi:10.1088/1757-899X/581/1/012032

Inhomogeneities and Extreme Fluctuations of Strains in Grains of Polycrystalline Materials

V E Shavshukov and A A Tashkinov

Perm National Research Polytechnic University, 29 Komsomolsky Av., Perm 614990, Russia

E-mail: shavshukov@pstu.ru

Abstract. Inhomogeneity of strains in individual grains are calculated using the field-theoretical approach in the mechanics of polycrystalline materials. It is shown that the microstructure surrounding the grain strongly influences the amplitude of the non-uniformity of strains in the grain. In the random microstructure of a polycrystal, there are specific clusters of grains in which very large strain fluctuations are realized. The strains in these clusters exceed macrostrains several times. The patterns of such extreme clusters are presented. Extreme strain fluctuations in bulk grains exceed fluctuations in surface grains by 40%. This may be one of the reasons for the movement of sites of damage initiation in gygacycle fatigue mode from the surface of a polycrystalline sample into the bulk.

1. Introduction

Determining microstructure-property relationships is an essential engineering problem and an important area of experimental and computational materials science. Due to the stochastic structure of polycrystals, the strain and stress fields at the meso-level are random, highly fluctuating inhomogeneous fields. Numerous studies described, for example, in review articles [1, 2, 3, 4], indicate the importance of estimating the magnitudes of strains fluctuations and their connection with the microstructure, especially in matters of gygacycle fatigue, where macro stresses and strains are very small. Fluctuations of strains in a grain depend on its shape, orientation, and interaction with the surrounding microstructure. Experiments of recent years using X-ray microscopy [5, 6, 7] revealed the existence of specific clusters of grains in polycrystals, which have a strong influence on the damage initiation and other localized critical phenomena. Extremely large strains and stresses are realized in these clusters. It was found that the interaction of grains plays a crucial role in the appearance of high strains and stresses [8, 9, 10].

Extreme clusters have characteristic microstructure patterns specific for different types of grain anisotropy and conditions of macro loading. The probability of the formation of these clusters is small; therefore, an experimental study of possible cluster patterns and extreme fluctuations of the fields in the cluster grains is difficult. As a result, the development of effective theoretical and computational tools for finding configurations of extreme clusters, estimating the magnitudes of fluctuations in random structures of polycrystals is relevant. In the present work, we carried out such estimates using the field-theoretical approach in the mechanics of polycrystals [11].

2. Field theory method for calculating strains in polycrystal grains

Let us consider a macroscopically isotropic polycrystalline body of volume V, surface S, and denote by, $u_i(\vec{r})$, $\varepsilon_{ij}(\vec{r})$, $C_{ijkl}(\vec{r})$ the displacement vector, the strain tensor, and the elastic moduli tensor, respectively. These variables are global fields. The elastic tensor is decomposed into homogenized and fluctuating parts

$$C_{ijkl}(\vec{r}) = \langle C_{ijkl}\rangle + C'_{ijkl}(\vec{r}), \qquad (1)$$

where $\langle C_{ijkl}\rangle \equiv \frac{1}{V}\int\limits_V C_{ijkl}(\vec{r})d\vec{r}$ is a homogeneous isotropic tensor, and $C'_{ijkl}(\vec{r})$ - a rapidly oscillating tensor function of arbitrary anisotropy and zero mean value. The isotropic tensor has the form

$$\langle C_{ijkl}\rangle = 3\langle K\rangle V_{ijkl} + 2\langle\mu\rangle D_{ijkl}\ , \qquad (2)$$

where $\langle K\rangle$ and $\langle\mu\rangle$ are the homogenized bulk modulus and shear modulus, V_{ijkl} and D_{ijkl} are the spherical and deviatoric parts of the identity tensor $I_{ijkl} = (\delta_{ik}\delta_{jl} + \delta_{il}\delta_{jk})/2$, and δ_{ik} is the Kronecker symbol.

We set two boundary problems for this body. The first problem is posed for a body with a uniform tensor of elastic moduli $\langle C_{ijkl}\rangle$. The second problem is set for the same body with an inhomogeneous tensor $C_{ijkl}(\vec{r})$. The boundary conditions for both bodies are the same. We use the integral form of boundary value problems in displacements.

The vector of displacements in a homogeneous body with an homogenized tensor is denoted as $u^*_m(\vec{r})$. The integral equation for the displacement vector in such a body has the form [12]

$$u^*_m(\vec{r}) = \int f_i(\vec{r}')G_{im}(\vec{r}-\vec{r}')d^3\vec{r}' + \oint\limits_S \langle C_{ijkl}\rangle\left[u^*_{k,l}(\vec{r}')G_{im}(\vec{r}-\vec{r}') - u^*_i(\vec{r}')G_{km,l}(\vec{r}-\vec{r}')\right]dS'_j\ , \qquad (3)$$

where $f_i(\vec{r})$ is the vector of volume forces, $G_{im}(\vec{r})$ is the Green's function of an isotropic medium with an averaged tensor of elastic moduli $\langle C_{ijkl}\rangle$, the index after the comma means differentiation by the corresponding coordinate.

The displacement vector for the second problem in an inhomogeneous body with a tensor of elastic moduli $C_{ijkl}(\vec{r})$ is denoted by $u_m(\vec{r})$. It satisfies the integral equation, which is derived similarly to (2) and has the form [11]

$$\begin{aligned} u_m(\vec{r}) &= \int \left[C'_{ijkl}(\vec{r}')u_{k,l}(\vec{r}')\right]_{,j} G_{im}(\vec{r}-\vec{r}')d^3\vec{r}' + \\ &+ \int f_i(\vec{r}')G_{im}(\vec{r}-\vec{r}')d^3\vec{r}' + \oint\limits_\Gamma \langle C_{ijkl}\rangle\left[u_{k,l}(\vec{r}')G_{im}(\vec{r}-\vec{r}') - u_i(\vec{r}')G_{km,l}(\vec{r}-\vec{r}')\right]dS'_j \end{aligned} \qquad (4)$$

Differentiating (3) and (4) and subtracting one from the other, as a result we obtain the equation for the global strain tensor $\varepsilon_{mn}(\vec{r})$

$$\varepsilon_{mn}(\vec{r}) = \varepsilon^*_{mn}(\vec{r}) + \int\limits_V g_{mnij}(\vec{r}-\vec{r}')C'_{ijkl}(\vec{r}')\varepsilon_{kl}(\vec{r}')d^3\vec{r}'\ , \qquad (5)$$

where $g_{mnij}(\vec{r}) = [(G_{im,nj}(\vec{r}) + G_{in.mj}(\vec{r})]/2$ is the modified Green tensor. Equation (5) was obtained by many authors [13, 14, 15] and used in various aspects. It is similar to the Lippmann-Schwinger equation in quantum field theory [14].We use equation (5) to calculate inhomogeneous strains in the grains. To do this, we discretize the equation. First, we decompose the global strain fields and the elastic modulus tensor into a sum of local variables in the grains using the shape functions $\lambda_\xi(\vec{r})$ of the grains (equal to one inside the grain and zero outside it)

$$\varepsilon_{ij}(\vec{r}) = \sum_{\xi=1}^N \lambda_\xi(\vec{r})\varepsilon^{(\xi)}_{ij}(\vec{r}) \text{ and } C_{ijmn}(\vec{r}) = \sum_{\xi=1}^N \lambda_\xi(\vec{r})C^{(\xi)}_{ijmn}(\vec{r}), \qquad (6)$$

where N is the total number of grains in a polycrystal.

Substituting (6) into (5) leads to a system of interrelated integral equations for local fields in the grains

$$\varepsilon_{ij}^{(\xi)}(\vec{r}_\xi) = \varepsilon_{ij}^* + \int_{\omega_\xi} d\vec{r}_\xi'\, g_{ijkl}(\vec{r}_\xi - \vec{r}_\xi')C_{klmn}'^{(\xi)}\varepsilon_{mn}^{(\xi)}(\vec{r}_\xi') + \sum_{\eta\neq\xi}\int_{\omega_\eta} d\vec{r}_\eta' g_{ijkl}(\vec{r}_\xi - \vec{r}_\eta')C_{klmn}'^{(\eta)}\varepsilon_{mn}^{(\eta)}(\vec{r}_\eta') \quad . \tag{7}$$

The subscript of the radius vector $\vec{r}_\xi$ indicates that it changes in the region ω_ξ of the ξ-th grain.

Next, we divide each grain into a large number of small subgrains and decompose the strains in the grains into sums of strains in the subgrains $\varepsilon_{ij}^{(\xi)(a)}(\vec{r})$ using the indicator functions of the subgrains $\mu_a^\xi(\vec{r})$ in which the upper Greek index enumerates the grains and the lower Latin index enumerates the subgrains inside the grain

$$\varepsilon_{ij}^{(\xi)}(\vec{r}) = \sum_{a=1}^{n}\mu_a^\xi(\vec{r}_\xi)\varepsilon_{ij}^{(\xi)(a)}(\vec{r}) \,. \tag{8}$$

This provides exact system of equations for $\varepsilon_{ij}^{(\xi)(a)}(\vec{r}_a)$

$$\begin{aligned}\varepsilon_{ij}^{(a)(\xi)}(\vec{r}_a) &= \varepsilon_{ij}^* + \int_{\omega_a} d\vec{r}_a'\, g_{ijkl}(\vec{r}_a - \vec{r}_a')C_{klmn}'^{(\xi)}\varepsilon_{mn}^{(a)(\xi)}(\vec{r}_a') + \\ &+\sum_{b\neq a}^{n}\int_{\omega_b} d\vec{r}_b\, g_{ijkl}(\vec{r}_a - \vec{r}_b)C_{klmn}'^{(\xi)}\varepsilon_{mn}^{(b)(\xi)}(\vec{r}_b) + \sum_{\eta\neq\xi}^{N}\sum_{e=1}^{n}\int_{\omega_e} d\vec{r}_e g_{ijkl}(\vec{r}_a - \vec{r}_e)C_{klmn}'^{(\eta)}\varepsilon_{mn}^{(e)(\eta)}(\vec{r}_e)\end{aligned} \tag{9}$$

,

where term related to a-th subgrain of ξ-th grain extracted from under the sum. To solve (9) we use perturbation theory. Let us consider both sums as perturbation and expand solution into series upon it

$$\varepsilon_{ij}^{(a,b,e)(\xi,\eta)}(\vec{r}) = \varepsilon_{ij}^{(0)(a,b,e)(\xi,\eta)}(\vec{r}) + \varepsilon_{ij}^{(1)(a,b,e)(\xi,\eta)}(\vec{r}) + \ldots \tag{10}$$

Iterating of (9) gives equations for corrections of all orders of magnitudes

$$\varepsilon_{ij}^{(0)(a)(\xi)}(\vec{r}_a) - \int_{\omega_a} d\vec{r}_a'\, g_{ijkl}(\vec{r}_a - \vec{r}_a')C_{klmn}'^{(\xi)}\varepsilon_{mn}^{(0)(a)(\xi)}(\vec{r}_a') = \varepsilon_{ij}^*$$

$$\begin{aligned}\varepsilon_{ij}^{(1)(a)(\xi)}(\vec{r}_a) - \int_{\omega_a} d\vec{r}_a'\, g_{ijkl}(\vec{r}_a - \vec{r}_a')C_{klmn}'^{(\xi)}\varepsilon_{mn}^{(1)(a)(\xi)}(\vec{r}_a') &= \sum_{b\neq a}^{n}\int_{\omega_b} d\vec{r}_b\, g_{ijkl}(\vec{r}_a - \vec{r}_b)C_{klmn}'^{(\xi)}\varepsilon_{mn}^{(0)(b)(\xi)}(\vec{r}_b) + \\ &+\sum_{\eta\neq\xi}\sum_{e=1}^{n}\int_{\omega_e} d\vec{r}_e g_{ijkl}(\vec{r}_a - \vec{r}_e)C_{klmn}'^{(\eta)}\varepsilon_{mn}^{(0)(e)(\eta)}(\vec{r}_e)\end{aligned} \tag{11}$$

.... and so on.

In the following we restrict solution by first order on perturbation and make last step of problem discretization. We will take strains be constant inside subgrain (due to it's small sizes) and varied by step-wise manner when go to adjacent subgrain. Then system of integral equations (11) transforms into system of linear algebraic ones

$$\varepsilon_{ij}^{(0)(a)(\xi)} - B_{ijkl}^{(aa)(\xi)}\varepsilon_{kl}^{(0)(a)(\xi)} = \varepsilon_{ij}^*$$

$$\varepsilon_{ij}^{(1)(a)(\xi)} - B_{ijkl}^{(aa)(\xi)}\varepsilon_{mn}^{(1)(a)(\xi)} = \sum_{b\neq a}^{n}B_{ijkl}^{(ab)(\xi)}\varepsilon_{kl}^{(0)(b)(\xi)} + \sum_{\eta\neq\xi}\sum_{e=1}^{n}B_{ijkl}^{(ae)(\eta)}\varepsilon_{kl}^{(0)(e)(\eta)} \,, \tag{12}$$

where constant coefficients $B_{ijmn}^{(ab)(\xi)} \equiv \int_{\omega_b} d\vec{r}_b' g_{ijkl}(\vec{r}_a - \vec{r}_b')\cdot C_{klmn}'^{(\xi)}$ and $B_{ijmn}^{(ae)(\eta)} \equiv \int_{\omega_b} d\vec{r}_b' g_{ijkl}(\vec{r}_a - \vec{r}_b')\cdot C_{klmn}'^{(\eta)}$ describe interactions of subgains inside the same grain and in different grains. These coefficients depend on microstructure of polycrystal and can be found numerically.

3. Numerical results for model polycrystals

We perform calculations on model polycrystals with spherical and cubic grains of equal volume. Consider a spherical ξ-th grain immersed in the averaged medium of all other polycrystal grains. In this case, as shown in [16], all corrections of higher order in exact equations (11) vanish, and corrections of zero order are the same for all subgrains of a spherical grain and are found from the algebraic equation

$$\left\{ I_{ijmn} + \frac{1}{3\langle\mu\rangle}\left[(1-\chi)V_{ijkl} + (1-\frac{2}{5}\chi)D_{ijkl}\right]\cdot C'_{klmn} \right\}\varepsilon^{(a)}_{mn} = \varepsilon^*_{ij}, \quad (13)$$

where $\chi = (3\langle K\rangle + \langle\mu\rangle)/(3\langle K\rangle + 4\langle\mu\rangle)$. For an isotropic grain in an isotropic matrix, the solution of (13) coincides with the Eshelby solution [17]. Inhomogeneity of strains in a spherical grain can occur only due to interaction with a non-uniform environment.

In a cubic single grain inside the inscribed sphere with a diameter of 0.9 from the cube edge, the strains are also almost homogeneous and coincide with strains in a spherical grain with an accuracy of several percent. Continuous singularities take place near the edges of the cube. The interaction with the inhomogeneous environment of the cubic grain makes the deformations inhomogeneous inside this spherical region. Numerical solution of equations (12) gives the deformation at any point of the grain under consideration, depending on the surrounding microstructure. In this model, this microstructure is reduced to the orientation of the crystallographic axes of each neighboring grain. An algorithm has been developed that searches for extremes of deformations at any point of the grain and the corresponding microstructures of the environment. The algorithm sets the array of random orientations of the surrounding grains, for each orientation solves system (12), from the array of solutions it finds the maximum (or minimum) value and the corresponding microstructure of the environment. Table 1 shows the results of calculations for a zinc polycrystal under uniaxial tension along the vertical axis X_3.

Table 1. Extreme deformation at grain points.

Macro strain $\varepsilon^*_{33} = 3\cdot10^{-4}$		Strain at given point ε_{33}
Grain center	maximum	$10.66\cdot10^{-4}$
	minimum	$-3.66\cdot10^{-4}$
Grain zenith	maximum	$11.34\cdot10^{-4}$
	minimum	$-2.79\cdot10^{-4}$

The absolute maximum is observed on the periphery of the grain at the zenith of the spherical part, the absolute minimum in the center of the grain. The maximum strain in the grain exceeds the macrostrain 3.78 times. The maximum (or minimum) strain at any point corresponds to a well-defined microstructure of the surrounding grains. Figure 1 shows a microstructure that generates a strain maximum in the center of the grain under consideration.

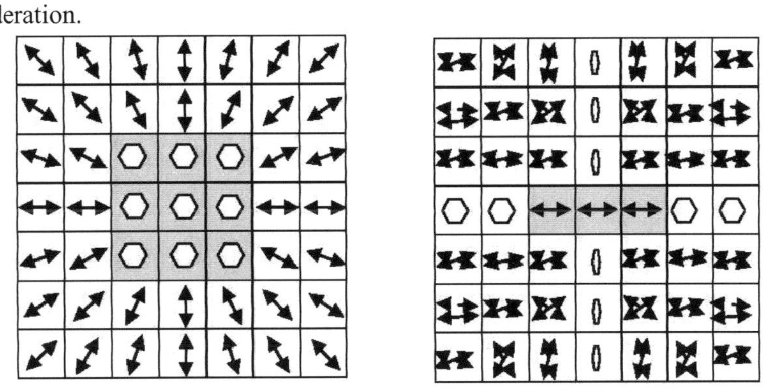

(a) (b)

Figure 1. Extreme grain orientation of neighbors: horizontal (a) and vertical (b) section. Conventions in the text.

The hexagon conventionally depicts the plane of the hexagonal elastic symmetry of single crystal grains. The icon corresponds to the position of the symmetry plane in the drawing plane, the icon

↔ depicts the position perpendicular to the drawing plane, and its inclination approximately shows the angle between the normal to the plane of elastic symmetry and the X_3 axis, the icon () corresponds to the inclined position to the drawing plane at some angle, but at the same time the normal of the plane of symmetry is perpendicular to the axis X_3. In an extreme configuration, clusters of grain-neighbors form symmetrical patterns. The nearest neighbors form a flat disk-shaped cluster of 9 grains of the same orientation around the central grain (they are darkened in Figure 1). The plane of the disk is perpendicular to the direction of the load. The tensile stiffness of these 9 grains is minimal in the direction of the load, while the surrounding grains are maximal. They form a kind of weakening zone. The diameter of this zone is equal to three grain diameters. If we introduce into the model a weakening zone of 5 grain diameters, then the strain in the central grain will decrease. It was experimentally found that in ultra-high cycle fatigue regime, cracks originate just in clusters of equally oriented grains [5], and the cluster size is equal to 3-4 grain diameters.

The magnitude of the fluctuations is significantly different for the bulk and surface grains of the polycrystal. The grain on the surface interacts with almost half as many neighbors as the grains deep inside the sample. As a result, the maximum possible strains in the surface grains are less by 40% compared to deep bulk grains. The range of fluctuations also decreases. Figure 2 shows the dependence of the maximum strain in the grain center on the distance from the surface of the sample. Thus, the surface grains form a special layer of material. In a sense, this is the discharge area of a polycrystalline body. The thickness of the surface layer is approximately equal to four average grain diameters.

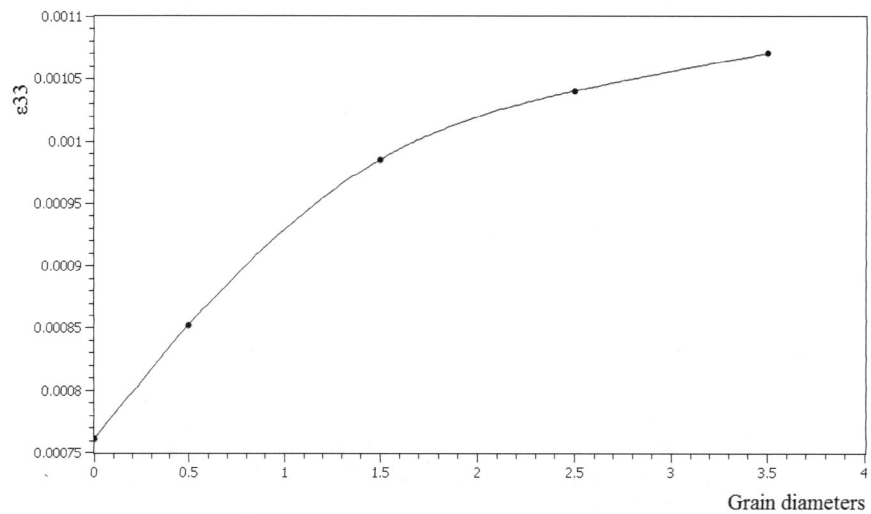

Figure 2. The dependence of the maximum strain on the distance to the surface.

4. Conclusions

A characteristic feature of polycrystalline materials is a high degree of heterogeneity of kinematic and force fields in combination with a stochastic internal structure. This heterogeneity is insignificant in some cases, but it becomes crucial in the study of various critical phenomena, the occurrence of which strongly depends on the microstructure of the material. In this work, extreme strains in grains of polycrystals are investigated using the field theory approach. In a zinc polycrystal, the grains of which have a high anisotropy, the maximum strains in the grain are 3.78 times greater than the macrostrain. Such a concentration is due exclusively to the elastic interaction of structure inhomogeneities. Maximum deformations in deep bulk grains are 40% higher than in surface ones. This may be the cause of the movement of sites of damage initiations from the surface inside the samples during the transition from high-cycle to ultra-high-cycle fatigue.

The calculations were performed on model polycrystals with cubic grains. Calculations on polycrystals with a different grain shape give similar results. Against the background of the geometric symmetry of the models, patterns of extreme microstructure of the environment of the grains generating extreme deformations clearly manifest themselves. The specific type of patterns depends on the anisotropy of the grains and the type of loading. In particular, it is theoretically predicted that maximum deformations under uniaxial tension arise in the center of the cluster from the nearest neighbors with the same orientation, that is, in fact, a large grain of a specific shape and orientation.

Acknowledgment
This work was supported by the Russian Foundation for Basic Research under grants 18-01-00675, 17-41-590433 and 19-41-590021.

References
[1] Pineau A, Benzerga A A, Pardoen T 2016 *Acta Materialia* **107** 424-483
[2] Pineau A, McDowell D L, Busso E P, Antolovich S D 2016 *Acta Materialia* **107** 484-507
[3] Das A. 2017 *Philosophical Magazine* **97** 867-916
[4] Fullwood D T, Niezgoda S R, Adams B L, Kalidindi S R 2010 *Prog. Mater. Sci.* **55** 477–562
[5] Oja M, Chandran K S and Tyron R J 2010 *Int. J. Fatigue* **32** 551-556
[6] Zimmermann M 2012 *Int. Mater. Revews* **57** 73-91
[7] Davidson D L, Tryon R G, Oja M, Matthews R and RaviChandra K S 2007 *Met. Mat. Trans.* **38A** 2214-25
[8] Abdolvand H, Wright J and Wilkinson A 2018 *Nature Communications* **9** 171
[9] Schuren J C, et all 2015 *Current Opinion in Solid State and Materials Science* **19** 235–244
[10] Guilhem Y, Basseville S, Gurtit F, Stephan J, Cailletaud G 2010 *Int. J. Fatigue* **32** 1748-76
[11] Shavshukov V E 2018 *Physical Mesomechanics* **21** 67-75
[12] deWit R 1960 *Solid State Physics* **10** 249-292
[13] Dederichs P H and Zeller R 1973 *Z.Physik* **259** 103-116
[14] Kroner E 1977 *J. Mech. Phys. Solids* **25** 137-155
[15] Mura T 1987 *Micromechanics of Defects in Solids* (Martinus Nijhoff Publishers)
[16] Tashkinov A A and Shavshukov V E 2018 *PNRPU Mechanics Bulletin* **1** 87-105
[17] Eshelby J D 1957 *Proc. R. Soc. A* **241** 376–396

IOP Conf. Series: Materials Science and Engineering **581** (2019) 012033 doi:10.1088/1757-899X/581/1/012033

Investigation of the Extrusion Process Influence on the Thermophysical, Rheological Properties and Operational Characteristics of Insulating Polymers

S V Ershov, N M Trufanova

Perm National Research Polytechnic University, 29, Komsomolsky ave., Perm, 614990, Russia

E-mail: Ershov_sv@bk.ru

Abstract. In this work, we carried out complex experimental studies of the thermophysical and rheological properties of the initial and secondary (after extrusion) polymer compositions based on low, medium and high density polyethylene, intended for the cable insulation production. On the basis of the obtained experimental data, a comparative analysis of the numerical study results of the flow processes in the extruder metering zone with a constant technological regime was carried out. Changes in the graphics of maximum and average temperature values along the length of the channel are received.

1. Introduction

At present, pure polymers such as polyethylene are rarely used in industry. They are replaced by polymer compositions with various fillers, which make it possible to give the finished product necessary performance characteristics. At the same time, depending on the composition, their thermophysical properties may differ significantly. Among the known methods of processing such materials, extrusion is the most frequently used method, which allows to continuously obtain products of a sufficiently large length [1].

In the modern cable industry during extrusion, along with the well-known problems caused by the occurrence of local overheating, there are also difficulties with utilization of the polymer materials that remain in the transition from one brand-size cable to another [2]. In this regard, it is important to study the extrusion process influence on the thermophysical, rheological properties and operatonal characteristics of polymer materials and to evaluate the possibility of their reuse in cable production.

2. Experiments

At the first stage, a study was conducted of the polymer materials thermal stability range on Discovery TGA manufactured by the TA Instruments [3]. Determination of the thermal stability range was carried out for polymer compositions based on low density (PE 153-10K) and high density polyethylene (PE 273-81K) of domestic production, and medium density polyethylene (Borealis ME 6052). The maximum heating temperature in experiment was 600 ˚C, at a heating rate of 10 ˚C /min. The results of the study are shown in the figure 1 (a).

The next step of the thermal paramete research was determination of the polymer compositions specific heat capacity (Figure 1,b) [4]. The study was conducted on the differential scanning calorimeter DSC Q2000 at a heating rate of 20 ˚C /min, in temperature range (50–200) °˚C.

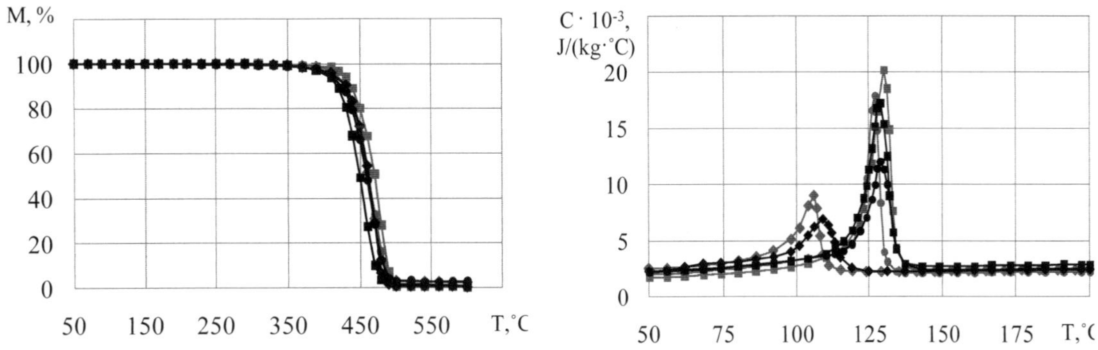

a b

Figure 1. Weight loss curves (a) and specific heat capacity temperature dependence (b) of the original polymer compositions: ● – Borealis ME 6052; ♦ – PE 153 - 10К; ■ – PE 273 - 81К; and secondary (after processing):● –Borealis ME 6052; ♦ – PE 153 - 10К; ■ – PE 273 - 81К.

Analysis of the curves in Figure 1,a showed that the main mass loss of all studied materials occurs at a temperature (400-500) °C. However, it is worth to note that the decomposition of the low molecular weight portion of each sample, which is approximately 5%, begins when the temperature value reaching 300 °C. When samples are heated to temperatures above 400 °C, decomposition of the remaining high molecular weight structure begins. That part of the material that did not burn at a temperature above 600 °C consists of mineral fillers, which are less in domestic polymer compositions (1%) than in foreign material Borealis ME 6052 (3%). It is also worth to note that the processing process does not affect on the thermal decomposition beginning temperature.

From the graph of specific heat capacity temperature dependence (Figure 1,b) it is clear that the phase transition to the molten state in case of low density polyethylene composition occurs at lower temperatures than the other two materials. The magnitude of the maximum value, as well as the curve shape, depends on the polymer material brand and the fillers included in its base. In the case of materials after processing, an increase in the maximum value of specific heat capacity is noted. Herewith in the melt state, all polymer compositions, both the original materials and the secondary materials, have similar average heat capacity values and low temperature dependence.

At the next stage of experimental studies, temperature dependences of thermal conductivity (Figure 2,a) and density (Figure 2,b) were obtained. The first parameter was determined on the FOX 50 analyzer, which is based on the stationary method of heat flux measuring [5]. To determine the temperature dependence of density, a horizontal type dilatometer DIL 802 was used. [6].

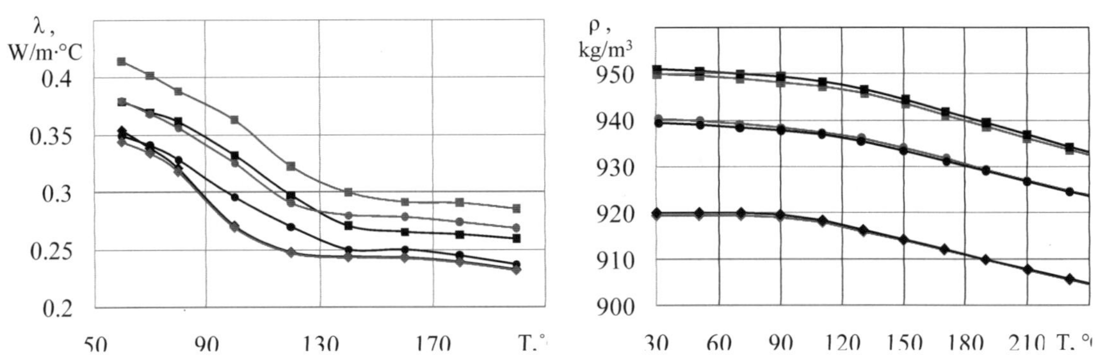

a b

Figure 2. Temperature dependence of thermal conductivity (a) and density (b) the original polymer compositions: ● – Borealis ME 6052; ♦ – PE 153 - 10К; ■ – PE 273 - 81К; and secondary (after processing):● –Borealis ME 6052; ♦ – PE 153 - 10К; ■ – PE 273 - 81К.

The difference in the thermal conductivity values of the materials in solid state is more significant (the maximum difference is 20%). In the melt state, the curve slope is changing, and with increasing temperature the difference in values decreases to 12%. It is also worth to note that the composition based on high density polyethylene in the entire temperature range has the highest value of thermal conductivity. A comparison of the obtained data allows us to conclude that in the case of medium and high density polyethylenes the processing process has a significant effect on the investigated parameters, in contrast to low density polyethylene.

Analyzing the curves of Figure 2,b, it can be noted that all materials in the solid state have a small temperature dependence of density (the value is less than 0.5%). During transition to the molten state, the slope of the curve increases, and the dependence becomes more pronounced (from 120 °C to 240 °C the density value decreases by 1.5%). It is also worth to note that all the studied polymer compositions in the liquid state have a linear temperature dependence of density. At the same time, the extrusion process practically does not affect on the value of studied parameter.

Analysis of the polymer materials rheological properties was carried out on a Discovery HR-2 rotary type rheometer manufactured by TA Instruments [7].

On Figure 3 the dependence of effective viscosity on the shear rate for the studied polymer compositions at a temperature of 200 °C are presented.

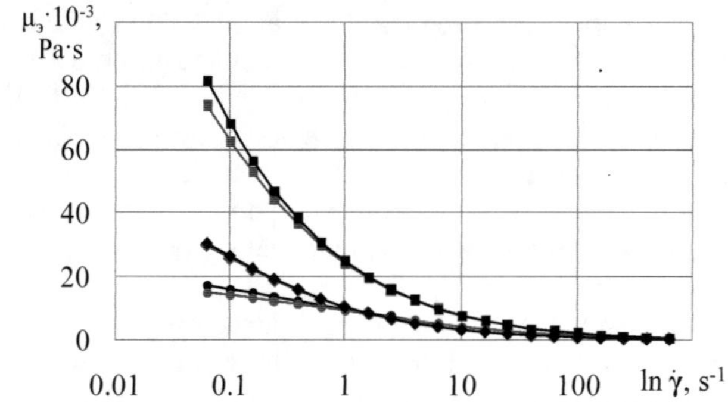

Figure 3. The dependence of the effective viscosity on the shear rate of the original polymer compositions at T = 200 °C: ● – Borealis ME 6052; ♦ – PE 153 - 10K; ■ – PE 273 - 81K; and secondary (after processing): ● –Borealis ME 6052; ♦ – PE 153 - 10K; ■ – PE 273 - 81K.

The Figure 3 shows that the extrusion process in the case of low density polyethylene (153-10K) has less influence on rheological parameters, in contrast to other two polymer compositions. The maximum difference is observed in the region of low shear rates.

Since the investigated polymer compositions are used as electrical cables insulation, their main operational parameters are the specific volume resistance R , dielectric constant ε and dielectric loss tangent $tg\delta$ [8,9].

Experimental determination of the dielectric constant and dielectric loss tangent was carried out using the MEP-5SA high-voltage AC bridge.

Table 1 presents the experimental data of the measured dielectrical parameters for the studied compositions before and after processing.

Table 1. Dielectrical properties of polymer compositions before and after processing.

Material	Measurements	R, TOhm·cm	$tg\delta$, $1\cdot10^{-4}$	ε
153-10K	before processing	>24000	6.0	2.24
	after processing	>24000	6.0	2.17
ME 6052	before processing	16890	5.3	2.41
	after processing	20438	4.3	2.26
273-81K	before processing	>24000	2.6	1.54
	after processing	>24000	2.0	2.21

Analyzing the obtained data, we can note that the processing process does not have a significant effect on the electrical properties of polymer compositions. All parameters are within acceptable limits, in accordance with the regulatory documentation for these materials.

In addition to the dielectric properties, the cables and wires polymer insulation must be resistant to mechanical stress, and at the same time provide the product necessary flexibility. For these purposes tensile tests are carried out.

An experimental study of the materials mechanical properties was carried out on a Walter Bai LFM-50-T tensile testing machine. [10]. For testing were used samples in the form of "blades" with a length of 115mm. The resulting stress-strain curves are shown below (Figure 4).

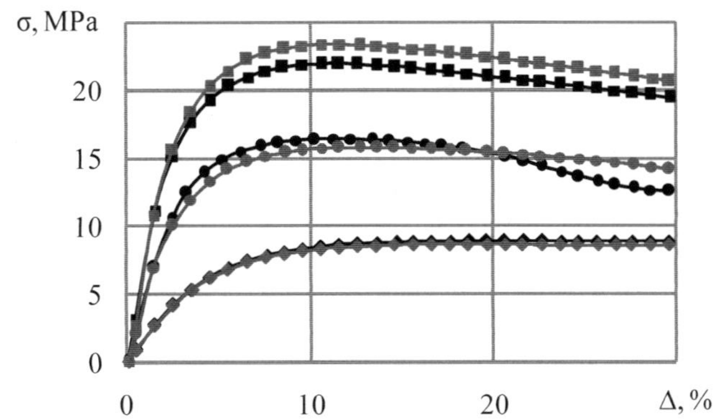

Figure 4. Experimental dependence of stress on strain of the original polymer:
● – Borealis ME 6052; ♦ – PE 153 - 10K; ■ – PE 273 - 81K;
and secondary (after processing): ● –Borealis ME 6052; ♦ – PE 153 - 10K; ■ – PE 273 - 81K.

A comparative analysis of the investigated materials mechanical properties suggests that the processing process has almost no effect on the polymers strength characteristics. The greatest difference in the maximum stress value was 4% for the polymer composition of the brand 273-81K.

3. Mathematical modeling

An experimental study of the extruded material properties showed that the processing process for a number of polymer compositions leads to a significant change in the thermal characteristics, especially thermal conductivity. In this regard, it is important to evaluate the behavior of the material during reprocessing and determine the rational mode of the extrusion.

To study the processes of flow and heat transfer in the metering zone of an extruder with an internal cylinder diameter of 160 mm, a screw channel model unrolled onto a plane was used. In the construction of spatial mathematical models of flow in the extruder metering zone following assumptions were made: the process is stationary and steady at constant mass flow, the polymer melt is considered to be a purely viscous, incompressible liquid, mass forces are equal to zero [11]. A more detailed mathematical formulation of the problem and the extruder channel geometrical parameters are described in the article [12].

Modeling was performed for the following technological processing mode: the body temperature was set at 200 °C; the temperature of the screw is 210°C; the temperature of the input material is 225°C; screw rotation speed was 60 rpm. This mode was used both for original polymer and for the material after first processing cycle.

Table 2 presents the values of temperature, viscosity, shear rate and pressure drop for all materials studied during both cycles primary and reprocessing.

IOP Conf. Series: Materials Science and Engineering **581** (2019) 012033 doi:10.1088/1757-899X/581/1/012033

Table 2. Changes in pressure, maximum and average temperatures, shear rates and viscosities for various polymer compositions

Material	Processing cycle	T, [°C]		γ, [s^{-1}]	η, [Pa·s]	Δ P, [MPa]
		Max.	Ave.	Ave.	Ave.	
153-10K	first	253.0	231.6	134.3	759.5	40.1
	second	249.3	229.5	134.5	684.9	35.3
ME 6052	first	273.4	243.5	135.2	1038	68.0
	second	257.3	233.6	136.0	841.9	53.8
273-81K	first	309.0	263.1	135.2	1890	112.7
	second	312.5	263.6	135.2	1938	116.6

Analyzing the obtained results, it can be noted that despite the unchanged technological mode parameters during the materials processing, the values of temperature, viscosity and pressure, that are realized in the extruder metering zone, differ significantly. Whereas the values of the shear rate in the channel remain unchanged. The greatest decrease in polymer temperature at the exit from the dosing zone was observed when reprocessing polymer Borealis ME 6052. The value of the maximum temperature at the channel exit decreases by 16 °C, and the average – by 10 °C.

4. Conclusions

As a result of experimental studies, it was found that the processing process does not lead to a significant change in the materials operational characteristics. It can be concluded that polyethylene-based polymer compositions can be reused in the manufacture of cables and wires electrical insulation. However, there are a number of important conditions, without which it is impossible to obtain high-quality finished product: All material should be collected in clean metal containers, excluding the possibility of various impurities, and the second – is ensuring the correct extrusion technological mode, without exceeding the temperature of thermal destruction.

Obtained in the course of numerical studies values of temperature, shear rate, viscosity and pressure drop allow us to conclude, that it is possible to re-process the investigated polymer compositions at an unchanged technological mode. At the same time, for polymer brand ME 6052, a decrease in the heaters temperature in extruder metering zone is allowed.

References

[1] Rauwendaal C 2014 *Polymer Extrusion.*(Hanser Publications) 950

[2] Maria del Pilar Noriega E end Rauwendaal C 2010 *Troubleshooting the Extrusion Process: A Systematic Approach to Solving Plastic Extrusion Problems.* (Hanser Publications) 194

[3] ISO 7111-87. Plastics. Thermogravity of polymers. Temperature scanning method

[4] ISO 11357-4:2005 Plastics – Differential scanning calorimetry (DSC) – Part 4: Determination of Specific Heat Capacity

[5] Callister W D and Rethwiisch D G 2007 *Fundamentals of materials science and engineering : an integrated approach* (Wiley) 912

[6] Shah V 2007 *Handbook of Plastics Testing and Failure Analysis, 3rd Edition* (Wiley) 648

[7] Malkin A Y and Isaev A I 2007 *Rheology: concepts, methods, and applications*. (St. Petersburg: Profession) 184

[8] GOST 22372-77. Dielectric materials. Methods of determination of permittivity and powerfactor with in a frequency range of 100 to 5x10 in 6 degree Hz

[9] GOST R 50499-93 (IEC 93-80) Solid electrical insulating materials. Methods of test for volume resistivity and surface resistivity

[10] ISO 527-2:2012, Plastics — Determination of tensile properties — Part 2: Test conditions for moulding and extrusion plastics

[11] Rauwendaal C 2004 Finite element studies of flow and temperature evolution in single screw extruders *Plastics, Rubbers and Composites* **33 (9/10)** 390–369

[12] Trufanova N M and Ershov S V 2017 Comparative analysis of heat and mass transfer processes in the extruder metering zone with the use of different spatial mathematical models and rheological law *Computational continuum mechanics* **10 (2)** 154-164

IOP Conf. Series: Materials Science and Engineering **581** (2019) 012034 doi:10.1088/1757-899X/581/1/012034

Geometrically Nonlinear Constitutive Equations of the Plastic Flow Theory in Terms of Asymmetric Stress and Strain Measures

A Yu Yants [a], P V Trusov
Perm National Research Polytechnic University, Komsomolsky pr. 29a, Perm, Russia

E-mail: [a] maximus5.59@gmail.com

Abstract. A formulation of the geometrically nonlinear plastic flow theory (PFT) based on asymmetric measures of stress and strain states is proposed. A main emphasis is placed on the physically reasonable decomposition of the deformation gradient into three components: elastic distortions, which determine stresses, an orthogonal tensor characterizing the quasi-rigid motion of a material and the plastic strain gradient. The quasi-rigid motion of the material is defined by introducing for a representative volume element a generalized lattice, which represents its symmetry elements. The hypoelastic anisotropic law is introduced in terms of the movable coordinate system associated with the material. The rate of plastic deformations is determined by the associated law of plastic flow. As a result, the closed system of constitutive equations of the geometrically nonlinear PFT of is obtained.

1. Introduction

Currently, there is a high need for physically correct description of severe deformation processes (SPD). The improvement of SPD technologies leads to the need to solve optimization problems that require the implementation of a variety of calculations with different process parameters. To the solve of corresponding boundary value problems, it is necessary to use computational effective mathematical models. At the same time, these material models must be physically relevant. Special attention should be paid to the correct description of the geometric nonlinearity arising in the study of SPD processes using the methods of deformable solid body mechanics. Usually, various modifications of classical plastic flow theories based on symmetric of stress and strain measures are used to describe of SPD. The use of a symmetric measure of the strain state (or its velocity) excludes any information about the quasi-rigid material motion and the effects associated with the material rotation. In the present work, a modification of the classical plastic flow theory based on asymmetric stress and strain measures. Correct geometric nonlinearity description is physically ground separation movement on quasirigid (rotational) and deformation. Separating the quasi rigid motion is necessary to introduce the correct constraint between the stress and the strain in terms of a rigid moving coordinate system (MCS). The complete motion of the continuum is represented as a combination of a quasi-rigid motion of the material (together with the MCS) and the deformation one (with respect to the MCS). The stress measure is the Cauchy stress tensor, defined in terms of the MSC. The transposed velocity gradient is the total strain measure. The quasi-rigid material rotation should not cause changes in the response of the material (for the observer in the MCS), so quasi-rigid motion must be excluded from the complete motion; only the strain movement effects on the material response. Determining the quasi-solid motion of the continuum is a difficult due to it is impossible to select an undeformable (throughout the whole process) aggregate of any material elements, preferably connected with axes or symmetry planes.

IOP Conf. Series: Materials Science and Engineering **581** (2019) 012034 doi:10.1088/1757-899X/581/1/012034

2. Continuum kinematic

2.1. Multiplicative decomposition of the deformation gradient

The classical multiplicative decomposition of the deformation gradient into elastic $\mathbf{F}^e$ and plastic $\mathbf{F}^p$ components [1] is represented by the relation:

$$\mathbf{F} = \mathbf{F}^e \cdot \mathbf{F}^p, \tag{1}$$

where the inelastic component $\mathbf{F}^p$ connects the basis vectors of the Lagrangian coordinate system in the reference and the plastically deformed configurations.

In [2, 3], for crystalline materials with an internal structure, a multiplicative decomposition different from the classical one (1) was proposed. The motion of deformable crystalline body, which is described by the deformation gradient $\mathbf{F}$, is represented by the sequence of plastic deformations (it is assumed that plastic deformations do not rotate the MCS associated with the lattice of the crystallite), a rotation of the MCS (together with the material) and an elastic distortion of the material with respect to the MCS:

$$\mathbf{F} = \bar{\mathbf{F}}^e \cdot \mathbf{R} \cdot \mathbf{F}^p, \tag{2}$$

where the inelastic component $\mathbf{F}^p$ connects the reference and plastically deformed (intermediate unloaded) configurations, $\mathbf{R}$ is the orthogonal tensor, which converts the reference basis of the rigid moving coordinate system into the current (rotating tensor of the MCS from the reference to actual configuration), $\bar{\mathbf{F}}^e$ – the deformation gradient, which transforms the plastically deformed configuration, which has experienced the quasi-rigid motion, into the actual configuration. It should be noted that determining the relationship between the MCS motion and the continuum (orthogonal transformation $\mathbf{R}$ at any time) and the initial orientation of the MCS for the continuum that does not have an internal structure is difficult, moreover, it has many possible solutions.

It is known that polycrystalline materials, in particular, metals, consist of separate crystallites with a certain structure. For a single crystallite, inelastic deformations realized by the motion of dislocations do not lead to the rotation of the crystal lattice [2, 3]. In the case of a polycrystalline material, the actual plastic deformations of each individual crystallite, realized by yields along crystal slip systems (SS), also do not lead to rotate of lattice. For the polycrystalline, it is possible to introduce a "generalized crystal lattice" (GL), which characterizes the average (by representative volume) symmetry properties of the material. Thus, the plastic deformation of a polycrystalline, described by the plastic component $\mathbf{F}^p$ determined by the plastic deformations of crystallites of the representative volume. It is assumed, that $\mathbf{F}^p$ (by analogy with the crystallites) not to turn the "generalized crystal lattice" of the representative volume element (RVE). The MCS is associated with the introduced "generalized lattice" of the representative volume element. The MCS is responsible for the quasi-rigid motion of the representative volume element. As a result, relation (2) represents the movement of the polycrystalline by the sequence of plastic deformation, described by the plastic component $\mathbf{F}^p$, carried out without distortion and rotation of the "generalized lattice" (in other words, it is assumed that during plastic deformation the material "flows" through the "generalized lattice", leaving the latter invariant), and the affine transformation of $\mathbf{F}^e$, which includes the rotation of $\mathbf{R}$ and the distortion of this lattice, described by $\bar{\mathbf{F}}^e$.

We note that the "generalized lattice" is experiencing both rotation and distortion, which are collectively described by the elastic component of the position gradient $\mathbf{F}^e$ (1), while the MCS is experiencing only rotation (the translation motion of the MCS do not lead to a change of basis, and therefore are not considered further). The decomposition $\mathbf{F}^e = \bar{\mathbf{F}}^e \cdot \mathbf{R}$ differs significantly from the classical polar decomposition, since $\bar{\mathbf{F}}^e$ is not symmetric in the general case and is calculated after the determination of the quasi-rigid motion of the MCS. By analogy with the deformation of single crystals [2], when determining the quasi-hard rotation, it is proposed to exclude purely plastic deformation.

2.2. Additive decomposition of the strain rate measure in the current configuration

The introduced multiplicative decomposition (2) is used to obtain the additive decomposition of the transposed gradient of total velocity in the current configuration:

$$\mathbf{L} = \hat{\nabla}\mathbf{V}^T = [illegible] + \bar{\mathbf{F}}^e \cdot \mathbf{\Omega} \cdot \bar{\mathbf{F}}^{e-1} + \bar{\mathbf{F}}^e \cdot \mathbf{R} \cdot \mathbf{L}^p \cdot \mathbf{R}^T \cdot \bar{\mathbf{F}}^{e-1}, \tag{3}$$

where $\mathbf{V}$ – is velocity of continuum points, $\hat{\nabla}$ – the operator nabla in the current configuration, $\bar{\mathbf{L}}^e = [illegible]$, $\mathbf{L}^p = [illegible]$. From decomposition (2) it follows that $\bar{\mathbf{F}}^e$ reflects only elastic distortions of the GL. In the case of large gradients of displacements and severe deformations of metals, the elastic distortions are small compared to the rotations **R** and the plastic deformations $\mathbf{F}^{\mathrm{p}}$. In this regard, $\bar{\mathbf{F}}^{\mathrm{e}}$ is close to unity, and the unloaded configuration consists with current configuration. However, the rate $\dot{\mathrm{i}}$ cannot be neglected; it is comparable to the values of the other terms in the decomposition (3). The additive decomposition is introduced:

$$\mathbf{Z} = \mathbf{Z}^{\mathrm{e}} + \mathbf{Z}^{\mathrm{in}}, \tag{4}$$

where, based on [2], $\mathbf{Z} = \mathbf{L} - \boldsymbol{\Omega}$, $\mathbf{Z}^{\mathrm{in}} = \mathbf{R} \cdot \mathbf{L}^{\mathrm{p}} \cdot \mathbf{R}^{\mathrm{T}}$, $\mathbf{Z}^{\mathrm{e}} = \dot{\mathrm{i}} \cdot \cdot$ – are measures of velocities of complete, inelastic and elastic distortions, defined in terms of the actual (unloaded) configuration, $\boldsymbol{\Omega}$ is the spin tensor of MCS tensor, $\mathbf{L}^{\mathrm{p}}$ is the plastic component of the displacement velocity gradient, defined in terms of the reference configuration.

Following [2], the defining relation is introduced:

$$\boldsymbol{\Sigma}^{\mathrm{CR}} = \mathrm{d}\boldsymbol{\Sigma} / \mathrm{d}t + \boldsymbol{\Sigma} \cdot \boldsymbol{\Omega} - \boldsymbol{\Omega} \cdot \boldsymbol{\Sigma} = \bar{\boldsymbol{\Pi}} : \mathbf{Z}^{\mathrm{e}}. \tag{5}$$

where $\boldsymbol{\Sigma}$ is the Cauchy stress tensor, $\bar{\boldsymbol{\Pi}} = \bar{\Pi}_{ijkl}\mathbf{k}^i\mathbf{k}^j\mathbf{k}^k\mathbf{k}^l$ is a fourth-rank tensor of elastic properties, symmetric with respect to pairs of indices, whose components are constant in the MCS basis $\{\mathbf{k}^i\}$.

3. The plastic flow theory based on asymmetric measures

In this section, the constitutive equations of the plastic flow theory based on asymmetric strain and stress rate measures are introduced. The previously obtained additive decomposition of the strain rate measure into the elastic and inelastic components (4) is used. The deviators of stress and strain measures and their velocities are used to derive the equations of the plasticity theory. The deviators of the Cauchy stress tensor $\boldsymbol{\Sigma}$ and the velocity gradient of relative displacements $\mathbf{Z}$ are introduced:

$$\begin{aligned} \mathbf{Z}' &= \mathbf{Z} - 1/3\mathrm{I}_1(\mathbf{Z})\mathbf{E} = \mathbf{L} - \boldsymbol{\Omega} - 1/3\mathrm{I}_1(\mathbf{L} - \boldsymbol{\Omega})\mathbf{E} = \mathbf{L} - \boldsymbol{\Omega} - 1/3\mathrm{I}_1(\mathbf{L})\mathbf{E} = \mathbf{L}' - \boldsymbol{\Omega}, \\ \boldsymbol{\Sigma}' &= \boldsymbol{\Sigma} - 1/3\mathrm{I}_1(\boldsymbol{\Sigma})\mathbf{E}, \end{aligned} \tag{6}$$

where **E** is the isotropic second rank tensor.

As the intensities of measures, the classical expressions are used, extended for the asymmetric case:

$$\begin{aligned} \Sigma_u &= \sqrt{3/2\boldsymbol{\Sigma}' : \boldsymbol{\Sigma}'^{\mathrm{T}}}, \\ \mathrm{Z}_u &= \sqrt{2/3\mathbf{Z}' : \mathbf{Z}'^{\mathrm{T}}}, \\ \lambda &= \int \sqrt{2/3\mathbf{Z}^{\mathrm{in}} : \mathbf{Z}^{\mathrm{in}\,\mathrm{T}}}\,\mathrm{d}\tau. \end{aligned} \tag{7}$$

It should be noted that the scalar measures of stresses and the accumulated strain λ need further refinement in the case of severe deformations in order to correctly introduce the single curve hypothesis. In numerous articles [4-6], there are experimental data indicating that it is impossible to introduce a single dependence $\Sigma_u(\lambda)$ for all loading types without significant errors (up to 20%).

As a basis for constructing the asymmetric plasticity theory, the associated law of flow is used [7]:

$$\mathbf{Z}^{\mathrm{in}} = \frac{\boldsymbol{\Sigma}'}{\Sigma_u} \dot{\cdot}, \tag{8}$$

where Σ_u=R(λ) during an active loading and R(λ) determines the single dependence of the intensity of the flow stresses on the accumulated plastic strain [7].

3.1. The derivation of the constitutive equation in the modified plastic flow theory

To derive the constitutive equation connecting the stress and strain rates, we will follow the scheme outlined in [8]. The aim of the subsequent transformations is to determine the tensor of elastoplastic properties, which connects the measures of these rates. Using the yield criterion Σ_u=R(λ), we determine the change rate of λ during an active plastic deformation $\left(\boldsymbol{\Sigma}' : \mathbf{Z}^{\mathrm{in}\,\mathrm{T}} \geq 0\right)$:

$$\frac{d}{dt}\mathrm{R}(\lambda)=\mathrm{R}_\lambda(\lambda)\dot{\lambda}=\frac{\left(3/2\,\boldsymbol{\Sigma}':\boldsymbol{\Sigma}'^{\mathrm{T}}\right)^{\bullet}}{2\sqrt{3/2\,\boldsymbol{\Sigma}':\boldsymbol{\Sigma}'^{\mathrm{T}}}}=\frac{1}{2\Sigma_u}3/2\left(\dots\right)$$

$$=\frac{1}{2\Sigma_u}3/2\left(\dots\right)\quad\frac{}{2\Sigma_u}\left(\dots\right)\quad\frac{}{2\;\Sigma_u}, \tag{9}$$

$$\dot{\lambda}\ \ \frac{}{\mathrm{R}_\lambda(\lambda)}=\frac{}{2\,\Sigma_u\mathrm{R}_\lambda(\lambda)},$$

where $\dot{\Sigma}$ is the total material derivative of the Cauchy stress tensor deviator. The elastic relation (5) determines the rate of the total stresses, and not their deviator part, relationship is needed:

$$(\boldsymbol{\Sigma}')^{\bullet}=\left(\boldsymbol{\Sigma}-1/3(\boldsymbol{\Sigma}:\mathbf{E})\mathbf{E}\right)^{\bullet}=\dot{\Sigma}\ \dots \overset{def}{=}\dots$$

$$=\dot{\Sigma}\dots(\mathbf{C}_{\mathrm{III}}\dots\mathbf{C}_{\mathrm{I}}), \tag{10}$$

where **E**= $\mathbf{e}_i\mathbf{e}^i$ is the isotropic tensor of the second rank, $\mathbf{C}_{\mathrm{I}}=\mathbf{EE}$, $\mathbf{C}_{\mathrm{III}}=\mathbf{e}_i\mathbf{e}^j\mathbf{e}_j\mathbf{e}^i$ are isotropic tensors of the fourth rank. From relations (8)-(10), the relationship between the rate of plastic strain and the stress rate is obtained:

$$\mathbf{Z}^{\mathrm{in}}=\frac{\boldsymbol{\Sigma}'}{\Sigma_u}\dot{\lambda}=\frac{\boldsymbol{\Sigma}'}{\Sigma_u}\frac{\dots}{2\,\Sigma_u\mathrm{R}_\lambda(\lambda)}=\frac{3}{2}\frac{\boldsymbol{\Sigma}'}{\Sigma_u}\frac{\boldsymbol{\Sigma}'^{\mathrm{T}}:\dot{\Sigma}}{\Sigma_u\mathrm{R}_\lambda(\lambda)}=\mathbf{P}:\dot{\Sigma},$$

$$\mathbf{P}\overset{o6}{=}\frac{3}{2}\frac{\boldsymbol{\Sigma}'\boldsymbol{\Sigma}'^{\mathrm{T}}}{\Sigma_u^2}\frac{1}{\mathrm{R}_\lambda(\lambda)}=\frac{3}{2}\frac{\boldsymbol{\Sigma}'\boldsymbol{\Sigma}'^{\mathrm{T}}}{\Sigma_u^2}\frac{1}{\mathrm{R}_\lambda(\lambda)}, \tag{11}$$

where **P** is the fourth rank tensor, which completely determines the inelastic properties of the material when the flow stress is reached with active loading; this tensor includes the tensor product of the stress tensor $\boldsymbol{\Sigma}'/\Sigma_u$ on itself and the value of the instantaneous hardening modulus $\mathrm{R}_\lambda(\lambda)$.

The rate of elastic deformations is expressed from the elastic relation (5):

$$\mathbf{Z}^{\mathrm{e}}=\bar{\boldsymbol{\Pi}}^{-1}:\boldsymbol{\Sigma}^{\mathrm{CR}}. \tag{12}$$

Considering the additivity of strain rates (4) and the expressions for the component strain rates (11), (12):

$$\mathbf{Z}=\bar{\boldsymbol{\Pi}}^{-1}:\boldsymbol{\Sigma}^{\mathrm{CR}}+\mathbf{P}:\dot{\Sigma}, \tag{13}$$

whence, using the expression for the corotational derivative (5), the total rate of change of stresses from the point of view of a fixed observer in the laboratory coordinate system (LCS) is expressed:

$$\mathbf{Z}=\bar{\boldsymbol{\Pi}}^{-1}:\left(\dot{\Sigma}\dots\right)\dots\dot{\Sigma},$$

$$\mathbf{Z}=\bar{\boldsymbol{\Pi}}^{-1}:\left(\dot{\Sigma}\dots\right)\dots(\mathbf{C}_{\mathrm{III}}\dots\mathbf{C}_{\mathrm{I}})\dots, \tag{14}$$

$$\mathbf{Z}-\bar{\boldsymbol{\Pi}}^{-1}:\boldsymbol{\Sigma}^{\Omega}=\left(\bar{\boldsymbol{\Pi}}^{-1}+\mathbf{P}:\left(\mathbf{C}_{\mathrm{III}}-1/3\mathbf{C}_{\mathrm{I}}\right)\right):\dot{\Sigma},$$

where the designation of the term $\boldsymbol{\Sigma}^{\Omega}\overset{def}{=}\boldsymbol{\Sigma}\cdot\boldsymbol{\Omega}-\boldsymbol{\Omega}\cdot\boldsymbol{\Sigma}$ characterizing the geometric nonlinearity is introduced. From the relation (14), the final expression for the stress rate with active loading is obtained, i.e. when the imaging stress point is found on the yield surface and the condition $\boldsymbol{\Sigma}':\mathbf{Z}^{\mathrm{inT}}\geq 0$ [9] is fulfilled:

$$\dot{\Sigma}\ \left(\dots\mathbf{P}:\left(\mathbf{C}_{\mathrm{III}}-1/3\mathbf{C}_{\mathrm{I}}\right)\right)^{-1}:\left(\mathbf{Z}-\bar{\boldsymbol{\Pi}}^{-1}:\boldsymbol{\Sigma}^{\Omega}\right),$$

$$\text{by conditional: } \Sigma_u=\mathrm{R}(\lambda),\ \boldsymbol{\Sigma}:\mathbf{Z}^{\mathrm{inT}}\geq 0. \tag{15}$$

In the case of elastic deformation, when the stress vector is located inside the yield surface, the plastic deformations are identically zero, then the connection between the stress rate and the total strain rate is performed in the following form:

$$\boldsymbol{\Sigma}^{\mathrm{CR}}=\bar{\boldsymbol{\Pi}}:\mathbf{Z}^{\mathrm{e}}=\bar{\boldsymbol{\Pi}}:\mathbf{Z},$$

$$\text{by conditional: } \Sigma_u<\mathrm{R}(\lambda). \tag{16}$$

$$\text{by conditional: } \Sigma_u=\mathrm{R}(\lambda),\ \boldsymbol{\Sigma}:\mathbf{Z}^{\mathrm{inT}}<0.$$

The final set of relations of the geometrically non-linear plastic flow theory is:

$$
\begin{aligned}
&\mathbf{Z} = \mathbf{L} - \mathbf{\Omega},\\
&\mathbf{\Sigma}^{\Omega} \overset{def}{=} \mathbf{\Sigma}\cdot\mathbf{\Omega} - \mathbf{\Omega}\cdot\mathbf{\Sigma},\\
&\Sigma_u = \sqrt{3/2\,\mathbf{\Sigma}' : \mathbf{\Sigma}'^{\mathrm{T}}},\\
&\mathbf{Z}^{\mathrm{in}} = \mathbf{P} : \mathbf{\Sigma}'^{\mathrm{CR}}, \quad \mathbf{P} = \frac{3}{2}\frac{\mathbf{\Sigma}'\mathbf{\Sigma}'^{\mathrm{T}}}{\Sigma_u^2}\frac{1}{\mathrm{R}_{\lambda}(\lambda)},\\
&(\mathbf{\Sigma}')^{\mathrm{CR}} = \dot{\Sigma}\ \text{[illegible]}\ /3\mathbf{C}_{\mathrm{I}}) + \mathbf{\Sigma}^{\Omega},\\
&\lambda = \int \sqrt{2/3\,\mathbf{Z}^{\mathrm{in}} : \mathbf{Z}^{\mathrm{in}}}\,\mathrm{d}\tau,\\
&\begin{cases}
\Sigma_u < \mathrm{R}(\lambda) \text{ or } \Sigma_u = \mathrm{R}(\lambda), \mathbf{\Sigma} : \mathbf{Z}^{\mathrm{inT}} < 0 \ : \ \dot{\Sigma}\ \text{[illegible]}\ \mathbf{\Sigma}^{\Omega},\\
\Sigma_u = \mathrm{R}(\lambda), \mathbf{\Sigma} : \mathbf{Z}^{\mathrm{inT}} \geq 0 \ : \ \dot{\Sigma}\ \text{[illegible]}\ \mathbf{P} : (\mathbf{C}_{\mathrm{III}} - 1/3\mathbf{C}_{\mathrm{I}}))^{-1} : (\mathbf{Z} - \bar{\mathbf{\Pi}}^{-1} : \mathbf{\Sigma}^{\Omega}).
\end{cases}
\end{aligned}
\tag{17}
$$

3.2. On quasi-solid motion

It should be noted that the main problem in the construction of the proposed model is the determination of the value of the spin tensor $\mathbf{\Omega}$ of MCS of representative volume element of the material at any time. As a solution of this problem, it is proposed to rely on the introduction of a generalized lattice of representative volume element, in which the MCS is rigidly associated with one direction and the plane of the generalized lattice. However, this does not take into account the orientation of the axes of symmetry at the reference configuration (the basis GL is assumed to coincide with the LCS), as well as at an arbitrary moment of process. It is obvious that the symmetry properties of the representative volume of the polycrystalline are determined by the law of distribution of orientations (changing in the process of deformation) in the current configuration and by the symmetry properties and orientations of its individual elements, i.e., crystallites. At the current stage of development of a modification of the PFT model, it is proposed an alternative method for determining the MCS spin tensor, assuming the use of the two-level crystal plasticity model [3]. The developing approach [10] is used to determine symmetry properties of the representative volume element. This approach makes it possible to determine compliance with the selected classes of symmetry of the representative volume element of the polycrystalline at any time during the process of arbitrary treatment. However, if at the initial moment the polycrystalline does not possess anisotropy of elastic properties, for which it is not possible to select character directions, the MCS is chosen to coincide with the LCS. In the future, it is planned to develop a macromodel to describe the evolution of symmetry properties. For identifying and verifying the parameters of the macromodel the two-level crystal plasticity model will used. The advantage of this approach is binding to the symmetry elements of the material rather than the objects induced by the loading kinematics, e.g., the polar decomposition [11] or the skew-symmetric part of the velocity gradient [12, 13], where the PCS is associated with the eigenvectors of certain strain measures that are not related to the material. The widely used logarithmic spin [14] allows to avoid the dissipation of the elastic energy on closed cycles (in the elastic region), as well as stress oscillations. However, the derivation of the logarithmic spin is based on the assumption of elastic isotropy, and thus it cannot be associated with the symmetric elements of the material. An approach, which is the most similar to the proposed one, is described in [15, 16], where the concept of a reference generalized lattice with corresponding material vectors is introduced. The evolution of these vectors is determined by the elastic component of the velocity gradient, the definition of which in the framework of the symmetrized measures raises questions undiscussed in the cited articles.

4. Conclusion

In this work, the structure of the macro-phenomenological geometrically non-linear plastic flow theory based on asymmetric stress and strain measures is presented. By adopting the Mises-Hyber-Hencky yield criterion, the associated plasticity law and the anisotropic hypoelastic law, conclusive resolving equations are obtained for the change rate of stresses. The developed model of elastoplastic material will be further used in solving boundary-value problems corresponding to real technological processes.

5. Acknowledgement
The work was supported by the Russian Federation Ministry of Science and Higher Education (basic part of the PNRPU state assignment, No. 9.7434.2017/8.9) and the Russian Foundation for Basic Research (grant No. 17-01-00379-a).

References
[1] Lee E H 1969 *ASME J. Appl. Mech.* **36** 1–6
[2] Trusov P V, Shveykin A I, Kondratev N S 2017 *NTIJ* **8** 133–166
[3] Shveykin A I, Trusov P V 2019 *NTIJ* **10** 1–20
[4] Anand L 2004 *Comput. Methods Appl. Mech. Engrg.* **193** 5359–83.
[5] Bronkhorst C A, Kalidindi S R, Anand L 1992 *Phil. Trans. Roy. Soc. Lond. A* **341** 443–477
[6] Tomé C, Canova G R, Kocks U F, Christodoulou N, Jonas J J 1984 *Acta Metall.* **32** 1637–53
[7] Malinin N N 1975 *Applied theory of plasticity and creep. Textbook for students and universities* (Moscow, Mechanical Engineering)
[8] Yamada Y, Yoshimura N, Sakurai T 1968 *Int. J. Mech Sci.* **10** 343–354.
[9] Birger I A, Shorr B F, Demyanushko I V 1975 *Thermal strength of machine parts. Theory. Experimental studies. Calculation.* (Moscow, Mechanical Engineering)
[10] Trusov P V, Ostapovich K V 2017 *Symmetry* **9** 1-28
[11] Green A E, Naghdi P M 1965 *Arch. Ration. Mech. Analysis.* **18** 251–281
[12] Zaremba S 1903 *Bull. Int. Acad. Sci. Cracovie.* 595–614.
[13] Jaumann G *Sitzber. Akad. Wiss. Wien, Abt. IIa.* 1911 B **120** 385-530
[14] Xiao H, Bruhns O T, Meyers A 1997 *Acta Mechanica.* **124** 88-105
[15] Rubin M B 1994 Int J Solids Struct. **31** 2615-34
[16] Rubin M B 1994 Int J Solids Struct. **31** 2635-52

Heat Transfer by Steady Streaming in a Horizontal Annulus with Inner Heating at Boundary Vibration

N V Kozlov

Institute of Continuous Media Mechanics UrB RAS,
1, Ac. Koroleva St., Perm, 614013, Russia

E-mail: kozlov.n@icmm.ru

Abstract. Convection and heat transfer are studied experimentally in a horizontal annulus, the outer wall of which is formed by an elastic silicone tube, and the inner one – by a copper cylindrical heat exchanger. The working volume is filled with a viscous fluid. The temperature of the outer boundary is kept constant. At the inner boundary, a constant heat generation power is maintained. The elastic outer wall is driven in a periodic motion by two linear motors, installed symmetrically to the left and to the right of the container and performing synchronous oscillations. Thermal measurements are carried out at steady state temperature. In the absence of vibrations, free convection is observed. Vibrations lead to the generation of steady averaged flows in the form of a system of eight pairs of rolls, parallel to the cylinder axis. Vibrational convection enhances heat transfer across the annulus. The structure of steady flows and the rate of heat transfer are determined by the interaction between two convective mechanisms.

1. Introduction

One of the actual challenges for the modern industry is the intensification of heat and mass transfer in fluids. Succeeding in this suggests opportunities in product quality enhancement or in production costs' reduction. Traditional methods, for example mechanical stirring, may have limitations in quite a number of hydrodynamic systems, which include droplets, microfluidic devices and others. An outstanding approach to fluid mixing is steady streaming [1] that is generated in dynamic viscous boundary layers, also referred to as Stokes layers, due to fluid oscillations [2].

It is known that at free oscillations of rising (falling) droplets in extraction columns, the mass transfer coefficient increases with the dimensionless oscillation frequency, see for example [3, 4]. Oscillations of a drop lead to the generation of steady flows both inside it and in the surrounding fluid [5]. This leads to the intensification of mass transfer [6–8]. For example, flow stagnation inside a droplet due to surfactant accumulation on its surface [9] can be successfully compensated by steady streaming. Another example is an original experiment [10] whose authors studied the heat transfer from a hot upper boundary of a vertical cylinder by steady streaming from an oscillating disk: the vibrational mixing produced an efficient heat transfer and a homogeneous temperature distribution in the fluid bulk. As found in theoretical studies [5, 7], steady flows in an oscillating droplet have the shape of toroidal vortices and lead to the interface renewal by tangential time-average flows. In recent experiments on fluid oscillations in cylindrical and spherical containers with elastic boundaries, similar steady flows were modeled [11, 12]. The experiments were systematically conducted in a large range of dimensionless frequencies, and the impact of the latter on the flow structure was theoretically analyzed.

The purpose of this study is to find out how the structure and velocity of convection are related to the contribution of the steady streaming to heat transfer under the condition of the simultaneous existence of free convection. A horizontal annulus filled with a viscous fluid is chosen as experimental model. Its inner boundary is rigid and hot, while the outer one is elastic and is kept at a constant temperature via the water cooling. It is known that in a horizontal annulus with a hot inner cylinder, natural thermal convection develops in the form of two symmetric rolls with a hot plume in the region of their contact, above the inner cylinder [13–16]. If the annulus is vibrated as a whole perpendicular to its axis, in the fluid the thermal vibrational convection develops, which alters the fluid motion and enhances the heat transfer [17, 18]. In contrast to [17, 18], in the present experiments, the annulus remains immobile as a whole, and only its outer boundary is brought into oscillations by external electrodynamic actuators. Consequently the boundary takes alternately the oblate and prolate shape, while its average boundary shape remains circular. The dimensionless frequency of vibrations is set relatively low, so as to be of interest for the application to droplet-size systems (see the argumentation in [11]). The impact of the competing convection mechanisms is analyzed.

2. Experimental setup and techniques

The experimental model consisted of the following principal parts: an elastic silicone tube (*1*) (figure 1a,b), a copper cylindrical heat exchanger (*2*), and transparent flanges (*3*), that all together formed a hermetically closed annulus. The flanges of the container were integrated with the front and rear walls of the cooling jacket (*4*). The annulus was filled with a working fluid – an aqueous solution of glycerol with concentration 0.90. Its kinematic viscosity varied in the range $\nu = (73.0 - 190)$ cSt depending on temperature. Other properties of the fluid were the thermal diffusivity $\chi = (9.7 - 9.9) \cdot 10^{-2}$ mm^2/s, the thermal conductivity $\lambda = 3.09 \cdot 10^{-4}$ W/(mmK), the thermal expansion coefficient $\beta = 4.70 \cdot 10^{-4}$ K^{-1}. The radius of the copper cylinder was $R_1 = 7.5$ mm, its length $l_h = 194$ mm. The interior dimensions of the silicone cylinder were $R_2 = 25.0$ mm and $L = 200.0$ mm, the thickness of its walls made 2.0 mm. The gap between the boundaries made $h = R_2 - R_1 = 17.5$ mm. On the silicone boundary, two rectangular activators (*5*) were installed symmetrically. Each activator consisted of two aluminum plates that squeezed the silicone wall in a sandwich-like configuration, their dimensions being the following: length $l = 170.0$ mm, height $d = 20.0$ mm, and thickness 1.2 mm.

The temperature of the outer boundary was maintained constant, $T_2 = (20.0 \pm 0.1)$ °C, by means of pumping water from a thermostatic bath trough the cooling jacket. The copper cylinder was heated up to the temperature T_1 by a nichrome wire with resistance 2.4 ± 0.1 Ohm installed in a narrow shaft inside it and powered by a DC supply. The heating power Q was calculated with the values of current I and voltage U measured from the indicators of the power supply, $Q = IU$. The temperature T_2 was controlled by two thermistors (*6*) of Pt100 type installed in the front and the rear sides of the jacket. The temperature difference between the inner and outer boundaries of the annulus, $\Theta = T_1 - T_2$, was measured directly by a differential thermocouple of type T. Its hot junction was put in a narrow groove in the heat exchanger, while the cold junction was immersed in the cooling water. The thermometers' data was automatically collected by a multichannel measuring device Termodat 22K5. Temperature measurement error did not exceed 0.1°C.

The activators were mechanically connected with the rods of electromagnetic actuators Dunkermotoren STA1112 (*7*) (figure 1b). Each actuator was controlled by an amplifier Copley Accelnet ACJ-055-18-S and fed by a DC power supply MASTECH HY5020E. The position of the actuators was defined by the harmonic signal from a generator GSPF-052. The actuators performed synchronous periodic motion and induced vibrations of the activators. As a result, the elastic boundary was periodically contracted and extended along the horizontal axis. The vibration frequency was $f = 3.00$ Hz, and the amplitude varied in the range $b = (0 - 11)$ mm. The oscillations of the opposite activators were symmetric, and the oscillation amplitude of the whole boundary was equal to the span of one activator. In experiments, b was found as an average of the values obtained for the left and the right activators, the discrepancy between them did not exceed 0.1 mm.

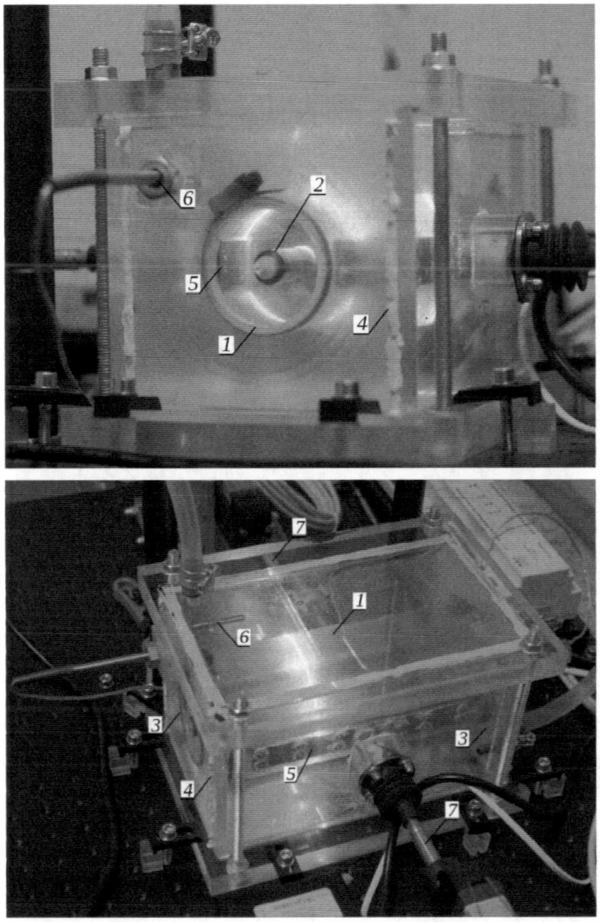

a

b

Figure 1. Experimental setup: *1* – silicone tube, *2* – copper cylinder, *3* – flanges, *4* – cooling jacket, *5* – activators, *6* – thermometer, *7* – linear motors.

The flow structure was visualized by tracers. For this, the working fluid was seeded with Rilsan light-scattering powder (particle size 40 μm and density 1030 gr/mm^3). A cross-section of the annulus was illuminated with a light-sheet from a solid state continuous laser (figure 1). The time-average flow structure and velocity were studied using PIV (Particle Image Velocimetry). The images were taken all in the same oscillation phase with the timestep equal to a multiple of the oscillation period. This way, the time-average velocity was calculated directly. The test experiments demonstrated that the flow structure reproduced itself in different cross-sections along the cylinder axis, except the regions immediately near the flanges. Hence, within the considered range of experimental parameters, the steady flows under the study may be considered effectively two-dimensional. All flow studies were done at $\Theta = \text{const}$.

3. Experimental results

3.1. Steady flows. Isothermal case

Oscillations of the container boundary lead to the generation of steady time-averaged flows in the form of two-dimensional rolls elongated parallel to the cylinder axis. In the cross section, the structure consists of eight pairs of vortices, regularly spaced along the curvilinear boundary (figure 2). The large vortices are generated near the outer boundary, and the small ones – near the inner boundary. The angular coordinates of both coincide. The four vortices in the bottom of the container fall into the shadow of the copper cylinder and are only partially visible, but the regular structure of the flows indicates their presence.

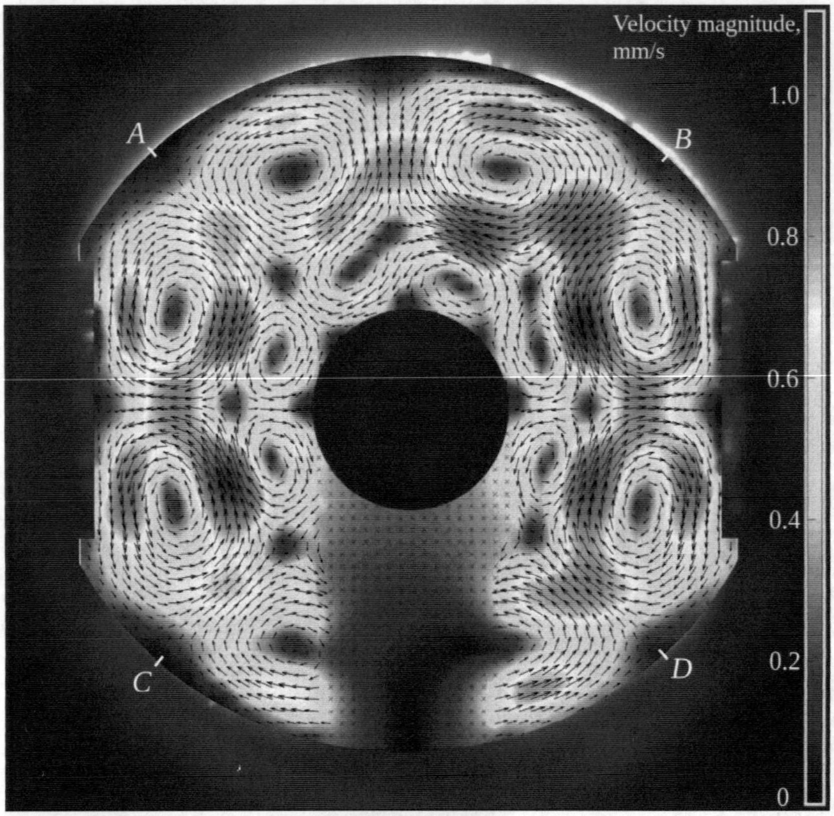

a

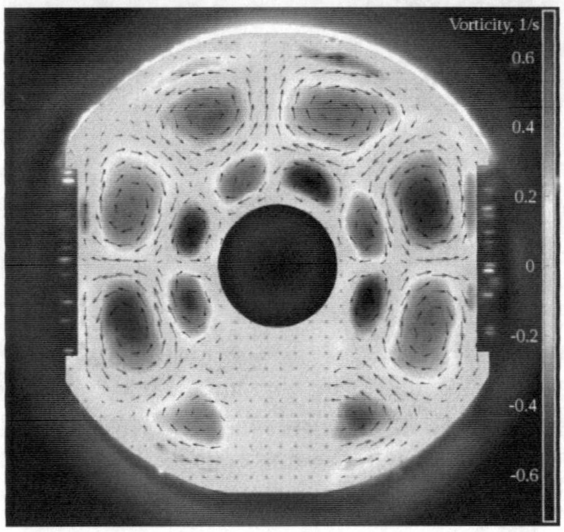

b

Figure 2. Structure of the steady flows at $f = 3.0$ Hz, $b = 8.3$ mm, $\mathrm{Re_p} \cdot \omega = 209$. The vector plot of the time-average fluid velocity is obtained by PIV. The colour maps represent the average velocity magnitude (a), and the vorticity of the time-average velocity (b).

At oscillations, distinctly localized regions of the boundary remain almost motionless. In figure 2a, they are indicated by points A, B, C, D. In these points, steady flows are directed perpendicular to the average position of the solid boundary, which coincides with the radial direction. This applies both to the external and the internal boundaries. Since the fluid oscillates with respect to the internal and external boundaries, a viscous boundary layer forms on each of the surfaces. In turn, the averaged vorticity is generated in each boundary layer. This explains the two-level structure of steady flows.

As a quantitative measure of the intensity of steady flows, a characteristic velocity is often chosen at some point [10–12]. In the present problem, it is convenient to take the velocity of flows directed along the radius from points A, B, C, D towards the container center (figure 2a). Let us denote the arithmetical average value of these velocities as v_j. It increases with the oscillation amplitude according to a quadratic law (figure 3a), which is typical for steady flows generated in Stokes layers [2]. Such approach to mathematical description is justified in the case of self-similar flows. It is worth noting, however, that in the case of simultaneous existence of the vibrational and the free convection, the flow structure transforms making it difficult to choose the characteristic velocity. Due to this, it is more convenient to evaluate the convection intensity by the vorticity of the averaged velocity (figure 2b). Let us introduce as a characteristic the average vorticity $Y = N^{-1}\sum_{i=1}^{N}(\mathrm{rot}\,\mathbf{v})_i$, where $(\mathrm{rot}\,\mathbf{v})_i$ is the maximal value of vorticity in the center of some vortex, $N = (12-14)$ is the number of measured vortices. As it is shown in figure 3b (symbols *1*), for the average vorticity, the quadratic scaling law is also satisfied. At the same time, Y may be used as a characteristic to compare the steady flows that have differences in the structure (compare figures 2b and 4a–d).

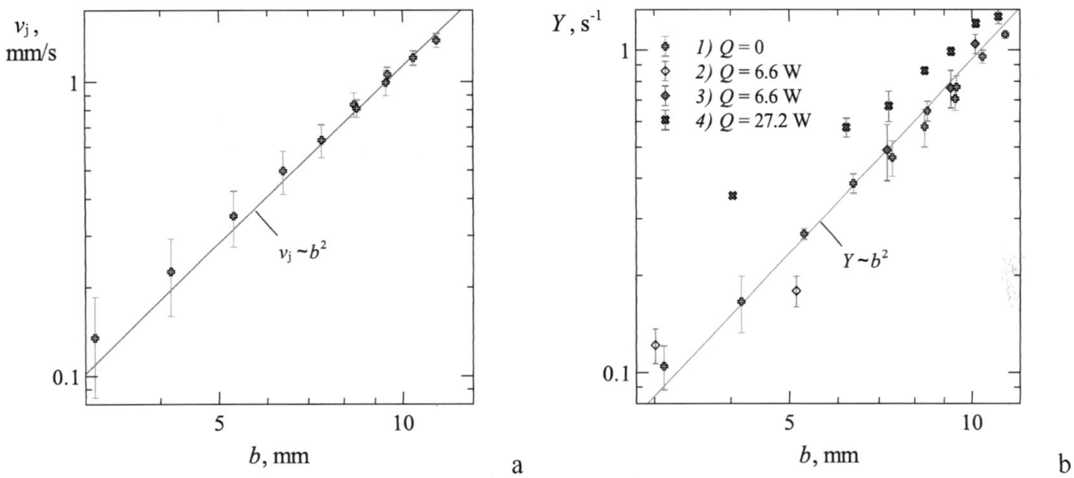

Figure 3. Intensity of steady flows as a function of the oscillation amplitude: a) the graph for v_j in the isothermal situation; b) the graph for Y at various heating power.

3.2. Steady flows at heating

At heating, in the layer, thermal convection emerges, which has been studied in detail in the absence of vibrations, see for example [13–16]. The free convection affects the flow structure (figure 4). Now, in the annulus, two motion mechanisms are present: the free convection excited by the heating of the inner cylinder, and the forced convection – steady streaming excited by the oscillations of the outer boundary. The relative contribution of each of the mechanisms to the fluid motion can be judged by the structure of steady flows.

At relatively low vibration amplitudes, the impact of the free convection is the most pronounced: above the hot cylinder, an intensive upward flow is formed that resembles the convective plume [14], while two descending flows are formed near the outer boundary (figure 4a,c). In this case, the return radial flow in the upper half of the container is absent (compare to figure 2, points *A* and *B*). The deviation of the $Y(b)$ dependence from the quadratic law is characteristic for such convection regimes (figure 3b, points *2* and *4*).

At $\mathrm{Ra} \sim 10^4$, and lower values, it is possible to suppress the free convection by increasing the amplitude of vibrations. Here, the Rayleigh number is calculated as $\mathrm{Ra} = g\beta\Theta h^3/(\nu\chi)$. In this situation, the flow structure takes the form that is typical for the vibrational convection (figure 4b). Simultaneously, the average vorticity approaches to the values observed in the absence of heating (figure 3b, points *3*). At $\mathrm{Ra} \sim 10^5$, in the whole studied range of vibrational parameters the impact of the free convection remains noticeable. This is evidenced by the flow structure (figure 4d), as well as by experimental points *4* (figure 3b), which have a discrepancy with the isothermal case.

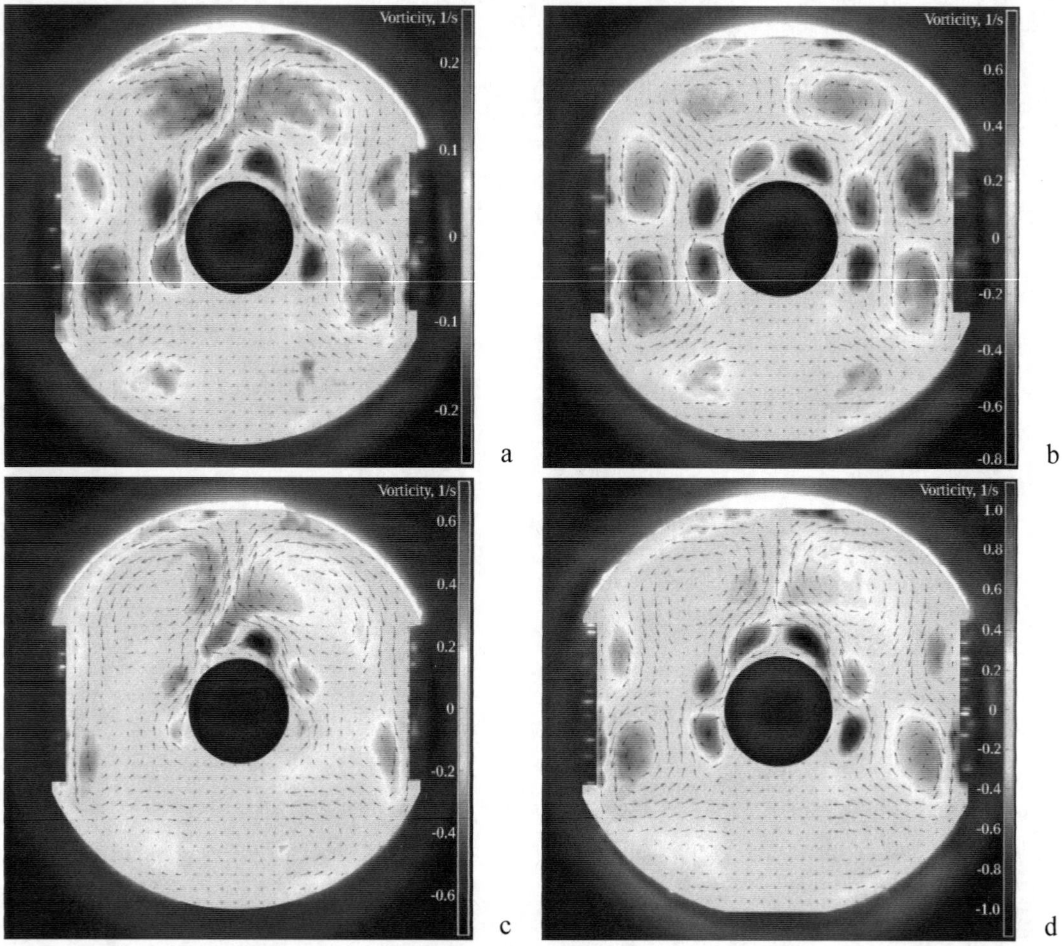

Figure 4. Steady-flow structures at simultaneous heating and vibration: a) $Q = 6.6$ W, $b = 5.1$ mm, $\mathrm{Ra} = 13.2 \cdot 10^3$, $\mathrm{Re_p} \cdot \omega = 136$; b) $Q = 6.6$ W, $b = 9.2$ mm, $\mathrm{Ra} = 9.12 \cdot 10^3$, $\mathrm{Re_p} \cdot \omega = 383$; c) $Q = 27.2$ W, $b = 6.2$ mm, $\mathrm{Ra} = 81.8 \cdot 10^3$, $\mathrm{Re_p} \cdot \omega = 683$, d) $Q = 27.2$ W, $b = 9.2$ mm, $\mathrm{Ra} = 64.3 \cdot 10^3$, $\mathrm{Re_p} \cdot \omega = 1.22 \cdot 10^3$. The vector plots represent the average velocity (the scale of arrows varies between figures), and the colour maps – the average vorticity.

4. Discussion

As it is shown in the equations of thermal vibrational convection [19], the action of vibrations on hydrodynamic systems is described by two control parameters: the dimensionless frequency $\omega = \Omega h^2 / \nu$ and the pulsation Reynolds number $\mathrm{Re_p} = b^2 \Omega / \nu$, where $\Omega = 2\pi f$. In experiments [11, 12], it was shown that ω determines the structure of steady flows, while $\mathrm{Re_p}$ is responsible for the intensity of vibrational convection at a given dimensionless frequency. The analysis of experimental results carried on in works [11, 12] revealed that the intensity of steady flows is determined on the parameters plane $\left(\omega, V / \mathrm{Re_p}\right)$, where $V = vh / \nu$, v and h are the characteristic average velocity and container size. In the low-frequency region, at $\omega \sim 1-10$, the following scaling law holds:

$$V / \mathrm{Re_p} \sim \omega . \tag{1}$$

In the present experiments, in the case of heating, due to the dependence of fluid viscosity on the temperature, the dimensionless frequency varies in the range $\omega = 30-80$, despite the experiments have been performed at a constant vibration frequency. In terms of the present problem, taking into account that $v \sim Yh$, we obtain $V \sim Yh^2 / \nu$. By renorming the expression (1) we obtain

$$Yh^2 / \nu \sim \mathrm{Re_p}\, \omega \sim b^2\Omega^2 h^2 / \nu^2 . \tag{2}$$

As shown in figure 5, the law (2) is found in the case of purely vibrational convection (points *1*). It is also satisfied when the free convection can be neglected compared to the steady streaming (points *3*). The latter is evidenced by the flow structure (figure 4b). In the situations when the vibration intensity is not sufficiently high compared to the Rayleigh number, the dependence law between Yh^2 / ν and $\mathrm{Re_p}\, \omega$ is different from (2) (figure 5, points *2*, *4*).

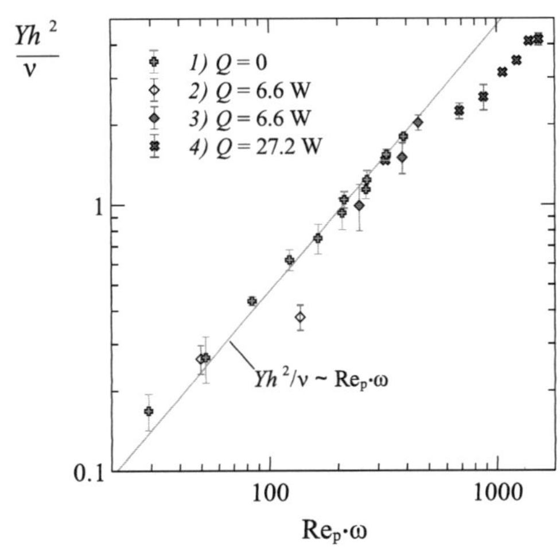

Figure 5. The steady flows' intensity versus the vibration intensity; $\omega = 30.4$ (*1*), $37-41$ (*2*, *3*), $62-79$ (*4*).

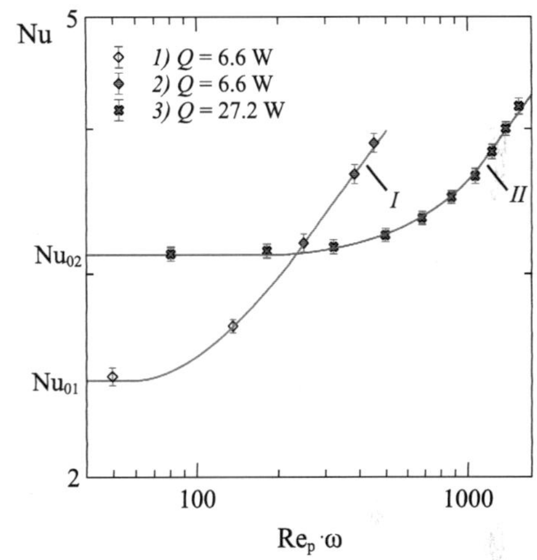

Figure 6. The heat transfer rate versus the vibration intensity; $\mathrm{Ra} = (8.5-15.2)\cdot 10^3$ (*1*, *2*), $(54.9-94.2)\cdot 10^3$ (*3*).

The heat transfer rate is characterized by the Nusselt number, $\mathrm{Nu} = Q / Q_\mathrm{m}$. Here, the molecular heat transfer is calculated as $Q_\mathrm{m} = 2\pi\lambda L\Theta / \ln(R_2 / R_1)$. At relatively low values of $\mathrm{Re_p}\, \omega$ (< 60 at $\mathrm{Ra} \sim 10^4$ and < 200 at $\mathrm{Ra} \sim 10^5$), vibrations do not intensify the heat transfer (figure 6). When the vibrational convection dominates over the free convection, the dependence $\mathrm{Nu} \sim (\mathrm{Re_p}\, \omega)^{1/3}$ is found (points *2*). It is interesting to notice that, despite the pronounced contribution of the free convection to the flow structure at $\mathrm{Ra} \sim 10^5$, in the region $\mathrm{Re_p}\, \omega > 10^3$ points *3* approach the 1/3 power law (compare curves *I* and *II* in figure 6). Let us find the limit between the parameters domains, which determine what type of convection dominates. For this, from the graph in figure 6 we obtain the marginal values of $(\mathrm{Re_p}\, \omega)^*$ by tracing the points of intersection of the laws $\mathrm{Nu} = \mathrm{const}$ and $\mathrm{Nu} \sim (\mathrm{Re_p}\, \omega)^{1/3}$. Then from the experimental data we find the corresponding values of Ra^*. This procedure yields the following:

$\left(\mathrm{Re_p}\,\omega\right)^* = 110$, if $\mathrm{Ra}^* = 14\cdot 10^3$, and $\left(\mathrm{Re_p}\,\omega\right)^* = 650$, if $\mathrm{Ra}^* = 82\cdot 10^3$. In both cases, the ratio $\left(\mathrm{Re_p}\,\omega\right)^* / \mathrm{Ra}^* \approx 8\cdot 10^{-3}$ is obtained.

5. Conclusion

Convective heat transfer has been studied experimentally in the horizontal annulus whose elastic outer boundary performed harmonic oscillations. Two experimental techniques have been applied independently allowing to study the flow velocity and the heat transfer simultaneously. Comparative analysis of the steady flows structure, the average vorticity and the heat transfer rate reveals a correlation between them. Namely, the vibration induced steady flows have quite a particular structure, regular with respect to the azimuthal direction, that is characterized by the scaling law $\mathrm{Nu} \sim \left(\mathrm{Re_p}\,\omega\right)^{1/3}$. This law gives the maximal growth rate of the heat transfer rate with the increase in the vibration intensity and is valid in the situation when the steady streaming clearly dominates over the free convection. A marginal ratio exists between the vibrational parameter $\mathrm{Re_p}\,\omega$ and Ra that allows defining two regions, in which either of two convective mechanisms dominates. In the limit between two regions $\left(\mathrm{Re_p}\,\omega\right)^* \sim 0.01\mathrm{Ra}^*$, and this means that the condition of steady streaming dominance should be: $\left(\mathrm{Re_p}\,\omega\right) \sim 0.1\mathrm{Ra}$ or greater. The considered experimental model allows successfully studying the rate of heat transfer by steady streaming in the presence of free thermal gravitational convection.

6. Acknowledgements

The research was supported by the Russian Science Foundation (project 17-71-10189).

References

[1] Riley N 2001 *Ann. Rev. Fluid Mech.* **33** 43–65
[2] Schlichting H 1968 *Boundary Layer Theory* (McGraw-Hill, New York (USA))
[3] Yamaguchi M, Fujimoto T and Katayama T 1975 *J. Chem. Engng Japan* **8 (5)** 361–6
[4] Kumar A, Hartland S 1999 *Chem. Eng. Res. Des.* **77 (5)** 372–84
[5] Murtsovkin V A and Muller V M 1992 *J. Colloid Interf. Sci.* **151** 150–6
[6] Yarin A L, Brenn G, Kastner O, Rensink D and Tropea C 1999 *J. Fluid Mech.* **399** 151–204
[7] Yarin A L 2001 *J. Fluid Mech.* **444** 321–42
[8] Muginstein A, Fichman M and Gutfinger C 2005 *Int. J. Multiph. Flow* **31** 263–84
[9] Wegener M, Paul N and Kraume M 2014 *Int. J. Heat Mass Transf.* **71** 475–95
[10] Ivanova A A, Kozlov V G, Polezhaev D A, Pareau D and Stambouli M 2009 *Fluid Dyn.* **44** 481–9
[11] Kozlov V G, Kozlov N V and Schipitsyn V D. 2017 *Physical Review Fluids* **2** 094501
[12] Kozlov V G, Sabirov R R and Subbotin S V 2018 *Fluid Dynamics* **53** 189–99
[13] Itoh M, Fujita T, Nishivaki N and Hirata M 1970 *Int. J. Heat Mass Transf.* **13** 1364–8
[14] Kuehn T H and Goldstein R J 1976 *J. Fluid Mech.* **74 (4)** 695–719
[15] Kuehn T H and Goldstein R J 1976 *Int. J. Heat Mass Transf.* **19** 1127–34
[16] Tsui Y T and Tremblay B 1984 *Int. J. Heat Mass Transf.* **27** 103–11
[17] Ivanova A A and Kozlov V G 1985 *Fluid Dynamics* **20** 989–92
[18] Ivanova A A and Kozlov V G 1998 *Fluid Dynamics* **33** 324–30
[19] Gershuni G Z and Lyubimov D V 1998 *Thermal Vibrational Convection* (New York, USA: Wiley)

IOP Conf. Series: Materials Science and Engineering **581** (2019) 012036 doi:10.1088/1757-899X/581/1/012036

Velocity Distribution in Rough Pipe: the Model Based on the Analytical Description of Resistance Curves in Nikuradse's Experiments

O Dudar

Applied Mathematics and Mechanics Faculty, Perm National Research Polytechnic University, 29, Komsomolsky Ave. Perm, 614990, Russia

E-mail: olegdudar@yandex.ru

Abstract. Calculation and optimization of such complex systems as ventilating networks, water supply systems, etc. taking into account roughness of a surface and possible processes of heat and mass transfer demands use simple hydrodynamic models, which are at the same time sufficiently exact for engineering calculations. The goal of this paper is to develop such model. Prandtl's model of a two-layer flow in the smooth pipe with the linear law of velocity distribution for a near-wall laminar layer and the logarithmic law for a turbulent core is used as a basis, which spreads to a case of rough pipes. The velocity profile equation and also the equations for the coordinate Y_b of the laminar and turbulent border and its velocity U_b are derived. This paper is based on the use of the analytical description of Nikuradse's experimental curves $\lambda(\mathrm{Re},\delta)$ (λ is the resistance coefficient, δ is the relative roughness) which earlier was found by the author. The dependence $\lambda(\mathrm{Re}^*,\delta)$ (Re^* is the dynamic Reynolds number) is included into all above-mentioned equations. Results of numerical calculations of Y_b and U_b as functions of Re^* for the values of δ which are the same as in Nikuradse's experiments are given. The calculation of velocity profiles for some values of Re^* and δ is also fulfilled. Comparisons with the experimental velocity profiles obtained by Nikuradse demonstrate a good agreement.

1. Introduction

Modern universal packages allow solving the most varied problems of hydrodynamics. However, their use for calculation and optimization of such complex systems as ventilating networks, water supply systems, etc. taking into account roughness of a surface and possible processes of heat and mass transfer is impossible because of significant time expenditure. The goal of this paper is to develop a simple, but at the same time exact model of a turbulent flow in a rough pipe, suitable for the solution of the above-mentioned difficult engineering problems. Prandtl's model [1-3] of a two-layer flow with the linear law of velocity distribution for a near-wall laminar layer and the logarithmic law for a turbulent core is used as a basis. The systematic verification of the assumptions which are the cornerstone of the Prandtl's mixing length theory was carried out in both real [4, 5] and numerical [6-9] experiments and proved their sufficient acceptability. However, for the analysis of a flow in rough pipes papers [1-3] used the single-layer model which is acceptable for the complete roughness (square-law) regime and gives a serious error for the regimes transitional from the smooth-pipe to the square-law regime. Most of later papers used the same approach [10-13]. At the same time a serious problem is the absence of analytical dependence $\lambda\left(\mathrm{Re},\frac{k}{r_p}\right)$ for transient regimes (λ is the resistance coefficient, k is the height of the uniform roughness elements, r_p is the pipe radius) entering into a velocity profile expression whereas for the limit (smooth-pipe and square-law) flow regimes such dependence is available. This paper is

IOP Conf. Series: Materials Science and Engineering **581** (2019) 012036 doi:10.1088/1757-899X/581/1/012036

based on the use of the analytical dependence $\lambda\left(\text{Re},\frac{k}{r_p}\right)$ found earlier by the author [14] as a result of the analysis of Nikuradse's experiments [1, 2] which can be applied for the description of all flow regimes. In total, a velocity profile for the laminar and turbulent layers, the thickness of a laminar layer and the velocity on the laminar - turbulent border were found as a function of Reynolds number Re and relative roughness $\frac{k}{r_p}$.

2. Model of a liquid flow in a rough pipe

Let's assume that the roughness of a surface is uniform (sand type) and is characterized by the height of the roughness elements k.

Following Prandtl's model of a flow in a smooth pipe we presume that there is a two-layer flow, namely: for a laminar near-wall layer the Newton's friction law is applied, and for a turbulent core Prandtl's hypothesis of mixing length is used, wherein the mixing length l is proportional to the distance from a wall y ($l = \kappa y$, where κ is the von Karman's constant).

The presence of roughness requires to assume the occurrence of a virtual wall on which the no-slip condition for longitudinal velocity u is satisfied. Its location is defined by a dimensionless parameter γ according to expression $u(\gamma k) = 0$ (about virtual wall see also [15]).

Then a velocity profile can be described by the following differential equations [1, 3, 10]

$$\frac{du}{dy} = \begin{cases} \dfrac{\tau_w}{\mu}, & \gamma k \le y \le y_b \\ \dfrac{1}{l}\sqrt{\dfrac{\tau_w}{\rho}}, & y_b \le y \le r_p \end{cases}, \tag{1}$$

where τ_w is the viscous shear on a wall, μ and ρ are the dynamic viscosity and the density of a liquid, r_p is the pipe radius, y_b is the coordinate of a border of the laminar and turbulent layers. Making transformation to dimensionless variables, we will get

$$\frac{dU}{dY} = \begin{cases} \dfrac{r_p}{l^*}, & \gamma\dfrac{k}{r_p} \le Y \le Y_b \\ \dfrac{1}{\kappa Y}, & Y_b \le Y \le 1 \end{cases}, \tag{2}$$

where $U = \frac{u}{V^*}$ is the dimensionless velocity, $V^* = \sqrt{\frac{\tau_w}{\rho}}$ is the friction velocity, $Y = \frac{y}{r_p}$ is the dimensionless distance from a wall, $l^* = \frac{\nu}{V^*}$ is the dynamic length, ν is the kinematic viscosity.

3. Law of the velocity distribution

After integration (2) and use of boundary conditions we obtain the law of velocity distribution

$$U(Y) = \begin{cases} \dfrac{r_p}{l^*}\left(Y - \gamma\dfrac{k}{r_p}\right), & \gamma\dfrac{k}{r_p} \le Y \le Y_b \\ U_{\max} + \kappa^{-1}\ln Y, & Y_b \le Y \le 1 \end{cases} \tag{3}$$

Keeping in mind that on the border of the laminar and turbulent layers the condition $U(Y_b) = U_b$ has to be satisfied, where U_b is the velocity on this border, (3) is possible to write down in a form

$$U(Y)=\begin{cases}\dfrac{U_b}{Y_b-\gamma\dfrac{k}{r_p}}\left(Y-\gamma\dfrac{k}{r_p}\right), & \gamma\dfrac{k}{r_p}\le Y\le Y_b\\ U_b+\kappa^{-1}\ln\dfrac{Y}{Y_b}, & Y_b\le Y\le 1\end{cases} \tag{4}$$

In the equation (4) Y_b, U_b, γ and κ remain unknown. From comparison (3) and (4) we can write down the system of the equations in Y_b and U_b

$$\begin{cases}\dfrac{U_b}{Y_b-\gamma\dfrac{k}{r_p}}=\dfrac{\mathrm{Re}^*}{2}\\ U_b=U_{\max}+\kappa^{-1}\ln Y_b\end{cases} \tag{5}$$

In (5) it is used that

$$\frac{2r_p}{l^*}=\frac{2r_pV^*}{\nu}=\mathrm{Re}^*, \tag{6}$$

where Re^* is the dynamic Reynolds number.

For determination of $U_{\max}$ we will apply the relation established by Prandtl between the law of resistance and the velocity distribution. Finding for a velocity profile (4) the average velocity according to the formula

$$\bar{U}=2\int_{\gamma\frac{k}{r_p}}^{1}U(1-Y)dY,$$

after transformations we will get

$$\bar{U}=U_b\left[1-\left(Y_b+\gamma\frac{k}{r_p}\right)+\frac{1}{3}\left(Y_b^2+Y_b\gamma\frac{k}{r_p}+\left(\gamma\frac{k}{r_p}\right)^2\right)\right]-\kappa^{-1}\left(\frac{3}{2}+\ln Y_b-2Y_b+\frac{Y_b^2}{2}\right). \tag{7}$$

Though the analysis (7) shows that the linear and square summands have significant effect at small values of Reynolds number (for a smooth surface discarding of the linear and square summands leads to an error of about 9%), in this paper we will be only limited to the initial approximation for $\bar{U}$

$$\bar{U}\approx U_b-\kappa^{-1}\left(\frac{3}{2}+\ln Y_b\right). \tag{8}$$

Taking into account (5) and (8), we can write down

$$U_{\max}=\bar{U}+\frac{3}{2}\kappa^{-1}. \tag{9}$$

Then, using Darci's formula in the form [1, 3, 10]

$$\bar{U}=\sqrt{\frac{8}{\lambda}}, \tag{10}$$

where λ is the resistance coefficient, we come to the following expression

$$U_{\max} = \sqrt{\frac{8}{\lambda}} + \frac{3}{2}\kappa^{-1}. \tag{11}$$

As a result we derive from (3) taking into account (6) and (11) the final expression for a velocity profile as a function of the resistance coefficient λ

$$U(Y) = \begin{cases} \dfrac{\mathrm{Re}^*}{2}\left(Y - \gamma\dfrac{k}{r_p}\right), & \gamma\dfrac{k}{r_p} \le Y \le Y_b \\ \sqrt{\dfrac{8}{\lambda}} + \kappa^{-1}\left(\dfrac{3}{2} + \ln Y\right), & Y_b \le Y \le 1 \end{cases} \tag{12}$$

4. Equations for a coordinate of the laminar and turbulent border and its velocity

Substitution of (11) in (5) allows entering a parameter λ into the system of equations in Y_b and U_b:

$$\begin{cases} \dfrac{U_b}{Y_b - \gamma\dfrac{k}{r_p}} = \dfrac{\mathrm{Re}^*}{2} \\ U_b = \sqrt{\dfrac{8}{\lambda}} + \kappa^{-1}\left(\dfrac{3}{2} + \ln Y_b\right) \end{cases} \tag{13}$$

Excluding U_b from a system (13), we get the transcendental equation in Y_b

$$Y_b = \gamma\frac{k}{r_p} + \frac{2}{\mathrm{Re}^*}\left(\sqrt{\frac{8}{\lambda}} + \kappa^{-1}\left(\frac{3}{2} + \ln Y_b\right)\right). \tag{14}$$

On the contrary, excluding Y_b, we derive the transcendental equation in U_b

$$U_b = \sqrt{\frac{8}{\lambda}} + \kappa^{-1}\left(\frac{3}{2} + \ln\left(\gamma\frac{k}{r_p} + \frac{2U_b}{\mathrm{Re}^*}\right)\right). \tag{15}$$

To use formulas (12, 14, 15), it is necessary to know dependence $\lambda\left(\mathrm{Re}^*, \frac{k}{r_p}\right)$. For the smooth-pipe and complete roughness limit regimes of flow such dependences are known from the experiments [1-3]. For the full roughness regime the resistance coefficient λ depends only on the relative roughness $\frac{k}{r_p}$ [1, 3]

$$\frac{1}{\sqrt{\lambda}} = -2\lg\left(\frac{k}{r_p}\right) + 1.74 = -0.869\ln\left(\frac{k}{r_p}\right) + 1.74 \tag{16}$$

and for the smooth-pipe flow regime it depends only on the Reynolds number [2, 3]

$$\frac{1}{\sqrt{\lambda}} = 2\lg\left(\mathrm{Re}\sqrt{\lambda}\right) - 0.8. \tag{17}$$

Taking into account that Darcy's formula (10) can be rewritten in a form

$$\bar{U} = \frac{\bar{u}}{V^*} = \frac{\mathrm{Re}}{\mathrm{Re}^*} = \sqrt{\frac{8}{\lambda}} \tag{18}$$

we come to a form of record of the smooth pipe resistance law through the dynamic Reynolds number

$$\frac{1}{\sqrt{\lambda}} = 2\lg\left(\mathrm{Re}^*\right) + 0.103 = 0.869\ln\left(\mathrm{Re}^*\right) + 0.103. \tag{19}$$

Let's use the (16) and (19) formulas for the determination of the remaining not found constants.

4.1. A smooth-pipe flow regime

In this case $\frac{k}{r_p} \to 0$. Then from (16) we derive:

$$U_b - \kappa^{-1}\ln U_b - \kappa^{-1}\left(\frac{3}{2} + \ln 2\right) + \kappa^{-1}\ln \mathrm{Re}^* = \sqrt{\frac{8}{\lambda}}. \tag{20}$$

Substituting in the right part of (20) the experimental formula (19) we will come to the expression

$$U_b - \kappa^{-1}\ln U_b - \kappa^{-1}\left(\frac{3}{2} + \ln 2\right) + \kappa^{-1}\ln \mathrm{Re}^* = 2.46\ln\left(\mathrm{Re}^*\right) + 0.291,$$

from which we get the following numerical equalities

$$\kappa^{-1} = 2.46, \tag{21}$$

$$U_b - \kappa^{-1}\ln U_b - \kappa^{-1}\left(\frac{3}{2} + \ln 2\right) = 0.291. \tag{22}$$

The last means that in the case of the smooth-pipe regime the dimensionless velocity on the laminar and turbulent border is constant. From (21) and (22) we obtain values of the constants

$$\kappa = 0.407, \quad U_b = 11.7. \tag{23}$$

This value of the von Karman's constant is in a good agreement with [2-4, 16]

4.2. A complete roughness regime

In this case $\mathrm{Re}^* \to \infty$. Then it follows from the first equation of the system (13)

$$Y_b - \gamma\frac{k}{r_p} = \frac{2U_b}{\mathrm{Re}^*} \to 0 \Rightarrow Y_b \to \gamma\frac{k}{r_p}.$$

It means that the thickness of a viscous layer tends to zero, and

$$U_b = U(Y_b) \to U\left(\frac{k}{r_p}\right) = 0.$$

Then the equation (15) takes the following form

$$\kappa^{-1}\left(\frac{3}{2} + \ln\gamma\frac{k}{r_p}\right) = -\sqrt{\frac{8}{\lambda}}. \tag{24}$$

Substituting in the right part of (24) the experimental formula (16) we will come to the expression

$$\kappa^{-1}\left(\frac{3}{2} + \ln\gamma + \ln\frac{k}{r_p}\right) = 2.46\ln\left(\frac{k}{r_p}\right) - 4.92,$$

wherefrom

$$\ln\gamma = -4.92\kappa - 1.5 = -3.503$$

and therefore

IOP Conf. Series: Materials Science and Engineering **581** (2019) 012036 doi:10.1088/1757-899X/581/1/012036

$$\gamma = 0.0301. \tag{25}$$

In paper [17] $\gamma = \frac{1}{3}$.

5. Numerical results

To find the velocity profile $U(Y)$, the coordinate Y_b of the laminar and turbulent border and its velocity U_b using formulas (12, 14, 15) it is necessary to have the analytical dependence $\lambda\left(\text{Re}^*, \frac{k}{r_p}\right)$. In paper [14] the procedure of obtaining these dependences as a result of the analysis of Nikuradse's experimental data on a turbulent flow in rough pipes is described. The derived analytical dependence $\lambda\left(\text{Re}^*, \frac{k}{r_p}\right)$ allows to find $U(Y)$, Y_b and U_b for any combination of Re^* and $\frac{k}{r_p}$. In figure 1 the dependences $Y_b(\text{Re}^*)$ and $U_b(\text{Re}^*)$ are provided for the same values of relative roughness $\frac{k}{r_p}$ presented in Nikuradse's paper [1] for experimental curves $\lambda\left(\text{Re}^*, \frac{k}{r_p}\right)$. In figure 2 the velocity profiles obtained experimentally [1] and with the help of the analytical dependence $\lambda\left(\text{Re}^*, \frac{k}{r_p}\right)$ for two particular cases are compared. Such comparison was carried out for all experimental points presented in [1]. It showed that in 953 experimental points from 963 points the deviation did not exceed 4.5%.

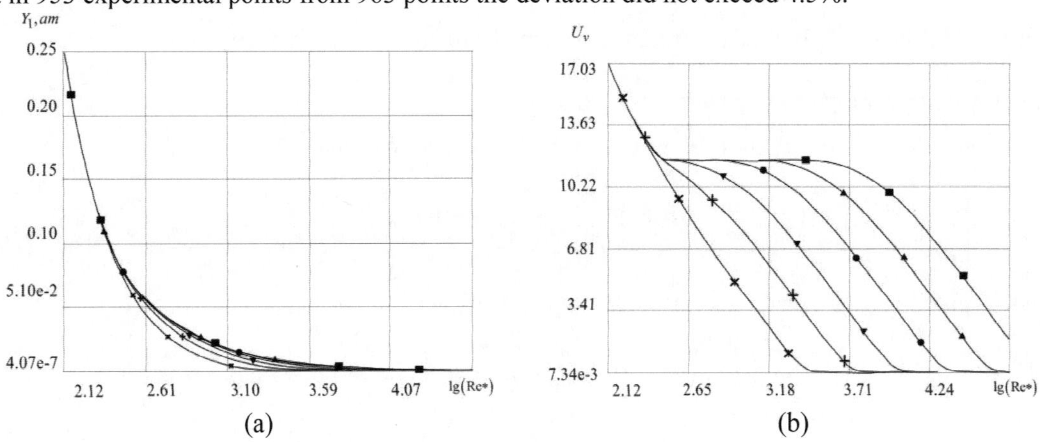

Figure 1. The coordinate Y_b (a) and the dimensionless velocity U_b (b) on the laminar and turbulent border. Legend ■ – $k/r_p = 507$; ▲ – $k/r_p = 252$; ● – $k/r_p = 126$; ▼ – $k/r_p = 60$; + – $k/r_p = 30.6$; * – $k/r_p = 15$.

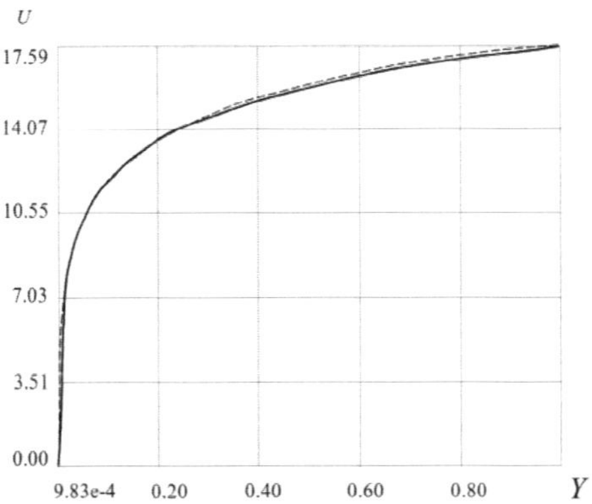

(a)

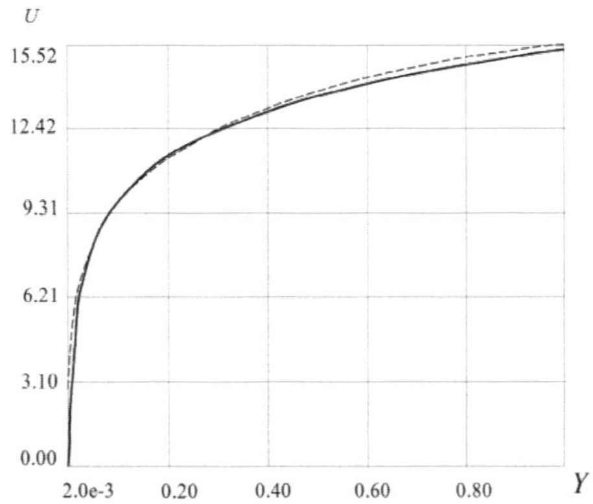

(b)

Figure 2. Velocity profile $k/r_p = 30.6$, $\mathrm{Re}^* = 1642$ (a); $k/r_p = 15$, $\mathrm{Re}^* = 1894$ (b) ______ – calculation; _ _ _ _ _ – experiment.

6. Conclusions

According to the results of Nikuradse's experiments on a turbulent flow in the pipes with sand roughness analytical dependences $\lambda\left(\mathrm{Re}^*, \frac{k}{r_p}\right)$ are derived.

The formula of velocity distribution for any regime of a turbulent flow (smooth-pipe, square-law, transitional) expressed in terms of the resistance law for this regime is obtained.

The equations describing dependence of the coordinate of a laminar and turbulent border and its velocity on the dynamic Reynolds number and relative roughness are derived and solved.

References

[1] Nikuradse J 1933 *Forschung auf dem Gebiete des Ingenieurwesens* **361 (B/4)** 1
[2] Nikuradse J 1932 *Forschung auf dem Gebiete des Ingenieurwesens* **356 (B/3)** 1
[3] Schlichting H 1979 *Boundary Layer Theory* (New York: McGraw-Hill) p. 817
[4] Bailey S C C, Vallikivi M, Hultmark M and Smits A J 2014 *J. Fluid Mech.* **749** 79
[5] Marusic I, Monty J P, Hultmark M and Smits A J 2013 *J. Fluid Mech.* **716 (R3)** 1
[6] Orlandi P, Bernardini M and Pirozzoli S 2015 *J. Fluid Mech.* **770** 424
[7] Pirozzoli S 2014 *J. Fluid Mech.* **745** 378
[8] Chan L, MacDonald M, Chung D, Hutchins N and Ooi A 2015 *J. Fluid Mech.* **771** 743
[9] Chung D, Chan L, MacDonald M, Hutchins N and Ooi A 2015 *J. Fluid Mech.* **773** 418
[10] Loitsyanskii L G 1987 *Liquids and gases mechanics* (Moscow: Nauka) p. 840
[11] Hramtsov I V, Bulbovich R V and Pavlohradskyi V V 2013 *Volga scientific bulletin* **6 (22)** 5
[12] Bryanskaya Yu V 2013 *Magazine of Civil Engineering* **6** 31
[13] Kondratyev A S, Nya T L and Shvydko P P 2017 *Fundamental research* **1** 74
[14] Dudar O I 2018 Influence model of roughness on a turbulent flow in a pipe based on the analytical description of Nikuradze experiments *Nonlinear problem of the theory of hydrodynamic stability and turbulence (Electronic Materials)* ed N V Nikitin and N V Popelenskaya (Moscow: MAX Press) pp. 127 – 133
[15] Herwig H, Gloss D and Wenterodt T 2008 *J. Fluid Mech.* **613** 35
[16] Bernardini M, Pirozzoli S and Orlandi P 2014 *J. Fluid Mech.* **742** 171
[17] Colebrook C F and White C M 1937 *Proc.Royal Soc. London* A **161 (960)** 367

IOP Conf. Series: Materials Science and Engineering **581** (2019) 012037 doi:10.1088/1757-899X/581/1/012037

Dynamics of the Gas Bubbles in the Magnetic Fluid in the Non-Uniform Magnetic Field

V M Polunin, P A Ryapolov[a], E ASokolov

South-West State University, Kursk, Russia

E-mail: [a] r-piter@yandex.ru

Abstract. This work presents a study of the hydrodynamics of gas bubbles in a magnetic fluid which is in a flat channel with a thickness of 2 mm, placed in the non-uniform field of the rare-earth annular permanent magnet. The peculiarity of the magnetic field configuration of this magnet is as follows: at a distance of 9 mm from the surface of the magnet there are two symmetric regions where the magnetic field is zero, the so-called 'magnetic vacuum' regions. The specified configuration of the magnetic field allows capturing and holding gas cavities when the magnet moves; while pressing the cavity to the bottom of the channel, gas bubbles separate. An experimental setup was created to study the hydrodynamics of gas inclusions; using this setup studies on the concentration series of four magnetic fluids based on kerosene with a volume concentration of 10.56 %; 6.32 %; 3.93 % and 2.08% were conducted. The mechanism of gas bubbles capturing and gas cavity formation in a magnetic fluid in the magnetic vacuum region of an annular magnet is investigated. The process of the separation of the bubbles from the air cavity when it is pressed down to the bottom of the channel by a non-uniform magnetic field is studied.

1. Introduction

Studies of gas inclusions (bubbles and cavities) in a magnetic fluid is of great interest due to the possibility of controlling by means of an external magnetic field. The scope of experimental research is the influence of the magnetic field on the bubbles shape, speed of movement, and the gas jets decay [1-3]. In theoretical terms, the popular technique for interpreting such systems is a method called the volume of fluid (VOF) [4-6] which was applied to theoretically simulate the shapes of gas bubbles in a magnetic field, including external uniform magnetic fields. A theoretical model of nonlinear movement and oscillations of a bubble wall in a magnetic fluid in a uniform magnetic field when exposed to an acoustic signal and a variable magnetic field is presented in [7]. Most of these works consider interaction of magneto-fluid systems with gas inclusions with uniform magnetic fields. Only [1] studies the influence of a non-uniform magnetic field created by an annular electromagnet on the deformation and velocity of the bubbles in a magnetic fluid. From the practical viewpoint, it is of interest to use such systems for gas flow control systems. In [8–9], a gas microflow rate sensor was proposed; in this sensor gas bubbles float onto the surface in the magnetic fluid in a uniform magnetic field, passing through a cylindrical measuring coil; when passing through it a gas bubble perturbs magnetic field, which leads to the EMF induced in the coil.

In the previous work [10], the dynamics of gas inclusions in the magnetic fluid in the cylindrical tube with the diameter of 12 mm in a non-uniform magnetic field was studied. An original acousto-magnetic mechanism for recording oscillations of the gas bubble wall in a magnetic fluid is proposed. Estimation of the sizes of gas bubbles was carried out. A theoretical interpretation of the mechanism of gas cavity capturing, transporting and pressing in the magnetic fluid in the non-uniform magnetic field of the annular magnet was presented. However, the experiments were carried out in an optically non-transparent system, which made it difficult to experimentally interpret the results obtained.

2. Experimental results and their interpretation

The idea of the experiment is as follows: if an annular magnet is brought from below into the tube filled with a magnetic fluid, then at a certain distance between the surface of the fluid and the surface of the annular magnet a portion of the gas is trapped into the fluid (Figure 1a). A further slow rise of the magnet (~ 0.02 mm/s) leads to the capture of gas bubbles and the formation of a larger cavity, the sealing of which is then broken (Fig. 1b). This process is similar to the process of breakdown of the magneto-liquid sealer. Jumper recovery (sealing cavity) occurs during the subsequent lowering of the magnet. In the process of lowering the magnet at a constant speed of ~ 0.8 mm/s, the jumper turns into a solid column due to the flow of the MF (Fig. 1c). Upon further lowering of the magnet to the level 'below the tube with the MF', the air cavity is pressed by the non-uniform magnetic field to the bottom (Fig. 1d) and when lowering below it, the gas bubbles separate [10].

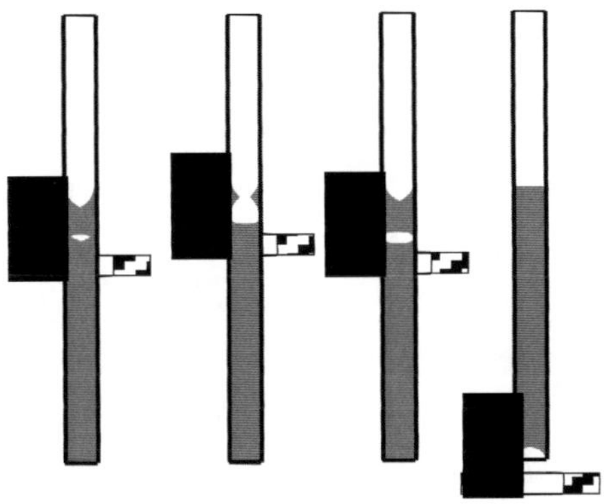

a b c d

Figure 1. The process of capturing air cavity by the magnetic fluid

The experimental setup is shown in Figure 2. A flat channel 1 with a thickness of 2 mm and a width of 10 mm filled with magnetic fluid 2 is rigidly fixed to an aluminum structure with fixtures made of plexiglas. The annular magnet 3 and the inductor coil 4, used to register magnetic oscillations, are mounted coaxially to the tube. The signals from it are amplified by means of the amplifier 5 and transmitted by means of the digital oscilloscope 6 to PC for further processing. A helical gear with a stepper motor 7 is used to move the annular magnet of the inductor coil at a speed of 0.05-45 mm/s. In the initial position, the ring magnet is located below the bottom of the tube. A cavity 8 is formed while lifting the magnet. For video recording, a LED controlled luminaire 9 and a high-speed camera 10 configured for recording in the transmitted light are used. Figure 3 shows the result of video recording of the air cavity in the magnetic fluid MF-1 in the flat channel. The air cavity is near the bottom of the flat channel and is superimposed on the isoline pattern of the module of the magnetic field intensity. The figure demonstrates that the upper part of the gas-fluid interphase boundary follows the isolines of the magnetic field intensity module equal to 35 kA/m.

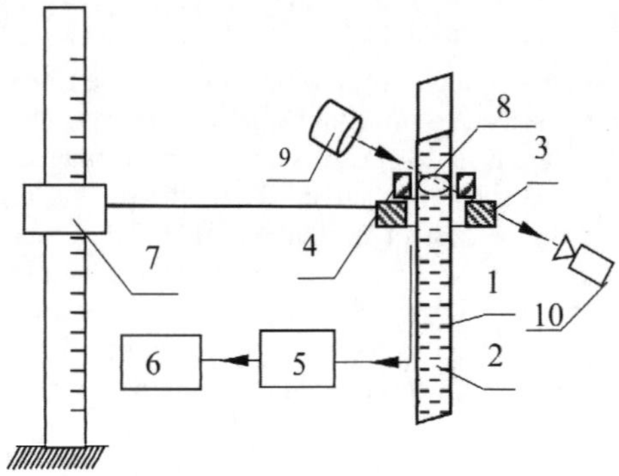

Figure 2. Block diagram of the experimental setup.

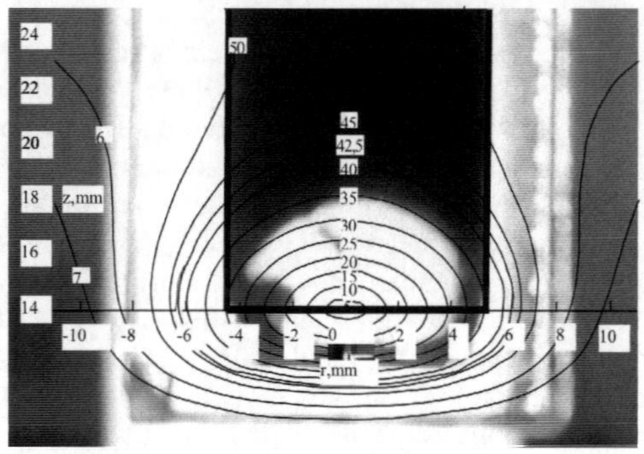

Figure 3. Air cavity in the magnetic fluid MF-1

A concentration series obtained by diluting MF-1 sample based on kerosene was studied. The physical parameters of the samples are presented in table 1.

Table 1. Physical parameters of the samples

	MF-4	MF-3	MF-2	MF-1
MF density	870	952	1,058	1,245
Volume concentration φ, %	2.08	3.93	6.32	10.56
Saturation magnetization M_s, kA/m	9.2	12.9	20.7	43.3
Magnetic concentration φ_m, %	1.93	2.70	4.34	9.08
MF viscosity, mPa·s	1.8	2.45	4.15	31.8

To analyze video recorded data in the NI LabView environment, a program for controlling and processing machine vision data and determining physical parameters of gas inclusions in the magnetic fluid (volume, surface shape evolution and their velocity) was developed. Figure 4 shows a graph of the dependence of the air cavity volume on the time; the time when the air cavity touches the bottom of the flat channel is taken as a reference point.

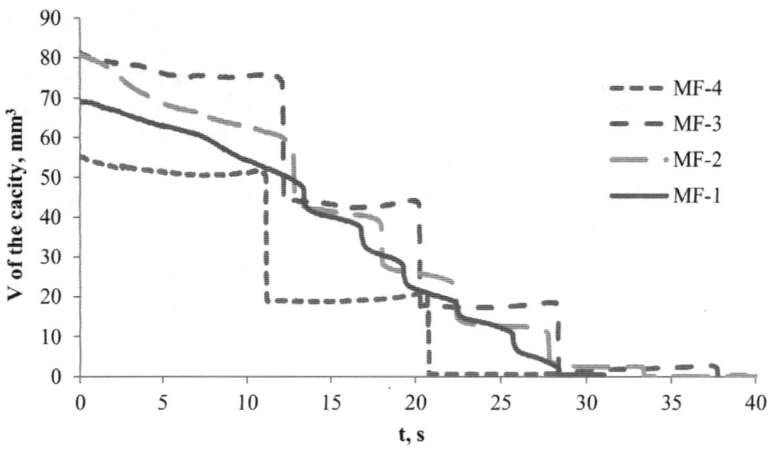

Figure 4. Dependence of the air cavity volume on the time when pressing the cavity to the bottom of the flat channel by non-uniform field of the annular magnet

On the graph, the vertical segments demonstrate the moment of a sharp change in the gas cavity volume, associated with the gas bubble separation. It can be seen from the graph that for magnetic fluids with lower concentration, bubble separation occurs earlier. Based on the graph data, table 2 was created; the table presents the sizes of the bubbles in the order of their numbers for each sample of magnetic fluid. The table shows that in low concentrated MFs a smaller number of gas bubbles of larger diameter is formed; on the contrary, concentrated MFs are characterized by the formation of a large number of the bubbles of small diameter.

Table 2. Bubble sizes

Number of the bubble	MF-4		MF-3		MF-2		MF-1	
	time of separation	bubble volume	time of separation	bubble volume	time of separation	bubble volume	time of separation	bubble volume
1	11.14	28.89	12.21	29.95	12.97	14.31	13.41	2.29
2	20.79	20.83	20.28	26.27	18.03	9.43	16.83	2.32
3	-	-	28.38	17.49	22.41	6.73	19.31	2.40
4	-	-	-	-	27.79	5.50	22.48	2.36
5	-	-	-	-	33.38	1.92	25.72	2.23
6	-	-	-	-	-	-	28.41	1.69

The data obtained for the MF-1 sample, coincide with the data obtained in [10] using acousto-magnetic display for the setup with a cylindrical tube filled with MF.

3. Conclusions

The dynamics of bubble trapping by magnetic fluid in the 'magnetic vacuum' region of the annular magnet is investigated. The data conerning the influence of concentration on the strength of magnetofluid bridges under the influence of external pressure are obtained. This information may be useful for the development of a test bench for magnetic fluids used in sealers, where fluids experience similar effects.

To obtain a detailed mechanism of air cavity formation in the 'magnetic vacuum' region of the annular magnet, an experiment with video recording of the gas-magnetic fluid interface in the flat thin channel was performed. An experiment on the air cavity pressing was performed, as a result of which gas bubbles separated from the MF samples. It is shown that in low-concentrated MFs, bubbles separation occurs earlier, with the formation of a small number of gas bubbles of large diameter. The data obtained for the most concentrated MF-1 sample coincide with the data obtained in [10] using an acoustomagnetic display for the setup with a cylindrical tube filled with MF.

Acknowledgements

The work was carried out with the financial support of the grant from the President of the Russian Federation MK-1393.2019.8 and the basic part of the state task of the Ministry of Education and Science of the Russian Federation. Project code is 3.8949.2017/BP.

References

[1] Ishimoto J, Okubo M, Kamiyama S and Higashitani M 1995 *JSME Int. J. Ser. B Fluids Therm. Eng.* **38** 382-387
[2] He Y Q, Bi Q C and Shi D X 2011 *FDMP* **7** 357-370
[3] Bashtovoi V, Kovalev M and Reks A 2005 *JMMM* **289** 350-352
[4] Korlie M S, Mukherjee A, Nita B G, Stevens J G, Trubatch A D and Yecko P 2008 *J. Phys. Condens. Matter* **20** 204143
[5] Tian X H, Shi W Y, Tang T and Feng L 2015 *ISIJ INT.* ISIJINT-2015-493
[6] Dizaji A S, Mohammadpourfard M and Aminfar H 2018 *JMMM* **449** 185-196
[7] Malvar S, Gontijo R G and Cunha F R *J. 2018 Engrg. Math.* **108** 143-170
[8] Popa N C, Potencz I, Anton I and Vékás L 1997 *Sens Actuators A Phys.* **59** C. 307-310
[9] Popa N C, De Sabata I, Anton I, Potencz I and Vékás L 1999 *JMMM* **201** 385-390
[10] Polunin V M, Ryapolov P A, Ryabtsev K S, Postnikov E B, Bridskiy A V, Sokolov E A and Khudyakov A N 2018 *MHD* **54** 211-224

Peculiarities of the Thermal Regime of the Magnitogorsk Reservoir in the Zone of Water Use by PJSC "Magnitogorsk Iron and Steel Works"

A P Lepikhin[1] Yu S Lyakhin[1], T P Lyubimova[2, a], Ya N Parshakova[2]
[1]Mining Institute UB RAS, Sibirskaya 78a, Perm 614007, Russia
[2] Institute of Continuous Media Mechanics UB RAS, Acade. Koroleva 1, Perm 614013, Russia

E-mail: [a] lubimova@psu.ru

Abstract. The solution of a wide range of tasks of protection and rational use of water bodies requires, first of all, an assessment of the levels of impact of industrial complexes. The solution of such problems for extreme technological and hydrometeorological conditions is carried out on the basis of appropriate computational experiments. However, verification of the underlying mathematical models can be performed only on the basis of correct input data. The paper presents materials of field studies of the temperature distribution of water both in the water area and in the depth of the Magnitogorsk reservoir, in the zone of influence of thermal effluents of PJSC "Magnitogorsk Iron and Steel Works" (PJSC MMK). It is shown that in the warmest limiting period, a significant vertical temperature inhomogeneity is formed.

1. Introduction

Magnitogorsk reservoir on the Ural River (cooling pond) was commissioned in 1939 according to the design of the Vodokanalproekt Institute, Moscow. The reservoir is of the channel type, the type of regulation is seasonal [1]. The area of the mirror is 21.1 km^2, the volume at NPU is 148 million m^3, the average depth is ~ 7 m.

In 1971-1972, separating and diverting dams were built on the Magnitogorsk reservoir (cooling pond), which formed the circulating part of the Magnitogorsk reservoir.

Industrial water supply of the main industrial site of PJSC "MMK" is carried out according to the circulating scheme through the circulating part of the Magnitogorsk reservoir and local revolving cycles, with repeated use of water. The total length of the waterway from the discharge to the common industrial septic tanks to the water intake facilities is 15–16 km (Figure 1).

The recharges of the circulating water supply system through the circulating part of the Magnitogorsk reservoir are:

- wastewater entering the circulating part of the Magnitogorsk reservoir, after using the intaken water from the Magnitogorsk reservoir by pumping stations of the first rise No. 16, 16a (wastewater after cooling the process equipment of the oxygen department (oxygen station No. 4) and TTP (northern discharge in the summer period);
- river water entering the circulating part of the Magnitogorsk reservoir through the northern regulating water meter during the flood period (during the low-flow period the tray is closed).

Recharge is used to cover irretrievable losses at the industrial site: losses in production, feeding of local circulating cycles (evaporation from the surface of sewage treatment plants, dripping and venting, from cooling towers).

2. The main factors of formation of the thermal regime of the reservoir

The thermal regime of reservoirs is formed under the influence of various factors that can be divided into two main groups. The first group includes various types of income and heat loss by water. This includes the heat exchange of water mass with the environment (atmosphere and bottom soil), as well as heat from internal sources (heat generation and ice melting, internal friction, biological processes, etc.) [2].

The second group includes factors due to which the incoming heat is redistributed within the water mass: turbulent, wind and dynamic mixing and free convective mixing. These processes occur simultaneously and are in interaction. The morphometric characteristics of the reservoir and, first of all, the depth of the reservoir and the area of the water surface have a great influence on the thermal regime. Obviously, at the same specific heat input to the water surface, the temperature value and its variability in the vertical direction depend on the water depth. As for the size of the water surface, they largely determine the intensity of wind action on thermal regime.

It should be taken into account that each individual reservoir may have its own individual "features" and peculiarities in the formation of the thermal regime. They can influence both at a sufficiently local level, by changing the temperature in certain areas, and play dominating role in the formation of temperature fields. In modern conditions of operation of reservoirs, such influence is often associated with anthropogenic impact.

For most part of the Magnitogorsk Reservoir, the general conditions and principles of the formation of the temperature regime in different seasons of the year are the common. For the assessment of technogenic impacts and the development of measures to minimize them, the analysis of thermal fields generated by diverting heated effluents from PJSC MMK is fundamentally important. If in work [3], the processes of formation of zones of thermal effects of the energy complex with a direct-flow cooling system were studied, in the present work the features of the formation of zones with a re-sequential cooling system are considered. Field studies were conducted in May 2015 and 2017, as well as in July 2013, in compliance with the requirements of the governing documents [4] and the use of modern equipment.

3. Features of technological and meteorological conditions during studies

At all stages of the investigations, the operation of PJSC “MMK” and the water use system was carried out in a regular mode. Water balance, thermal and hydrochemical regimes in the central part of the site under consideration, is mainly formed under the influence of inflow of the Ural river, as well as wastewater of the outlet number 1.

During studies performed on July 11, 2013 the flow rate in the Ural river (Magnitogorsk Reservoir) in the area of the "Southern passage" was ~ 5 m3 / s, the surface water temperature was 25.2 °C. The discharge flow rate by discharge No. 1 is 11.6 m3 / s with a temperature of 29.1 °C. When mixing of 2 streams of different densities occurs in the water area between passages, a water temperature distribution is formed, characterized by the presence of a jump layer, which is located at a depth of 2-4 m. The difference between surface and bottom temperatures reaches 5-7 °C (Figure 1). Characteristics of meteorological conditions are given in Table 1.

During surveys of May 24 in 2015 and 2017, due to the passage of different phases of the flood during this period, the flow rate in the Ural river was, respectively, ~ 110 and 12 m3 / s. The surface water temperature was 13 and 13.3 ,3C, respectively. The discharge flow rate of issue No. 1 was about 8.5 m3 / s with temperatures of 23.1 and 21.4 ° C, respectively. The difference between surface and bottom temperatures did not exceed 3 ° C (Figure 1).

The most detailed study of water temperature fields was performed on May 24, 2017. The measurement results are presented in Figure 2 and Figure 3.

IOP Conf. Series: Materials Science and Engineering **581** (2019) 012038 doi:10.1088/1757-899X/581/1/012038

Table 1. Meteorological conditions during field studies (www.rp5.ru)

Date	Local time	Air temperature, °C	Direction of the wind	Wind speed, m/s	Cloudiness,%
11.07.2013	02:00	17.3	Northeast	3	60
	08:00	20.0	Northeast	2	20–30.
	14:00	28.5	South-South-West	3	70 – 80.
	20:00	26.1	South	3	60
24.05.2015	02:00	14.5	Southeast	1	70 – 80.
	08:00	19.4	West-North-West	3	40
	14:00	27.0	West-North-West	6	70 – 80.
	20:00	20.4	Northwest	3	90
24.05.2017	02:00	3.8	Calm	0	100
	08:00	7.7	Western	3	20–30
	14:00	12.4	Western	4	60
	20:00	13.3	Western	3	No clouds

4. Main results of temperature surveys

Attention is drawn to the fact that the water temperature distribution curves in depth are close to each other in the measurements of 2015 and 2017, with a significant difference in the flow rates of water in the Ural river, as well as air temperature. These results quite well confirm the statement that the flow rate in the Ural river is not the governing factor in the formation of the thermal regime of the central part of the Magnitogorsk reservoir. The role of daily air temperatures, due to the very high heat capacity of the water, is also small. Thus, in particular, according to Table 1, the average daily air temperature on May 24, 2015 was by 11 ° C higher than on May 24, 2017, while temperature distributions, as follows from Figure 1, are very close.

The presented materials make it possible to get a fairly "solid" idea about the formation of temperature fields in the central part of the Magnitogorsk reservoir at the time of the studies, to provide input data for verifying hydrodynamic models for calculating the heat affected zones of large energy complexes for both 2D and 3D approaches. The water temperature distribution was characterized by substantial inhomogeneity both in the water area and in the depth of the water body, and also by the presence of a fairly stable jump layer, especially during the warmest period (July-August). The presence of a significant difference in water temperature on the surface and in the near-bottom horizons allows us to work out possible variants for selective water intake in order to increase the sustainability of the enterprise's water supply system in fairly short-term periods with extremely high air temperatures.

5. Conclusions

Field studies of thermal fields in the Magnitogorsk reservoir, formed in the active impact zone of PJSC MMK, showed their stability and inertia with respect to both hydrological and meteorological factors. In the warmest limiting period, a substantial vertical temperature inhomogeneity is formed. The obtained data are a necessary and obligatory component for verification and binding of hydrodynamic models for the conditions of the Magnitogorsk reservoir.

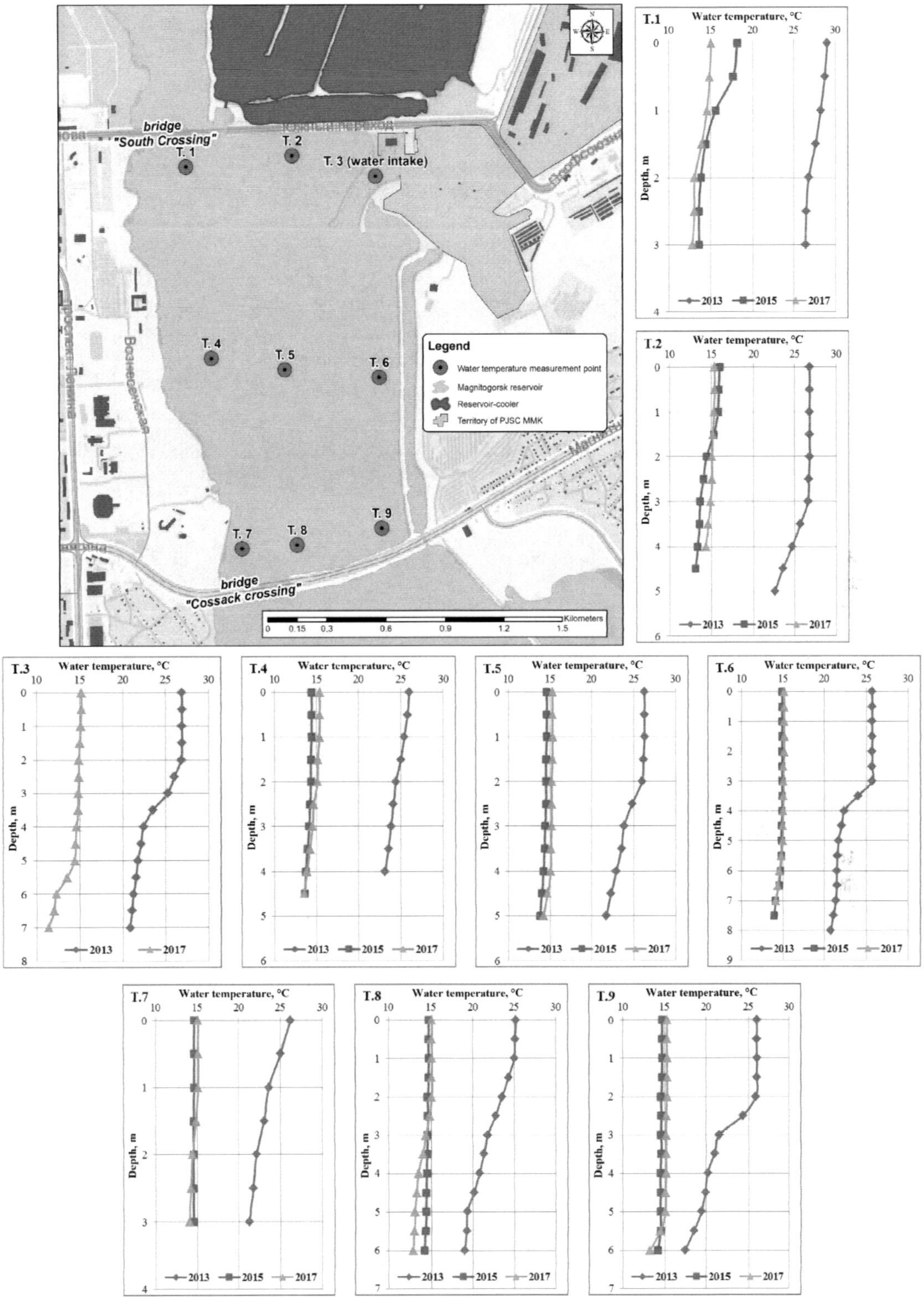

Figure 1 Water temperature distribution in depth at different verticals in the central part of the Magnitogorsk reservoir (area of activity of PJSC "MMK")

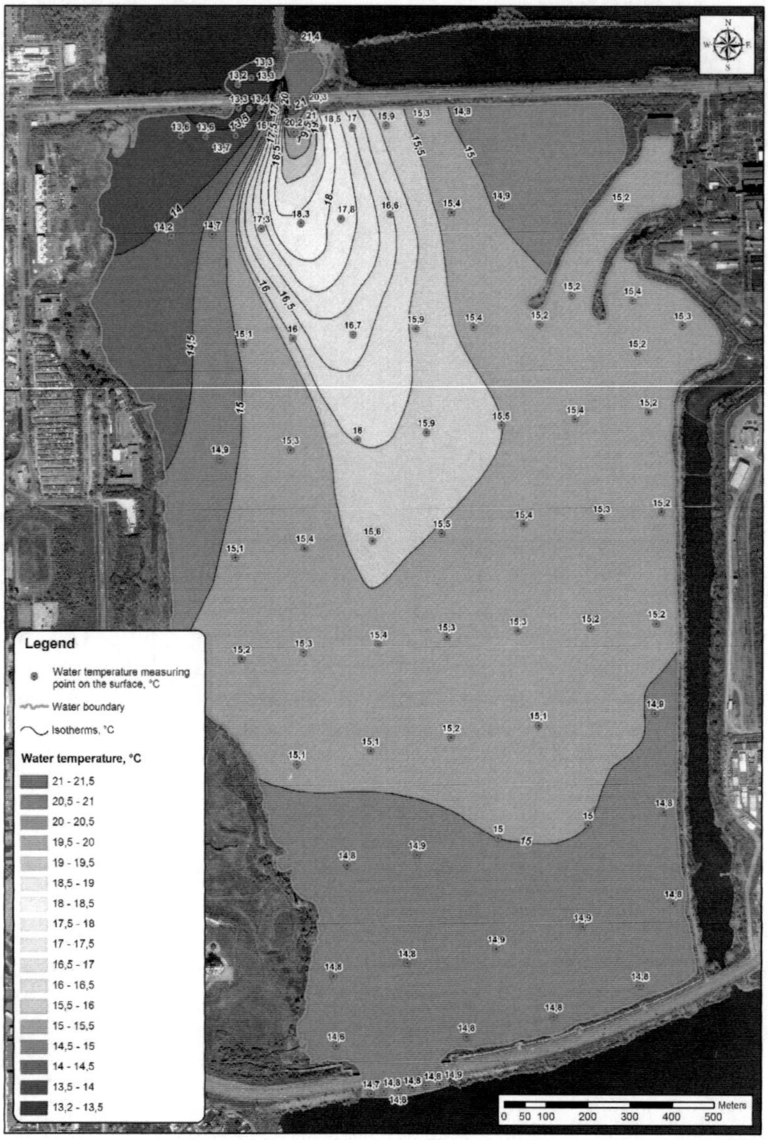

Figure 2 Water temperature distribution as measured by May 24, 2017, at the surface

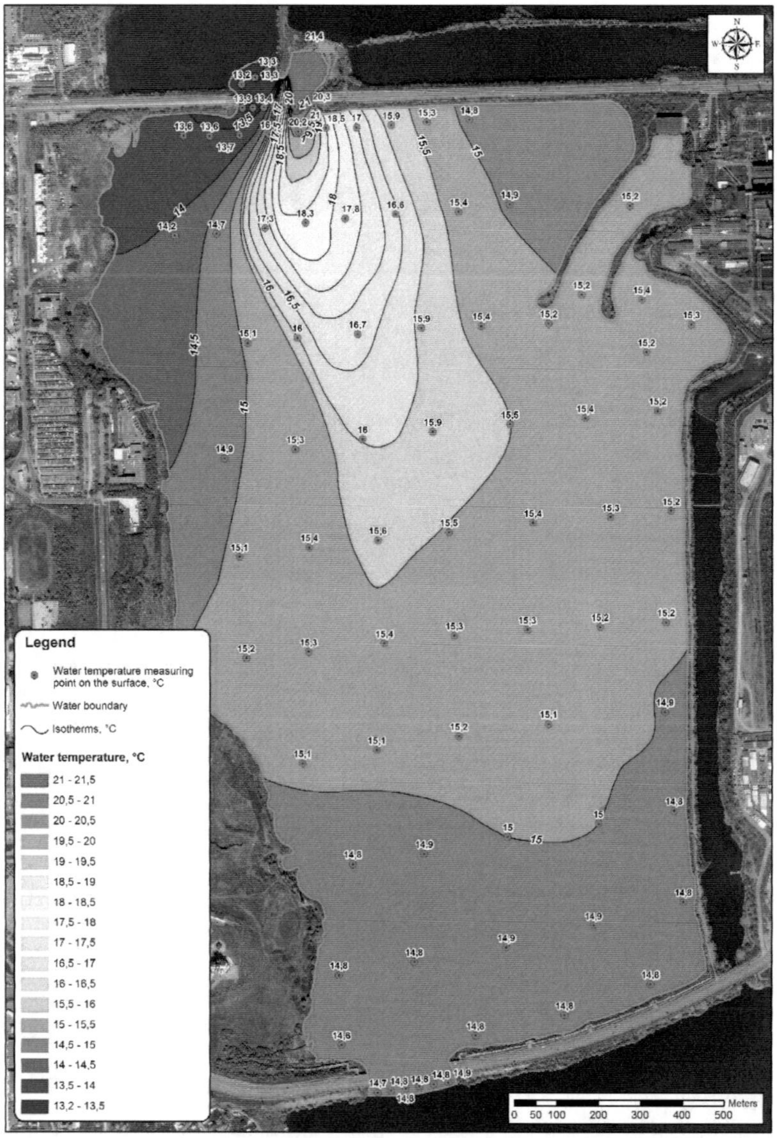

Figure 3 Water temperature distribution as measured by May 24, 2017, at a depth of 2 m

Acknowledgments
The work was supported by Russian Scientific Foundation (grant No. 17-77-20093).

References

[1] Technical and economic substantiation of water supply in Sverdlovsk 1983 *M: Soyuzgiprovodkhoz*, (in Russian).
[2] Gottlieb J.L. Thermal regime of hydroelectric reservoirs 1976 *Leningrad: Gidrometeoizdat* (in Russian).
[3] Lyubimova T, Lepikhin A, Parshakova Ya, Lyakhin Yu, Tiunov A 2018 *International Journal of Heat and Mass Transfer* **126** 342
[4] RD 52.10.842-17 Manual of hydrometeorological stations and posts 2017 **9** Hydrometeorological observations at sea stations and posts. Part 1. Hydrological observations at coastal stations and posts *M .: Publishing house ITRK*, (in Russian).

IOP Conf. Series: Materials Science and Engineering **581** (2019) 012039 doi:10.1088/1757-899X/581/1/012039

Simulation of Uniaxial Deformation of a Ferrogel Sample Exposed by the External Magnetic Field

P Melenev [1, 2] **and A Ryzhkov** [2, 3]

[1] Institute of Natural Sciences and Mathematics, Ural Federal University, 620026, Ekaterinburg, Russia
[2] Institute of Continuous Media Mechanics, Perm Federal Research Centre, 614013, Perm, Russia
[3] Applied Mathematics and Mechanics Faculty, Perm National Research Polytechnical University, 614000, Perm, Russia

E-mail: melenev@icmm.ru

Abstract. Ferrogels are magnetoactive composite materials with very soft hydrogel matrix, filled by ferromagnetic particles. This work is devoted to a numerical investigation of magneto-mechanical behaviour of a small sample of ferrogel. Our model is based on a coarse grained molecular dynamics approach: polymeric matrix is emulated by lattice of bead-strings chains and magnetic single-domain particles are considered as rigid spheres and placed in a nodes of the lattice. This model was previously successfully used for a research of equilibrium magnetization of the material. Here we examine of the reaction of ferrogel sample on uniaxial tension in presence of an external magnetic field. Modelling proves that key role in a changes of mechanical response during the magnetization plays an internal structure of material, specifically, the existence of particles aggregates, induced by forces of dipolar magnetic interaction. Also the influence of magnetic anisotropy on a process is considered.

1. Introduction

Soft magnetic composites (SMC) are interested for researches due to pronounced interrelation between mechanical and magnetic characteristics of SMC, caused by the ability of sufficient magnetically induced rearrangements of magnetic filler inside the soft polymeric matrix. Ferrogels may be considered as an "extreme case" of SMC with respect to matrix softness. These materials consist of hydrogels with very low elastic modules (characteristic value is less than 10 kPa) filled by ferromagnetic particles which usually have sizes within tens of nanometers and so have single-domain magnetic structure. This combination allows ferrogels to demonstrate noticeable response to magnetic field even of relatively low strengths (less then 0.1 T) which is substantial from the perspective of utilization of ferrogels in a bio-medical applications. Beyond the mechanical properties magnetic action could change the shape of the ferrogel samples [1-3].

The complex composition of SMC accounts for their uncial magneto-mechanical behavior, but simultaneously it is "headache" for the researches who try to investigate theoretically (and numerically) of these materials. Indeed, a mechanical response of composites is determined to a great extent by the local structure of their filler. Which in case of SMC may be altered by magnetic forces, induced by the external field and/or particle dipolar interaction. In turn the distribution of the magnetic field in the sample may also depends on filler allocation. This strong linkage between macrocharacteristics of the material and its internal structure complicates application of conventional continuous methods for the SMC theoretical study.

One of the possible way of direct examination of the micro-structure of magnetic composite is application of particle-based methods, where both polymeric matrix and dispersed filler are modelled by discrete objects, namely, particles and bonds. This simplifies sufficiently the description of composite components and because of corresponding decrease of computation cost allows to consider in simulations systems which contain tens or even hundreds of thousands of magnetic particles [4, 5]. This approach is especially prospective for examination of ferrogel since real objects made of these materials developed for bio-medical applications may have submicron size. So it is possible to model the whole material sample [6]. On the other hand, there are examples of multiscales models [7-10] where data, obtained on the base, structural level of magnetic composite, are used for an evaluation of macroscopic material characteristics.

A nice review paper [11] contains description of various approaches of theoretical treatment of ferrogels, including several particle-based models. They could be divided into two types regarding way of account of polymer matrix. In the first one, matrix is represented exclusively by bonds between the filler particles, thus polymeric excluded volume is not considered (e.g. in [12]). This approach allows to perform a calculations of systems with larger amount of particles, but simultaneously complicates the procedure of identification of bonds parameters based on experimental and theoretical data regarding microstate of real polymers. Also it appears that in this case examination of composite with low or moderate filler concentration would not efficiently due to oversimplified model of the matrix. The second type of models presents polymer network as a set of chains of weakly deformable or rigid particles, which could be associated with polymer segments or blobs. This approach allows to model various types of matrix structure and variants of inclusion of filler particles into the polymer. The model, employed in our work here, belongs to this method with bead-strings description of polymer network.

We consider the behavior of small sample of ferrogel contains up to 1000 single-domain magnetic particles, arranged in a nodes of polymer network. Our simulation is based on coarse grained molecular dynamics approach. Earlier this model was used in simulations of equilibrium state of the samples with various values of filler concentration and particle characteristics in presence of permanent magnetic field. Particularly we considered dependence on the field and material properties of the sample volume and filler structure [13-14]. Development and studies of real ferrogels which may be utilized, among other, as micromachines (or their details) require ability to model a material response to mechanical loading. Also, the following calculations may be necessary for the identification of model parameters based on experimental data. This work is devoted to numerical examination of mechanical response of small ferrogel sample to uniaxial tension and examination of dependence of material reaction on the external magnetic field.

2. Simulation

2.1. Ferrogel sample model

Here we briefly describe the model of ferrogel, used in calculations. More details may be found in [13]. Filler particles are presented as rigid spheres of diameter d_p, each of which has magnetic moment of constant magnitude m_p. Dipolar magnetic interaction operates between the particles. Also each particles possesses uniaxial magnetic anisotropy with finite value of energy barrier E_A.

Polymer macromolecules is modelled by the chains of identical rigid spheres – beads (with diameter $d_m < d_p$), linked by practically non-deformable bonds. Chains may bend freely – flexural stiffness is absent. Chains is formed to quasi-cubic lattice some of nodes of which are occupied by magnetic particles. The portion of occupied nodes depends on filler concentration.

The system may be exposed by the external permanent and uniform magnetic field with strength H_0. The presence of constant temperature T, essential for the simulation of processes on submicron scales, is accounted by introduction to the model of Langevin thermostat. Thus the mechanical reaction of matrix is determined by the mechanism of entropic elasticity. Then, characteristic energy of interaction of dipoles with each other and with the external field as well as the energy of particle magnetic anisotropy are estimated relatively to thermal energy and specified in terms of corresponding dimensionless parameters:

$$\lambda = \frac{\mu_0}{4\pi} \frac{m_p^2}{d_p^3 k_B T}; \quad \xi = \mu_0 \frac{m_p H_0}{k_B T}; \quad \sigma = \frac{E_A}{k_B T}, \tag{1}$$

IOP Conf. Series: Materials Science and Engineering **581** (2019) 012039 doi:10.1088/1757-899X/581/1/012039

here μ_0 denotes the vacuum permittivity.
The calculations are performed using ESPResSo software package [15] with our modifications, charged with magnetic anisotropy simulations.

2.2. Simulation of deformation process
At the start of each numerical experiment the thermostating of the sample is performed. Earlier it was found that in samples with high filler concentration and strength of dipolar interactions in absence of the external magnetic field elongated and chaotically oriented chain-like aggregates of particles are formed (see figure 2 in [13]). External field attempts to turn these chains along its direction, but degree of such an orientation depends on the field magnitude and material properties, particularly magnetic anisotropy energy. In the samples where dipolar forces are not able to form particles aggregates without external field, magnetization practically does not lead to filler clustering – usually, only small number of new particle pairs appears.

When system has reached equilibrium an uniaxial tension of the sample starts. Because of irregular relief of the sample's sides for performing of hard loading we use the following trick. On two opposite sides the layers are chosen, each with thickness of several d_p. Model elements – magnetic particles and polymer beads – inside these layers are fixed, which means their position do not change according to equation of motion integration. During the calculation one of the layer – "base" – remains immobile. And all of elements of the second one were moved in direction of tension to new positions which ensured the deformation ε of the sample in limit of 5%. Then the system has time to reach the new equilibrium state. And this operation repeats until achievement $\varepsilon = 50\%$ or fracture of the sample. The reaction of the sample is described by the net force f_b acts on the magnetic particles in "base" level. The presence of thermofluctuations of model elements causes to fluctuate the instantaneous value of this force. Therefore the time averaging of f_b is performed during given number (in practice – several tens) of "relaxation" numerical steps. Obtained value of force $\langle f_b \rangle_t$ is used in calculation of mechanical module of material:

$$G = S_0^{-1} \frac{\Delta \langle f_b \rangle_t}{\Delta \varepsilon}, \tag{2}$$

where S_0 is initial area of sample cross-section.

It is significant to note that used in present calculations values of model parameters do not correspond to some concrete examples of real ferrogels. And in this work we are focused on obtaining of qualitative results only. Particularly, we were not interested to find absolute values of mechanical module but try to examine it changes, caused by different factors: deformation, external magnetic field, magnetic anisotropy and strength of dipolar interaction of the particles.

3. Results

Simulation proves the key role which plays the internal filler structure in mechanical behaviour of the ferrogel. Thus, in absence of the external field the samples where dipolar forces able to form particle aggregates have higher value of G in comparison with analogous (in respect of particle concentration and anisotropy) samples with weaker dipolar interaction. Samples of the latter type, characterised by $\lambda < 3$, in absence of the field demonstrate practically linear elastic response for whole range of tension. For the materials with stronger magnetic forces the growth of mechanical module during the deformation also is not dramatic and does not exceed 10%, see dashed lines on figure 1. Magnetisation leads to noticeable increase of G in case of deformation along the external field. Also mechanical response of the samples with high values of λ becomes sufficiently nonlinear. Meanwhile in materials with weaker magnetic interactions much lower increase of mechanical module is observed, see solid lines on figure 1. The found difference of field dependence of G may be described by the origin of anisotropy of filler structure in the samples, where the external filed orients already existed particle aggregates. This assumption is supported by the simulation of deformation perpendicular to the direction of magnetisation, where only a small growth of value of G was observed – for all considered samples it was several times lower, than in case of deformation along the field, see figures 2-3. It is interesting that we obtained results similar to the experimental data for deformations of samples of soft magnetic elastomers with polydimethylsiloxane matrix, filled by microparticles of carbonyl iron [16], which emphasizes the similar sense of dipolar interaction in forming of mechanical response of different types of SMC.

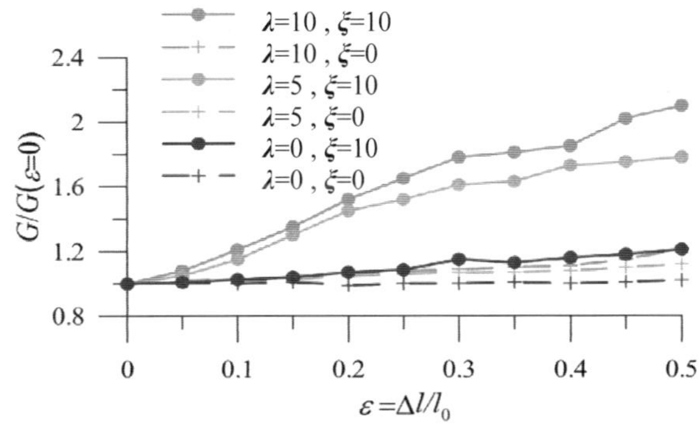

Figure 1. The dependence of relative value of mechanical module on strain. Presented results obtained for particles with uniaxial magnetic anisotropy, $\sigma = 2$.

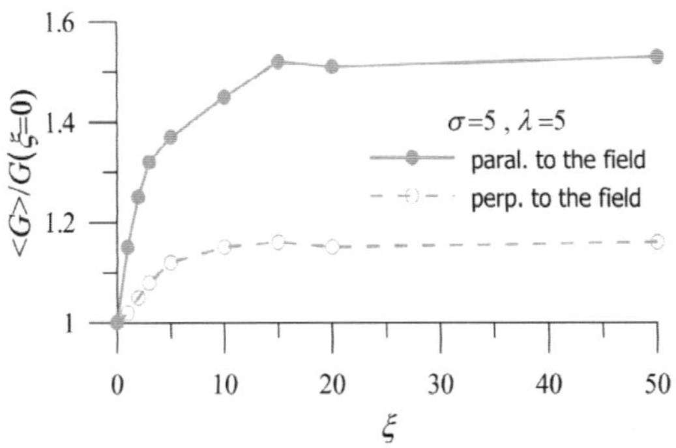

Figure 2. The dependence of the relative value of mechanical module on the external field. Presented results obtained for the system with $\sigma = 5, \lambda = 5$. Dashed line corresponds deformations perpendicular to field direction and solid line – parallel case.

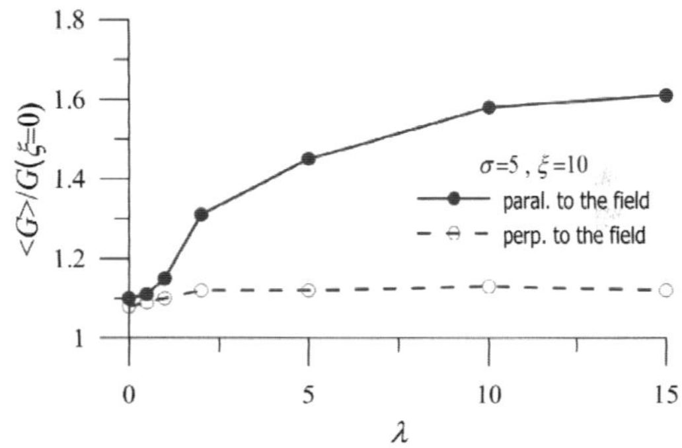

Figure 3. The dependence of the relative value of mechanical module on dipolar interaction strength. Presented results obtained for the system with $\sigma = 5, \xi = 10$. Dashed line corresponds deformations perpendicular to field direction and solid line – parallel case.

The distinguish feature of our ferrogel model is capability to consider particles with finite energy of magnetic anisotropy. The simulation of the samples with various E_A values shows that in materials with intense dipolar interactions and noticeable magnetic anisotropy (i.e. $\sigma \geq 1$) the increase of E_A entails some growth of mechanical module, which is evidently connected with the presence of local stresses occurred in matrix due to rotations of particles induced by reorientations of their magnetic moments. The magnetization, as it was already described, leads to increase of mechanical module (see figure 2). But for samples with intense dipolar interactions relative value of field-induced variation of G decreases within growth of σ, see figure 4. We suppose that this happens because above-mentioned initial hardening of the material hampers the rearrangements of the filler in the external field. So lesser changes of internal structure leads to lower growth of the mechanical module during the magnetization.

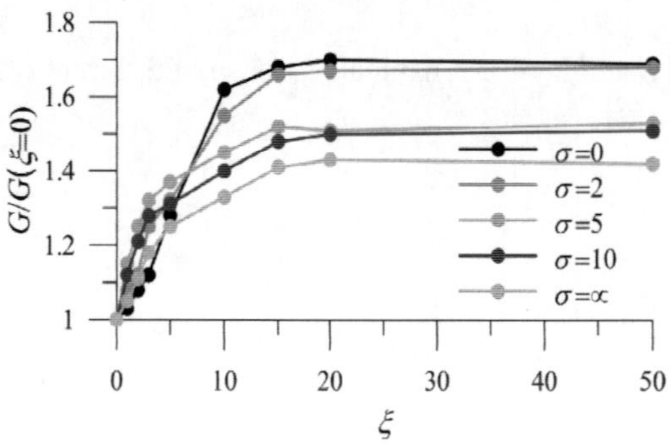

Figure 4. The dependence of relative value of mechanical module on the external for the systems with different values of anisotropy energy. Presented results obtained for samples with $\lambda = 10$ at strain $\varepsilon = 25\%$.

4. Conclusions

The results of numerical investigations on the process of uniaxial tension of small ferrogel sample in presence of permanent magnetic field, performed with employ of coarse grained molecular dynamics approach, confirm the key role of internal structure in forming of material response. In cases, when dipolar magnetic forces form chain-like aggregates of particles even in absence of the external field, the magnetisation leads to orientation of these clusters along the field. Occurred structure anisotropy provides the much larger growth of mechanical module than one, observed for the samples with uniform distribution of particles. Also it was found that the samples with higher energy of magnetic anisotropy are stiffer and, at the same time, demonstrate lower field-induced variation of mechanical module.

The performed calculations prove the applicability of chosen model of ferrogel for simulation of deformation processes. Now we can point out the following primary prospects of prolongation of this work. Firstly, it is necessary to clarify the influence of temperature on the mechanical response, because in our model thermofluctions alongside the polymer network structure determine the elastic properties of the matrix. Secondly, it would be interested to consider material compression, during which in the sample the regions of high particle concentration could occur. It may favour the formation of new particle aggregates and thus sufficient changes of mechanical properties of the sample.

Acknowledgments

The work was supported by Russian Foundation for Basic Research, project 17-41-590123. The calculations were performed using computer clusters of the Institute for Computational Physics of University of Stuttgart (Germany) and the Institute of Continuous Media Mechanics of Perm Federal Research Centre (Russia).

References

[1] Mitsumata T, Ikeda K, Gong J P, Osada Y, Szabó D and Zrínyi M 1999 *J. Appl. Phys.* **85** 8451-8455
[2] Gollwitzer C, Turanov A, Krekhova M, Lattermann G, Rehberg I and Richter R 2008 *J. Chem. Phys.* **128** 164709
[3] Datta P 2018 Magnetic gels *Polymeric Gels: Characterization, Properties and Biomedical Applications* (*Woodhead Publishing Series in Biomaterials*) ed K Pal and I Banerjee (Woodhead Publishing) chapter 17 pp 441–465
[4] Dudek M, Grabiec B and Wojciechowski K 2007 *Rev. Adv. Mater. Sc.* **14** 167-173
[5] Weeber R, Kantorovich S and Holm C 2012 *Soft Matter* **8** 9923–32
[6] Ryzhkov A and Raikher Y 2018 *Nanomaterials* **10** 763
[7] Metch P, Kalina K and Kästner M 2016 *Int. J. Solids and Struct.* **102-103** 286-296
[8] Keip M-A and Rambausek M 2017 *Int. J. Solids and Struct.* **121** 1-20
[9] Danas K 2017 *J. Mech. Phys. of Solids* **105** 25-53
[10] Zabihyan R, Merghiem J, Javili A and Steinmann P 2018 *Int. J. Solids and Struct.* **130-131** 105-121
[11] Weeber R 2018 *J. Phys.: Cond. Matter* **30** 063002
[12] Sánchez P, Minina E, Kantorovich S and Kramarenko E 2019 *Soft Matter* **15** 179-189

[13] Ryzhkov A, Melenev P, Balasoiu M and Raikher Y 2016 *J. Chem. Phys.* **145** 074905
[14] Ryzhkov A and Raikher Y 2017 *J. Magn. Magn. Mater.* **431** 192-195
[15] Arnold A, Lenz O, Kesselheim S, Weeber R, Fahrenberger F, Röhm D, Košovan P and Holm C 2013 ESPResSo 3.1 - Molecular dynamics software for coarse-grained models *Meshfree Methods for Partial Differential Equations VI* (*Lecture Notes in Computational Science and Engineering* vol 89) ed M Griebel and M Schweitzer (Springer) pp 1–23
[16] Varga Z, Filipcsei G and Zrínyi M 2006 *Polymer* **47** 227-233

IOP Conf. Series: Materials Science and Engineering **581** (2019) 012040 doi:10.1088/1757-899X/581/1/012040

Two-Phase System in a Rotating Cylindrical Cavity under the Transverse Vibrations

I E Karpunin [1], N V Kozlov [2, a]
[1]Perm State Humanitarian Pedagogical University, Sibirskaya St. 24, 614990, Russia, Perm
[2]Institute of Continuous Media Mechanics of the Ural Branch of Russian Academy of Science, Academika Koroleva St. 1, 614013, Russia, Perm

E-mail: vankarpunin@yandex.ru, [a] kozlov.n@icmm.ru

Abstract. The effect of large-amplitude translational vibrations on the dynamics of a two-phase system (light cylinder in liquid or two immiscible liquids, placed in a rotating cylindrical cavity) is studied experimentally. The experiments are run at high rotation rate, when under the action of centrifugal force the light phase is located near the rotation axis. Vibrations are perpendicular to the rotation axis and their frequency is close to the rotation rate. When the frequencies coincide, the vibrations change the centrifugal field configuration. The light phase column shifts stationary in the rotating frame of reference.

1. Introduction

Vibrations are an efficient instrument of control of mechanical systems and make new technological solutions possible [1]. The effect of vibrations on density-inhomogeneous hydrodynamic systems leads to a number of averaged effects. One of the bright effects connected with vibrations is steady streaming [2], which finds its application in the most various areas: from geophysics and industrial technologies to the description of the principles of physiology of a human body. Separately, it is worth mentioning rotating systems, for which a large number of natural oscillations is specific that can be excited by variable force fields [3]. Examples of the rotating two-phase systems: a centrifuged liquid layer with a free surface [4–6], bodies in liquid [7, 8], systems of immiscible liquids of different density [9–11]. It is known that the field of gravity leads to a radial displacement of the phase inclusion in the rotating cavity [4, 8]. The oscillations of the liquid excited in this case lead to the differential rotation of the interface [5, 8]. Independently of the action of the gravity field, vibrations lead to the excitation of an additional differential rotation of the light phase due to the excitation of inertial oscillations [6, 7, 9, 10]. In the case of a light solid body, its averaged motion depends on the frequency ratio between cavity rotation and vibrations, and its dynamics is complex [9–11]. In the case of a centrifuged two-liquid system, the stability of the interface strongly depends on the relative radius of the surface, determined by the volume filling, and the relative length of the cavity [5, 8]. Interestingly, the effects described above also manifest themselves in rotating systems with a temperature distribution that is not uniform along the radius [12]. Experiments showed that resonant excitation of oscillations of the rotating hydrodynamic systems by means of vibrations with eigenfrequencies can serve as an efficient instrument of control of flows and of heat and mass transfer. In this work, we consider the case of vibrations applied to a rotating two-phase system with the frequency close to the frequency of rotation.

2. Experimental setup and methods

The experimental model is a transparent cylindrical cavity *1* (figure 1) made of acrylic glass tube with a transparent flange *2*. In the first case, a light (whose density is less than that of the surrounding liquid) cylindrical body *3* is placed inside, and the space between it and the walls of the cavity is filled with a working liquid. In the second case, the cavity is filled with a pair of immiscible liquids of different densities. Dimensions of the working cavity are the following: the length $L_1 = 26.5$ cm (with a body) and $L_1 = 13.5$ cm (with two liquids), the inner radius $R_1 = 3.5$ cm. The solid body is made of polyamide and has the length $L_2 = 24.5$ cm, the radius $R_2 = 2.0$ cm. Its density is 0.88 g/cm^3. The length of the light liquid column is $L_3 = 13.5$ cm, and its radius varies in the range $R_3 = (0.7 - 2.3)$ cm. In experiments with a two-liquid system, a shorter cavity is used. This allows to weaken the effects associated with the occurrence of centrifugal waves at the interface of liquids during vibrations. The physical characteristics of the working liquids used in the experiments are given in Table 1. The cavity is fixed in pillars with bearings *4*, which are tightened with metal studs *5* in order to prevent the poles from tilting relative to the cavity. The cavity is brought into rotation by a FL86STH80-4208A stepper motor *6*, fixed on a textolite platform *7*. The motor is controlled by an SMD-4.2 driver paired with a ZET 210 ADC module (*8*), and the power is supplied by a NES-350-48 power supply (*9*). On the motor, a fan *10* of the HAF-71M type is fixed, aimed to maintain the required thermal operation mode. Rotation from the motor is transmitted to the cavity through the coupling *11*.

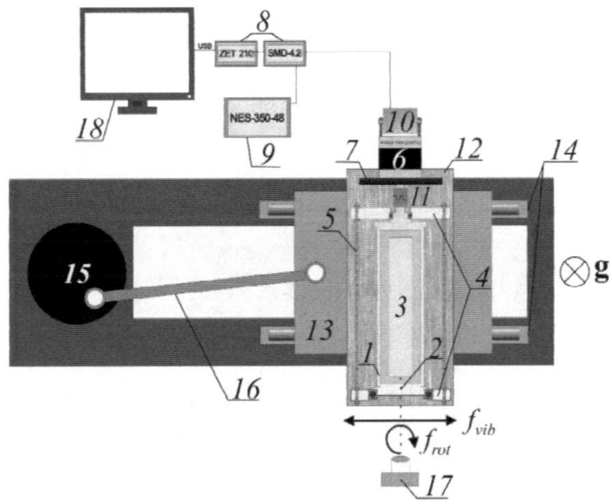

Figure 1. Experimental setup (top view).

Table 1. Physical characteristics of the working liquids.

	Density, g/cm^3	Viscosity, cSt	Surface tension coefficient, dyn/cm
Water	1	1	54
Oil I-5A	0.83	10.3	27
Water+glycerol (50%)	1.12	5	37
Water+glycerol (75%)	1.2	21	44

The entire experimental model is fixed on an aluminum platform *12*. Without rotation, the light cylindrical body (light liquid) is located in the upper part of the cavity. Experimental study is conducted for the case of centrifuged state when the light phase inclusion is located near the axis of cavity rotation (figure 2). In this case, the light liquid takes the form of an axisymmetric column, elongated parallel to the axis of rotation.

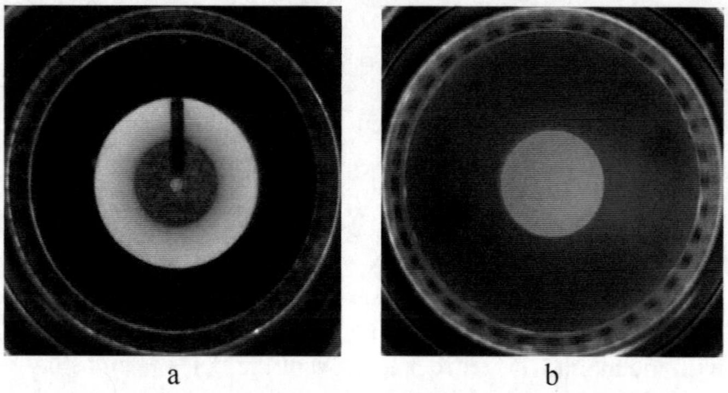

Figure 2. Centrifuged state of the light phase in the absence of vibrations: (a) the solid body; (b) the liquid column.

The experimental setup is located on a vibrator platform *13* which performs horizontal vibrations (figure 1). The vibrator platform is installed on linear ball bearings, which slide on horizontal parallel rails *14*. Rotational motion of the rotor of a motor *15* is transformed into translational motion of the vibrator platform by means of a crank-and-rod mechanism *16*. The direction of vibrations is perpendicular to the cavity rotation axis. Observation of the body's behavior, as well as its photo and video recording, is carried out from the side of the transparent end of the cell under continuous illumination. High-speed video recording is made with an Optronis CamRecord CL600x2 (*17*) camera. Capturing images and processing the results of photo and video recordings is performed using a computer *18*. To study the dynamics of the phase inclusion, frame-by-frame video recording is performed. The rotation rate of the body in the laboratory reference frame f_s is calculated from the displacement of a label stuck on its end and from the time between frames, which is known from the camera records. The rotation rate in the cavity reference system is $\Delta f = f_s - f_r$, where f_r is the rotation rate of the cavity. In the case of liquid inclusion, the definition of Δf by this method is not possible. To find the trajectory of the light body and of the light-liquid column, the images are processed using the programme ImageJ. The coordinates of the cavity center and of the phase inclusion are obtained. The axis of the cavity retains a fixed position relative to the vibrator platform. Subtracting the coordinates of the cavity center from the coordinates of the center of inclusion, a transition is made to the reference system associated with the vibrator platform, relative to which the inclusion motion is considered.

In order to determine the working ranges of the frequency of cavity rotation and the frequency of vibrations, experiments were conducted in the gravity field in the absence of vibrations. Thus, the solid body goes from the cavity wall to the middle at f_r =4.88 rps. For the solid phase inclusion, the parameters of experiments are: the frequency of rotation (vibrations) $f_r = f_{vib} = 5-7$ Hz, the amplitude of vibrations is $A_{vib} = 0-3.0$ cm. For a two-liquid system, the frequency of centrifugation depends on the parameters of the liquids: viscosity, density, relative filling of the cavity $q = R_3^2 / R_1^2$. The studied pairs of liquids and their relative characteristics are presented in Table 2. The parameters of experiments: $f_r = f_{vib} = 5.75-7.75$ Hz, $A_{vib} = 0-1.5$ cm. The amplitude of vibrations is lower than for the solid body, due to the appearance of waves at the interface and, as a result, destruction of the steady centrifuged state of the light-liquid column.

Table 2. Physical characteristics of liquid pairs.

	Density ratio ρ_{rel}	Viscosity ratio ν_{rel}	Coefficient of interfacial tension, dyn/cm
Water – Oil I-5A	0.83	10.3	31
Water+glycerol (50%) – Oil I-5A	0.74	2.1	17
Water+glycerol (75%) – Oil I-5A	0.69	0.49	8.6

3. Experimental results and discussion

The impact on the cylindrical phase inclusion in the rotating cavity is defined by the superposition of gravity and the oscillating force of inertia related to vibrations. Both of these forces excite oscillations in the cavity reference frame (with different frequencies), causing additive averaged effects. The intensity of vibrations can be characterized by the dimensionless vibration acceleration a_{vib}, describing the inertial forces in the reference frame of the vibrator platform due to the transverse horizontal vibrations relative to the rotation axis. If the unit of measurement is the characteristic centrifugal acceleration in the cavity $(2\pi f_r)^2 R_2$, then this parameter takes the form

$$a_{vib} = \frac{A_{vib}}{R_2}\left(\frac{f_{vib}}{f_r}\right)^2. \quad (1)$$

In the following section, the dynamics of a solid light body and a column of light liquid in a rotating cavity under vibrations with a frequency extremely close to the frequency of rotation, $f_{vib} = f_r$, is considered. In this regard, discussing the variation in the rotation frequency implies the variation in both of them.

3.1. Solid body dynamics

In the absence of vibrations under the action of gravity, the body shifts along the radius, and the differential rotation rate $\Delta f < 0$. The impact of vibrations with a frequency close to the frequency of rotation causes two-dimensional circular body oscillations, which are summed with the stationary displacement caused by the action of gravity. This is evidenced by the trajectory of the body (figure 3(a, b)). It performs circular oscillations (point 1) along a trajectory of radius b_1 relative to its average position (point 2), which in turn is displaced from the axis of cavity rotation by the distance b_2. This means that the body performs both azimuthal and radial motion in the cavity. At the same time, the body rotates around its axis.

Qualitatively, two types of body motion can be distinguished. The choice between them does not depend on the frequency of cavity rotation (vibrations) and is determined by the amplitude. The first type is characteristic for vibration amplitudes $A_{vib} < 1.5$ cm. The trajectory of the center of the body is characterized by a periodic change in its radius b_1 (figure 3 (a)). The second type of body motion is observed with an increase in the amplitude of vibrations and, accordingly, an increase in the vibration acceleration. At the same time, the body moves along a path with the constant radius b_1 (figure 3(b)). Considering the trajectory shown in figure 3(b) in the reference frame rotating with the cavity, the light body occupies a stable position at the distance b_1 from the axis of rotation. Relative to it, the body oscillates at a small amplitude b_2. The angular coordinate of the body relative to the rotating cavity does not change with time. Comparing figures 3(a) and (b) reveals that b_1 increases with the amplitude of vibrations. In the extreme case, when $b_1 = R_1 - R_2$, the body is in contact with the wall of the cell (figure 3(c)).

Displacement and oscillations of the body relative to the cavity lead to oscillations of the liquid and, as a result, to the generation of an average mass force causing differential rotation of the body [9]. At a fixed frequency, the differential rotation rate of the body Δf increases almost monotonically with the increase in vibration amplitude A_{vib} (figure 4(a)). It should be noted that the series of experimental points split according to the rotation frequency; this is due to the contribution of the gravity field to the differential rotation of the body [11]. In the range $A_{vib} = 1.0 - 1.5$ cm, non-monotonous behavior of Δf is observed. This phenomenon is under study. Presumably, it may be associated with a change in the type of body motion. When $A_{vib} = 3.0$ cm, the body is in contact with the wall of the cavity and the differential rotation stops, $\Delta f = 0$. With a fixed amplitude in the studied range of parameters, an increase in the rotation frequency leads to a slight increase in the differential rotation of the body (figure 4(b)). It should be noted that the dimensionless amplitude a_{vib} does not change for some given value of A_{vib}.

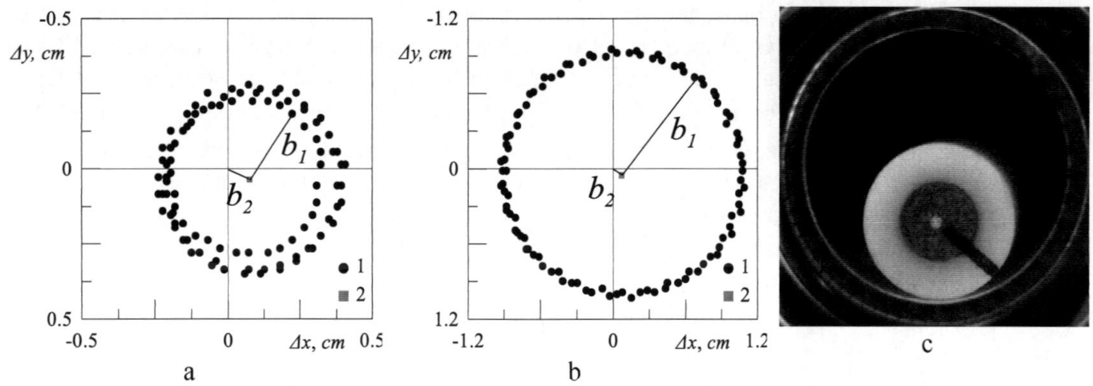

Figure 3. The trajectory of the center of the body in the reference frame of the vibrator platform at $f_r = f_{vib} = 5$ Hz: (a) $A_{vib} = 0.53$ cm; (b) $A_{vib} = 2.05$ cm; (c) body position at $A_{vib} = 3$ cm.

The data from figures 4(a) and (b) is generalized in figure 4(c) on the plane of dimensionless parameters. As it is shown [10], the differential rotation rate of the body is a linear superposition of two components, caused independently by the action of gravity and vibrations: $\Delta f = \Delta f_{vib} + \Delta f_g$. Consequently, the vibration component of the differential rotation rate $\Delta f_{vib} = (\Delta f - \Delta f_g)$. Figure 4(c) shows the graph of dependence of $\Delta f_{vib} / f_r$ on A_{vib} / R_2.

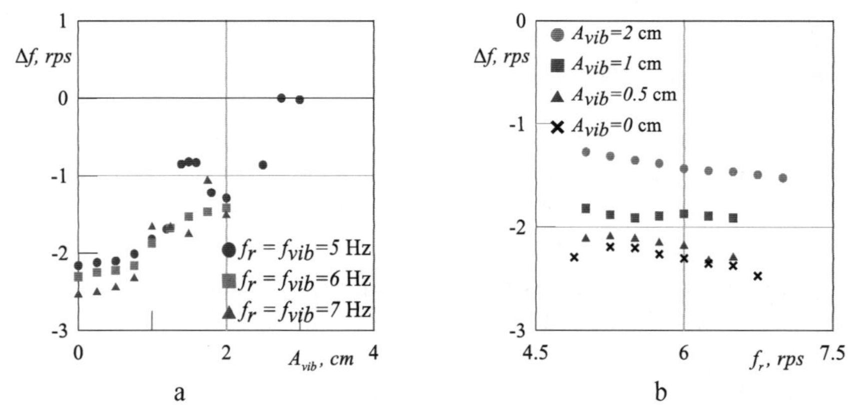

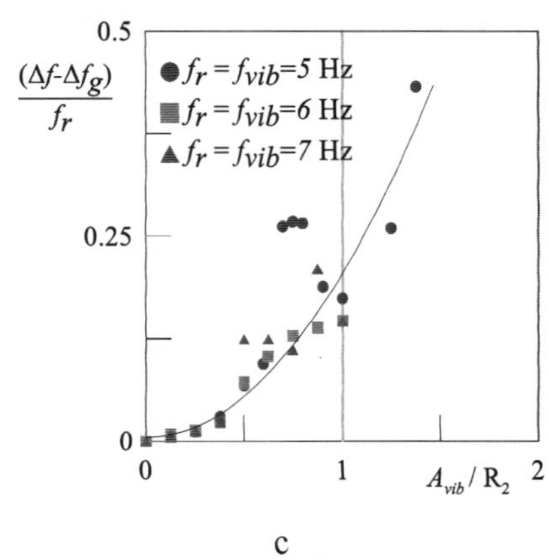

Figure 4. Experimental dependence of Δf: (a) on A_{vib} in the dimensional form; (b) on f_r in the dimensional form; (c) on A_{vib} in the dimensionless form.

The amplitude of vibrations is nondimensionalized by the radius of the body, and Δf_{vib} – by the frequency of cavity rotation. The gravitational component of the body rotation rate, Δf_g, was found in the absence of vibrations ($f_{vib} = 0$ Hz, $A_{vib} = 0$ cm). It can be noted that at $A_{vib} = 0 - 0.75$ cm the response of the body to a disturbance with a different frequency is the same. In general, the three dependences are qualitatively consistent with each other and can be approximated by a quadratic dependence: $(\Delta f - \Delta f_g) / f_r \sim (A_{vib} / R_2)^2$ (solid line in figure 4(c)). As it can be seen from the graph, the vibrations excite the leading body rotation ($\Delta f_{vib} > 0$) whose rate increases with the amplitude.

3.2. Light liquid column

In general, the dynamics of the light liquid-column is similar to the case of the light cylinder: under vibrations, the column shifts along the radius as a whole and occupies a stable position in the rotating reference frame. However, the system of two liquids also has a number of features. First, this system is characterized by the regular motion along a closed circular trajectory, b_1 does not oscillate (figure 5). The figure shows the trajectories for two cases: (a) $\nu_{rel} = 2.1$, $\rho_{rel} = 0.74$, $R_{rel} = 0.36$; (b) $\nu_{rel} = 0.49$, $\rho_{rel} = 0.69$, $R_{rel} = 0.49$. Here, $R_{rel} = R_3 / R_1$. All relative parameters are normalized to the parameters of the heavy liquid (solution of glycerin in water). In the reference frame of the vibrator platform, the center of the liquid column performs circular oscillations relative to the average shifted position. This dynamics does not depend on such parameters as the relative radius of the column R_{rel} and the properties of liquids ρ_{rel}, ν_{rel}.

Since the interface between two liquids is deformable, at high-amplitude vibrations, the centrifuged state is destroyed, and the column is broken. However, this is preceded by the appearance of steady waves at the interface. Two characteristic modes of wave propagation can be distinguished. The first type is marked by the occurrence of an azimuthal wave (figure 6 (a)). The liquid column is distorted, however, these disturbances are two-dimensional, and the wave crests are elongated parallel to the axis of the column. The second type is recognized by the simultaneous appearence of axial and azimuthal waves (figure 6 (b)). These disturbances are already three-dimensional and precede the regime when the liquid column collapses. At the same time, in the lateral projection, a stable pattern of clusters of visualizer particles is observed, forming lines in the wave nodes (figure 6 (c)). Despite the fact of the interface perturbation in both cases, the column of light liquid under the action of vibrations performs circular oscillations similar to figure 5.

3.3. Discussion

In [12], the dynamics of a rotating nonisothermal liquid in a field of transverse vibrations was experimentally and theoretically considered. It was shown for the case of small density gradients ($1 - \rho_{rel} \sim 10^{-3}$) that, if the frequency of cavity rotation and the frequency of transverse vibrations in the rotating reference frame coincide, a stationary force field is created that breaks the symmetry of the centrifugal force field. In the present experimental study, the case of coincidence of the frequencies of rotation and vibrations in systems with an interface is considered, where the density profile is step-like and $(1 - \rho_{rel}) \sim 10^{-1}$. Despite this difference, we observe a qualitatively similar dynamics. The averaged effect of vibrations at $f_r = f_{vib}$ is equivalent to a parallel translation of the rotation axis to a distance equal to half A_{vib}. This is evidenced by the experimental values of the displacement relative to the axis of cavity rotation of the solid (figure 7(a)) and the liquid column (figure 7(b)). The ratio $b_1 / A_{vib} \approx 0.5$ is maintained in all experiments and does not depend on what kind of light inclusion is considered. This ratio does not depend on the values of f_r and f_{vib} provided that $f_r / f_{vib} \approx 1$.

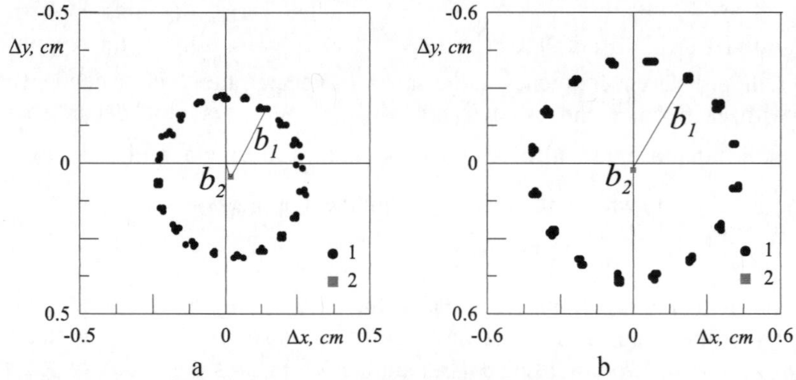

Figure 5. The trajectory of the center of the liquid column in the reference frame of the vibrator platform at (a) $f_r = f_{vib} = 7.75$ Hz, $A_{vib} = 0.55$ cm; (b) $f_r = f_{vib} = 7.5$ Hz, $A_{vib} = 0.88$ cm.

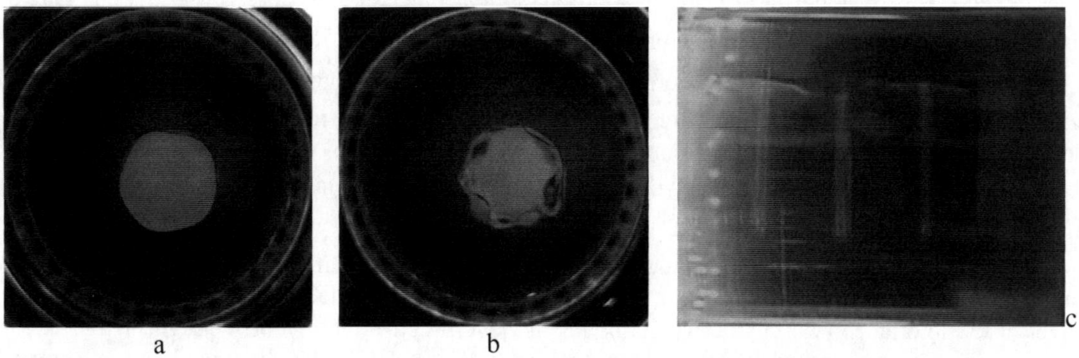

Figure 6. Wave visualization at the interface of two liquids at $f_r = f_{vib} = 7.75$ Hz: (a) $A_{vib} = 0.75$ cm; (b,c) $A_{vib} = 1.1$ cm.

Also, in the case of a two-liquid system, with the variation in the relative radius R_{rel}, relative density ρ_{rel} and viscosity ratio ν_{rel}, the value $b_1 / A_{vib} \approx 0.5$ remains constant. This confirms the result of the theory [12] that the shift of the light inclusion from the symmetry axis of the cavity is caused by the averaged inertial field. The results also indicate that the theory [12] can be generalized for a wide class of rotating systems that are inhomogeneous in density. An interesting feature of the interface between liquids is that centrifugal waves propagate in the reference system associated with the column of light liquid. This means that the wave inertia axis is also shifted at $0.5A_{vib}$ relative to the cavity axis.

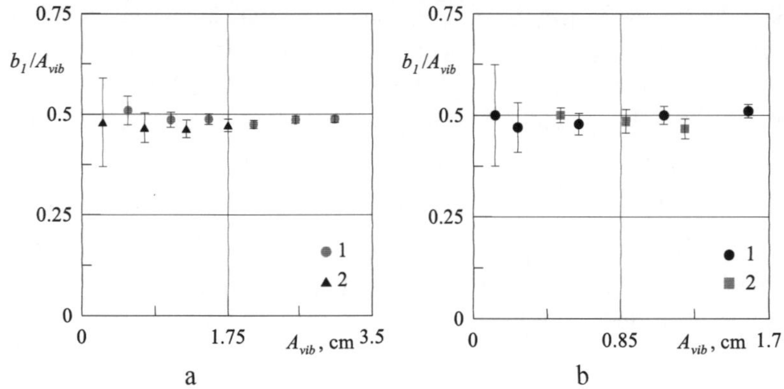

Figure 7. The dependence of the dimensionless radial displacement on A_{vib}: (a) the solid body, $f_r = f_{vib} = 5$ (1), 7 Hz (2); (b) the light liquid column, $(f_r, \nu_{rel}, \rho_{rel}, R_{rel}) = (7.75$ Hz, $2.1, 0.74, 0.36)$ (1) and $(7.5$ Hz, $0.49, 0.69, 0.49)$ (2).

4. Conclusion

At the coincidence of f_r and f_{vib}, according to the conducted experimental study, translational vibrations lead to the creation of a stationary force field in the cavity reference system. In this case, the action of vibrations is equivalent to a parallel translation of the axis of rotation, both of the solid body and the column of light liquid, to the distance equal to half the amplitude of perpendicular vibrations. A slight mismatch of the frequency of vibrations and the frequency of cavity rotation leads to an average azimuthal drift of the phase inclusion (light body or column of light liquid) relative to the cavity. The latter is determined by the azimuthal displacement of the direction of the averaged force field and can be both lagging and leading. The studies have shown that the vibrations perpendicular to the rotation axis with a frequency close to the frequency of rotation can be used for azimuthal and radial positioning of light phase inclusions in rotating cavities. The positioning is possible at varying parameters of the system.

5. Acknowledgments

The work was supported by the Russian Science Foundation (grant No. 18-71-10053).

References

[1] Blekhman I I 2000 *Vibrational mechanics: nonlinear dynamic effects, general approach, applications* (Singapore: World Scientific)
[2] Riley N 2001 *Ann. Rev. Fluid Mechanics* **33** 43
[3] Greenspan H P 1968 *The Theory of Rotating Fluids* (Great Britain: Cambridge University Press)
[4] Phillips O M 1960 *Journal of Fluid Mechanics* **7(3)** 340
[5] Ivanova A A, Kozlov V G and Chigrakov A V 2004 *Fluid Dynamics* **39(4)** 594
[6] Kozlov V G and Polezhaev D A 2008 *Microgravity Sci. Technol.* **21** 79
[7] Kozlov N V and Shuvalova D A 2016 *Acta Astronaut.* **127** 561–71
[8] Kozlov N V, Kozlova A N and Shuvalova D A 2016 *Phys. Fluids* **28** 112102
[9] Kozlov V G and Kozlov N V 2007 *Doklady Phys.* **52(8)** 458–61
[10] Kozlov V G, Kozlov N V and Subbotin S V 2015 *Phys. Fluids* **27(12)** 124101
[11] Karpunin I E, Kozlova A N and Kozlov N V 2018 *Microgravity Sci. Technol.* **30(4)** 399
[12] Vjatkin A A, Kozlov V G and Sabirov R R 2019 *Int. Journal of Thermal Sciences* **137** 560

IOP Conf. Series: Materials Science and Engineering **581** (2019) 012041 doi:10.1088/1757-899X/581/1/012041

Dynamics of Vapor-Gas Bubbles in a Liquid Near Solid Surfaces with Different Properties

T P Lyubimova[1,2], K A Rybkin[2], O O Fattalov[2], L O Filippov[1]
[1] Institute of Continuous Media Mechanics UB RAS, Acad. Koroleva 1, Perm 614013, Russia
[2] Perm State University, Bukireva Str 15, Perm 614990, Russia

E-mail: lubimova@psu.ru

Abstract. In this paper, the dynamics of vapor-gas bubbles arising in a distilled water and water-salt solutions as a result of coalescence of cavitation bubbles formed under the action of ultrasound (US) near solid surfaces with different properties has been experimentally investigated. Experiments have shown a significant effect of salt concentration in water and properties of solid surface on the behavior of bubbles, both under the action of ultrasonic and after its shutdown. In distilled water, mobile bubbles were observed on hydrophilic surface of wood, most of which are detached from the surface and rise after US shutdown. Bubbles on the hydrophobic surface of fluoroplastic, on the contrary, remain motionless both during the exposure to ultrasonic and after its shutdown. In a 20% aqueous solution of NaCl under the action of ultrasound on the hydrophilic surface, the bubbles remain stationary, and at the time of the shutdown of the ultrasound only a small part of them rise. In similar conditions, mobile bubbles are observed on the hydrophobic surface that rise when the ultrasonic is switched off.

1. Introduction

Under the influence of ultrasound cavitation bubbles are formed in water. Their further behavior depends on various factors, there can be observed both the collapse of cavitation bubbles in a liquid, and their coalescence and long existence of relatively large bubbles [1,2]. The interaction of cavitation bubbles with solid surfaces is also a subject of discussion [3], in particular, due to the formation of nanobubbles under the action of ultrasonic, which can change the surface properties of a solid surface, for example, increasing the likelihood of large bubbles attaching to a solid surface when they touch [4] . Interest in such studies, in particular, is due to the fact that under the action of ultrasound erosion of solid surfaces occurs, and its intensity in salt solutions is higher than in pure water [5].

The problem of the interaction of bubbles with ore particles in the process of their enrichment by the flotation method deserves special attention. According to the available theoretical and experimental data, it can be concluded that the effect of ultrasound on the flotation pulp can increase the efficiency of flotation [6]. This behavior may be associated with the following effects: additional generation of a larger number of cavitation microbubbles in the pulp during its irradiation with ultrasonic, an increase in the trajectory of ascent of bubbles, the action of averaged forces arising in vibration fields [7], generation of vapor-gas nanobubbles formed in microcracks of solids [8-10], the change in the value of the Zeta potential on the surface of the bubble [11,12]. Some authors also point out the negative impact of ultrasonic exposure, in particular on the flotation process, since at a certain critical value of sound pressure there is a decrease in the percentage of valuable ore recovery [13].

One of the reasons for the positive or negative effect of ultrasound on the flotation process is that it affects the interaction of bubbles with solid surfaces due to the appearance of additional acoustic forces [14,15]. In [16], the interaction of two bubbles caused by the action of the Bjerknes force is discussed in the case when two bubbles in an acoustic field are attracted to each other. As a result of convergence, under certain parameters, their coalescence occurs. Also, ultrasonic impact can contribute to the retention of a bubble on a solid surface due to the formation of gas nanobubbles in microcracks of a solid surface. In [17], it was experimentally observed how nanobubbles covered the microcracks of solids, and in [18], the change in the structure of a solid surface under the action of ultrasound was shown. All these effects can influence the dynamics of vapor-gas bubbles near the solid surface.

In this paper, we experimentally investigate the effects observed when bubbles approach each other with a solid surface, followed by their fixation and retention. In addition, the features of coalescence of vapor-gas bubbles in water and water-salt solutions of different concentrations arising on the surface of solids with different surface properties under the action of ultrasonic are discussed.

2. Experimental technique

The experiments were carried out in a parallelepiped-shaped cuvette with dimensions of 110 × 116 × 160 mm3 (Fig. 1). The cuvette was made of 3 mm thick plexiglas. In the central part of the cuvette, the samples under investigation were fixed at an angle of 45 to the horizon; a laser knife created by a cylindrical lens and a KLM-532 laser using a DPSS green laser was used to visualize the bubbles. A high-speed camera was used to record the bubble dynamics. Degassed distilled water and aqueous solutions of chemically pure NaCl, in various mass concentrations, were used as the studied liquids.

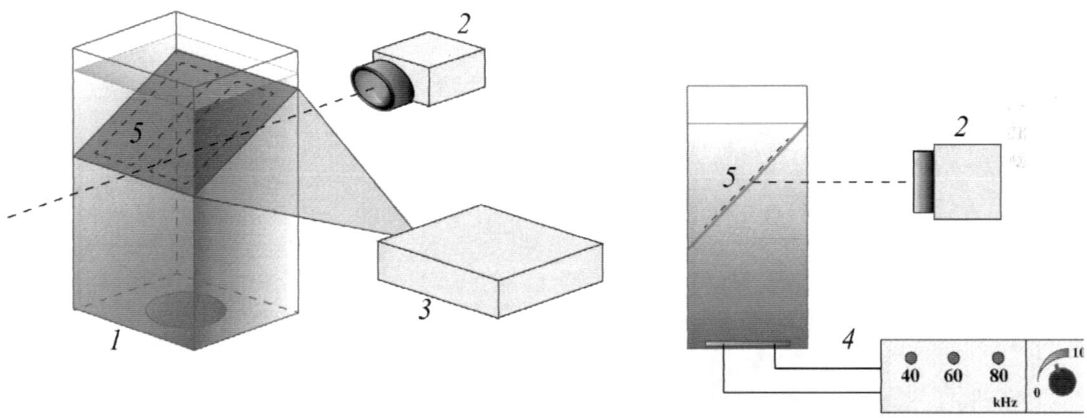

Figure 1. Laboratory setup: 1) cavity 110×116×160 mm³; 2) high speed camera or reflex camera, 3) laser, 532nm; 4) ultrasound source *f*=40 kHz, *P*=100 w; 5) wood or fluoroplastic plates

A metal disk with a diameter of 88 mm, placed at the bottom of the cell, was used as the source of ultrasound so that the center of the radiator coincided with the center of the bottom of the cell. It was connected to an ultrasonic oscillator with a frequency of f = 40 kHz and a maximum power of P = 100w.

From the photographs obtained in the course of the experiments, the critical diameter of the bubbles retained on solid surfaces was tracked. The dynamics of bubbles near solid surfaces with different surface properties were also investigated: the surface of wood with a high degree of wettability by water and the surface of a fluoroplastic with much more pronounced hydrophobic properties.

3. Results of the experiments

Distilled water

As shown by experiments conducted with distilled water and solid plates of wood and fluoroplast, under the action of ultrasound in a liquid cavitation bubbles are formed, the diameter of which increases with time due to coalescence, in the future some of the bubbles formed can be fixed on a solid surface. Depending on the surface properties and the composition of the liquid, the bubbles either adhere to the surface or drift along it.

Fig. 2a, b present photographs obtained in experiments with plates made of wood and fluoroplastic in distilled water. The photos clearly demonstrate the difference in fixing vapor-gas bubbles on surfaces with different wettability properties. Figure 2c shows the diasetter distribution of bubbles attached to the surface, 2 minutes after the start of the ultrasonic effect.

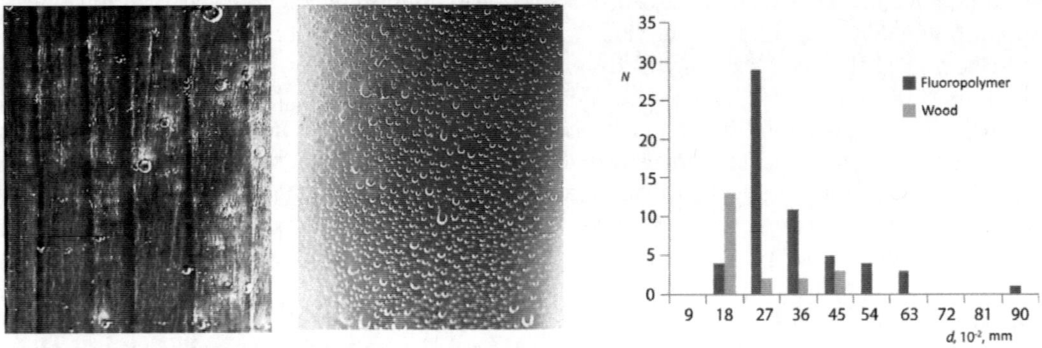

a b c

Figure 2. Photo of surface a - wood, b - fluoroplastic in distilled water under the action of ultrasonic; c is the distribution of the diameter of the bubbles attached to the surface, depending on its properties, where N is the number of bubbles

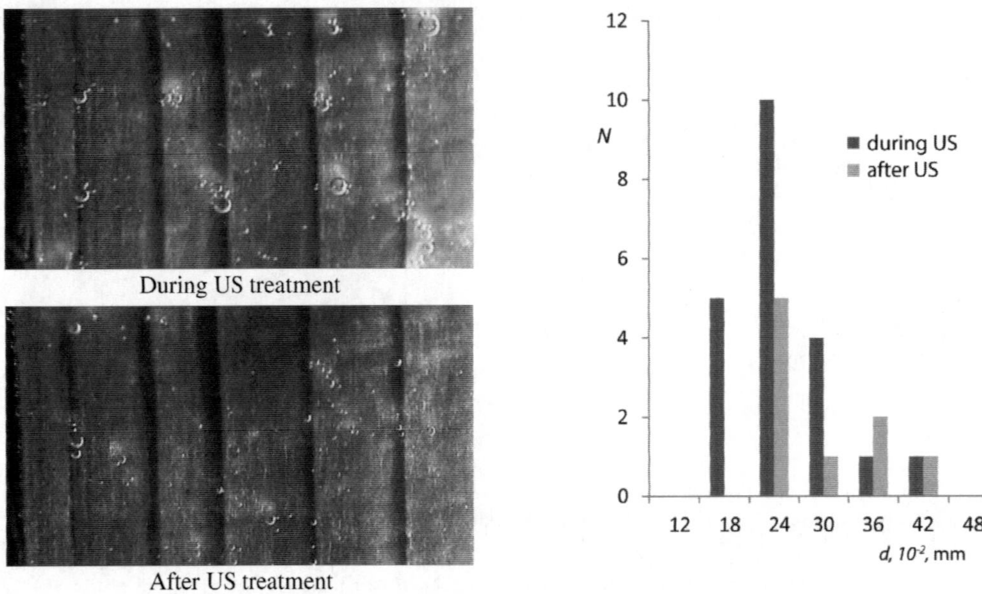

Figure 3. Size distribution of bubbles attached to the wood surface at the time of the impact of ultrasound and 5 seconds after it is switched off

On the hydrophilic surface of wood, bubbles drift, passing for 1 second a distance of about 5 to 10 of its diameter, quickly increasing in size due to coalescence and float up. In the case of the hydrophobic surface of the fluoroplast, stable attachment of bubbles is observed. Thus, the force of interaction of a bubble with a solid surface in distilled water depends on the surface properties of the solid surface, as a result of which the total number of bubbles on the wood surface is much less than on the surface of the fluoroplastic. Such features in the dynamics of bubbles can be due to the balance of electrostatic and hydrodynamic forces, namely the Archimedes force, the surface tension force, the pressure gradient arising from the averaged flows caused by the action of ultrasonic, and the secondary Bjorkens force. One of the reasons for the mobility of bubbles on wood is the predominance of hydrodynamic forces over the surface ones; in the case of fluoroplast, the main contribution is made by forces of an electrostatic nature.

Fig.3 on the left shows a photograph of the wood surface 5 seconds after the shutdown of the ultrasound, the right shows the size distribution of bubbles attached to the surface during the ultrasound impact and 5 seconds after it was turned off. As you can see, after turning off the ultrasound bubbles attached to the surface of the wood float. In the case of a fluoroplastic plate, the bubbles attached to the surface, unlike the bubbles attached to the surface of the wood, almost do not float, but their subsequent growth stops.

The different behavior of the bubbles when turning off the ultrasound in the case of solid surfaces of wood and fluoroplastic, in our opinion, is explained as follows. The averaged vibration forces of attraction between a solid surface and a bubble, acting under the influence of ultrasound [19], cease to exist when the ultrasound is turned off. In the case of wood, these forces exceed the surface ones, so the bubbles tend to float. In the case of fluoroplast, the surface forces exceed the hydrodynamic forces, while the bubbles remain frozen into the surface.

Water salt solutions

The size distributions of bubbles attached to the surfaces of wood and fluoroplastic when subjected to ultrasound, depending on the salt concentration are shown in Fig.4. Histograms show that for both wood and fluoroplastic, an increase in salt concentration leads to a decrease in the size variation of bubbles. Thus, as the salt concentration increases, the diameter of the bubbles decreases simultaneously and their size distribution becomes narrower.

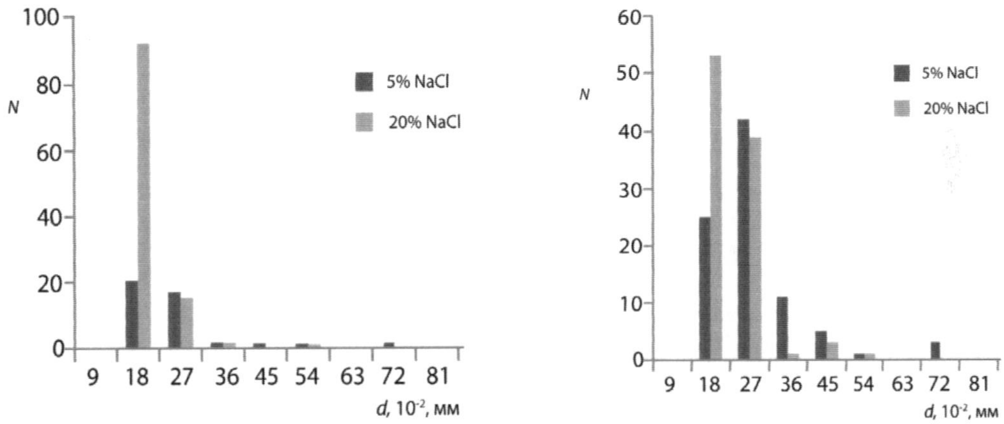

a b

Figure 4. Comparison of histograms of the distribution of the number of bubbles from their diameter for wood (a) and fluoroplast (b) with increasing salt concentration in water from 5% to 20%

The pattern of bubble drift also changes dramatically with changes in salt concentration. In contrast to the case of distilled water (zero salt concentration), at a salt concentration of 20%, the bubbles are frozen into the surface of the wood, but drift along the surface of the fluoroplastic. When drifting over the surface of the fluoroplast, they additionally coalesce with the bubbles already attached to the surface, in which case their ascent occurs more intensively. As a result, by the time of the shutdown of UZ, there are practically no adherent bubbles on the fluoroplastic, while only a small amount of bubbles detach from the surface of the wood. This behavior is due to the fact that in such a system the bubbles are energetically advantageous to remain on the surface of the wood, which makes it impossible for them to coalesce. The change in the number of bubbles attached to a solid surface with time under the action of ultrasound is shown in Fig.5a for wood surface in 25% NaCl solution and in Fig.5b for fluoroplastic surface in distilled water.

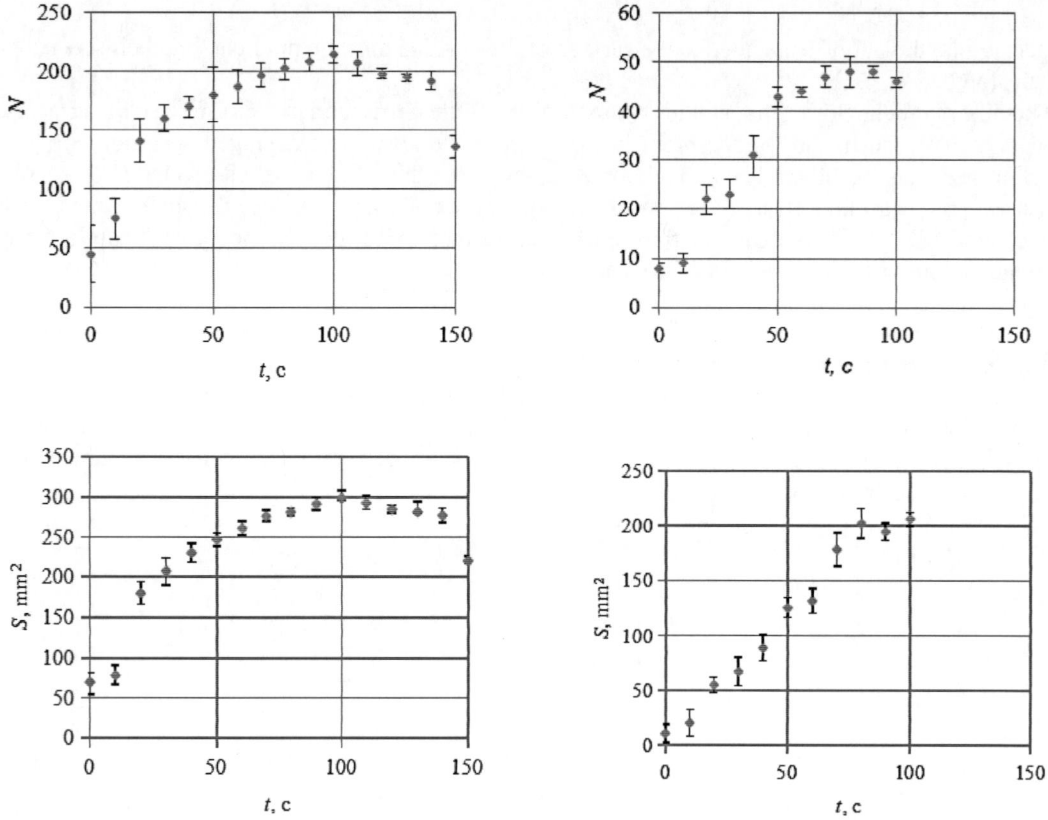

Figure 5 Changes in the number of attached bubbles and the total area of attachment of bubbles over time when exposed to ultrasound: left - to the surface of the fluoroplast, in distilled water, right - to the surface of wood in a 20% solution of salt NaCl

Fig.5 shows the time dependence of the number of bubbles and the total surface area occupied by the bubbles in distilled water and a 20% aqueous solution of NaCl salt. It is important to note that the dynamics of vapor-gas bubbles near the surface with hydrophobic properties in distilled water is qualitatively similar to the dynamics of vapor-gas bubbles near the hydrophilic surface in a 20% aqueous solution of NaCl. In both cases, the number of bubbles increases over time, however, there is a point at which the total number of bubbles near the surface does not change. The same can be said about the surface area covered by bubbles, it grows only to a certain critical value, after which the system enters the stationary mode. In distilled water, the ultrasonic effect made it possible to additionally hold 35% of the total number of bubbles on the hydrophobic surface.

4. Conclusions

As experiments have shown, in distilled water, under the action of ultrasound, moving bubbles appeared on the hydrophilic surface, most of which broke off from the surface and surfaced when the ultrasonic was switched off. However, the bubbles on the hydrophobic surface remained frozen at all stages of the experiment.

It was found that with an increase in salt concentration in water, in the case of a hydrophilic surface, the total surface area covered with bubbles first grows according to a law close to linear, and then enters the saturation stage, and the number of bubbles on both surfaces changes in the same way over time.

With a high concentration of NaCl under the influence of ultrasound on a hydrophilic surface, the bubbles remained stationary, in contrast to the hydrophobic surface, where there are moving bubbles that pop up when the ultrasonic is turned off.

Acknowledgments

The work was supported by Russian Foundation for Basic Research (grant No. 17-41-590974).

IOP Conf. Series: Materials Science and Engineering **581** (2019) 012041 doi:10.1088/1757-899X/581/1/012041

References

[1] Bremond, N., Arora, M., Ohl, C. D., & Lohse, D. (2006). Controlled multibubble surface cavitation. Physical review letters, 96(22), 224501.
[2] Rybkin, K. A., Bratukhin, Y. K., Lyubimova, T. P., Fatallov, O., & Filippov, L. O. (2017, July). Experimental study of formation and dynamics of cavitation bubbles and acoustic flows in NaCl, KCl water solutions. In Journal of Physics: Conference Series (Vol. 879, No. 1, p. 012026). IOP Publishing.
[3] Bremond, N., Arora, M., Dammer, S. M., & Lohse, D. (2006). Interaction of cavitation bubbles on a wall. Physics of fluids, 18(12), 121505.
[4] Teshima, H., Nishiyama, T., & Takahashi, K. (2017). Nanoscale pinning effect evaluated from deformed nanobubbles. The Journal of Chemical Physics, 146(1), 014708.
[5] Gou, W., Zhang, H., Li, H., Liu, F., & Lian, J. (2018). Effects of silica sand on synergistic erosion caused by cavitation, abrasion, and corrosion. Wear, 412, 120-126.
[6] Ozkan, Safak G., and Halit Z. Kuyumcu. "Investigation of mechanism of ultrasound on coal flotation." International Journal of Mineral Processing 81.3 (2006): 201-203.
[7] Hassan, S., Lyubimova, T. P., Lyubimov, D. V., & Kawaji, M. (2006). Motion of a sphere suspended in a vibrating liquid-filled container. Journal of applied mechanics, 73(1), 72-78.
[8] Yang, S., Dammer, S. M., Bremond, N., Zandvliet, H. J., Kooij, E. S., & Lohse, D. (2007). Characterization of nanobubbles on hydrophobic surfaces in water. Langmuir, 23(13), 7072-7077.
[9] Belova, V., Gorin, D. A., Shchukin, D. G., & Möhwald, H. (2010). Selective ultrasonic cavitation on patterned hydrophobic surfaces. Angewandte Chemie International Edition, 49(39), 7129-7133.
[10] Belova, V., Gorin, D. A., Shchukin, D. G., & Möhwald, H. (2011). Controlled effect of ultrasonic cavitation on hydrophobic/hydrophilic surfaces. ACS applied materials & interfaces, 3(2), 417-425.
[11] Elmahdy, A. M., Mirnezami, M., & Finch, J. A. (2008). Zeta potential of air bubbles in presence of frothers. International Journal of Mineral Processing, 89(1-4), 40-43.
[12] Cho, S. H., Kim, J. Y., Chun, J. H., & Kim, J. D. (2005). Ultrasonic formation of nanobubbles and their zeta-potentials in aqueous electrolyte and surfactant solutions. Colloids and Surfaces A: Physicochemical and Engineering Aspects, 269(1-3), 28-34.
[13] Ca, G., Ozdemir, O., Wang, X., Ozkan, S. G., & Miller, J. D. (2018). Effect of ultrasound on bubble-particle 2 interaction in quartz-amine flotation system.
[14] Crum, Lawrence A. "Bjerknes forces on bubbles in a stationary sound field." The Journal of the Acoustical Society of America 57.6 (1975): 1363-1370.
[15] Mettin, R., et al. "Bjerknes forces between small cavitation bubbles in a strong acoustic field." Physical review E 56.3 (1997): 2924.
[16] Ida, Masato. "Alternative interpretation of the sign reversal of secondary Bjerknes force acting between two pulsating gas bubbles." Physical Review E 67.5 (2003): 056617.
[17] Berkelaar, Robin P., Harold JW Zandvliet, and Detlef Lohse. "Covering surface nanobubbles with a NaCl nanoblanket." Langmuir 29.36 (2013): 11337-11343.
[18] Belova, Valentina, et al. "Selective ultrasonic cavitation on patterned hydrophobic surfaces." Angewandte Chemie International Edition 49.39 (2010): 7129-7133.
[19] Lyubimov, D. V., Klimenko, L. S., Lyubimova, T. P., & Filippov, L. O. (2016). The interaction of a rising bubble and a particle in oscillating fluid. Journal of Fluid Mechanics, 807, 205-220.

IOP Conf. Series: Materials Science and Engineering **581** (2019) 012042 doi:10.1088/1757-899X/581/1/012042

Droplet Sizing in the Spray of a Fuel Injector Using Wavelet Analysis

R Stepanov [a], V Batalov, A Sukhanovskii

Institute of Continuous Media Mechanics
Academic Korolyov, 1, Perm, 614013, Russia

E-mail: [a] rodion@icmm.ru

Abstract. The paper discusses the applied methods for studying the structure of the two-phase flow. Measurements of the size of liquid droplets in a spray of fuel injector were carried out using two methods: Interferometric Particle Imaging (IPI) and Glare Point Technique (GPT). The combined use of these methods allows to significantly expand the range of measured droplet sizes, and also provides an opportunity for cross-validation of methods and determining the level of errors. In this paper, we tested the IPI method using wavelet processing of interference drop images. As a result, it was shown that IPI images can have a highly inhomogeneous structure. This leads to a considerable variation in the period of the IPI-image and, as a consequence, to an increase of the error in determining the drop size. The proposed error estimation algorithm allows one to introduce an additional selection criterion and obtain statistically comparable results obtained by the IPI and GPT methods.

1. Introduction

Measurement of distribution of spray droplets by size, velocity, and spatial density is an important applied problem. Spraying processes are used in the medical, fuel and energy industries. Atomization is a complex multi-phase mechanical process that is carried out using nozzles and is characterized by the flow of liquid inside the nozzle, disintegration at the edge of the nozzle, splitting and breaking into small droplets when mixed with the gaseous phase, followed by evaporation or combustion. The desire to determine the true distribution of liquid droplets (particles) motivates many researchers. There is a variety of optical methods that are commonly used for measurements in sprays [1, 2, 3]. In simple cases, when the particle flux density is low and their images do not intersect, it is possible to obtain a distribution of liquid droplets sufficiently close to the true distribution by various optical methods, in particular, using Interferometric Particle Imaging (IPI) [4,5] and Glare Point Technique (GPT) methods [6]. With an increase in particle density, the results obtained by the IPI and GPT methods begin to diverge. In this regard, we make efforts to determine the causes of this discrepancy. Some questionable results obtained by the GPT method are easier to verify than the results obtained by the IPI method. In the case of individual distorted drops one can deal with them manually, because GPT method operates with focused high resolved images of droplets. On the other hand, the IPI method can be used for a much larger area than the GPT method, and developed IPI software is much less time consuming (e.g. Actual Flow software with IPI Kit).

Serious limitation of GPT concerns measurements of small droplets. In case of one-to-one imaging system droplets should be larger than 40-50 μm. In opposite IPI measurements are limited for large droplet sizing because spacing between fringes with increasing of droplets size becomes smaller. Another intrinsic problem of IPI is overlap of interferometric images in case of highly concentrated sprays [13]. In this paper we suggest an approach to verify the IPI method using wavelet transforms at the stage of processing IPI particle images. Comparison with GPT method is done for the case when GPT gives reliable results.

2. Results

2.1. IPI and GPT methods

The process of registering IPI and GPT droplet images is as follows. At some distance from the nozzle, drops of a transparent sprayed liquid acquire a spherical shape. In this area one can estimate the size of the droplets by their diameters. Drops (2) (see in figure 1) are illuminated with a short (10 ns) laser flash (1) formed in the form of a narrow band of light ("knife") about 1 mm thick. The laser beam, reflected and refracted by a drop of liquid, forms two glares. The distance between them is proportional to the diameter of the drop with a certain coefficient depending on the angle of illumination [1, 2]. CCD cameras, synchronized with a laser, register the images of glares.

The difference between the IPI and GPT methods is that in the GPT method the camera captures focused glare images (3). The light glares (point light sources) on the drops are in focus, and the diameters of the drops are calculated by multiplying the number of pixels between the glares by a scale factor and a coefficient depending on the angle of illumination. In the case of the IPI method, a defocused image of the same glares is recorded, or rather, the pattern of interferometric fringes, and the drop diameters are calculated from the frequency of this interference pattern. To correctly understand why interference patterns are seen in the form of narrow strips, it should be noted that the IPI method uses cylindrical lenses that compress these images along one of the coordinates (in our case, vertically) to reduce their spatial overlap.

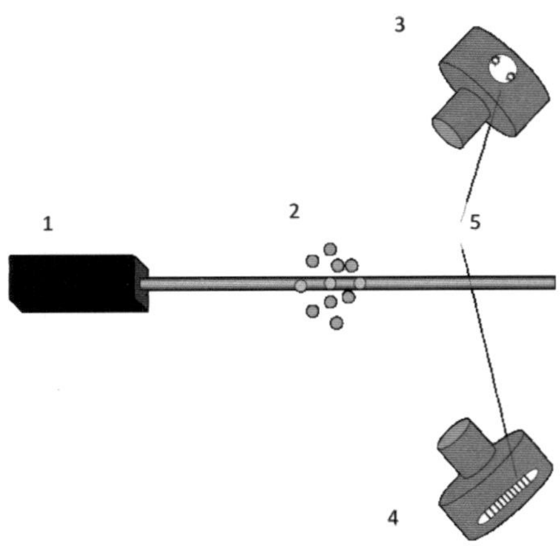

Figure 1. Scheme of combined IPI-GPT measurements. 1 - laser, 2 - spray, (3) - CCD camera configured for GPT method, 4 - CCD camera configured for IPI method, 5 - IPI and GPT primitive images of droplets.

In papers [7, 8, 9, 10, 11, 12], these methods were compared for dense and sparse flows and it was shown that it is possible to measure droplets diameter simultaneously using the IPI and the GPT methods. Typical IPI and GPT images are shown in figure 2. The results of droplet sizing are close within the limits of error with a significant and rough filtering of non-valid particles only. Therefore considerable parts of distribution responsible for drops with largest and smallest sized are not detected.

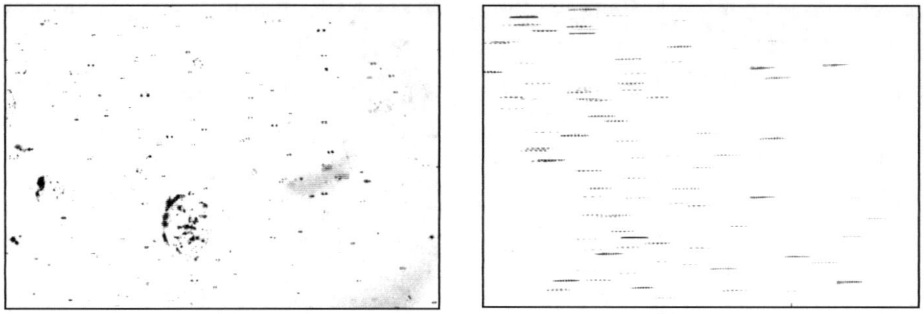

(a) (b)

Figure 2. Images of droplets: (a) GPT images (b) IPI images. Inverted intensity is used for better image perception.

We aimed in this work to recognize cases when droplets sizes, determined by the IPI and the GPT methods are substantially different. For this purpose, the flow was considered with a small density of about 3-7 drops per matrix frame of 24 by 36 mm and a resolution of 4004 by 2671 pixels (see figure 3). Also, we used a continuous two-dimensional wavelet transform to determine the dominant frequency in vicinity of each droplet.

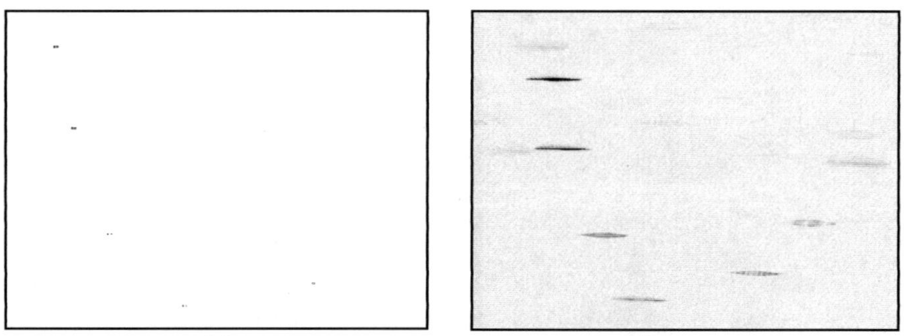

(a) (b)

Figure 3. Simultaneously captured images of individual droplets: (a) GPT images and (b) IPI images.

2.2. Wavelet analysis

Wavelet transform was used for GPT image pre-processing in [18] and proved its efficiency. Some specific details should be emphasized in case of using wavelet analysis for IPI images. The continuous wavelet transform of a 2D map $f(x,y)$ is defined by

$$w(s,x,y) = \frac{1}{s^{3/2}} \iint_{-\infty}^{-\infty} f(x,y)\psi\left(\frac{x'-x}{s},\frac{y'-y}{\delta}\right) dx'dy', \qquad (1)$$

where s is the spatial scale and the function $\psi(x, y)$ is 2D Morlet wavelet function with a satisfactory spatial resolution ($\omega = 2\pi, \sigma = 2$). The partiality of (1) is that we don't dilate $\psi(x, y)$ in y direction (δ is fixed). Wavelet coefficients are normalized by a factor $s^{3/2}$ in order to get the maximum intensity at scale corresponding scale of dominate interferometric component.

The result of wavelet analysis is shown in figure 5. One can see that of the seven images, the size of the droplets is determined only in four. The two drops in the bottom are outlined in the same color contours, indicating that the size is unambiguously defined in the image area.

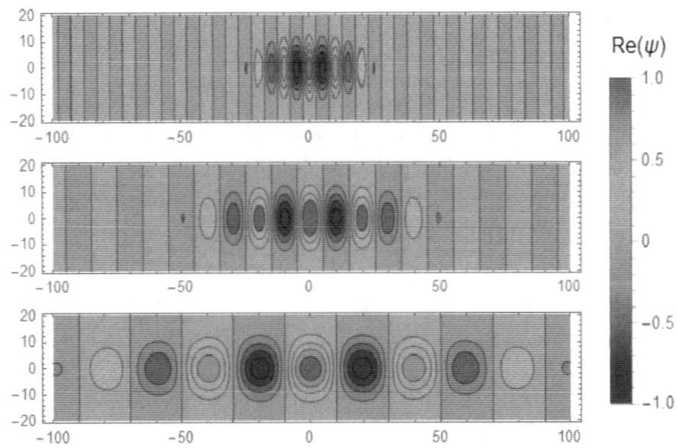

Figure 4. Real part of wavelet function for different scale s=10, 20, 40.

The two upper images of droplets include areas dominated by wavelet coefficients corresponding to different sizes of droplets. It means that there are several spatial frequencies in interferometric pattern. It can be a result of parasitic glares or complex non-spherical shape of the droplet. In that case it should be treated as invalid droplet. There is another source which can lead to the variation of spatial frequency and to the systematic error. The optical distortions due to imperfection of lens and interferometric images compression can also provide the variation of spatial frequency.

Figure 5. IPI images (gray) and isolines of intensity of wavelet coefficients of various scales (color).

Results of droplet sizing obtained by different methods are summarized in table 1.
Using of wavelets can help to localize the distorted parts of the interferometric image and seriously improve the quality of measurements. Thus, it is possible to determine the value of the effective size and determine its confidence interval, which would include the value of the size obtained on the basis of GPT method.

Table 1. Sizes (in pixels) of droplets from figure 5 which are determined by different methods. Non-valid determining are highlighted by gray backcolor.

	GPT	IPI by wavelet	IPI by AF IPI Tit
1	9.3±0.1	10±1.1	22
2	8.0±0.1	9.5±1.2	23
3	12.3±0.1	12.5±1.1	13
4	12.9±0.1	13.5±1.8	14
5	7.7±0.1	8.0±0.8	18

IOP Conf. Series: Materials Science and Engineering **581** (2019) 012042 doi:10.1088/1757-899X/581/1/012042

3. Conclusions

As a result, it was shown that IPI images can have a highly inhomogeneous structure. This leads to a significant variation in the period of the interference image and, as a consequence, to an increase in the error in determining the drop size. The proposed error estimation algorithm allows one to introduce an additional selection criterion and obtain statistically comparable results obtained by the IPI and GPT methods.

Acknowledgments

This work was supported by the Russian Foundation for Basic Research and by the Perm regional government under project No. 19-08-00574.

References

[1] Tropea C 2011 *Annual Review of Fluid Mechanics* **43** 399-426
[2] Fansler T D and Parrish S E 2014 *Measurement Science and Technology* **26** 012002
[3] Gouesbet G and Gréhan G 2015 *International Journal of Multiphase Flow* **72** 288-297
[4] Maeda M, Kawaguchi T and Hishida K 2000 *Measurement Science and Technology* **11** L13
[5] Kawaguchi T, Akasaka Y and Maeda M 2002 *Measurement Science and Technology* **13** 308
[6] Van de Hulst H and Wang R 1991 *Applied optics* **30** 4755-4763
[7] Zama Y, Kawahashi M and Hirahara H 2004 *Optical review* **11** 358-364
[8] Zama Y, Kawahashi M and Hirahara H 2005 *Measurement Science and Technology* **16** 1977-1986
[9] Hess C F and LEsperance D 2009 *Experiments in fluids* **47** 171-182
[10] Lacagnina G, Grizzi S, Falchi M, Di Felice F and Romano G P 2011 *Experiments in fluids* **50** 1153-1167
[11] Qieni L, Kan H, Baozhen G and XiangW 2016 *Optics express* **24** 16530-16543
[12] Qieni L, Xiaoxue Y, Baozhen G and Tingting C 2018 *Optics Express* **26** 1038-1048
[13] Bocanegra Evans H, Dam N, van der Voort D, Bertens G and van de Water W 2015 *Review of Scientic Instruments* **86** 023709
[14] Sukhanovskii A, Batalov V and Stepanov R 2016 *AIP Conference Proceedings* **1770** 030023
[15] Sukhanovskii A, Batalov V and Stepanov R 2019 *Exper. Thermal and Fluid Science* **103** 29-36

IOP Conf. Series: Materials Science and Engineering **581** (2019) 012043 doi:10.1088/1757-899X/581/1/012043

Registration of the Creep Behavior by Embedded and Surface Mounted FOSS

N Kosheleva, G Serovaev
Institute of Continuous Media Mechanics UB RAS, Perm, Russia

E-mail: kosheleva.n@icmm.ru, serovaev@icmm.ru

Abstract. Epoxy resins are widely used to connect various structural elements made of polymer composite materials. The integrity of the critical components depends on the strength and durability of such compounds. At the same time, epoxy resins have pronounced viscoelastic properties. An experimental study was conducted of the possibility of using fiber-optic strain sensors to register the viscoelastic behavior of materials. The fiber Bragg grating (FBG) was used as a sensor. The main objectives of this study were: registration of material creep (growth of strain with time at constant external load) and subsequent (after removal of load) recovery of strain using surface mounted and embedded fiber-optic strain sensors.

1. Introduction

The use of polymer composite materials as structural materials requires long-term stability of their size and strength. Under various operating temperature and mechanical loads, polymer composite materials can exhibit pronounced viscoelastic behavior, which requires additional research [1]. One of the most common ways to experimentally study the viscoelastic behavior is to test materials at high temperatures, when materials exhibit their viscoelastic properties more vividly, which can significantly reduce the test time [2–4]. Material creep failure is the result of a combination of such phenomena as the breaking of bonds of material fibers (for fiber composites), shearing and cracking of the binder, the formation and growth of voids. In addition, in polymer composites, such parameters as polymer / fiber or polymer / particles interfacial strength affect the material creep failure. Temperature, stress level and time are also crucial factors affecting the viscoelastic behavior of polymer composite materials. The influence of these factors is used to describe the behavior and prediction of the polymer composite materials failure [5–8]. Significant amount of studies of the viscoelastic behavior of composite materials, in particular, of layered composites, have been conducted in the military and aerospace industries [9]. In [10], a literature review is presented about the most relevant studies of the creep behavior of materials, specifically designed for use in civil engineering infrastructure. These studies focus both on the experimental characterization of materials and on the development of analytical models. Special attention is paid to flexural and compression loads, which were used in most studies on the creep phenomenon of composites with different reinforcement structures. Using of the materials exhibiting viscoelastic properties for structural design, requires to take into account a number of features [11].

Epoxy resins are widely used to connect various structural elements and the integrity of the critical components depends on the strength and durable characteristics of such compounds. At the same time, epoxy resins exhibit viscoelastic behavior that require determination, both in laboratory tests and during operation. Fiber optic strain sensors (FOSS) have a great prospect of use for registration of the mechanical behavior of laboratory samples and joints in structures made from epoxy resins during long-term testing. They are small in size, can be embedded in the samples or junctions, or mounted to the surface. It is also possible to place several sensors on one fiber and to record data for a long period of time.

This paper presents the results of an experimental study of the possibility of using fiber-optic strain sensors to register the viscoelastic behavior of materials. The fiber Bragg grating (FBG) was used as a sensor. Experiments on creep and strain recovery were carried out on an epoxy sample with an embedded and surface mounted FBGs. The main objectives of this study were: registration of material creep (growth of strain with time at a constant external load) and subsequent (after removal of the load) recovery of strain using fiber optic strain sensors and comparing the data for two types of sensor placement on the object.

2. Description of the experiment

For testing, epoxy resin sample was made. Two Bragg grating strain sensors were used to measure the strains of the sample. One of the sensors was embedded into the sample during its manufacture. The second was mounted to the surface of the sample using adhesive bonding. The sensor layout is shown in Figure 1.

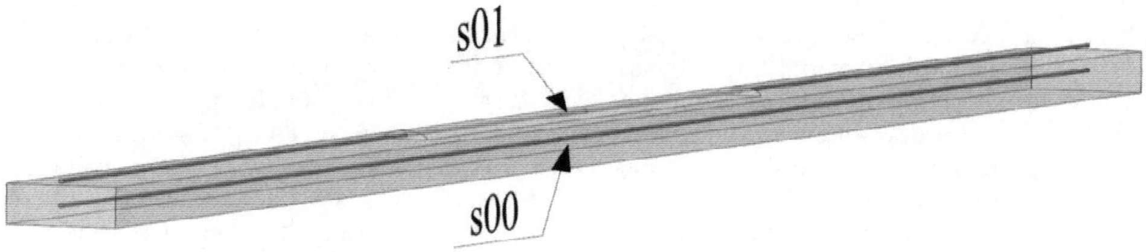

Figure 1. The scheme of location of optical fibers with FBG on a epoxy resin sample.

The technology of making a sample from epoxy resin was as follows. A silicone mold with a 200×10×5 mm hole, for the subsequent pouring of epoxy resin with a hardener, was made (Figure 1.). Two technological cuts along the edges of silicone mold were created. They are designed for centering the position and fixing the optical fiber with FBG in the mold. After positioning of the optical fiber in this silicone mold, a mixture of epoxy resin and hardener is poured in proportions determined by the manufacturer of the epoxy resin. This mixture is left for 24 hours at a temperature of 23°C for curing.

Figure 2. Silicone mold for making samples.

It is known that when a Bragg grating is subjected to a transverse load, as a result of the birefringence phenomenon, the peak of the reflected spectrum splits. The Bragg wavelength, which is subsequently used to calculate the strain, is determined by the maximum value of the spectrum. Thereby if the peak is bifurcated, difficulties may arise in the reliability of strain determination. This problem is common for FBGs embedded in the composite material, when during the technological process the optical fiber in the zone of Bragg grating loses a cylindrical shape. During experiments with epoxy samples, carried out in the framework of this study, such problem was not observed. Further, Figure 3 shows the reflected Bragg grating spectra: before embedding into the sample (black line) and after embedding under the tensile load (blue line).

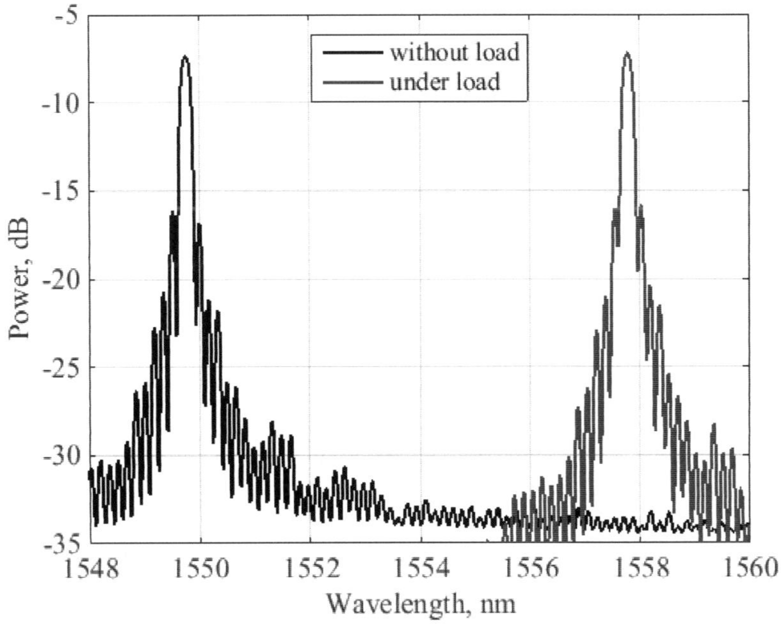

Figure 3. Reflected FBG spectra before embedding and after embedding under the load.

The second fiber with FBG was glued to the sample surface after the epoxy resin had hardened using cyanoacrylate glue. Patch pieces were made for fixation of the sample, the scheme and photos are shown in Figure 4 and 5. The patch pieces have two technological holes for attaching the sample and for hanging the load.

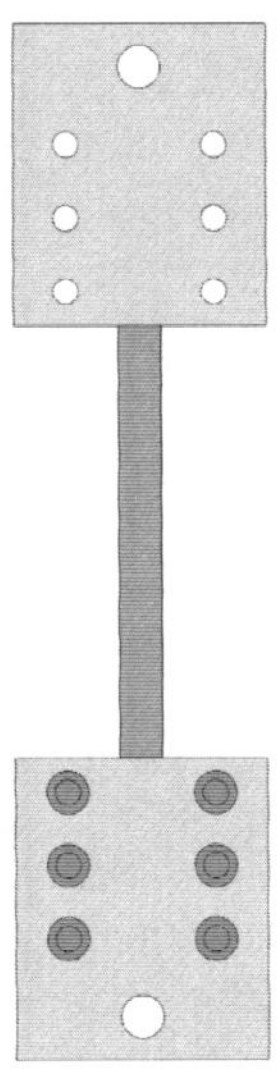

Figure 4. Experimental diagram of the sample with mounts.

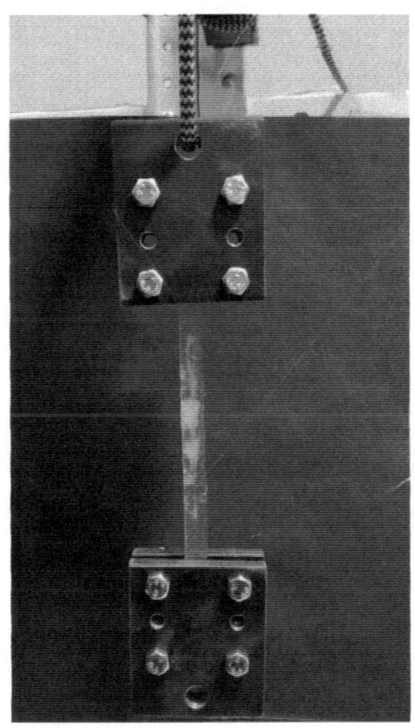

Figure 5. Sample mounted for testing.

In the experiment several stages of loading and unloading were carried out in order to study the reaction of the Bragg grating, as well as the sample itself in time. The scheme of the experiment stages of the viscoelastic behavior registration of an epoxy sample is presented in Figure 6.

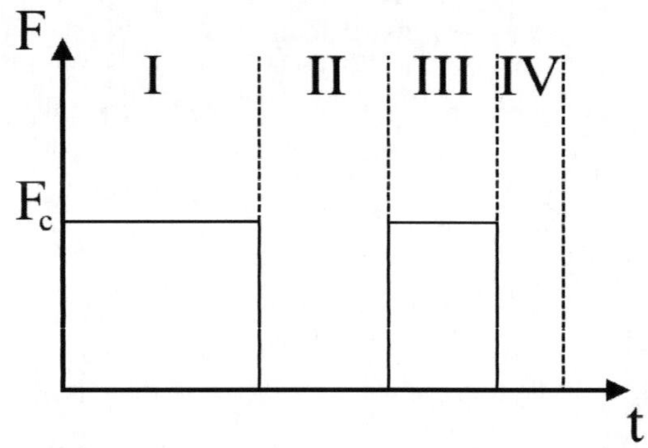

Figure 6. The scheme of the experiment stages of the viscoelastic behavior registration of an epoxy sample.

The experiment was carried out for 171 hours. During this time, the sensor readings were taken with a frequency of 1 measurement per second using an interrogator FS 2100. At the first stage (I), a constant load F_c was applied to the sample for 121 hours. Strains recorded by the sensors at this stage are presented in Figure 7. For convenience, the sensor readings of each of the experiment stages are presented from the zero time value corresponding to the beginning of this stage.

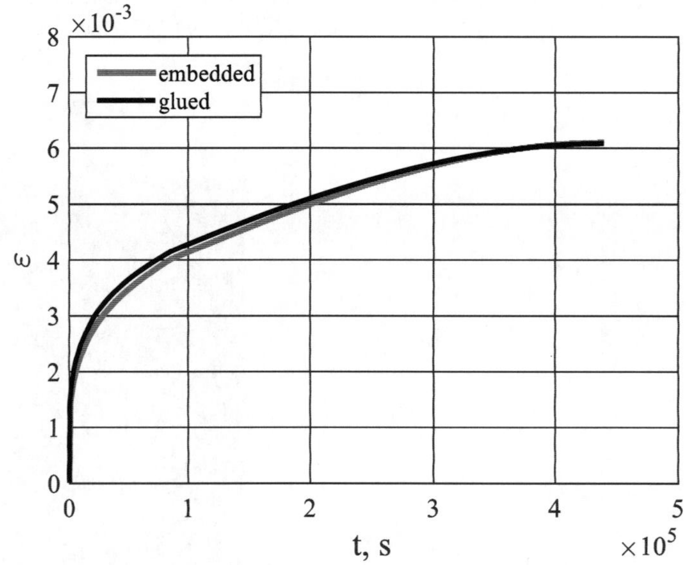

Figure 7. Strain measurements depending on the time under constant load of the sample for embedded FBG (s00) and glued on the surface (s01).

Then, at the second stage (II) the load was removed and the sample was in a free state for 25 hours. The process of decreasing of the strain in time, recorded by the sensors is shown in Figure 8.

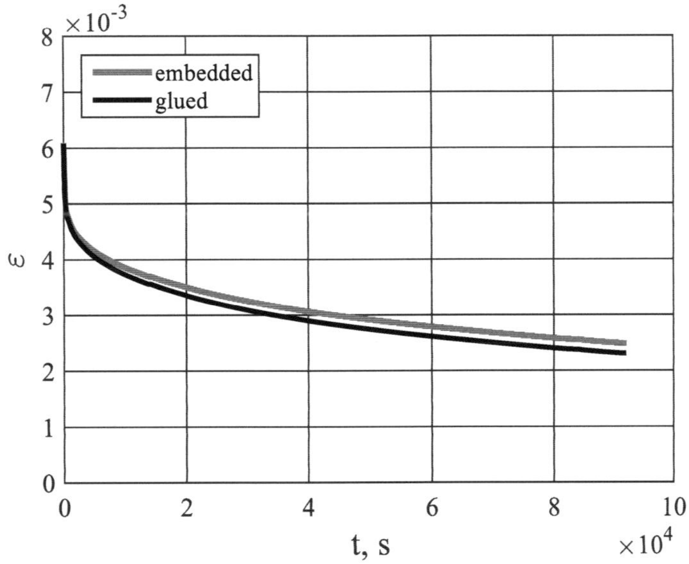

Figure 8. Strains depending on the time after removing the load for FBGs embedded in the sample (s00) and glued on the surface (s01).

At the third stage (III), a constant load F_c was applied to the sample again for 22 hours (Figure 9).

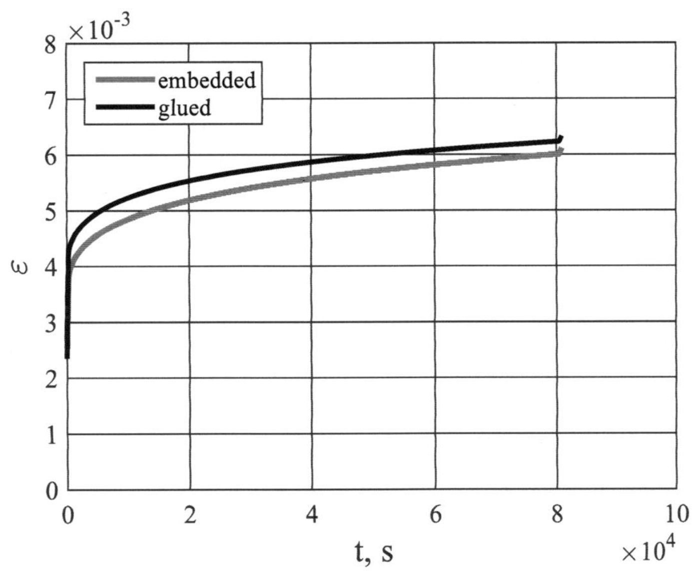

Figure 9. Strains depending on the time under constant load for FBG embedded in the sample (s00) and glued on the surface (s01).

And at the final stage of the experiment (IV), the sample was in the free state for 3 hours (Figure 10).

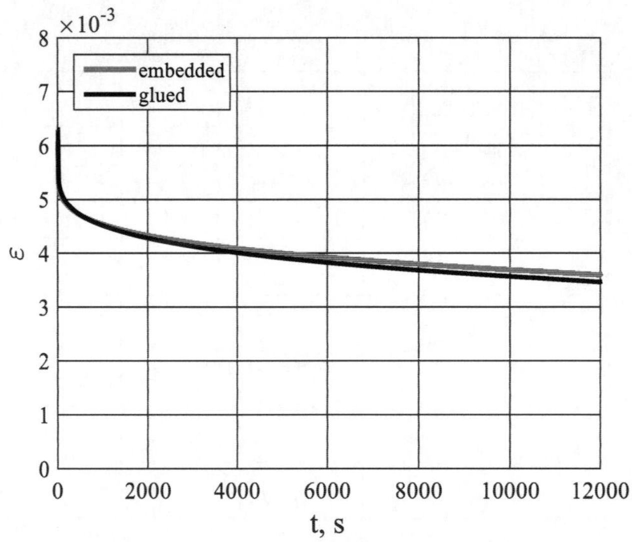

Figure 10. Strains depending on the time after removing the load for FBGs embedded in the sample (s00) and glued on the surface (s01).

Analysis of the results showed that the embedded fiber-optic sensor (s00) and placed on the surface (s01) demonstrate minor differences in the readings. Thus, the maximum difference in the strains recorded by the sensors in the first stage was 13%; in the second - 7%; in the third - 13.2%; in the fourth - 4%.

3. Conclusions

The experimental study of the viscoelastic behaviour registration using fiber-optic strain sensors based on the Bragg gratings was done. A sample of epoxy resin with FOSS embedded and placed on the surface was fabricated for experiment.

The experiments made it possible to track the strain growth with time under the constant load which is characteristic for creep material behaviour and the strain reduction over time after the load was removed using the sensitive elements. The maximum difference between the readings of the embedded sensor and glued on the surface was 13.2%.

Acknowledgments

The research was supported by RFBR (project No. 17-41-590684 r_ural_a).

References

[1] Bank L C 2007 *Composites for Construction: Structural Design with FRP Materials*

[2] Raghavan J and Meshii M 1997 Creep rupture of polymer composites *Compos. Sci. Technol.* **57** 375–88

[3] Sá M F, Gomes A M, Correia J R and Silvestre N 2011 Creep behavior of pultruded GFRP elements – Part 1: Literature review and experimental study *Compos. Struct.* **93** 2450–9

[4] Yeol C and L. Y R 2003 Time-Dependent Deformation of Pultruded Fiber Reinforced Polymer Composite Columns *J. Compos. Constr.* **7** 356–62

[5] Al-Haik M, Vaghar M R, Garmestani H and Shahawy M 2001 Viscoplastic analysis of structural polymer composites using stress relaxation and creep data *Compos. Part B Eng.*

[6] Gates T S and Sun C T 1991 Elastic/Viscoplastic Constitutive Model for Fiber Reinforced Thermoplastic Composites *AIAA J.*

[7] Guedes R M 2010 *Creep and Fatigue in Polymer Matrix Composites*

[8] Guedes R M 2007 Durability of polymer matrix composites: Viscoelastic effect on static and fatigue loading *Compos. Sci. Technol.*
[9] Keller T 2003 *Use of fibre reinforced polymers in bridge construction* (Zurich : IABSE)
[10] Scott D W, Lai J S and Zureick A-H 1995 Creep Behavior of Fiber-Reinforced Polymeric Composites: A Review of the Technical Literature *J. Reinf. Plast. Compos.* **14** 588–617
[11] Spathis G and Kontou E 2012 Creep failure time prediction of polymers and polymer composites *Compos. Sci. Technol.* **72** 959–64

Institute of Physics
Dirac House, Temple Back
Bristol BS1 6BE UK

ISSN: 1757-8981
ISBN 978-1-5108-9449-5

CURRAN ASSOCIATES INC.
proceedings.com